THE SECOND WORLI

Volume II: Their Fine

D0176659

Winston Churchill (1874–1965) was the elder son of Lord Randolph Churchill and his American wife, Jennie Jerome. In 1908, he married Clementine Ogilvy, who gave him life-long support, and they had four daughters and one son.

Churchill entered the army in 1895, served in Cuba, India, Egypt, and the Sudan and his first publications were *The Story of the Malakand Field Force* (1898), *The River War* (1899) and *Savrola* (1900), his only novel. On a special commission for the *Morning Post*, he became involved in the Boer War, was taken prisoner and escaped. His experiences led to the writing of *London to Ladysmith, via Pretoria* and *Ian Hamilton's March*, both published in 1900.

He began his erratic political career in October 1900, when he was elected Conservative M.P. for Oldham. Four years later, however, he joined the Liberal party. In 1906, he became Under-Secretary of State for the colonies and showed his desire for reform in such writings as *My African Journey* (1908). He became President of the Board of Trade in 1908 and Home Secretary in 1910 and, together with Lloyd George, introduced social legislation which helped form much of the basis for modern Britain. Because he foresaw the possibilities of war with Germany after the Agadir crisis, he was made First Lord of the Admiralty in October 1911. He achieved major changes, including that of modernising and preparing the Royal Navy for war, despite unpopularity in the Cabinet because of the cost involved. In May 1915, however, pressurised by the Opposition, he left the Admiralty and served for a time as Lieutenant-Colonel in France. Lloyd George appointed him Minister of Munitions in July 1917 and Secretary of State for War and Air the following year. In 1924, he rejoined the Conservative party and was made Chancellor of the Exchequer by Baldwin. He resigned in January 1931 and, during the 1930s, wrote numerous books, amongst which were *My Early Life* (1930), *Thoughts and Adventures* (1932) and *Great Contemporaries* (1937). Churchill was again asked to take office in September 1939, after the German invasion of Poland and, when Chamberlain was forced to retire because of the Labour party's refusal to serve under him, Churchill became Prime Minister (May 1940–May 1945). From 1945, he spent most of his time writing *The Second World War* and returned to office in 1951. In 1953, he accepted the garter and also won the Nobel prize for literature. In April 1955, however, owing to increasing illness, he resigned as Prime Minister, although he continued to write. *A History of the English-speaking Peoples* (1956–8) is his major work of this time. He died at the age of ninety.

WINSTON S. CHURCHILL

# THE SECOND WORLD WAR

## VOLUME II

# THEIR FINEST HOUR

HOUGHTON MIFFLIN COMPANY BOSTON

Copyright © 1949 by Houghton Mifflin Company
Copyright renewed © 1976 by Lady Spencer Churchill,
the Honourable Lady Soames, and the Honourable Lady Sarah Audley
Introduction copyright © 1985 by John Keegan

All rights reserved

For information about permission to reproduce selections from this book,
write to trade.permissions@hmhco.com or to Permissions,
Houghton Mifflin Harcourt Publishing Company, 3 Park Avenue,
19th Floor, New York, New York 10016.

ISBN 0-395-41056-8 (pbk.)

ISBN 978-0-395-41056-1

Printed in the United States of America
DOC 30 29 28 27 26 25
4500736280

The quotations from *Blood, Sweat and Tears,* by Winston S. Churchill,
are reprinted by the courtesy of G. P. Putnam's Sons.
The quotation from *The Memoirs of Cordell Hull,* copyright, 1948,
by Cordell Hull, is reprinted by the courtesy of The Macmillan Company.
The quotations from *The Ciano Diaries* are reprinted
by the courtesy of Countess Ciano.
The quotation from *Hamewith and Other Poems,*
by Dr. Charles Murray, is reprinted
by the courtesy of Constable & Company Limited.

# Moral of the Work

IN *WAR:* RESOLUTION

IN *DEFEAT:* DEFIANCE

IN *VICTORY:* MAGNANIMITY

IN *PEACE:* GOODWILL

# ACKNOWLEDGMENTS

I MUST again acknowledge the assistance of those who helped me with the previous volume, namely, Lieutenant-General Sir Henry Pownall, Commodore G. R. G. Allen, Colonel F. W. Deakin, Sir Edward Marsh, Mr. Denis Kelly, and Mr. C. C. Wood. I have also to thank the very large number of others who have kindly read these pages and commented upon them.

Lord Ismay has continued to give me his aid, as have my other friends.

I again record my obligation to His Majesty's Government for permission to reproduce the text of certain official documents of which the Crown Copyright is legally vested in the Controller of His Majesty's Stationery Office. At the request of His Majesty's Government, on security grounds, I have paraphrased some of the telegrams published in this volume. These changes have not altered in any way the sense or substance of the telegrams.

# INTRODUCTION

WINSTON CHURCHILL began to write the first of what were to be the six volumes of *The Second World War* in 1946. It was a work he had expected to postpone to a later stage of his life, since he had looked forward in 1945 to extending his wartime leadership into the peace. The rejection of his party by the electorate was a heavy blow, which might have dulled his urge to write. But resilience was perhaps the most pronounced of his traits of character, and he had already written the history of another great war in which he had been a principal actor. Once committed to the task, he attacked it with an energy, enthusiasm and power of organisation which would have been remarkable in a professional historian of half his age.

His five-volume history of the First World War, *The World Crisis*, had drawn heavily on the evidence he had submitted to the Dardanelles Committee and on episodic accounts written for newspapers. Its origins were therefore in political debate and in journalism. He set about composing *The Second World War* in an entirely different manner—different, too, from the way in which he had written his great life of Marlborough. Then his technique had been to dictate long passages of narrative, later correcting points of detail in consultation with experts. Now he began by assembling a team of advisers and collecting the documents on which the writing was to be based. The documents were set up in print by the publishers, Cassell, while the advisers worked on the chronologies into which they would fit. Churchill meanwhile prepared himself by dictating recollections of what he had identified as key episodes. They consisted partly of firm impressions and partly of queries to his team about dates, times, places, and personalities. He also wrote copiously to fellow-actors, begging of them their own papers and recollections, and inviting their comments on what he proposed to say. When documents, chronologies, corrections, and comments were collated, he began to write. The bulk of the writing, which was completed in 1953, was done by his normal method of dictation; however, long passages of the first volume, which is very much an *apologia pro vita sua*, were composed in his own hand.

Churchill was not, did not aspire to be, and would very probably have despised the label of, a scientific historian. Like Clarendon and Macaulay, he saw history as a branch of moral philosophy. Indeed, he gave his history a Moral. Its phrases have become some of the most famous words he pronounced—"In War: Resolution; In Defeat: Defiance; In Victory: Magnanimity; In Peace: Goodwill." Each of the component volumes was also given a Theme—"How the English-speaking peoples through their unwisdom, carelessness and good nature allowed the wicked to rearm" is that of the first— which the author believed encapsulated the period with which the volume dealt, but which he also organised his material to illustrate. He justified this method by comparing it to that of Defoe's *Memoirs of a Cavalier*, "in which the author hangs the chronicle and discussion of great military and political events upon the thread of the personal experience of an individual."

The history is, indeed, intensely personal. Explicitly so, because Churchill asks the reader to regard it as a continuation of *The World Crisis*, the two together forming both "an account of another Thirty Years War" and an expression of his "life-effort" on which he was "content to be judged." Implicitly so, because he related many of the major episodes of the war autobiographically. An excellent example is his account of the air fighting on September 15, 1940, which is regarded as the crisis of the Battle of Britain. He was lunching at Chequers and decided, since the weather seemed to favour a German attack, to spend the afternoon at the Headquarters of the R.A.F. No. 11 Group. He and his wife at once drove there, were given seats in the command room from which the British fighters were controlled, and watched the development of the action:

Presently the red bulbs showed that the majority of our squadrons were engaged. A subdued hum arose from the floor, where the busy plotters pushed their discs to and fro in accordance with the swiftly-changing situation. . . . In a little while all our squadrons were fighting, and some had already begun to return for fuel. . . . I became conscious of the anxiety of the Commander. Hitherto I had watched in silence. I now asked, "What other reserves have we?" "There are none," said Air Marshal Park. In an account which he wrote afterwards he said that at this I "looked grave." Well I might. The odds were great; our margins small; the stakes infinite. . . . Then it appeared that the enemy were going home. No new attack appeared. In another ten minutes the action was ended. We climbed again the stairways that led to the surface, and almost as we emerged the "All Clear" sounded. . . . It was 4.30 p.m. before I got back to Chequers, and I immediately went to bed [an unvarying wartime habit]. I did not wake till eight. When I rang my Principal Private Secretary came in with the evening budget of news from all over the world. It was repellent. . . . "However," he said, as he finished his account, "all is redeemed by the air. We have shot down one hundred and eighty-three for the loss of under forty."

This account is both unique—neither Roosevelt, Stalin, nor Hitler left any first-hand narrative of their involvement in the direction of the war—and acutely revealing. Churchill was fascinated by military operations and followed their progress very closely. But he forbore absolutely to intervene in their control at the hour-by-hour and unit-by-unit level adopted by Hitler. He warned and advised, encouraged and occasionally excoriated. He appointed and removed commanders. But he did not presume to do their job. Another chapter conveys the extent of his forbearance. It comes in Volume IV and concerns the fall of Singapore in February 1942. Very properly, Churchill was not merely disheartened but outraged by the failure of the Malaya garrison, under its commander, General Percival, to halt a Japanese invading force which it outnumbered. When it became clear that Percival was about to be defeated, outrage mingled with desperation and disbelief. Breaking a rule, he signalled Wavell, the Supreme Commander, to urge that the newly arrived 18th Division fight "to the bitter end" and that "commanders and senior officers should die with their troops." In the event, the 18th Division was captured by the Japanese almost intact and General Percival marched into enemy lines under a white flag. By not one immoderate word does the author convey in his narrative how deeply he—and, he felt, his country—were wounded by this humiliating and disastrous episode.

The restraint shown in the Singapore chapter was determined by another principle which he had adopted: that of "never criticising any measure of war or policy after the event unless I had before expressed publicly or formally my opinion or warning about it." The effect is to invest the whole history with those qualities of magnanimity and good will by which he set such store, and the more so as it deals with personalities. The volumes are not only a chronicle of events. They are a record of meetings, debate, and disagreements with a world of people. Some were friends with whom he was forced to differ. Some were with opponents or future enemies with whom he nevertheless succeeded in making common cause, Stalin foremost among them. The descriptions of his personal relationships with these men would alone assure the permanent value of this history to our understanding of the Second World War.

But the value of these volumes is assured in a host of other ways. They have their defects: they take no account, because they could not, of the then still secret Ultra intelligence available throughout the war to the Prime Minister (though a no longer cryptic reference on page 295 of Volume I alludes to its significance); and the first volume, in particular, may be judged excessively personal in its interpretation of the policies of the author's opponents. But these deficiencies do not detract from the history's monumental quality. It is an extraordinary achievement, extraordinary in its sweep and comprehensiveness, balance and literary effect; extraordinary in the singularity of its point of view; extraordinary as the labour of a man, already old, who still had ahead of him a career large enough to crown most other statesmen's lives; extraordinary as a contribution to the memorabilia of the English-speaking peoples. It is a great history and will continue to be read as long as Churchill and the Second World War are remembered.

JOHN KEEGAN

# PREFACE

DURING the period covered by this volume I bore a heavy burden of responsibility. I was Prime Minister, First Lord of the Treasury, Minister of Defence, and Leader of the House of Commons. After the first forty days we were alone, with victorious Germany and Italy engaged in mortal attack upon us, with Soviet Russia a hostile neutral actively aiding Hitler, and Japan an unknowable menace. However, the British War Cabinet, conducting His Majesty's affairs with vigilance and fidelity, supported by Parliament and sustained by the Governments and peoples of the British Commonwealth and Empire, enabled all tasks to be accomplished and overcame all our foes.

WINSTON SPENCER CHURCHILL

Chartwell,
    Westerham,
        Kent

*January 1, 1949*

# NOTE TO THE SECOND EDITION

A REPRINT having been called for, the opportunity has been taken to correct a few errors in detail. I am indebted to various correspondents who have drawn attention to these.

As in the case of the first volume, I have to express my appreciation of the generous reception given to the work, and I extend my cordial thanks to the many persons who have written to me concerning it.

WINSTON S. CHURCHILL

Chartwell
*January 10, 1950*

# Theme of the Volume

HOW THE BRITISH PEOPLE

HELD THE FORT

ALONE

TILL THOSE WHO HITHERTO HAD

BEEN HALF BLIND WERE

HALF READY

# THEIR FINEST HOUR

BOOK I

## The Fall of France

BOOK II

## Alone

# TABLE OF CONTENTS

## BOOK I

## THE FALL OF FRANCE

# BOOK II

# ALONE

## MAPS AND DIAGRAMS

BOOK I

# THE FALL OF FRANCE

# CHAPTER I

# THE NATIONAL COALITION

*The Beginning and the End – The Magnitude of Britain's Work for the Common Cause – Divisions in Contact with the Enemy throughout the War – The Roll of Honour – The Share of the Royal Navy – British and American Discharge of Air-bombs – American Aid in Munitions Magnifies Our War Effort – Formation of the New Cabinet - Conservative Loyalty to Mr. Chamberlain – The Leadership of the House of Commons – Heresy-hunting Quelled in Due Course – My Letter to Mr. Chamberlain of May 11 – A Peculiar Experience – Forming a Government in the Heat of Battle – New Colleagues: Clement Attlee, Arthur Greenwood, Archibald Sinclair, Ernest Bevin, Max Beaverbrook – A Small War Cabinet – Stages in the Formation of the Government, May 11–15 – A Digression on Power – Realities and Appearances in the New War Direction – Alterations in the Responsibilities of the Service Ministers – War Direction Concentrated in Very Few Hands – My Personal Methods – The Written Word – General Ismay - My Relations with the Chiefs of Staff Committee - Sir Edward Bridges – Kindness and Confidence Shown by the War Cabinet – The Office of Minister of Defence – Its Staff: Ismay, Hollis, Jacob – No Change for Five Years – Stability of Chiefs of Staff Committee – No Changes from 1941 till 1945 Except One by Death – Intimate Personal Association of Politicians and Soldiers at the Summit – The Personal Correspondence - My Relations with President Roosevelt – My Message to the President of May 15 – "Blood, Toil, Tears, and Sweat."*

NOW at last the slowly-gathered, long-pent-up fury of the storm broke upon us. Four or five millions of men met each other in the first shock of the most merciless of all the wars of which record has been kept. Within a week the front in France, behind which we had been accustomed to dwell

3

through the hard years of the former war and the opening phase of this, was to be irretrievably broken. Within three weeks the long-famed French Army was to collapse in rout and ruin, and our only British Army to be hurled into the sea with all its equipment lost. Within six weeks we were to find ourselves alone, almost disarmed, with triumphant Germany and Italy at our throats, with the whole of Europe open to Hitler's power, and Japan glowering on the other side of the globe. It was amid these facts and looming prospects that I entered upon my duties as Prime Minister and Minister of Defence and addressed myself to the first task of forming a Government of all parties to conduct His Majesty's business at home and abroad by whatever means might be deemed best suited to the national interest.

Five years later almost to a day it was possible to take a more favourable view of our circumstances. Italy was conquered and Mussolini slain. The mighty German Army had surrendered unconditionally. Hitler had committed suicide. In addition to the immense captures by General Eisenhower, nearly three million German soldiers were taken prisoners in twenty-four hours by Field-Marshal Alexander in Italy and Field-Marshal Montgomery in Germany. France was liberated, rallied, and revived. Hand in hand with our Allies, the two mightiest empires in the world, we advanced to the swift annihilation of Japanese resistance. The contrast was certainly remarkable. The road across these five years was long, hard, and perilous. Those who perished upon it did not give their lives in vain. Those who marched forward to the end will always be proud to have trodden it with honour.

\* \* \* \* \*

In giving an account of my stewardship and in telling the tale of the famous National Coalition Government it is my first duty to make plain the scale and force of the contribution which Great Britain and her Empire, whom danger only united more tensely, made to what eventually became the Common Cause of so many States and nations. I do this with no desire to make invidious comparisons or rouse purposeless rivalries with our greatest Ally, the United States, to whom we owe immeasurable and enduring gratitude. But it is to the combined interest of the English-speaking world that the magnitude of the British war-making effort should be known and realised. I have therefore had a table made which I print on the following page, which covers the whole period of the war. This shows that until July 1944 Britain and her Empire had a substantially larger number of divisions *in contact with the enemy* than the United States. This general figure includes not only the European and African spheres, but also all the war in Asia against Japan. Till the arrival in Normandy in the autumn of 1944 of the great mass of the American Army, we had always the right to speak at least as an equal and usually as the predominant partner in every theatre of war except the Pacific and the Australasian; and this remains also true, up to the time mentioned, of the aggregation of all divisions in all theatres for any given month. From July 1944 the fighting front of the United States, as represented by divisions in contact with the enemy, became increasingly predominant, and so continued, mounting and triumphant, till the final victory ten months later.

Another comparison which I have made shows that the British and Empire sacrifice in loss of life was even greater than that of our valiant Ally. The British total dead, and missing, presumed dead, of the armed forces amounted to 303,240, to which should be added over 109,000 from the Dominions, India, and the Colonies, a total of over 412,240. This figure does not include 60,500 civilians killed in the air raids on the United Kingdom, nor the losses of our Merchant Navy and fishermen, which amounted to about 30,000. Against this figure the United States mourn the deaths in the Army and Air Force, the Navy, Marines, and Coastguard, of 322,188.* I cite these sombre Rolls of Honour in the confident faith that the equal comradeship sanctified by so much precious blood will continue to command the reverence and inspire the conduct of the English-speaking world.

On the seas the United States naturally bore almost the entire weight of the war in the Pacific, and the decisive battles which they fought near Midway Island, at Guadalcanal, and in the Coral Sea in 1942 gained for them the whole initiative in that vast ocean domain, and opened to them the assault of all the Japanese conquests, and eventually of Japan herself. The American Navy could not at the same time carry the main burden in the Atlantic and the Mediterranean. Here again it is a duty to set down the facts. Out of 781 German and 85 Italian U-boats destroyed in

* Eisenhower, *Crusade in Europe*, p. 1.

## LAND FORCES IN FIGHTING CONTACT WITH THE ENEMY
## "EQUIVALENT DIVISIONS"

| | BRITISH EMPIRE | | | U.S.A. | | |
| | Western Theatre | Eastern Theatre | Total | Western Theatre | Eastern Theatre | Total |
|---|---|---|---|---|---|---|
| Jan. 1, 1940 | $5\frac{3}{8}$ | — | $5\frac{3}{8}$ (a) | — | — | — |
| July 1, 1940 | 6 | — | 6 | — | — | — |
| Jan. 1, 1941 | $10\frac{1}{3}$ | — | $10\frac{1}{3}$ (b) | — | — | — |
| July 1, 1941 | 13 | — | 13 (b) | — | — | — |
| Jan. 1, 1942 | $7\frac{2}{3}$ | 7 | $14\frac{2}{3}$ | — | $2\frac{2}{3}$ | $2\frac{2}{3}$ (c) |
| July 1, 1942 | 10 | $4\frac{2}{3}$ | $14\frac{2}{3}$ | — | $8\frac{1}{3}$ | $8\frac{1}{3}$ |
| Jan. 1, 1943 | $10\frac{1}{3}$ | $8\frac{2}{3}$ | 19 | 5 | 10 | 15 |
| July 1, 1943 | $16\frac{2}{3}$ | $7\frac{2}{3}$ | $24\frac{1}{3}$ | 10 | $12\frac{1}{3}$ | $22\frac{1}{3}$ |
| Jan. 1, 1944 | $11\frac{1}{3}$ | $12\frac{1}{3}$ | $23\frac{2}{3}$ | $6\frac{2}{3}$ | $9\frac{1}{3}$ | 16 |
| July 1, 1944 | $22\frac{2}{3}$ | 16 | $38\frac{2}{3}$ | 25 | 17 | 42 |
| Jan. 1, 1945 | $30\frac{1}{3}$ | $18\frac{2}{3}$ | 49 | $55\frac{2}{3}$ | $23\frac{1}{3}$ | 79 |

NOTES AND ASSUMPTIONS

(a) B.E.F. in France.
(b) Excludes guerrillas in Abyssinia.
(c) Excludes Filipino troops.

The dividing line between the Eastern and Western theatres is taken as a north/south line through Karachi.

The following are NOT taken as operational theatres:

N.W. Frontier of India; Gibraltar; West Africa; Iceland; Hawaii; Palestine; Iraq; Syria (except on July 1, 1941).

Malta is taken as an operational theatre; also Alaska from Jan. 1942 to July 1943.

Foreign contingents—e.g., Free French, Poles, Czechs—are NOT included.

the European theatre, the Atlantic and Indian Oceans, 594 were accounted for by British sea and air forces, who also disposed of all the German battleships, cruisers, and destroyers, besides destroying or capturing the whole Italian Fleet.

The table of U-boat losses is as follows:

U-BOAT LOSSES

| Destroyed by | German | Italian | Japanese |
|---|---|---|---|
| *British Forces       ..      ..      .. | 525 | 69 | 9½ |
| *United States Forces   ..      .. | 174 | 5 | 110½ |
| Other and unknown causes   .. | 82 | 11 | 10 |
| Totals   ..      .. | 781 | 85 | 130 |

Grand total of U-boats destroyed: 996

In the air superb efforts were made by the United States to come into action, especially with their daylight Fortress bombers, on the greatest scale from the earliest moment after Pearl Harbour, and their power was used both against Japan and from the British Isles against Germany.    However, when we reached Casablanca in January 1943 it is a fact that no single American bomber plane had cast a daylight bomb on Germany. Very soon the fruition of the great exertions they were making was to come, but up till the end of 1943 the British discharge of bombs upon Germany had in the aggregate exceeded by eight tons to one those cast from American machines by day or night, and it was only in the spring of 1944 that the preponderance of discharge was achieved by the United States. Here, as in the armies and on the sea, we ran the full course from the beginning, and it was not until 1944 that we were overtaken and surpassed by the tremendous war effort of the United States.

It must be remembered that our munitions effort from the beginning of Lend-Lease in January 1941 was increased by over one-fifth through the generosity of the United States. With the

---

* The terms British and United States Forces include Allied Forces under their operational control. Where fractional losses are shown the "kill" was shared. There were many cases of shared "kills", but in the German totals the fractions add up to whole numbers.

materials and weapons which they gave us we were actually able to wage war *as if we were a nation of fifty-eight millions instead of forty-eight*. In shipping also the marvellous production of Liberty Ships enabled the flow of supplies to be maintained across the Atlantic. On the other hand, the analysis of shipping losses by enemy action suffered by all nations throughout the war should be borne in mind. Here are the figures:

| Nationality | Losses in Gross Tons | Percentage |
|---|---|---|
| British      ..      ..      ..      .. | 11,357,000 | 54 |
| United States      ..      ..      .. | 3,334,000 | 16 |
| All other nations (outside enemy control)      ..      ..      .. | 6,503,000 | 30 |
| | 21,194,000 | 100 |

Of these losses 80 per cent. were suffered in the Atlantic Ocean, including British coastal waters and the North Sea. Only 5 per cent. were lost in the Pacific.

This is all set down, not to claim undue credit, but to establish on a footing capable of commanding fair-minded respect the intense output in every form of war activity of the people of this small Island, upon whom in the crisis of the world's history the brunt fell.

\* \* \* \* \*

It is probably easier to form a Cabinet, especially a Coalition Cabinet, in the heat of battle than in quiet times. The sense of duty dominates all else, and personal claims recede. Once the main arrangements had been settled with the leaders of the other parties, with the formal authority of their organisations, the attitude of all those I sent for was like that of soldiers in action, who go to the places assigned to them at once without question. The Party basis being officially established, it seemed to me that no sense of Self entered into the minds of any of the very large number of gentlemen I had to see. If some few hesitated it was only because of public considerations. Even more did this high standard of behaviour apply to the large number of Conservative and National Liberal Ministers who had to leave their offices and break their careers, and at this moment of surpassing interest and excitement to step out of official life, in many cases for ever.

The Conservatives had a majority of more than one hundred

and twenty over all other parties in the House combined. Mr. Chamberlain was their chosen leader. I could not but realise that his supersession by me must be very unpleasant to many of them, after all my long years of criticism and often fierce reproach. Besides this, it must be evident to the majority of them how my life had been passed in friction or actual strife with the Conservative Party, that I had left them on Free Trade and had later returned to them as Chancellor of the Exchequer. After that I had been for many years their leading opponent on India, on foreign policy, and on the lack of preparations for war. To accept me as Prime Minister was to them very difficult. It caused pain to many honourable men. Moreover, loyalty to the chosen leader of the party is a prime characteristic of the Conservatives. If they had on some questions fallen short of their duty to the nation in the years before the war, it was because of this sense of loyalty to their appointed chief. None of these considerations caused me the slightest anxiety. I knew they were all drowned by the cannonade.

In the first instance I had offered to Mr. Chamberlain, and he had accepted, the Leadership of the House of Commons, as well as the Lord Presidency. Nothing had been published. Mr. Attlee informed me that the Labour Party would not work easily under this arrangement. In a Coalition the Leadership of the House must be generally acceptable. I put this point to Mr. Chamberlain, and, with his ready agreement, I took the Leadership myself, and held it till February 1942. During this time Mr. Attlee acted as my deputy and did the daily work. His long experience in Opposition was of great value. I came down only on the most serious occasions. These were, however, recurrent. Many Conservatives felt that their party leader had been slighted. Everyone admired his personal conduct. On his first entry into the House in his new capacity (May 13) the whole of his party—the large majority of the House—rose and received him in a vehement demonstration of sympathy and regard. In the early weeks it was from the Labour benches that I was mainly greeted. But Mr. Chamberlain's loyalty and support was steadfast, and I was sure of myself.

There was considerable pressure by elements of the Labour Party, and by some of those many able and ardent figures who had not been included in the new Government, for a purge of the "guilty men" and of Ministers who had been responsible

9

for Munich or could be criticised for the many shortcomings in our war preparation. Among these Lord Halifax, Lord Simon, and Sir Samuel Hoare were the principal targets. But this was no time for proscriptions of able, patriotic men of long experience in high office. If the censorious people could have had their way at least a third of the Conservative Ministers would have been forced to resign. Considering that Mr. Chamberlain was the Leader of the Conservative Party, it was plain that this movement would be destructive of the national unity. Moreover, I had no need to ask myself whether all the blame lay on one side. Official responsibility rested upon the Government of the time. But moral responsibilities were more widely spread. A long, formidable list of quotations from speeches and votes recorded by Labour, and not less by Liberal, Ministers, all of which had been stultified by events, was in my mind and available in detail. No one had more right than I to pass a sponge across the past. I therefore resisted these disruptive tendencies. "If the present," I said a few weeks later, "tries to sit in judgment on the past it will lose the future." This argument and the awful weight of the hour quelled the would-be heresy-hunters.

* * * * *

Early on the morning of May 11 I sent a message to Mr. Chamberlain: "No one changes houses for a month." This avoided petty inconveniences during the crisis of the battle. I continued to live at Admiralty House, and made its Map Room and the fine rooms downstairs my temporary headquarters. I reported to Mr. Chamberlain my talk with Mr. Attlee and the progress made in forming the new Administration. "I hope to have the War Cabinet and the Fighting Services complete to-night for the King. The haste is necessitated by the battle. . . . As we [two] must work so closely together, I hope you will not find it inconvenient to occupy once again your old quarters which we both know so well in No. 11."* I added:

I do not think there is any necessity for a Cabinet to-day, as the Armies and other Services are fighting in accordance with pre-arranged plans. I should be very glad however if you and Edward [Halifax] would come to the Admiralty War Room at 12.30 p.m., so that we could look at the maps and talk things over.

* The house in Downing Street usually occupied by the Chancellor of the Exchequer.

British and French advanced forces are already on the Antwerp-Namur line, and there seem to be very good hopes that this line will be strongly occupied by the Allied armies before it can be assailed. This should be achieved in about forty-eight hours, and might be thought to be very important. Meanwhile the Germans have not yet forced the Albert Canal, and the Belgians are reported to be fighting well. The Dutch also are making a stubborn resistance.

*　　*　　*　　*　　*

My experiences in those first days were peculiar. One lived with the battle, upon which all thoughts were centred, and about which nothing could be done. All the time there was the Government to form and the gentlemen to see and the party balances to be adjusted. I cannot remember, nor do my records show, how all the hours were spent. A British Ministry at that time contained between sixty and seventy Ministers of the Crown, and all these had to be fitted in like a jig-saw puzzle, in this case having regard to the claims of three parties. It was necessary for me to see not only all the principal figures, but, for a few minutes at least, the crowd of able men who were to be chosen for important tasks. In forming a Coalition Government the Prime Minister has to attach due weight to the wishes of the party leaders as to who among their followers shall have the offices allotted to the party. By this principle I was mainly governed. If any who deserved better were left out on the advice of their party authorities, or even in spite of that advice, I can only express regret. On the whole however the difficulties were few.

In Clement Attlee I had a colleague of war experience long versed in the House of Commons. Our only differences in outlook were about Socialism, but these were swamped by a war soon to involve the almost complete subordination of the individual to the State. We worked together with perfect ease and confidence during the whole period of the Government. Mr. Arthur Greenwood was a wise counsellor of high courage and a good and helpful friend.

Sir Archibald Sinclair, as official Leader of the Liberal Party, found it embarrassing to accept the office of Air Minister, because his followers felt he should instead have a seat in the War Cabinet. But this ran contrary to the principle of a small War Cabinet. I therefore proposed that he should join the War Cabinet when any matter affecting fundamental political issues or party union

II

was involved. He was my friend, and had been my second-in-command when in 1916 I commanded the 6th Royal Scots Fusiliers at Ploegsteert ("Plug Street"), and personally longed to enter upon the great sphere of action I had reserved for him. After no little intercourse this had been amicably settled. Mr. Bevin, with whom I had made acquaintance at the beginning of the war, in trying to mitigate the severe Admiralty demands for trawlers, had to consult the Transport and General Workers' Union, of which he was secretary, before he could join the team in the most important office of Minister of Labour. This took two or three days, but it was worth it. The union, the largest of all in Britain, said unanimously that he was to do it, and stuck solid for five years till we won.

The greatest difficulty was with Lord Beaverbrook. I believed he had services to render of a very high quality. I had resolved, as the result of my experiences in the previous war, to remove the supply and design of aircraft from the Air Ministry, and I wished him to become the Minister of Aircraft Production. He seemed at first reluctant to undertake the task, and of course the Air Ministry did not like having their Supply Branch separated from them. There were other resistances to his appointment. I felt sure however that our life depended upon the flow of new aircraft; I needed his vital and vibrant energy, and I persisted in my view.

★ ★ ★ ★ ★

In deference to prevailing opinions expressed in Parliament and the Press it was necessary that the War Cabinet should be small. I therefore began by having only five members, of whom one only, the Foreign Secretary, had a department. These were naturally the leading party politicians of the day. For the convenient conduct of business it was necessary that the Chancellor of the Exchequer and the Leader of the Liberal Party should usually be present, and as time passed the number of "constant attenders" grew. But all the responsibility was laid upon the five War Cabinet Ministers. They were the only ones who had the right to have their heads cut off on Tower Hill if we did not win. The rest could suffer for departmental shortcomings, but not on account of the policy of the State. Apart from the War Cabinet, any one could say: "I cannot take the responsibility for this or that." The burden of policy was borne at a higher level. This saved many people a lot of worry in the days which were immediately to fall upon us.

Here are the stages by which the National Coalition Government was built up day by day in the course of the great battle.

## THE WAR CABINET

### May 11, 1940

| | | |
|---|---|---|
| Prime Minister, First Lord of the Treasury, Minister of Defence, and Leader of the House of Commons | * Mr. Churchill | Conservative |
| Lord President of the Council | * Mr. Neville Chamberlain | Conservative |
| Lord Privy Seal | Mr. C. R. Attlee | Labour |
| Secretary of State for Foreign Affairs | * Lord Halifax | Conservative |
| Minister without Portfolio | Mr. Arthur Greenwood | Labour |

### MINISTERS OF CABINET RANK

| | | |
|---|---|---|
| First Lord of the Admiralty | Mr. A. V. Alexander | Labour |
| Secretary of State for War | * Mr. Anthony Eden | Conservative |
| Secretary of State for Air | Sir Archibald Sinclair | Liberal |

### May 12

| | | |
|---|---|---|
| Lord Chancellor | * Sir John Simon [becoming Lord Simon] | National Liberal |
| Chancellor of the Exchequer | * Sir Kingsley Wood | Conservative |
| Home Secretary and Minister of Home Security | * Sir John Anderson | Non-Party |
| Secretary of State for the Colonies | Lord Lloyd | Conservative |
| President of the Board of Trade | Sir Andrew Duncan | Non-Party |
| Minister of Supply | Mr. Herbert Morrison | Labour |
| Minister of Information | Mr. Alfred Duff Cooper | Conservative |

### May 13

| | | |
|---|---|---|
| Secretary of State for India and Burma | Mr. L. S. Amery | Conservative |
| Minister of Health | * Mr. Malcolm Macdonald | National Labour |
| Minister of Labour and National Service | Mr. Ernest Bevin | Labour |
| Minister of Food | * Lord Woolton | Non-Party |

* Members in previous Administration.

### May 14

| | | |
|---|---|---|
| Dominions Secretary and Leader of the House of Lords | * Viscount Caldecote | Conservative |
| Secretary of State for Scotland | * Mr. Ernest Brown | National Liberal |
| Minister of Aircraft Production | Lord Beaverbrook | Conservative |
| President of the Board of Education | * Mr. Herwald Ramsbotham | Conservative |
| Minister of Agriculture | * Mr. Robert Hudson | Conservative |
| Minister of Transport | * Sir John Reith | Non-Party |
| Minister of Shipping | * Mr. Ronald Cross | Conservative |
| Minister of Economic Warfare | Mr. Hugh Dalton | Labour |
| Chancellor of the Duchy of Lancaster | * Lord Hankey | Non-Party |

### May 15

| | | |
|---|---|---|
| Minister of Pensions | * Sir W. J. Womersley | Conservative |
| Postmaster-General | * Mr. W. S. Morrison | Conservative |
| Paymaster-General | Lord Cranborne | Conservative |
| Attorney-General | * Sir Donald Somervell, K.C. | Conservative |
| Lord Advocate | * Mr. T. M. Cooper, K.C. | Conservative |
| Solicitor-General | Sir William Jowitt, K.C. | Labour |
| Solicitor-General for Scotland | * Mr. J. S. C. Reid, K.C. | Conservative |

\* \* \* \* \*

In my long political experience I had held most of the great offices of State, but I readily admit that the post which had now fallen to me was the one I liked the best. Power, for the sake of lording it over fellow-creatures or adding to personal pomp, is rightly judged base. But power in a national crisis, when a man believes he knows what orders should be given, is a blessing. In any sphere of action there can be no comparison between the positions of number one and numbers two, three, or four. The duties and the problems of all persons other than number one are quite different and in many ways more difficult. It is always a misfortune when number two or three has to initiate a dominant plan or policy. He has to consider not only the merits of the policy, but the mind of his chief; not only what to advise, but

\* Members in previous Administration.

14

what it is proper for him in his station to advise; not only what to do, but how to get it agreed, and how to get it done. Moreover, number two or three will have to reckon with numbers four, five, and six, or maybe some bright outsider, number twenty. Ambition, not so much for vulgar ends, but for fame, glints in every mind. There are always several points of view which may be right, and many which are plausible. I was ruined for the time being in 1915 over the Dardanelles, and a supreme enterprise was cast away, through my trying to carry out a major and cardinal operation of war from a subordinate position. Men are ill-advised to try such ventures. This lesson had sunk into my nature.

At the top there are great simplifications. An accepted leader has only to be sure of what it is best to do, or at least to have made up his mind about it. The loyalties which centre upon number one are enormous. If he trips he must be sustained. If he makes mistakes they must be covered. If he sleeps he must not be wantonly disturbed. If he is no good he must be pole-axed. But this last extreme process cannot be carried out every day; and certainly not in the days just after he has been chosen.

* * * * *

The fundamental changes in the machinery of war direction were more real than apparent. "A Constitution", said Napoleon, "should be short and obscure." The existing organisms remained intact. No official personalities were changed. The War Cabinet and the Chiefs of Staff Committee at first continued to meet every day as they had done before. In calling myself, with the King's approval, Minister of Defence I had made no legal or constitutional change. I had been careful not to define my rights and duties. I asked for no special powers either from the Crown or Parliament. It was however understood and accepted that I should assume the general direction of the war, subject to the support of the War Cabinet and of the House of Commons. The key change which occurred on my taking over was of course the supervision and direction of the Chiefs of Staff Committee by a Minister of Defence with undefined powers. As this Minister was also the Prime Minister, he had all the rights inherent in that office, including very wide powers of selection and removal of all professional and political personages. Thus for the first time the

Chiefs of Staff Committee assumed its due and proper place in direct daily contact with the executive head of the Government, and in accord with him had full control over the conduct of the war and the armed forces.

The position of the First Lord of the Admiralty and of the Secretaries of State for War and Air was decisively affected in fact though not in form. They were not members of the War Cabinet, nor did they attend the meetings of the Chiefs of Staff Committee. They remained entirely responsible for their departments, but rapidly and almost imperceptibly ceased to be responsible for the formulation of strategic plans and the day-to-day conduct of operations. These were settled by the Chiefs of Staff Committee acting directly under the Minister of Defence and Prime Minister, and thus with the authority of the War Cabinet. The three Service Ministers, very able and trusted friends of mine whom I had picked for these duties, stood on no ceremony. They organised and administered the ever-growing forces, and helped all they could in the easy, practical English fashion. They had the fullest information by virtue of their membership of the Defence Committee, and constant access to me. Their professional subordinates, the Chiefs of Staff, discussed everything with them and treated them with the utmost respect. But there was an integral direction of the war to which they loyally submitted. There never was an occasion when their powers were abrogated or challenged, and anyone in this circle could always speak his mind; but the actual war direction soon settled into a very few hands, and what had seemed so difficult before became much more sin ole—apart of course from Hitler. In spite of the turbulence of events and the many disasters we had to endure the machinery worked almost automatically, and one lived in a stream of coherent thought capable of being translated with great rapidity into executive action.

*    *    *    *    *

Although the awful battle was now going on across the Channel, and the reader is no doubt impatient to get there, it may be well at this point to describe the system and machinery for conducting military and other affairs which I set on foot and practised from my earliest days of power. I am a strong believer in transacting official business by *the Written Word*. No doubt, surveyed in the after-time, much that is set down from hour to hour under the impact of events may be lacking in proportion or may not come true. I am willing to take my chance of that. It is always better, except in the hierarchy of military discipline, to express opinions and wishes rather than to give orders. Still, written directives coming personally from the lawfully-constituted Head of the Government and Minister specially charged with defence counted to such an extent that, though not expressed as orders, they very often found their fruition in action.

To make sure that my name was not used loosely, I issued during the crisis of July the following minute:

*Prime Minister to General Ismay, C.I.G.S., and*
*Sir Edward Bridges*                                        19.VII.40

Let it be very clearly understood that all directions emanating from me are made in writing, or should be immediately afterwards confirmed in writing, and that I do not accept any responsibility for matters relating to national defence on which I am alleged to have given decisions unless they are recorded in writing.

When I woke about 8 a.m. I read all the telegrams, and from my bed dictated a continuous flow of minutes and directives to the departments and to the Chiefs of Staff Committee. These were typed in relays as they were done, and handed at once to General Ismay, Deputy-Secretary (Military) to the War Cabinet, and my representative on the Chiefs of Staff Committee, who came to see me early each morning. Thus he usually had a good deal in writing to bring before the Chiefs of Staff Committee when they met at ten-thirty. They gave all consideration to my views at the same time as they discussed the general situation. Thus between three and five o'clock in the afternoon, unless there were some difficulties between us requiring further consultation, there was ready a whole series of orders and telegrams sent by me or by the Chiefs of Staff and agreed between us, usually giving all the decisions immediately required.

In total war it is quite impossible to draw any precise line between military and non-military problems. That no such friction occurred between the military staff and the War Cabinet staff was due primarily to the personality of Sir Edward Bridges, Secretary to the War Cabinet. Not only was this son of a former

Poet Laureate an extremely competent and tireless worker, but he was also a man of exceptional force, ability, and personal charm, without a trace of jealousy in his nature. All that mattered to him was that the War Cabinet Secretariat as a whole should serve the Prime Minister and War Cabinet to the very best of their ability. No thought of his own personal position ever entered his mind, and never a cross word passed between the civil and military officers of the Secretariat.

In larger questions, or if there were any differences of view, I called a meeting of the War Cabinet Defence Committee, which at the outset comprised Mr. Chamberlain, Mr. Attlee, and the three Service Ministers, with the Chiefs of Staff in attendance. These formal meetings got fewer after 1941.* As the machine began to work more smoothly I came to the conclusion that the daily meetings of the War Cabinet with the Chiefs of Staff present were no longer necessary. I therefore eventually instituted what came to be known among ourselves as the "Monday Cabinet Parade". Every Monday there was a considerable gathering—all the War Cabinet, the Service Ministers, and the Minister of Home Security, the Chancellor of the Exchequer, the Secretaries of State for the Dominions and for India, the Minister of Information, the Chiefs of Staff, and the official head of the Foreign Office. At these meetings each Chief of Staff in turn unfolded his account of all that had happened during the previous seven days; and the Foreign Secretary followed them with his story of any important developments in foreign affairs. On other days of the week the War Cabinet sat alone, and all important matters requiring decision were brought before them. Other Ministers primarily concerned with the subjects to be discussed attended for their own particular problems. The members of the War Cabinet had the fullest circulation of all papers affecting the war, and saw all important telegrams sent by me. As confidence grew the War Cabinet intervened less actively in operational matters, though they watched them with close attention and full knowledge. They took almost the whole weight of home and party affairs off my shoulders, thus setting me free to concentrate upon the main theme. With regard to all future operations of importance I always consulted them in good time; but, while they gave

* The Defence Committee met 40 times in 1940, 76 in 1941, 20 in 1942, 14 in 1943, and 10 in 1944

careful consideration to the issues involved, they frequently asked not to be informed of dates and details, and indeed on several occasions stopped me when I was about to unfold these to them.

I had never intended to embody the office of Minister of Defence in a department. This would have required legislation, and all the delicate adjustments I have described, most of which settled themselves by personal goodwill, would have had to be thrashed out in a process of ill-timed constitution-making. There was however in existence and activity under the personal direction of the Prime Minister the Military Wing of the War Cabinet Secretariat, which had in pre-war days been the Secretariat of the Committee of Imperial Defence. At the head of this stood General Ismay, with Colonel Hollis and Colonel Jacob as his two principals, and a group of specially-selected younger officers drawn from all three Services. This Secretariat became the staff of the office of the Minister of Defence. My debt to its members is immeasurable. General Ismay, Colonel Hollis, and Colonel Jacob rose steadily in rank and repute as the war proceeded, and none of them was changed. Displacements in a sphere so intimate and so concerned with secret matters are detrimental to continuous and efficient dispatch of business.

After some early changes almost equal stability was preserved in the Chiefs of Staff Committee. On the expiry of his term as Chief of the Air Staff, in September 1940, Air Marshal Newall became Governor-General of New Zealand, and was succeeded by Air Marshal Portal, who was the accepted star of the Air Force. Portal remained with me throughout the war. Sir John Dill, who had succeeded General Ironside in May 1940, remained C.I.G.S. until he accompanied me to Washington in December 1941. I then made him my personal Military Representative with the President and head of our Joint Staff Mission. His relations with General Marshall, Chief of Staff of the United States Army, became a priceless link in all our business, and when he died in harness some two years later he was accorded the unique honour of a resting-place in Arlington Cemetery, the Valhalla hitherto reserved exclusively for American warriors. He was succeeded as C.I.G.S. by Sir Alan Brooke, who stayed with me till the end.

From 1941, for nearly four years, the early part of which was passed in much misfortune and disappointment, the only change made in this small band either among the Chiefs or in the Defence staff was due to the death in harness of Admiral Pound. This may well be a record in British military history. A similar degree of continuity was achieved by President Roosevelt in his own circle. The United States Chiefs of Staff—General Marshall, Admiral King, and General Arnold, subsequently joined by Admiral Leahy—started together on the American entry into the war, and were never changed. As both the British and Americans presently formed the Combined Chiefs of Staff Committee this was an inestimable advantage for all. Nothing like it between allies has ever been known before.

I cannot say that we never differed among ourselves even at home, but a kind of understanding grew up between me and the British Chiefs of Staff that we should convince and persuade rather than try to overrule each other. This was of course helped by the fact that we spoke the same technical language, and possessed a large common body of military doctrine and war experience. In this ever-changing scene we moved as one, and the War Cabinet clothed us with ever more discretion, and sustained us with unwearied and unflinching constancy. There was no division, as in the previous war, between politicians and soldiers, between the "Frocks" and the "Brass Hats"—odious terms which darkened counsel. We came very close together indeed, and friendships were formed which I believe were deeply valued.

The efficiency of a war Administration depends mainly upon whether decisions emanating from the highest approved authority are in fact strictly, faithfully, and punctually obeyed. This we achieved in Britain in this time of crisis owing to the intense fidelity, comprehension, and whole-hearted resolve of the War Cabinet upon the essential purpose to which we had devoted ourselves. According to the directions given, ships, troops, and aeroplanes moved, and the wheels of factories spun. By all these processes, and by the confidence, indulgence, and loyalty by which I was upborne, I was soon able to give an integral direction to almost every aspect of the war. This was really necessary, because times were so very bad. The method was accepted because everyone realised how near were death and ruin. Not only individual death, which is the universal experience, stood near, but, incomparably more commanding, the life of Britain, her message, and her glory.

* * * * *

Any account of the methods of government which developed under the National Coalition would be incomplete without an explanation of the series of personal messages which I sent to the President of the United States and the heads of other foreign countries and the Dominions Governments. This correspondence must be described. Having obtained from the Cabinet any specific decisions required on policy, I composed and dictated these documents myself, for the most part on the basis that they were intimate and informal correspondence with friends and fellow-workers. One can usually put one's thoughts better in one's own words. It was only occasionally that I read the text to the Cabinet beforehand. Knowing their views, I used the ease and freedom needed for the doing of my work. I was of course hand-in-glove with the Foreign Secretary and his department, and any differences of view were settled together. I circulated these telegrams, in some cases after they had been sent, to the principal members of the War Cabinet, and, where he was concerned, to the Dominions Secretary. Before dispatching them I of course had my points and facts checked departmentally, and nearly all military messages passed through Ismay's hands to the Chiefs of Staff. This correspondence in no way ran counter to the official communications or the work of the Ambassadors. It became however in fact the channel of much vital business, and played a part in my conduct of the war not less, and sometimes even more important than my duties as Minister of Defence.

The very select circle, who were entirely free to express their opinion, were almost invariably content with the drafts and gave me an increasing measure of confidence. Differences with American authorities for instance, insuperable at the second level, were settled often in a few hours by direct contact at the top. Indeed, as time went on the efficacy of this top-level transaction of business was so apparent that I had to be careful not to let it become a vehicle for ordinary departmental affairs. I had repeatedly to refuse the requests of my colleagues to address the President personally on important matters of detail. Had these intruded unduly upon the personal correspondence they would soon have destroyed its privacy and consequently its value.

My relations with the President gradually became so close that the chief business between our two countries was virtually conducted by these personal interchanges between him and me. In this way our perfect understanding was gained. As Head of the State as well as Head of the Government, Roosevelt spoke and acted with authority in every sphere; and, carrying the War Cabinet with me, I represented Great Britain with almost equal latitude. Thus a very high degree of concert was obtained, and the saving in time and the reduction in the number of people informed were both invaluable. I sent my cables to the American Embassy in London, which was in direct touch with the President at the White House through special coding machines. The speed with which answers were received and things settled was aided by clock-time. Any message which I prepared in the evening, at night, or even up to two o'clock in the morning, would reach the President before he went to bed, and very often his answer would come back to me when I woke the next morning. In all I sent him nine hundred and fifty messages, and received about eight hundred in reply. I felt I was in contact with a very great man, who was also a warm-hearted friend and the foremost champion of the high causes which we served.

*　　*　　*　　*　　*

The Cabinet being favourable to my trying to obtain destroyers from the American Government, I drafted during the afternoon of May 15 my first message to President Roosevelt since I became Prime Minister. To preserve the informality of our correspondence I signed myself "Former Naval Person", and to this fancy I adhered almost without exception throughout the war.

Although I have changed my office, I am sure you would not wish me to discontinue our intimate private correspondence. As you are no doubt aware, the scene has darkened swiftly. The enemy have a marked preponderance in the air, and their new technique is making a deep impression upon the French. I think myself the battle on land has only just begun, and I should like to see the masses engage. Up to the present Hitler is working with specialised units in tanks and air. The small countries are simply smashed up, one by one, like matchwood. We must expect, though it is not yet certain, that Mussolini will hurry in to share the loot of civilisation. We expect to be attacked here ourselves, both from the air and by parachute and air-borne troops, in the near future, and are getting ready for them. If necessary, we shall continue the war alone, and we are not afraid of that.

But I trust you realise, Mr. President, that the voice and force of the United States may count for nothing if they are withheld too long. You may have a completely subjugated, Nazified Europe established with astonishing swiftness, and the weight may be more than we can bear. All I ask now is that you should proclaim non-belligerency, which would mean that you would help us with everything short of actually engaging armed forces. Immediate needs are, first of all, the loan of forty or fifty of your older destroyers to bridge the gap between what we have now and the large new construction we put in hand at the beginning of the war. This time next year we shall have plenty. But if in the interval Italy comes in against us with another hundred submarines we may be strained to breaking-point. Secondly, we want several hundred of the latest types of aircraft, of which you are now getting delivery. These can be repaid by those now being constructed in the United States for us. Thirdly, anti-aircraft equipment and ammunition, of which again there will be plenty next year, if we are alive to see it. Fourthly, the fact that our ore supply is being compromised from Sweden, from North Africa, and perhaps from Northern Spain, makes it necessary to purchase steel in the United States. This also applies to other materials. We shall go on paying dollars for as long as we can, but I should like to feel reasonably sure that when we can pay no more you will give us the stuff all the same. Fifthly, we have many reports of possible German parachute or air-borne descents in Ireland. The visit of a United States squadron to Irish ports, which might well be prolonged, would be invaluable. Sixthly, I am looking to you to keep the Japanese quiet in the Pacific, using Singapore in any way convenient. The details of the material which we have in hand will be communicated to you separately.

With all good wishes and respect . . .

On May 18 a reply was received from the President welcoming the continuance of our private correspondence and dealing with my specific requests. The loan or gift of the forty or fifty older destroyers, it was stated, would require the authorisation of Congress, and the moment was not opportune. He would facilitate to the utmost the Allied Governments obtaining the latest types of United States aircraft, anti-aircraft equipment, ammunition, and steel. In all this the representations of our agent, the highly competent and devoted Mr. Purvis (presently to give his life in an air accident), would receive most favourable consideration. The President would consider carefully my suggestion that a United States squadron might visit Irish ports. About the Japanese, he merely pointed to the concentration of the American fleet at Pearl Harbour.

\*　　\*　　\*　　\*　　\*

On Monday, May 13, I asked the House of Commons, which had been specially summoned, for a vote of confidence in the new Administration. After reporting the progress which had been made in filling the various offices, I said: "I have nothing to offer but blood, toil, tears, and sweat." In all our long history no Prime Minister had ever been able to present to Parliament and the nation a programme at once so short and so popular. I ended:

You ask, What is our policy? I will say: It is to wage war, by sea, land, and air, with all our might and with all the strength that God can give us: to wage war against a monstrous tyranny, never surpassed in the dark, lamentable catalogue of human crime. That is our policy. You ask, What is our aim? I can answer in one word: Victory—victory at all costs, victory in spite of all terror; victory, however long and hard the road may be; for without victory there is no survival. Let that be realised: no survival for the British Empire; no survival for all that the British Empire has stood for, no survival for the urge and impulse of the ages, that mankind will move forward towards its goal. But I take up my task with buoyancy and hope. I feel sure that our cause will not be suffered to fail among men. At this time I feel entitled to claim the aid of all, and I say, "Come, then, let us go forward together with our united strength."

Upon these simple issues the House voted unanimously, and adjourned till May 21.

\*　　\*　　\*　　\*　　\*

Thus then we all started on our common task. Never did a British Prime Minister receive from Cabinet colleagues the loyal and true aid which I enjoyed during the next five years from these men of all parties in the State. Parliament, while maintaining free and active criticism, gave continuous, overwhelming support to all measures proposed by the Government, and the nation was united and ardent as never before. It was well indeed that this should be so, because events were to come upon us of an order more terrible than anyone had foreseen.

# CHAPTER II

# THE BATTLE OF FRANCE

The First Week. Gamelin
May 10–May 16

*Plan D – The German Order of Battle – German and French Armour
– French and British Advance through Belgium – Holland Overrun –
The Belgian Problem – Accepted Primacy of France in the Military Art
– The Gap in the Ardennes – British Difficulties during the Twilight
War Phase – Progress of Plan D – Bad News of 13th and 14th –
Kleist's Group of Armies Break the French Front – Heavy British Air
Losses – Our Final Limit for Home Defence – Reynaud Telephones
Me Morning of 15th – Destruction of the French Ninth Army Opposite
the Ardennes Gap – "Cease Fire" in Holland – The Italian Menace –
I Fly to Paris – Meeting at the Quai d'Orsay – General Gamelin's
Statement – No Strategic Reserve: "Aucune" – Proposed Attacks on
the German "Bulge" – French Demands for More British Fighter
Squadrons – My Telegram to the Cabinet on the Night of May 16 –
Cabinet Agrees to Send Ten More Fighter Squadrons.*

AT the moment in the evening of May 10 when I became
responsible no fresh decision about meeting the German
invasion of the Low Countries was required from me or
from my colleagues in the new and still unformed Administration. We had long been assured that the French and British
staffs were fully agreed upon General Gamelin's Plan D, and
it had already been in action since dawn. In fact, by the morning
of the 11th the whole vast operation had made great progress.
On the seaward flank General Giraud's Seventh French Army
had already begun its adventurous dash into Holland. In the
centre the British armoured-car patrols of the 12th Lancers were
upon the river Dyle, and to the south of our front all the rest of

General Billotte's First Group of Armies were hastening forward to the Meuse. The opinion of the Allied military chiefs was that Plan D, if successful, would save anything from twelve to fifteen divisions by shortening the front against Germany; and then of course there was the Belgian Army of twenty-two divisions, besides the Dutch Army of ten divisions, without which our total forces in the West were numerically inferior. I did not therefore in the slightest degree wish to interfere with the military plans, and awaited with hope the impending shock.

Nevertheless, if in the after-light we look back upon the scene, the important paper written by the British Chiefs of Staff on September 18, 1939, becomes prominent.* In this it had been affirmed that unless the Belgians were effectively holding their front on the Meuse and the Albert Canal it would be wrong for the British and French to rush to their aid, but that they should rather stand firm on the French frontier, or at the most swing their left hand slightly forward to the line of the Scheldt. Since those days of September 1939 agreement had been reached to carry out General Gamelin's Plan D. Nothing had however happened in the interval to weaken the original view of the British Chiefs of Staff. On the contrary, much had happened to strengthen it. The German Army had grown in strength and maturity with every month that had passed, and they now had a vastly more powerful armour. The French Army, gnawed by Soviet-inspired Communism and chilled by the long, cheerless winter on the front, had actually deteriorated. The Belgian Government, staking their country's life upon Hitler's respect for international law and Belgian neutrality, had not achieved any effective joint planning between their Army chiefs and those of the Allies. The anti-tank obstacles and defensive line which were to have been prepared on the front Namur-Louvain were inadequate and unfinished. The Belgian Army, which contained many brave and resolute men, could hardly brace itself for a conflict for fear of offending neutrality. The Belgian front had been, in fact, overrun at many points by the first wave of German assault, even before General Gamelin gave the signal to execute his long-prepared plan. The most that could now be hoped for was success in that very "encounter battle" which the French High Command had declared itself resolved to avoid.

* See Volume I, first edition, p. 378; second edition, p. 431.

On the outbreak of the war eight months before, the main power of the German Army and Air Force had been concentrated on the invasion and conquest of Poland. Along the whole of the Western front, from Aix-la-Chapelle to the Swiss frontier, there had stood forty-two German divisions without armour. After the French mobilisation France could deploy the equivalent of seventy divisions opposite to them. For reasons which have been explained, it was not deemed possible to attack the Germans then. Very different was the situation on May 10, 1940. The enemy, profiting by the eight months' delay and by the destruction of Poland, had armed, equipped, and trained about 155 divisions, of which ten were armoured ("Panzer"). Hitler's agreement with Stalin had enabled him to reduce the German forces in the East to the smallest proportions. Opposite Russia, according to General Halder, the German Chief of Staff, there was "no more than a light covering force, scarcely fit for collecting customs duties". Without premonition of their own future, the Soviet Government watched the destruction of that "Second Front" in the West for which they were soon to call so vehemently and to wait in agony so long. Hitler was therefore in a position to deliver his onslaught on France with 126 divisions and the whole of the immense armour-weapon of ten Panzer divisions, comprising nearly three thousand armoured vehicles, of which a thousand at least were heavy tanks.

These mighty forces were deployed from the North Sea to Switzerland in the following order:

> *Army Group B*, comprising 28 divisions, under General von Bock, marshalled along the front from the North Sea to Aix-la-Chapelle, was to overrun Holland and Belgium, and thereafter advance into France as the German right wing.
>
> *Army Group A*, of 44 divisions, under General von Rundstedt, constituting the main thrust, was ranged along the front from Aix-la-Chapelle to the Moselle.
>
> *Army Group C*, of 17 divisions, under General von Leeb, held the Rhine from the Moselle to the Swiss frontier.

The O.K.H. (Supreme Army Command) Reserve consisted of about forty-seven divisions, of which twenty were in immediate reserve behind the various Army Groups and twenty-seven in general reserve.

Opposite this array, the exact strength and disposition of which was of course unknown to us, the First Group of Armies, under General Billotte, consisting of fifty-one divisions, of which nine were held in G.Q.G. (Grand-Quartier-Général Réserve), and including nine British divisions, stretched from the end of the Maginot Line near Longwy to the Belgian frontier, and behind that frontier to the sea in front of Dunkirk. The Second and Third Groups of Armies, under Generals Prételat and Besson, consisting, with the reserves, of forty-three divisions, guarded the French frontier from Longwy to Switzerland. In addition the French had the equivalent of nine divisions occupying the Maginot Line—a total of 103 divisions. If the armies of Belgium and Holland became involved, this number would be increased by twenty-two Belgian and ten Dutch divisions. As both these countries were immediately attacked, the grand total of Allied divisions of all qualities nominally available on May 10 was therefore 135, or practically the same number as we now know the enemy possessed. Properly organised and equipped, well trained and led, this force should, according to the standards of the previous war, have had a good chance of bringing the invasion to a stop.

However, the Germans had full freedom to choose the moment, the direction, and the strength of their attack. More than half of the French Army stood on the southern and eastern sectors of France, and the fifty-one French and British divisions of General Billotte's Army Group No. 1, with whatever Belgian and Dutch aid was forthcoming, had to face the onslaught of upwards of seventy hostile divisions under Bock and Rundstedt between Longwy and the sea. The combination of the almost cannon-proof tank and dive-bomber aircraft which had proved so successful in Poland on a smaller scale was again to form the spearhead of the main attack, and a group of five Panzer and three motorised divisions under Kleist, included in German Army Group A, was directed through the Ardennes on Sedan and Monthermé.

To meet such modern forms of war the French deployed about 2,300 tanks, mostly light. Their armoured formations included some powerful modern types, but more than half their total armoured strength was held in dispersed battalions of light tanks, for co-operation with the infantry. Their six armoured divisions,*

* This figure includes the so-called light motorised divisions, which possessed tanks.

**THE FORWARD MOVEMENTS STARTING 10 May**

with which alone they could have countered the massed Panzer assault, were widely distributed over the front, and could not be collected together to operate in coherent action. Britain, the birthplace of the tank, had only just completed the formation and training of her first armoured division (328 tanks), which was still in England.

The German fighter aircraft now concentrated in the West were far superior to the French in numbers and quality. The British Air Force in France comprised the ten fighter squadrons (Hurricanes) which could be spared from vital Home Defence, eight squadrons of Battles, six of Blenheims, and five of Lysanders. Neither the French nor the British air authorities had equipped themselves with dive-bombers, which at this time, as in Poland, became prominent, and were to play an important part in the

29

demoralisation of the French infantry and particularly of their coloured troops.

During the night of May 9–10, heralded by widespread air attacks against airfields, communications, headquarters, and magazines, all the German forces in the Bock and Rundstedt Army Groups sprang forward towards France across the frontiers of Belgium, Holland, and Luxembourg. Complete tactical surprise was achieved in nearly every case. Out of the darkness came suddenly innumerable parties of well-armed, ardent storm troops, often with light artillery, and long before daybreak a hundred and fifty miles of front were aflame. Holland and Belgium, assaulted without the slightest pretext or warning, cried aloud for help. The Dutch had trusted to their water-line; all the sluices not seized or betrayed were opened, and the Dutch frontier guards fired upon the invaders. The Belgians succeeded in destroying the bridges of the Meuse, but the Germans captured intact two across the Albert Canal.

By Plan D, the First Allied Army Group, under General Billotte, with its small but very fine British army, was, from the moment when the Germans violated the frontier, to advance east into Belgium. It was intended to forestall the enemy and stand on the line Meuse–Louvain–Antwerp. In front of that line, along the Meuse and the Albert Canal, lay the main Belgian forces. Should these stem the first German onrush the Army Group would support them. It seemed more probable that the Belgians would be at once thrown back on to the Allied line. And this, in fact, happened. It was assumed that in this case the Belgian resistance would give a short breathing-space, during which the French and British could organise their new position. Except on the critical front of the French Ninth Army, this was accomplished. On the extreme left or seaward flank the Seventh French Army was to seize the islands commanding the mouth of the Scheldt, and, if possible, to assist the Dutch by an advance towards Breda. It was thought that on our southern flank the Ardennes constituted an impassable barrier, and south of that again began the regular fortified Maginot Line, stretching out to the Rhine and along the Rhine to Switzerland. All therefore seemed to depend upon the forward left-handed counter-stroke of the Allied Northern Armies. This again hung upon the speed with which Belgium could be occupied. Everything had been worked out in this way with the utmost detail, and only a signal was necessary to hurl forward the Allied force of well over a million men. At 5.30 a.m. on May 10 Lord Gort received a message from General Georges ordering "Alertes 1, 2, and 3"; namely, instant readiness to move into Belgium. At 6.45 a.m. General Gamelin ordered the execution of Plan D, and the long-prepared scheme of the French High Command, to which the British had subordinated themselves, came at once into action.

\* \* \* \* \*

Mr. Colijn, when as Dutch Prime Minister he visited me in 1937, had explained to me the marvellous efficiency of the Dutch inundations. He could, he explained, by a telephone message from the luncheon table at Chartwell press a button which would confront an invader with impassable water obstacles. But all this was nonsense. The power of a great State against a small one under modern conditions is overwhelming. The Germans broke through at every point, bridging the canals or seizing the locks and water-controls. In a single day all the outer line of the Dutch defences was mastered. At the same time the German Air Force began to use its might upon a defenceless country. Rotterdam was reduced to a blazing ruin. The Hague, Utrecht, and Amsterdam were threatened with the same fate. The Dutch hope that they would be by-passed by the German right-handed swing as in the former war was vain.

However, when the blow fell the Dutch nation instantly rallied to the cause with indomitable courage. Queen Wilhelmina, her family, and the members of her Government were safely brought to England by the Royal Navy, and continued from here to inspire their people and to manage their vast empire overseas. The Queen's Navy and her great merchant fleet were placed unreservedly under British control, and played a redoubtable part in Allied affairs.

The case of Belgium requires more searching statement. Several hundreds of thousands of British and French graves in Belgium mark the struggle of the previous war. The policy of Belgium in the years between the wars had not taken sufficient account of the past. The Belgian leaders saw with worried eyes the internal weakness of France and the vacillating pacifism of Britain. They clung to a strict neutrality. In the years before they were again invaded their attitude towards the two mighty arrays which confronted each other was, officially at any rate, quite impartial. Great allowance must be made for the fearful problems of a small State in such a plight, but the French High Command had for years spoken bitterly of the line taken by the Belgian Government. Their only chance of defending their frontier against a German attack lay in a close alliance with France and Britain. The line of the Albert Canal and other water fronts was highly defensible, and had the British and French armies, aided by the Belgian Army, after the declaration of war, been drawn up on the Belgian frontiers in good time a very strong offensive might have been prepared and launched from these positions against Germany. But the Belgian Government deemed that their safety lay in the most rigid neutrality, and their only hope was founded on German good faith and respect for treaties.

Even after Britain and France had entered into war it was impossible to persuade them to rejoin the old alliance. They declared they would defend their neutrality to the death, and placed nine-tenths of their forces on their German frontier, while at the same time they strictly forbade the Anglo-French army to enter their country and make effective preparations for their defence or for forestalling counter-strokes. The construction of new lines and the anti-tank ditch during the winter of 1939 by the British armies, with the French First Army on their right, along the Franco-Belgian frontier, had been the only measure open to us. It is a haunting question whether the whole policy of Plan D should not have been reviewed upon this basis, and whether we would not have been wiser to stand and fight on the French frontier, and amid these strong defences invite the Belgian Army to fall back upon them, rather than make the hazardous and hurried forward leap to the Dyle or the Albert Canal.

\* \* \* \* \*

No one can understand the decisions of that period without realising the immense authority wielded by the French military leaders and the belief of every French officer that France had the primacy in the military art. France had conducted and carried the main weight of the terrible land fighting from 1914 to 1918. She had lost fourteen hundred thousand men killed. Foch had been given the supreme command, and the great British and Imperial armies of sixty to seventy divisions had been placed, like the Americans, unreservedly under his orders. Now the British

Expeditionary Army numbered but three or four hundred thousand men, spread from the bases at Havre and along the coast forward to the line, compared with nearly a hundred French divisions, or over two million Frenchmen, actually holding the long front from Belgium to Switzerland. It was natural therefore that we should place ourselves under their command, and that their judgment should be accepted. It had been expected that General Georges would take full command of the French and British armies in the field from the moment when war was declared, and General Gamelin was expected to retire to an advisory position on the French Military Council. However, General Gamelin was averse from yielding his control as Generalissimo. He retained the supreme direction. A vexatious conflict of authority took place between him and General Georges during the eight months' lull. General Georges, in my opinion, never had the chance to make the strategic plan in its entirety and on his own responsibility.

The British General Staff and our headquarters in the field had long been anxious about the gap between the northern end of the Maginot Line and the beginning of the British fortified front along the Franco-Belgian frontier. Mr. Hore-Belisha, the Secretary of State for War, raised the point in the War Cabinet on several occasions. Representations were made through military channels. The Cabinet and our military leaders however were naturally shy of criticising those whose armies were ten times as strong as our own. The French thought that the Ardennes were impassable for large modern armies. Marshal Pétain had told the Senate Army Commission, "This sector is not dangerous." A great deal of field work was done along the Meuse, but nothing like a strong line of pill-boxes and anti-tank obstacles, such as the British had made along the Belgian sector, was attempted. Moreover, General Corap's Ninth French Army was mainly composed of troops who were definitely below the French standards. Out of its nine divisions, two were of cavalry, partly mechanised, one was a fortress division, two (61 and 53) belonged to a secondary category, two (22 and 18) were not much inferior to active divisions; only two were divisions of the permanent regular army. Here, then, from Sedan to Hirson, on the Oise, along a front of fifty miles, there were no permanent fortifications, and only two divisions of professional troops.

One cannot be strong everywhere. It is often right and necessary to hold long sectors of a frontier with light covering forces, but this of course should be only with the object of gathering larger reserves for counter-attacks when the enemy's striking-points are revealed. The spreading of forty-three divisions, or half the mobile French army, from Longwy to the Swiss frontier, the whole of which was either defended by the Maginot Line forts or by the broad, swift-flowing Rhine, with its own fortress system behind it, was an improvident disposition. The risks that have to be run by the defender are more trying than those which an assailant, who is presumably the stronger at the point of attack, must dare. Where very long fronts are concerned, they can only be met by strong mobile reserves which can rapidly intervene in a decisive battle. A weight of opinion supports the criticism that the French reserves were inadequate, and, such as they were, badly distributed. After all, the gap behind the Ardennes opened the shortest road from Germany to Paris, and had for centuries been a famous battleground. If the enemy penetrated here the whole forward movement of the Northern Armies would be deprived of its pivot, and all their communications would be endangered equally with the capital.

Looking back, we can see that Mr. Chamberlain's War Cabinet, in which I served, and for whose acts or neglects I take my full share of responsibility, ought not to have been deterred from thrashing the matter out with the French in the autumn and winter of 1939. It would have been an unpleasant and difficult argument, for the French at every stage could say: "Why do you not send more troops of your own? Will you not take over a wider sector of the front? If reserves are lacking, pray supply them. We have five million men mobilised.* We follow your ideas about the war at sea; we conform to the plans of the British Admiralty. Pray show a proper confidence in the French Army and in our historic mastery of the art of war on land."

Nevertheless we ought to have done it.

Hitler and his Generals were in little doubt as to the military views and general arrangements of their opponents. During this same autumn and winter the German factories had poured out tanks, the plants for making which must have been well

* The French "mobilisation" of five millions included many not under arms—*e.g.*, in factories, on the land, etc.

advanced at the Munich crisis in 1938, and bore abundant fruit in the eight months that had passed since war began. They were not at all deterred by the physical difficulties of traversing the Ardennes. On the contrary, they believed that modern mechanical transport and vast organised road-making capacity would make this region, hitherto deemed impassable, the shortest, surest and easiest method of penetrating France and of rupturing the whole French scheme of counter-attack. Accordingly, the German Supreme Army Command planned their enormous onrush through the Ardennes to sever the curling left arm of the Allied Northern Armies at the shoulder-joint. The movement, though on a far larger scale and with different speeds and weapons, was not unlike Napoleon's thrust at the Plateau of Pratzen in the Battle of Austerlitz, whereby the entire Austro-Russian turning move was cut off and ruined and their centre broken.

*    *    *    *    *

At the signal the Northern Armies sprang to the rescue of Belgium and poured forward along all the roads amid the cheers of the inhabitants. The first phase of Plan D was completed by May 12. The French held the left bank of the Meuse to Huy, and their light forces beyond the river were falling back before increasing enemy pressure. The armoured divisions of the French First Army reached the line Huy–Hannut–Tirlemont. The Belgians, having lost the Albert Canal, were falling back to the line of the river Gette and taking up their prescribed position from Antwerp to Louvain. They still held Liége and Namur. The French Seventh Army had occupied the islands of Walcheren and South Beveland, and were engaged with mechanised units of the German Eighteenth Army on the line Herenthals-Bergen-op-Zoom. So rapid had been the advance of the French Seventh Army that it had already outrun its ammunition supplies. The superiority in quality though not in numbers of the British Air Force was already apparent. Thus up till the night of the 12th there was no reason to suppose that the operations were not going well.

However, during the 13th Lord Gort's headquarters became aware of the weight of the German thrust on the front of the French Ninth Army. By nightfall the enemy had established themselves on the west bank of the Meuse, on either side of Dinant and Sedan. The French G.Q.G. (Grand-Quartier-Général) were not yet certain whether the main German effort was directed through Luxembourg against the left of the Maginot Line or through Maastricht towards Brussels. Along the whole front Louvain–Namur–Dinant to Sedan an intense, heavy battle had developed, but under conditions which General Gamelin had not contemplated, for at Dinant the French Ninth Army had no time to install themselves before the enemy was upon them.

*   *   *   *   *

During the 14th the bad news began to come in. At first all was vague. At 7 p.m. I read to the Cabinet a message received from M. Reynaud stating that the Germans had broken through at Sedan, that the French were unable to resist the combination of tanks and dive-bombing, and asking for ten more squadrons of fighters to re-establish the line. Other messages received by the Chiefs of Staff gave similar information, and added that both Generals Gamelin and Georges took a serious view of the situation and that General Gamelin was surprised at the rapidity of the enemy's advance. In fact, Kleist's Group, with its immense mass of armour, heavy and light, had completely scattered or destroyed the French troops on their immediate front, and could now move forward at a pace never before known in war. At almost all points where the armies had come in contact the weight and fury of the German attack was overpowering. They crossed the Meuse in the Dinant sector with two more armoured divisions. To the north the fighting on the front of the French First Army had been most severe. The Ist and IInd British Corps were still in position from Wavre to Louvain, where our 3rd Division, under General Montgomery, had had sharp fighting. Farther north the Belgians were retiring to the Antwerp defences. The French Seventh Army, on the seaward flank, was recoiling even quicker than it had advanced.

From the moment of the invasion we began Operation "Royal Marine", the launching of the fluvial mines into the Rhine, and in the first week of the battle nearly 1,700 were "streamed".* They produced immediate results. Practically all river traffic between Karlsruhe and Mainz was suspended, and extensive damage was done to the Karlsruhe barrage and a number of

* Operation "Royal Marine" was first planned in November 1939. The mines were designed to float down the Rhine and destroy enemy bridges and shipping. They were fed into the river from French territory upstream. See Vol. I, App. Q.

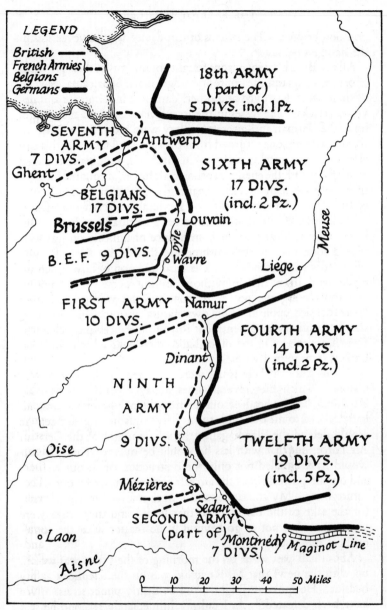

LEGEND
British
French Armies
Belgians
Germans

18th ARMY
(part of)
5 DIVS. incl. 1 Pz.

SEVENTH
ARMY
7 DIVS.

Ghent

Antwerp

SIXTH ARMY
17 DIVS.
(incl. 2 Pz.)

BELGIANS
17 DIVS.

Brussels

Louvain

Meuse

B.E.F. 9 DIVS.

Dyle

Wavre

Liége

FIRST ARMY
10 DIVS.

Namur

FOURTH ARMY
14 DIVS.
(incl. 2 Pz.)

Dinant

NINTH
ARMY

9 DIVS.

TWELFTH ARMY
19 DIVS.
(incl. 5 Pz.)

Oise

Mézières

Sedan

Laon

SECOND ARMY
(part of)
7 DIVS.

Montmédy

Maginot Line

Aisne

0   10   20   30   40   50 Miles

THE OPPOSING FORCES   13 MAY

pontoon bridges. The success of this device was however lost in the deluge of disaster.

All the British air squadrons fought continuously, their principal effort being against the pontoon bridges in the Sedan area. Several of these were destroyed and others damaged in desperate and devoted attacks. The losses in the low-level attacks on the bridges from the German anti-aircraft artillery were cruel. In one case, of six aircraft only one returned from the successful task. On this day alone we lost a total of sixty-seven machines, and, being engaged principally with the enemy's anti-aircraft forces, accounted for only fifty-three German aircraft. That night there remained in France of the Royal Air Force only 206 serviceable aircraft out of 474.

This detailed information came only gradually to hand. But it was already clear that the continuance of fighting on this scale would soon completely consume the British Air Force in spite of its individual ascendancy. The hard question of how much we could send from Britain without leaving ourselves defenceless and thus losing the power to continue the war pressed itself henceforward upon us. Our own natural promptings and many weighty military arguments lent force to the incessant, vehement French appeals. On the other hand, there was a limit, and that limit if transgressed would cost us our life.

At this time all these issues were discussed by the whole War Cabinet, which met several times a day. Air Chief Marshal Dowding, at the head of our metropolitan Fighter Command, had declared to me that with twenty-five squadrons of fighters he would defend the Island against the whole might of the German Air Force, but that with less he would be overpowered. Defeat would have entailed not only the destruction of all our airfields and our air-power, but of the aircraft factories on which our whole future hung. My colleagues and I were resolved to run all risks for the sake of the battle up to that limit—and those risks were very great—but not to go beyond it, no matter what the consequences might be.

About half-past seven on the morning of the 15th I was woken up with the news that M. Reynaud was on the telephone at my bedside. He spoke in English, and evidently under stress. "We have been defeated." As I did not immediately respond he said again: "We are beaten; we have lost the battle." I said: "Surely it can't have happened so soon?" But he replied: "The front is broken near Sedan; they are pouring through in great numbers with tanks and armoured cars"—or words to that effect. I then said: "All experience shows that the offensive will come to an end after a while. I remember the 21st of March, 1918. After five or six days they have to halt for supplies, and the opportunity for counter-attack is presented. I learned all this at the time from the lips of Marshal Foch himself." Certainly this was what we had always seen in the past and what we ought to have seen now. However, the French Premier came back to the sentence with which he had begun, which proved indeed only too true: "We are defeated; we have lost the battle." I said I was willing to come over and have a talk.

On this day the French Ninth Army, Corap's, was in a state of complete dissolution, and its remnants were divided up between General Giraud, of the Seventh French Army, who took over from Corap in the north, and the headquarters of the Sixth French Army, which was forming in the south. A gap of some fifty miles had in fact been punched in the French line, through which the vast mass of enemy armour was pouring. By the evening of the 15th German armoured cars were reported to be in Liart and Montcornet, the latter sixty miles behind the original front. The French First Army was also pierced on a 5,000-yards front south of Limal. Farther north all attacks on the British were repulsed. The German attack and the retirement of the French division on their right compelled the making of a British defensive flank facing south. The French Seventh Army had retreated into the Antwerp defences west of the Scheldt, and was being driven out of the islands of Walcheren and South Beveland.

On this day also the struggle in Holland came to an end. Owing to the capitulation of the Dutch High Command at 11 a.m., only a very few Dutch troops could be evacuated.

Of course this picture presented a general impression of defeat. I had seen a good deal of this sort of thing in the previous war, and the idea of the line being broken, even on a broad front, did not convey to my mind the appalling consequences that now flowed from it. Not having had access to official information for so many years, I did not comprehend the violence of the revolution effected since the last war by the incursion of a mass of fast-moving heavy armour. I knew about it, but it had not altered my inward convictions as it should have done. There was nothing I could have done if it had. I rang up General Georges, who seemed quite cool, and reported that the breach at Sedan was being plugged. A telegram from General Gamelin also stated that although the position between Namur and Sedan was serious he viewed the situation with calm. I reported Reynaud's message and other news to the Cabinet at 11 a.m.

On the 16th the German spearheads stood along the line La Capelle–Vervins–Marle–Laon, and the vanguards of the German XIVth Corps were in support at Montcornet and Neufchâtel-sur-Aisne. The fall of Laon confirmed the penetration of over sixty miles inward upon us from the frontier near Sedan. Under this threat and the ever-increasing pressure on their own front, the First French Army and the British Expeditionary Force were ordered to withdraw in three stages to the Scheldt. Although none of these details were available even to the War Office, and no clear view could be formed of what was happening, the gravity of the crisis was obvious. I felt it imperative to go to Paris that afternoon.

\*　　\*　　\*　　\*　　\*

We had to expect that the disastrous events on the front would bring new foes upon us. Although there were no indications of a change in Italian policy, the Minister of Shipping was given instructions to thin out the shipping in the Mediterranean. No more British ships were to come homewards from Aden. We had already diverted round the Cape the first convoy carrying Australian troops to England. The Defence Committee were instructed to consider action in the event of war with Italy, particularly with regard to Crete. Schemes for evacuating civilians from Aden and Gibraltar were put into operation.

\*　　\*　　\*　　\*　　\*

At about 3 p.m. I flew to Paris in a "Flamingo", a Government passenger plane, of which there were three. General Dill, Vice-Chief of the Imperial General Staff, came with me, and Ismay.

It was a good machine, very comfortable, and making about 160 miles an hour. As it was unarmed, an escort was provided, but we soared off into a rain-cloud and reached Le Bourget in little more than an hour. From the moment we got out of the "Flamingo" it was obvious that the situation was incomparably worse than we had imagined. The officers who met us told General Ismay that the Germans were expected in Paris in a few

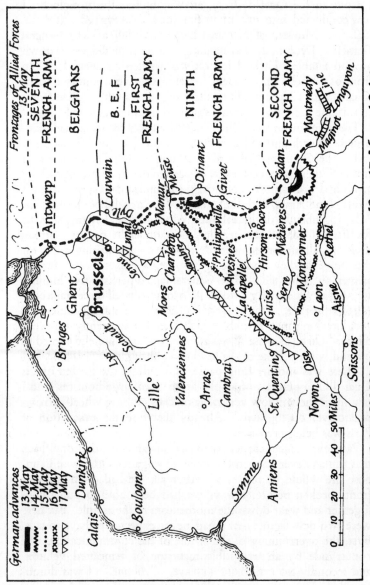

**GERMAN ADVANCES** on successive days 13—17 May 1940

German advances
13 May
14 May
15 May
16 May
17 May

Frontages of Allied Forces
13 May
SEVENTH FRENCH ARMY
BELGIANS
B.E.F.
FIRST FRENCH ARMY
NINTH FRENCH ARMY
SECOND FRENCH ARMY

Antwerp
Louvain
Dyle
Namur
Meuse
Dinant
Givet
Philippeville
Rocroi
Mézières
Sedan
Montmédy
Maginot Line
Longuyon

Brussels
Ghent
Bruges
Senne
Charleroi
Mons
Sambre
Avesnes
Hirson
Serre
Montcornet

Dunkirk
Scheldt
Lys
Lille
Valenciennes
Arras
Cambrai
La Capelle
Guise
Laon
Aisne
Rethel

Calais
Boulogne
St. Quentin
Noyon
Oise
Soissons

Somme
Amiens

0  10  20  30  40  50 Miles

41

days at most. After hearing at the Embassy about the position, I drove to the Quai d'Orsay, arriving at five-thirty o'clock. I was conducted into one of its fine rooms. Reynaud was there, Daladier, Minister of National Defence and War, and General Gamelin. Everybody was standing. At no time did we sit down around a table. Utter dejection was written on every face. In front of Gamelin on a student's easel was a map, about two yards square, with a black ink line purporting to show the Allied front. In this line there was drawn a small but sinister bulge at Sedan.

The Commander-in-Chief briefly explained what had happened. North and south of Sedan, on a front of fifty or sixty miles, the Germans had broken through. The French army in front of them was destroyed or scattered. A heavy onrush of armoured vehicles was advancing with unheard-of speed towards Amiens and Arras, with the intention, apparently, of reaching the coast at Abbeville or thereabouts. Alternatively they might make for Paris. Behind the armour, he said, eight or ten German divisions, all motorised, were driving onwards, making flanks for themselves as they advanced against the two disconnected French armies on either side. The General talked perhaps five minutes without anyone saying a word. When he stopped there was a considerable silence. I then asked: "Where is the strategic reserve?" and, breaking into French, which I used indifferently (in every sense): "*Où est la masse de manœuvre?*" General Gamelin turned to me and, with a shake of the head and a shrug, said: "*Aucune.*"

There was another long pause. Outside in the garden of the Quai d'Orsay clouds of smoke arose from large bonfires, and I saw from the window venerable officials pushing wheel-barrows of archives on to them. Already therefore the evacuation of Paris was being prepared.

Past experience carries with its advantages the drawback that things never happen the same way again. Otherwise I suppose life would be too easy. After all, we had often had our fronts broken before; always we had been able to pull things together and wear down the momentum of the assault. But here were two new factors that I had never expected to have to face. First, the overrunning of the whole of the communications and countryside by an irresistible incursion of armoured vehicles and secondly NO STRATEGIC RESERVE. "*Aucune.*" I was dumbfounded. What were we to think of the great French Army and its highest chiefs? It had never occurred to me that any commanders having to defend five hundred miles of engaged front would have left themselves unprovided with a mass of manœuvre. No one can defend with certainty so wide a front; but when the enemy has committed himself to a major thrust which breaks the line one can always have, one *must* always have, a mass of divisions which marches up in vehement counter-attack at the moment when the first fury of the offensive has spent its force.

What was the Maginot Line for? It should have economised troops upon a large sector of the frontier, not only offering many sally-ports for local counter-strokes, but also enabling large forces to be held in reserve; and this is the only way these things can be done. But now there was no reserve. I admit this was one of the greatest surprises I have had in my life. Why had I not known more about it, even though I had been so busy at the Admiralty? Why had the British Government, and the War Office above all, not known more about it? It was no excuse that the French High Command would not impart their dispositions to us or to Lord Gort except in vague outline. We had a right to know. We ought to have insisted. Both armies were fighting in the line together. I went back again to the window and the curling wreaths of smoke from the bonfires of the State documents of the French Republic. Still the old gentlemen were bringing up their wheelbarrows, and industriously casting their contents into the flames.

There was a considerable conversation in changing groups around the principals, of which M. Reynaud has published a detailed record. I am represented as urging that there should be no withdrawal of the Northern Armies, that on the contrary they should counter-attack. Certainly this was my mood. But here was no considered military opinion.* It must be remembered

---

* As other accounts of what passed have appeared, I asked Lord Ismay, who was at my side throughout, to give his recollection. He writes:

"We did not sit round a table, and much may have been said as we walked about in groups. I am positive that you did not express any 'considered military opinion' on what should be done. When we left London we considered the break-through at Sedan serious, but not mortal. There had been many 'break-throughs' in 1914–18, but they had all been stopped, generally by counter-attacks from one or both sides of the salient.

"When you realised that the French High Command felt that all was lost, you asked Gamelin a number of questions, with, I believe, the dual object, first of informing yourself as to what had happened and what he proposed to do, and secondly of stopping the panic. One of these questions was: 'When and where are you going to counter-attack the flanks of the Bulge? From the north or from the south?' I am sure that you did not press any particular strategical or tactical thought upon the conference. The burden of your song was: 'Things may be bad, but are certainly not incurable.' "

that this was the first realisation we had of the magnitude of the disaster or of the apparent French despair. We were not conducting the operations, and our army, which was only a tenth of the troops on the front, was serving under the French command. I and the British officers with me were staggered at the evident conviction of the French Commander-in-Chief and leading Ministers that all was lost, and in anything that I said I was reacting violently against this. There is however no doubt that they were quite right, and that the most rapid retreat to the south was imperative. This soon became obvious to all.

Presently General Gamelin was speaking again. He was discussing whether forces should now be gathered to strike at the flanks of the penetration, or "Bulge", as we called such things later on. Eight or nine divisions were being withdrawn from quiet parts of the front, the Maginot Line; there were two or three armoured divisions which had not been engaged; eight or nine more divisions were being brought from Africa and would arrive in the battle zone during the next fortnight or three weeks. General Giraud had been placed in command of the French army north of the gap. The Germans would advance henceforward through a corridor between two fronts on which warfare in the fashion of 1917 and 1918 could be waged. Perhaps the Germans could not maintain the corridor, with its ever-increasing double flank-guards to be built up, and at the same time nourish their armoured incursion. Something in this sense Gamelin seemed to say, and all this was quite sound. I was conscious however that it carried no conviction in this small but hitherto influential and responsible company. Presently I asked General Gamelin when and where he proposed to attack the flanks of the Bulge. His reply was: "Inferiority of numbers, inferiority of equipment, inferiority of method"—and then a hopeless shrug of the shoulders. There was no argument; there was no need of argument. And where were we British anyway, having regard to our tiny contribution—ten divisions after eight months of war, and not even one modern tank division in action?

This was the last I saw of General Gamelin. He was a patriotic, well-meaning man and skilled in his profession, and no doubt he has his tale to tell.*

\* \* \* \* \*

* His book entitled *Servir* throws little light either upon his personal conduct of events or generally upon the course of the war.

44

The burden of General Gamelin's, and indeed of all the French High Command's subsequent remarks, was insistence on their inferiority in the air and earnest entreaties for more squadrons of the Royal Air Force, bomber as well as fighter, but chiefly the latter. This prayer for fighter support was destined to be repeated at every subsequent conference until France fell. In the course of his appeal General Gamelin said that fighters were needed not only to give cover to the French Army, but also to stop the German tanks. At this I said: "No. It is the business of the artillery to stop the tanks. The business of the fighters is to cleanse the skies [*nettoyer le ciel*] over the battle." It was vital that our metropolitan fighter air force should not be drawn out of Britain on any account. Our existence turned on this. Nevertheless it was necessary to cut to the bone. In the morning, before I started, the Cabinet had given me authority to move four more squadrons of fighters to France. On our return to the Embassy, and after talking it over with Dill, I decided to ask sanction for the dispatch of six more. This would leave us with only the twenty-five fighter squadrons at home, and that was the final limit. It was a rending decision either way. I told General Ismay to telephone to London that the Cabinet should assemble at once to consider an urgent telegram which would be sent over in the course of the next hour or so. Ismay did this in Hindustani, having previously arranged for an Indian Army officer to be standing by in his office. This was my telegram:

9 p.m. 16th May, 1940

I shall be glad if the Cabinet could meet immediately to consider following. Situation grave in the last degree. Furious German thrust through Sedan finds French armies ill-grouped, many in north, others in Alsace. At least four days required to bring twenty divisions to cover Paris and strike at the flanks of the Bulge, which is now fifty kilometres wide.

Three [German] armoured divisions with two or three infantry divisions have advanced through gap and large masses hurrying forward behind them. Two great dangers therefore threaten. First that B.E.F. will be largely left in the air to make a difficult disengagement and retreat to the old line. Secondly, that the German thrust will wear down the French resistance before it can be fully gathered.

Orders given to defend Paris at all costs, but archives of the Quai d'Orsay already burning in the garden. I consider the next two, three, or four days decisive for Paris and probably for the French Army.

Therefore the question we must face is whether we can give further aid in fighters above four squadrons, for which the French are very grateful, and whether a larger part of our long-range heavy bombers should be employed to-morrow and the following nights upon the German masses crossing the Meuse and flowing into the Bulge. Even so results cannot be guaranteed; but the French resistance may be broken up as rapidly as that of Poland unless this Battle of the Bulge is won. I personally feel that we should send squadrons of fighters demanded (*i.e.*, six more) to-morrow, and, concentrating all available French and British aviation, dominate the air above the Bulge for the next two or three days, not for any local purpose, but to give the last chance to the French Army to rally its bravery and strength. It would not be good historically if their requests were denied and their ruin resulted. Also night bombardment by a strong force of heavy bombers can no doubt be arranged. It looks as if the enemy was by now fully extended both in the air and tanks. We must not underrate the increasing difficulties of his advance if strongly counter-attacked. I imagine that if all fails here we could still shift what is left of our own air striking force to assist the B.E.F. should it be forced to withdraw. I again emphasise the mortal gravity of the hour, and express my opinion as above. Kindly inform me what you will do. Dill agrees. I must have answer by midnight in order to encourage the French. Telephone to Ismay at Embassy in Hindustani.

The reply came at about eleven-thirty. The Cabinet said "Yes." I immediately took Ismay off with me in a car to M. Reynaud's flat. We found it more or less in darkness. After an interval M. Reynaud emerged from his bedroom in his dressing-gown and I told him the favourable news. Ten fighter squadrons! I then persuaded him to send for M. Daladier, who was duly summoned and brought to the flat to hear the decision of the British Cabinet. In this way I hoped to revive the spirits of our French friends, as much as our limited means allowed. Daladier never spoke a word. He rose slowly from his chair and wrung my hand. I got back to the Embassy about 2 a.m., and slept well, though the cannon fire in petty aeroplane raids made one roll over from time to time. In the morning I flew home, and, in spite of other preoccupations, pressed on with the construction of the second level of the new Government.

# CHAPTER III

# THE BATTLE OF FRANCE

## The Second Week. Weygand
## May 17-May 24

*The Battle Crisis Grows - The Local Defence Volunteers - Reinforcements from the East - My Telegrams to the President of May 18 and May 20 - General Gamelin's Final Order No. 12, May 19 - French Cabinet Changes - General Weygand Appointed - First Orders to the Little Ships, May 20 - Operation "Dynamo" - Weygand Tours the Front - Billotte Killed in a Motor Accident - French Failure to Grapple with German Armour - Ironside's Report, May 21 - Parliament Votes Extraordinary Powers to the Government - My Second Visit to Paris - Weygand's Plan - Peril of the Northern Armies - Fighting round Arras - Correspondence with M. Reynaud - Sir John Dill Chief of the Imperial General Staff.*

T HE War Cabinet met at 10 a.m. on the 17th, and I gave them an account of my visit to Paris, and of the situation so far as I could measure it.

I said I had told the French that unless they made a supreme effort we should not be justified in accepting the grave risk to the safety of our country that we were incurring by the dispatch of the additional fighter squadrons to France. I felt that the question of air reinforcements was one of the gravest that a British Cabinet had ever had to face. It was claimed that the German air losses had been four or five times our own, but I had been told that the French had only a quarter of their fighter aircraft left. On this day Gamelin thought the situation "lost", and is reported to have said: "I will guarantee the safety of Paris only for to-day, to-morrow [the 18th], and the night following." In Norway it appeared that Narvik was likely to be captured by us at any moment, but Lord Cork was informed that in the light of the news from France no more reinforcements could be sent to him.

The battle crisis grew hourly in intensity. At the request of General Georges the British Army prolonged its defensive flank by occupying points on the whole line from Douai to Péronne, thus attempting to cover Arras, which was a road-centre vital to any southward retreat. That afternoon the Germans entered Brussels. The next day they reached Cambrai, passed St. Quentin, and brushed our small parties out of Péronne. The French Seventh, the Belgian, the British, and the French First Army all continued their withdrawal to the Scheldt, the British standing along the Dendre for the day, and forming the detachment "Petreforce" (a temporary grouping of various units under Major-General Petre) for the defence of Arras.

At midnight (May 18–19) Lord Gort was visited at his head-quarters by General Billotte. Neither the personality of this French General nor his proposals, such as they were, inspired confidence in his allies. From this moment the possibility of a withdrawal to the coast began to present itself to the British Commander-in-Chief. In his dispatch published in March 1941 he wrote: "The picture was now [night of the 19th] no longer that of a line bent or temporarily broken, but of a besieged fortress."

As the result of my visit to Paris and the Cabinet discussions I already found it necessary to pose a general question to my colleagues.

*Prime Minister to Lord President*          17.V.40
I am very much obliged to you for undertaking to examine to-night the consequences of the withdrawal of the French Government from Paris or the fall of that city, as well as the problems which would arise if it were necessary to withdraw the B.E.F. from France, either along its communications or by the Belgian and Channel ports. It is quite understood that in the first instance this report could be no more than an enumeration of the main considerations which arise, and which could thereafter be remitted to the Staffs. I am myself seeing the military authorities at 6.30.

\*     \*     \*     \*     \*

The swift fate of Holland was in all our minds. Mr Eden had already proposed to the War Cabinet the formation of Local Defence Volunteers, and this plan was energetically pressed. All over the country, in every town and village, bands of determined men came together armed with shot-guns, sporting rifles, clubs, and spears. From this a vast organisation was soon to spring. But the need of Regulars was also vital.

*Prime Minister to General Ismay, for C.O.S.* 18.v.40

1. I cannot feel that we have enough trustworthy troops in England, in view of the very large numbers that may be landed from air-carriers preceded by parachutists. I do not consider this danger is imminent at the present time, as the great battle in France has yet to be decided.

I wish the following moves to be considered with a view to immediate action:

(i) The transports which brought the Australians to Suez should bring home eight battalions of Regular infantry from Palestine, properly convoyed, even at some risk, by whatever route is thought best. I hope it will be possible to use the Mediterranean.

(ii) The Australian fast convoy arrives early in June with 14,000 men.

(iii) These ships should be immediately filled with eight battalions of Territorials and sent to India, where they should pick up eight [more] Regular battalions. The speed of this fast convoy should be accelerated.

2. Everything must be done to carry out the recommendations for the control of aliens put forward by the Committee and minuted by me on another paper. Action should also be taken against Communists and Fascists, and very considerable numbers should be put in protective or preventive internment, including the leaders. These measures must, of course, be brought before the Cabinet before action.

3. The Chiefs of Staff must consider whether it would not be well to send only half of the so-called Armoured Division to France. One must always be prepared for the fact that the French may be offered very advantageous terms of peace, and the whole weight be thrown on us.

\* \* \* \* \*

I also thought it necessary, with the approval of my colleagues, to send the following grave telegrams to President Roosevelt in order to show how seriously the interests of the United States would be affected by the conquest and subjugation not only of France but of Great Britain. The Cabinet pondered over these drafts for a while, but made no amendment.

*Former Naval Person to President Roosevelt* 18.v.40

I do not need to tell you about the gravity of what has happened. We are determined to persevere to the very end, whatever the result

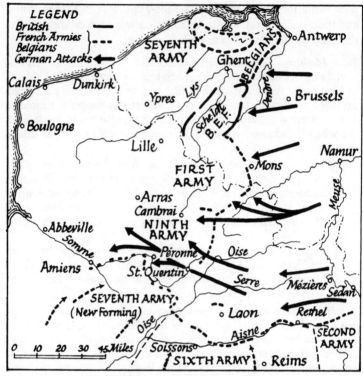

**LEGEND**
British
French Armies
Belgians
German Attacks

SEVENTH ARMY
Ghent
BELGIANS
Antwerp
Calais
Dunkirk
Ypres
Lys
Brussels
Boulogne
Dendre
Scheldt
B.E.F.
Lille
Mons
Namur
FIRST ARMY
Arras
Cambrai
NINTH ARMY
Meuse
Abbeville
Somme
Péronne
Oise
Amiens
St. Quentin
Serre
Mézières
Sedan
SEVENTH ARMY
(New Forming)
Oise
Laon
Rethel
Aisne
SECOND ARMY
0 10 20 30 45 Miles
Soissons
SIXTH ARMY
Reims

## SITUATION: EVENING 18 May

of the great battle raging in France may be. We must expect in any case to be attacked here on the Dutch model before very long, and we hope to give a good account of ourselves. But if American assistance is to play any part it must be available soon.

*Former Naval Person to President Roosevelt*                        20.V.40

Lothian has reported his conversation with you. I understand your difficulties, but I am very sorry about the destroyers. If they were here in six weeks they would play an invaluable part. The battle in France is full of danger to both sides. Though we have taken heavy toll of the enemy in the air and are clawing down two or three to one of their planes, they have still a formidable numerical superiority. Our most vital need is, therefore, the delivery at the earliest possible date of the largest possible number of Curtiss P.40 fighters, now in course of delivery to your Army.

With regard to the closing part of your talk with Lothian, our intention is, whatever happens, to fight on to the end in this Island, and, provided we can get the help for which we ask, we hope to run them very close in the air battles in view of individual superiority. Members of the present Administration would [be] likely [to] go down during this process should it result adversely, but in no conceivable circumstances will we consent to surrender. If members of the present Administration were finished and others came in to parley amid the ruins, you must not be blind to the fact that the sole remaining bargaining counter with Germany would be the Fleet, and if this country was left by the United States to its fate no one would have the right to blame those then responsible if they made the best terms they could for the surviving inhabitants. Excuse me, Mr. President, putting this nightmare bluntly. Evidently I could not answer for my successors, who in utter despair and helplessness might well have to accommodate themselves to the German will. However, there is happily no need at present to dwell upon such ideas. Once more thanking you for your goodwill . . .

*   *   *   *   *

Far-reaching changes were now made by M. Reynaud in the French Cabinet and High Command. On the 18th Marshal Pétain was appointed Vice-President of the Council. Reynaud himself, transferring Daladier to Foreign Affairs, took over the Ministry of National Defence and War. At 7 p.m. on the 19th he appointed Weygand, who had just arrived from the Levant, to replace General Gamelin. I had known Weygand when he was the right-hand man of Marshal Foch, and had admired his masterly intervention in the Battle of Warsaw against the Bolshevik invasion of Poland in August 1920—an event decisive for Europe at that time. He was now seventy-three, but was reported to be efficient and vigorous in a very high degree. General Gamelin's final Order (No. 12), dated 9.45 a.m. on May 19, prescribed that the Northern Armies, instead of letting themselves be encircled, must at all costs force their way southwards to the Somme, attacking the Panzer divisions which had cut their communications. At the same time the Second Army and the newly-forming Sixth were to attack northward towards Mézières. These decisions were sound. Indeed, an Order for the general retreat of the Northern Armies southward was already at least four days overdue. Once the gravity of the breach in the French centre at Sedan was apparent, the only hope for the Northern Armies lay in an immediate march to the Somme. Instead, under General Billotte,

51

they had only made gradual and partial withdrawals to the Scheldt and formed the defensive flank to the right. Even now there might have been time for the southward march.

The confusion of the northern command, the apparent paralysis of the First French Army, and the uncertainty about what was happening had caused the War Cabinet extreme anxiety. All our proceedings were quiet and composed, but we had a united and decided opinion, behind which there was silent passion. On the 19th we were informed (4.30 p.m.) that Lord Gort was "examining a possible withdrawal towards Dunkirk if that were forced upon him". The C.I.G.S. (Ironside) could not accept this proposal, as, like most of us, he favoured the southward march. We therefore sent him to Lord Gort with instructions to move the British Army in a south-westerly direction and to force his way through all opposition in order to join up with the French in the south, and that the Belgians should be urged to conform to this movement, or, alternatively, that we would evacuate as many of their troops as possible from the Channel ports. He was to be told that we would ourselves inform the French Government of what had been resolved. At the same Cabinet we sent Dill to General Georges' H.Q., with which we had a direct telephone line. He was to stay there for four days, and tell us all he could find out. Contacts even with Lord Gort were intermittent and difficult, but it was reported that only four days' supplies and ammunition for one battle were available.

\*     \*     \*     \*     \*

At the morning War Cabinet of May 20 we again discussed the situation of our Army. Even on the assumption of a successful fighting retreat to the Somme, I thought it likely that considerable numbers might be cut off or driven back on the sea. It is recorded in the minutes of the meeting: "The Prime Minister thought that as a precautionary measure the Admiralty should assemble a large number of small vessels in readiness to proceed to ports and inlets on the French coast." On this the Admiralty acted immediately and with ever-increasing vigour as the days passed and darkened. Operational control had been delegated on the 19th to Admiral Ramsay, commanding at Dover, whose resources at that time comprised thirty-six personnel vessels of various sorts based on Southampton and Dover. On the afternoon of the 20th, in consequence of the orders from London, the first conference of all concerned, including representatives of the Shipping Ministry, was held at Dover to consider "*the emergency evacuation across the Channel of very large forces*". It was planned if necessary to evacuate from Calais, Boulogne, and Dunkirk, at a rate of ten thousand men from each port every twenty-four hours. Thirty craft of passenger-ferry type, twelve naval drifters, and six small coasters were provided as a first instalment. On May 22 the Admiralty ordered forty Dutch *schuits* which had taken refuge with us to be requisitioned and manned with naval crews. These were commissioned between May 25 and May 27. From Harwich round to Weymouth sea-transport officers were directed to list all suitable ships up to a thousand tons, and a complete survey was made of all shipping in British harbours. These plans for what was called "Operation Dynamo" proved the salvation of the Army ten days later.

*   *   *   *   *

The direction of the German thrust had now become more obvious. Armoured vehicles and mechanised divisions continued to pour through the gap towards Amiens and Arras, curling westwards along the Somme towards the sea. On the night of the 20th they entered Abbeville, having traversed and cut the whole communications of the Northern Armies. These hideous, fatal scythes encountered little or no resistance once the front had been broken. The German tanks—the dreaded "*chars allemands*"—ranged freely through the open country, and, aided and supplied by mechanised transport, advanced thirty or forty miles a day. They had passed through scores of towns and hundreds of villages without the slightest opposition, their officers looking out of the open cupolas and waving jauntily to the inhabitants. Eye-witnesses spoke of crowds of French prisoners marching along with them, many still carrying their rifles, which were from time to time collected and broken under the tanks. I was shocked by the utter failure to grapple with the German armour, which, with a few thousand vehicles, was compassing the entire destruction of mighty armies, and by the swift collapse of all French resistance once the fighting front had been pierced. The whole German movement was proceeding along the main roads, which at no point seemed to be blocked.

Already on the 17th I had asked the Chief of the Air Staff:

**Is** there no possibility of finding out where a column of enemy armoured vehicles harbours during the dark hours, and then bombing? We are being ripped to pieces behind the front by these wandering columns.

I now telegraphed to Reynaud:

21.V.40

Many congratulations upon appointing Weygand, in whom we have entire confidence here.

It is not possible to stop columns of tanks from piercing thin lines and penetrating deeply. All ideas of stopping holes and hemming in these intruders are vicious. Principle should be, on the contrary, to punch holes. Undue importance should not be attached to the arrival of a few tanks at any particular point. What can they do if they enter a town? Towns should be held with riflemen, and tank personnel should be fired upon should they attempt to leave vehicles. If they cannot get food or drink or petrol, they can only make a mess and depart. Where possible, buildings should be blown down upon them. Every town with valuable cross-roads should be held in this fashion. Secondly, the tank columns in the open must be hunted down and attacked in the open country by numbers of small mobile columns with a few cannon. Their tracks must be wearing out, and their energy must abate. This is the one way to deal with the armoured intruders. As for the main body, which does not seem to be coming on very quickly, the only method is to drive in upon the flanks. The confusion of this battle can only be cleared by being aggravated, so that it becomes a *mêlée*. They strike at our communications; we should strike at theirs. I feel more confident than I did at the beginning of the battle; but all the armies must fight at the same time, and I hope the British will have a chance soon. Above is only my personal view, and I trust it will give no offence if I state it to you.

Every good wish.

Weygand's first act was to confer with his senior commanders. It was not unnatural that he should wish to see the situation in the north for himself, and to make contact with the commanders there. Allowances must be made for a general who takes over the command in the crisis of a losing battle. But now there was no time. He should not have left the summit of the remaining controls and have become involved in the delays and strains of personal movement. We may note in detail what followed. On the morning of the 20th Weygand, installed in Gamelin's place, made arrangements to visit the Northern Armies on the 21st.

After learning that the roads to the north were cut by the Germans he decided to fly. His plane was attacked, and forced to land at Calais. The hour appointed for his conference at Ypres had to be altered to 3 p.m. on the 21st. Here he met King Leopold and General Billotte. Lord Gort, who had not been notified of time and place, was not present, and no British officers were there. The King described this conference as "four hours of confused talking". It discussed the co-ordination of the three Armies, the execution of the Weygand plan, and if that failed the retirement of the British and French to the Lys, and the Belgians to the Yser. At 7 p.m. General Weygand had to leave. Lord Gort did not arrive till eight, when he received an account of the proceedings from General Billotte. Weygand drove back to Calais, embarked on a submarine for Dieppe, and returned to Paris. Billotte drove off in his car to deal with the crisis, and within the hour was killed in a motor collision. Thus all was again in suspense.

\* \* \* \* \*

On the 21st Ironside returned and reported that Lord Gort, on receiving the Cabinet instructions, had put the following points to him:

(i) That the southward march would involve a rearguard action from the Scheldt at the same time as an attack into an area already strongly held by the enemy armoured and mobile formations. During such a movement both flanks would have to be protected.

(ii) That sustained offensive operations were difficult in view of the administrative situation, and

(iii) That neither the French First Army nor the Belgians were likely to be able to conform to such a manœuvre if attempted.

Ironside added that confusion reigned in the French High Command in the north, that General Billotte had failed to carry out his duties of co-ordination for the past eight days and appeared to have no plans, that the B.E.F. were in good heart and had so far had only about five hundred battle casualties. He gave a vivid description of the state of the roads, crowded with refugees, scourged by the fire of German aircraft. He had had a rough time himself.

Two fearsome alternatives therefore presented themselves to

the War Cabinet. The first, the British Army at all costs, with or without French and Belgian co-operation, to cut its way to the south and the Somme, a task which Lord Gort doubted its ability to perform; the second, to fall back on Dunkirk and face a sea evacuation under hostile air attack, with the certainty of losing all the artillery and equipment, then so scarce and precious. Obviously great risks should be run to achieve the first, but there was no reason why all possible precautions and preparations should not be taken for the sea evacuation if the southern plan failed. I proposed to my colleagues that I should go to France to meet Reynaud and Weygand and come to a decision. Dill was to meet me there from General Georges' H.Q.

\* \* \* \* \*

This was the moment when my colleagues felt it right to obtain from Parliament the extraordinary powers for which a Bill had been prepared during the last few days. This measure would give the Government practically unlimited power over the life, liberty, and property of all His Majesty's subjects in Great Britain. In general terms of law the powers granted by Parliament were absolute. The Act was to "include power by Order in Council to make such Defence Regulations making provision for requiring persons to place themselves, their services and their property at the disposal of His Majesty as appear to him to be necessary or expedient for securing the public safety, the defence of the Realm, the maintenance of public order, or the efficient prosecution of any war in which His Majesty may be engaged, or for maintaining supplies or services essential to the life of the community".

In regard to persons, the Minister of Labour was empowered to direct anyone to perform any service required. The regulation giving him this power included a Fair Wages clause, which was inserted in the Act to regulate wage conditions. Labour supply committees were to be set up in important centres. The control of property in the widest sense was imposed in equal manner. Control of all establishments, including banks, was imposed under the authority of Government orders. Employers could be required to produce their books, and Excess Profits were to be taxed at 100 per cent. A Production Council, to be presided over by Mr. Greenwood, was to be formed, and a Director of Labour Supply to be appointed.

This Bill was accordingly presented to Parliament on the afternoon of the 22nd by Mr. Chamberlain and Mr. Attlee, the latter himself moving the Second Reading. Both the Commons and the Lords with their immense Conservative majorities passed it unanimously through all its stages in a single afternoon, and it received the Royal Assent that night.

> For Romans in Rome's quarrel
> Spared neither land nor gold,
> Nor son nor wife, nor limb nor life,
> In the brave days of old.

Such was the temper of the hour.

\*        \*        \*        \*        \*

When I arrived in Paris on May 22 there was a new setting. Gamelin was gone; Daladier was gone from the war scene. Reynaud was both Prime Minister and Minister of War. As the German thrust had definitely turned seaward, Paris was not immediately threatened. Grand-Quartier-Général (G.Q.G.) was still at Vincennes. M. Reynaud drove me down there about noon. In the garden some of those figures I had seen round Gamelin— one a very tall cavalry officer—were pacing moodily up and down. "*C'est l'ancien régime,*" remarked the aide-de-camp. Reynaud and I were brought into Weygand's room, and afterwards to the map room, where we had the great maps of the Supreme Command. Weygand met us. In spite of his physical exertions and a night of travel, he was brisk, buoyant, and incisive. He made an excellent impression upon all. He unfolded his plan of war. He was not content with a southward march or retreat for the Northern Armies. They should strike south-east from around Cambrai and Arras in the general direction of St. Quentin, thus taking in flank the German armoured divisions at present engaged in what he called the St. Quentin–Amiens pocket. Their rear, he thought, would be protected by the Belgian Army, which would cover them towards the east, and if necessary towards the north. Meanwhile a new French army under General Frère, composed of eighteen to twenty divisions drawn from Alsace, from the Maginot Line, from Africa, and from every other quarter, were to form a front along the Somme. Their left hand would push forward through Amiens to Arras, and thus by their utmost efforts establish contact with the armies

57

of the north. The enemy armour must be kept under constant pressure. "The Panzer divisions must not," said Weygand, "be allowed to keep the initiative." All necessary orders had been given so far as it was possible to give orders at all. We were now told that General Billotte, to whom he had imparted his whole plan, had just been killed in the motor accident. Dill and I were agreed that we had no choice, and indeed no inclination, except to welcome the plan. I emphasised that "it was indispensable to reopen communications between the armies of the north and those of the south by way of Arras". I explained that Lord Gort, while striking south-west, must also guard his path to the coast. To make sure there was no mistake about what was settled, I myself dictated a *résumé* of the decision and showed it to Weygand, who agreed. I reported accordingly to the Cabinet, and sent the following telegram to Lord Gort:

22.V.40

I flew to Paris this morning with Dill and others. The conclusions which were reached between Reynaud, Weygand, and ourselves are summarised below. They accord exactly with general directions you have received from the War Office. You have our best wishes in the vital battle now opening towards Bapaume and Cambrai.

It was agreed—

1. That the Belgian Army should withdraw to the line of the Yser and stand there, the sluices being opened.

2. That the British Army and the French First Army should attack south-west towards Bapaume and Cambrai at the earliest moment, certainly to-morrow, with about eight divisions, and with the Belgian Cavalry Corps on the right of the British.

3. That as this battle is vital to both armies and the British communications depend upon freeing Amiens, the British Air Force should give the utmost possible help, both by day and by night, while it is going on.

4. That the new French Army Group which is advancing upon Amiens and forming a line along the Somme should strike northwards and join hands with the British divisions who are attacking southwards in the general direction of Bapaume.

It will be seen that Weygand's new plan did not differ except in emphasis from the cancelled Instruction No. 12 of General Gamelin. Nor was it out of harmony with the vehement opinion which the War Cabinet had expressed on the 19th. The Northern Armies were to shoulder their way southwards by offensive

LEGEND
British
French Armies
Belgian
German Attacks

SITUATION: EVENING 22 May

action, destroying, if possible, the armoured incursion. They were to be met by a helpful thrust through Amiens by the new French Army Group under General Frère. This would be most important if it came true. In private I complained to M. Reynaud that Gort had been left entirely without orders for four consecutive days. Even since Weygand had assumed command three days had been lost in taking decisions. The change in the Supreme Command was right. The resultant delay was evil.

I slept the night at the Embassy. The air raids were trivial; the guns were noisy, but one never heard a bomb. Very different indeed were the experiences of Paris from the ordeal which London was soon to endure. I had a keen desire to go to see my friend General Georges at his headquarters at Compiègne. Our liaison officer with him, Brigadier Swayne, was with me for some time and gave me the picture of the French armies so far as he knew it, which was only part of the way. I was persuaded that it would be better not to intrude at this time, when this vast and complicated operation was being attempted in the teeth of every form of administrative difficulty and frequent breakdowns in communication.

In the absence of any supreme war direction events and the enemy had taken control. On the 17th Gort had begun to direct troops to the line Ruyaulcourt–Arleux and to garrison Arras, and was constantly strengthening his southern flank. The French Seventh Army, less the XVIth Corps, which had suffered heavily in the Walcheren fighting, had moved south to join the First French Army. It had traversed the British rear without serious disturbance. On the 20th Gort had informed both Generals Billotte and Blanchard that he proposed to attack southwards from Arras on May 21st with two divisions and an armoured brigade, and Billotte had agreed to co-operate with two French divisions from the First French Army. This army of thirteen divisions was gathered in an oblong some nineteen miles by ten: Maulde–Valenciennes–Denain–Douai. The enemy had crossed the Scheldt on the 20th around Oudenarde, and the three British corps, which still faced east, withdrew on the 23rd to the defences we had erected in the winter along the Belgian frontier, from which they had advanced so eagerly twelve days before. On this day the B.E.F. were put on half rations. The impression of French helplessness derived from many sources led me to protest to Reynaud.

*Prime Minister to M. Reynaud*        23.V.40
  (Copy to Lord Gort)

Communications of Northern Armies have been cut by strong enemy armoured forces. Salvation of these armies can only be obtained by immediate execution of Weygand's plan. I demand the issue to the French commanders in north and south and Belgian G.H.Q. of the most stringent orders to carry this out and turn defeat into victory. Time is vital as supplies are short.

I reported this message to the War Cabinet when they met at 11.30 a.m., pointing out that the whole success of the Weygand plan was dependent on the French taking the initiative, which they showed no signs of doing. We met again at 7 p.m.

And the next day:

*Prime Minister to M. Reynaud, for General Weygand*      24.V.40

General Gort wires that co-ordination of northern front is essential with armies of three different nations. He says he cannot undertake this co-ordination, as he is already fighting north and south and is threatened on his lines of communications. At the same time Sir Roger Keyes tells me that up to 3 p.m. to-day (23rd) Belgian Headquarters and King had received no directive. How does this agree with your statement that Blanchard and Gort are *main dans la main*? Appreciate fully difficulties of communication, but feel no effective concert of operations in northern area, against which enemy are concentrating. Trust you will be able to rectify this. Gort further says that any advance by him must be in the nature of sortie, and that relief must come from south, as he has not (*repeat* not) ammunition for serious attack. Nevertheless, we are instructing him to persevere in carrying out your plan. We have not here even seen your own directive, and have no knowledge of the details of your northern operations. Will you kindly have this sent through French Mission at earliest? All good wishes.

*   *   *   *   *

Some account of the small battle fought by the British around Arras must be given here. General Franklyn, who commanded, intended to occupy the area Arras–Cambrai–Bapaume. He had the 5th and 50th British Divisions and the 1st Army Tank Brigade. His plan was to attack with this armour and one brigade of each division, the whole under General Martel, round the western and southern sides of Arras, with an immediate objective on the river Sensée. The French were to co-operate with two divisions on the east to the Cambrai–Arras road. The British divisions consisted of only two brigades each, and the tanks numbered sixty-five Mark I and eighteen Mark II, all of whose tracks, the life of which was short, were wearing out. The attack began at 2 p.m. on May 21, and soon found itself engaged with much stronger opposition than was expected. French support on the eastern flank did not materialise, and on the western was limited to one light mechanised division. The enemy armour actually consisted of about four hundred tanks of the 7th and 8th Armoured Divisions, a general named Rommel commanding the former. At first the attack prospered, and four hundred prisoners were taken, but the line of the river Sensée was not reached, and the German counter-attack, in overwhelming numbers, with full air support, caused heavy casualties. The 12th Lancers presently reported strong enemy columns moving towards St. Pol and threatening to turn the western flank. During the night the Army Tank Brigade, the 13th Brigade of the 5th Division, and the 151st Brigade of the 50th Division gradually withdrew to the river Scarpe. Here three British brigades stood until the afternoon of the 22nd, and in this neighbourhood repulsed various attacks. We still held Arras, but the enemy gradually tended to swing round towards Béthune. The French light mechanised division guarding our western flank was forced from Mont St. Eloi, and the enemy tanks soon after approached Souchez. By 7 p.m. on the 23rd the British eastern flank was under heavy pressure, and the enemy reaching Lens had encircled the western flank. Thus the position was precarious. We were hopelessly outnumbered, beset by masses of armour, and almost surrounded. At 10 p.m. General Franklyn informed G.H.Q. that unless his force was withdrawn during the night its retirement would become impossible. He was told that orders to withdraw had been sent him three hours before. The operation had some temporary effect on the enemy; they recorded at the time "heavy British counterattacks with armour", which caused them considerable anxiety.

In pursuance of the Weygand plan Gort proposed to General Blanchard, who now commanded the northern group, that two British divisions, one French division, and the French Cavalry Corps should attack southwards between the Canal du Nord and the Scheldt Canal. Two French divisions had in fact twice previously reached the outskirts of Cambrai, but on each occasion they were bombed and withdrew. In all these days this was the only offensive action of the French First Army.

★　　★　　★　　★　　★

In London we had no knowledge of the progress of this forlorn attempt at Arras to break the encircling line. However, during the 24th very reproachful telegrams arrived from Reynaud. The shorter of his two messages tells the story.

You wired me [he said] this morning that you had instructed General Gort to continue to carry out the Weygand plan. General Weygand now informs me that, according to a telegram from General Blanchard, the British Army had carried out, on its own initiative, a retreat of twenty-five miles towards the ports at a time when our troops moving up from the south are gaining ground towards the north, where they were to meet their allies.

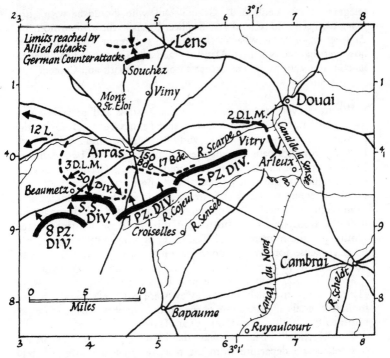

Limits reached by
Allied attacks — – –
German Counterattacks ▬▬

## BATTLE OF ARRAS   May 21/22

This action of the British Army is in direct opposition to the formal orders renewed this morning by General Weygand. This retreat has naturally obliged General Weygand to change all his arrangements, and he is compelled to give up the idea of closing the gap and restoring a continuous front. I need not lay any stress upon the gravity of the possible consequences.

Up to this time General Weygand had been counting on General Frère's army advancing northwards on Amiens, Albert, and Péronne. They had, in fact, made no noticeable progress, and were still forming and assembling. The following are my replies to M. Reynaud:

25.V.40

My telegram last night told you all we knew over here, and we have still heard nothing from Lord Gort to contradict it. But I must tell you that a Staff officer has reported to the War Office confirming the withdrawal of the two divisions from the Arras region, which your telegram to me mentioned. General Dill, who should be with Lord Gort, has been told to send a Staff officer by air at the earliest moment. As soon as we know what has happened I will report fully. It is clear however that the Northern Army is practically surrounded and that all its communications are cut except through Dunkirk and Ostend.

25.V.40

We have every reason to believe that Gort is still persevering in southward move. All we know is that he has been forced by the pressure on his western flank, and to keep communication with Dunkirk for indispensable supplies, to place parts of two divisions between himself and the increasing pressure of the German armoured forces, which in apparently irresistible strength have successively captured Abbeville and Boulogne, are menacing Calais and Dunkirk, and have taken St. Omer. How can he move southward and disengage his northern front unless he throws out this shield on his right hand? Nothing in the movements of the B.E.F. of which we are aware can be any excuse for the abandonment of the strong pressure of your northward move across the Somme, which we trust will develop.

Secondly, you complained of heavy materials being moved from Havre. Only materials moved away were gas shells, which it was indiscreet to leave. Also some of the stores have been moved from the north to the south side of the river at Havre.

Thirdly, should I become aware that extreme pressure of events has compelled any departure from the plan agreed I shall immediately inform you. Dill, who was this morning wholly convinced that the sole hope of any effective extrication of our Army lies in the southward move and in the active advance of General Frère, is now with Gort. You must understand that, having waited for the southward move for a week after it became obvious[ly necessary], we find ourselves now ripped from the coast by the mass of the enemy's armoured vehicles. We therefore have no choice but to continue the southward move, using such flank guard protection to the westward as is necessary.

General Spears will be with you to-morrow morning, and it will probably be quickest to send him back when the position is clear.

\* \* \* \* \*

There was a very strong feeling in Cabinet and high military circles that the abilities and strategic knowledge of Sir John Dill, who had been since April 23 Vice-Chief of the Imperial General Staff, should find their full scope in his appointment as our principal Army adviser. No one could doubt that his professional standing was in many ways superior to that of Ironside.

As the adverse battle drew to its climax I and my colleagues greatly desired that Sir John Dill should become C.I.G.S. We had also to choose a Commander-in-Chief for the British Island, if we were invaded. Late at night on May 25 Ironside, Dill, Ismay, myself, and one or two others in my room at Admiralty House were trying to measure the position. General Ironside volunteered the proposal that he should cease to be C.I.G.S., but declared himself quite willing to command the British Home Forces. Considering the unpromising task that such a command was at the time thought to involve, this was a spirited and selfless offer. I therefore accepted General Ironside's proposal; and the high dignities and honours which were later conferred upon him arose from my appreciation of his bearing at this moment in our affairs. Sir John Dill became C.I.G.S. on May 27. The changes were generally judged appropriate for the time being.

# CHAPTER IV

# THE MARCH TO THE SEA

## May 24–May 31

*Review of the Battle – General Halder's Account of Hitler's Personal Intervention – Halt of the German Armour – The Truth from the German Staff Diaries – A Separate Cause for the Halt at the Decisive Point – The Defence of Boulogne – The Drama of Calais – The Consequences of Prolonged Defence – Gort Abandons the Weygand Plan – His Decision of May 25 – Filling the Belgian Gap – Withdrawal of the British Army to the Dunkirk Bridgeheads – Extrication of the Four British Divisions from Lille – A Question to the Chiefs of Staff – Their Answer – My Message to Lord Gort – And to Admiral Keyes – General Pownall's Account of the Gort-Blanchard Meeting on the Morning of the 28th – Surrender of the Belgian Army, May 28 – Decisive Battle Fought by General Brooke and the IInd Corps, May 28 – Withdrawal to the Bridgehead – Escape by Sea of Half the French First Army.*

WE may now review up to this point the course of this memorable battle.

Only Hitler was prepared to violate the neutrality of Belgium and Holland. Belgium would not invite the Allies in until she was herself attacked. Therefore the military initiative rested with Hitler. On May 10 he struck his blow. The First Army Group, with the British in the centre, instead of standing behind their fortifications, leaped forward into Belgium on a vain, because belated, mission of rescue. The French had left the gap opposite the Ardennes ill fortified and weakly guarded. An armoured inroad on a scale never known in war broke the centre of the French line of armies, and in forty-eight hours threatened to cut all the Northern Armies alike from their southern communications and from the sea. By the 14th at the latest the French High Command should have given imperative orders to these armies to make a general retreat at full speed, accepting not only risks but heavy losses of *matériel*. This issue was not faced in its brutal realism by General Gamelin. The French commander of the northern group, Billotte, was incapable of taking the necessary decisions himself. Confusion reigned throughout the armies of the threatened left wing.

As the superior power of the enemy was felt they fell back. As the turning movement swung round their right they formed a defensive flank. If they had started back on the 14th they could have been on their old line by the 17th and would have had a good chance of fighting their way out. At least three mortal days were lost. From the 17th onwards the British War Cabinet saw clearly that an immediate fighting march southwards would alone save the British Army. They were resolved to press their view upon the French Government and General Gamelin, but their own commander, Lord Gort, was doubtful whether it was possible to disengage the fighting fronts, and still more to break through at the same time. On the 19th General Gamelin was dismissed, and Weygand reigned in his stead. Gamelin's "Instruction No. 12", his last order, though five days late, was sound in principle, and also in conformity with the main conclusions of the British War Cabinet and Chiefs of Staff. The change in the Supreme Command, or want of command, led to another three days' delay. The spirited plan which General Weygand proposed after visiting the Northern Armies was never more than a paper scheme. In the main it was the Gamelin plan, rendered still more hopeless by further delay.

In the hideous dilemma which now presented itself we accepted the Weygand plan, and made loyal and persistent, though now ineffectual, efforts to carry it out until the 25th, when, all the communications being cut, our weak counter-attack being repulsed, with the loss of Arras, the Belgian front breaking and King Leopold about to capitulate, all hope of escape to the southward vanished. There remained only the sea. Could we reach it, or must we be surrounded and broken up in the open field? In any case the whole artillery and equipment of our Army, irreplaceable for many months, must be lost. But what was that compared with saving the Army, the nucleus and structure upon which alone Britain could build her armies of the future? Lord Gort, who had from the 25th onwards felt that evacuation by sea was our only chance, now proceeded to form a bridgehead around Dunkirk and to fight his way into it with what strength remained. All the discipline of the British, and the qualities of their commanders, who included Brooke, Alexander, and Montgomery, were to be needed. Much more was to be needed. All that man could do was done. Would it be enough?

\* \* \* \* \*

A much-disputed episode must now be examined. General Halder, Chief of the German General Staff, has declared that at this moment Hitler made his only effective direct personal intervention in the battle. He became, according to this authority, "alarmed about the armoured formations, because they were in considerable danger in a difficult country, honeycombed with canals, without being able to attain any vital results". He felt he could not sacrifice armoured formations uselessly, as they were essential to the second stage of the campaign. He believed, no doubt, that his air superiority would be sufficient to prevent a large-scale evacuation by sea. He therefore, according to Halder, sent a message to him through Brauchitsch ordering "the armoured formations to be stopped, the points even taken back". Thus, says Halder, the way to Dunkirk was cleared for the British Army. At any rate we intercepted a German message sent in clear at 11.42 a.m. on May 24, to the effect that the attack on the line Dunkirk–Hazebrouck–Merville was to be discontinued for the present. Halder states that he refused, on behalf of Supreme Army Headquarters (O.K.H.), to interfere in the movement of Army Group Rundstedt, which had clear orders to prevent the enemy from reaching the coast. The quicker and more complete the success here, he argued, the easier it would be later to repair the loss of some tanks. The next day he was ordered to go with Brauchitsch to a conference.

The excited discussion finished with a definite order by Hitler, to which he added that he would ensure execution of his order by sending personal liaison officers to the front. Keitel was sent by plane to Army Group Rundstedt, and other officers to the front command posts. "I have never been able," says General Halder, "to figure how Hitler conceived the idea of the useless endangering of the armoured formations. It is most likely that Keitel, who was for a considerable time in Flanders in the First World War, had originated these ideas by his tales."

Other German generals have told much the same story, and have even suggested that Hitler's order was inspired by a political motive, to improve the chances of peace with England after France was beaten. Authentic documentary evidence has now come to light in the shape of the actual diary of Rundstedt's headquarters *written at the time*. This tells a different tale. At midnight on the 23rd orders came from Brauchitsch at O.K.H., confirming that the Fourth Army was to remain under Rundstedt for "the last act" of "the encirclement battle". Next morning Hitler visited Rundstedt, who represented to him that his armour, which had come so far and so fast, was much reduced in strength and needed a pause wherein to reorganise and regain its balance for the final blow against an enemy who, his staff diary says, was "fighting with extraordinary tenacity". Moreover, Rundstedt foresaw the possibility of attacks on his widely-dispersed forces from north and south; in fact, the Weygand Plan, which, if it had been feasible, was the obvious Allied counter-stroke. Hitler "agreed entirely" that the attack east of Arras should be carried out by infantry and that the mobile formations should continue to hold the line Lens–Béthune–Aire–St. Omer–Gravelines in order to intercept the enemy forces under pressure from Army Group B in the north-east. He also dwelt on the paramount necessity of conserving the armoured forces for further operations. However, very early on the 25th a fresh directive was sent from Brauchitsch as the Commander-in-Chief ordering the continuation of the advance by the armour. Rundstedt, fortified by Hitler's verbal agreement, would have none of it. He did not pass on the order to the Fourth Army Commander, Kluge, who was told to continue to husband the Panzer divisions. Kluge protested at the delay, but it was not till next day, the 26th, that Rundstedt released them, although even then he enjoined that Dunkirk was not yet itself to be directly assaulted. The diary records that the Fourth Army protested at this restriction, and its Chief of Staff telephoned on the 27th:

The picture in the Channel ports is as follows. Big ships come up the quayside, boards are put down, and the men crowd on the ships. All material is left behind. But we are not keen on finding these men, newly equipped, up against us later.

It is therefore certain that the armour was halted; that this was done on the initiative not of Hitler but of Rundstedt. Rundstedt no doubt had reasons for his view both in the condition of the armour and in the general battle, but he ought to have obeyed the formal orders of the Army Command, or at least told them what Hitler had said in conversation. There is general agreement among the German commanders that a great opportunity was lost.

*　　*　　*　　*　　*

There was however a separate cause which affected the movements of the German armour at the decisive point.

After reaching the sea beyond Abbeville on the night of the 20th, the leading German armoured and motorised columns had moved northward along the coast by Étaples towards Boulogne, Calais, and Dunkirk, with the evident intention of cutting off all escape by sea. This region was lighted in my mind from the previous war, when I had maintained the mobile Marine Brigade operating from Dunkirk against the flanks and rear of the German armies marching on Paris. I did not therefore have to learn about the inundation system between Calais and Dunkirk, or the significance of the Gravelines waterline. The sluices had already been opened, and with every day the floods were spreading, thus giving southerly protection to our line of retreat. The defence of Boulogne, but still more of Calais, to the latest hour, stood forth upon the confused scene, and garrisons were immediately sent there from England. Boulogne, isolated and attacked on May 22, was defended by two battalions of the Guards and one of our few anti-tank batteries, with some French troops. After thirty-six hours' resistance it was reported to be untenable, and I consented to the remainder of the garrison, including the French, being taken off by sea. The Guards were embarked by eight destroyers on the night of May 23–24, with a loss of only 200 men. The French continued to fight in the Citadel until the morning of the 25th. I regretted our evacuation.

Some days earlier I had placed the conduct of the defence of the Channel ports directly under the Chief of the Imperial General Staff, with whom I was in constant touch. I now resolved that Calais should be fought to the death, and that no evacuation by sea could be allowed to the garrison, which consisted of one battalion of the Rifle Brigade, one of the 60th Rifles, the Queen

Victoria Rifles, the 229th Anti-tank Battery, R.A., and a battalion of the Royal Tank Regiment, with twenty-one light and twenty-seven cruiser tanks, and an equal number of Frenchmen. It was painful thus to sacrifice these splendid, trained troops, of which we had so few, for the doubtful advantage of gaining two or perhaps three days, and the unknown uses that could be made of these days. The Secretary of State for War and the C.I.G.S. agreed to this hard measure. The telegrams and minutes tell the tale.

*Prime Minister to General Ismay, for C.I.G.S.*            23.V.40

Apart from the general order issued, I trust, last night by Weygand, for assuring the southward movement of the armies via Amiens, it is imperative that a clear line of supply should be opened up at the earliest moment to Gort's army by Dunkirk, Calais, or Boulogne. Gort cannot remain insensible to the peril in which he is now placed, and he must detach even a division, or whatever lesser force is necessary, to meet our force pushing through from the coast. If the regiment of armoured vehicles, including cruiser tanks, has actually landed at Calais, this should improve the situation, and should encourage us to send the rest of the 2nd Brigade of that Armoured Division in there. This coastal area must be cleaned up if the major operation of withdrawal is to have any chance. The intruders behind the line must be struck at and brought to bay. The refugees should be driven into the fields and parked there, as proposed by General Weygand, so that the roads can be kept clear. Are you in touch with Gort by telephone and telegraph, and how long does it take to send him a ciphered message? Will you kindly tell one of your Staff officers to send a map to Downing Street with the position, so far as it is known to-day, of the nine British divisions. Do not reply to this yourself.

*Prime Minister to General Ismay*            24.V.40

I cannot understand the situation around Calais. The Germans are blocking all exits, and our regiment of tanks is boxed up in the town because it cannot face the field guns planted on the outskirts. Yet I expect the forces achieving this are very modest. Why, then, are they not attacked? Why does not Lord Gort attack them from the rear at the same time that we make a sortie from Calais? Surely Gort can spare a brigade or two to clear his communications and to secure the supplies vital to his army. Here is a general with nine divisions about to be starved out, and yet he cannot send a force to clear his communications. What else can be so important as this? Where could a reserve be better employed?

This force blockading Calais should be attacked at once by Gort, by the Canadians from Dunkirk, and by a sortie of our boxed-up tanks.

Apparently the Germans can go anywhere and do anything, and their tanks can act in twos and threes all over our rear, and even when they are located they are not attacked. Also our tanks recoil before their field guns, but our field guns do not like to take on their tanks. If their motorised artillery, far from its base, can block us, why cannot we, with the artillery of a great army, block them? . . . The responsibility for cleansing the communications with Calais and keeping them open rests primarily with the B.E.F.

This did less than justice to our troops. But I print it as I wrote it at the time.

*Prime Minister to General Ismay*                              24.V.40

Vice-Chief of the Naval Staff informs me that [an] order was sent at 2 a.m. to Calais saying that evacuation was decided on in principle, but this is surely madness. The only effect of evacuating Calais would be to transfer the forces now blocking it to Dunkirk. Calais must be held for many reasons, but specially to hold the enemy on its front. The Admiralty say they are preparing twenty-four naval 12-pounders, which with S.A.P.* vill pierce any tank. Some of these will be ready this evening.

*Prime Minister to C.I.G.S.*                                    25.V.40

I must know at earliest why Gort gave up Arras, and what actually he is doing with the rest of his army. Is he still persevering in Weygand's plan, or has he become largely stationary? If the latter, what do you consider the probable course of events in the next few days, and what course do you recommend? Clearly, he must not allow himself to be encircled without fighting a battle. Should he [not] do this by fighting his way to the coast and destroying the armoured troops which stand between him and the sea with overwhelming force of artillery, while covering himself and the Belgian front, which would also curl back, by strong rearguards? To-morrow at latest this decision must be taken.

It should surely be possible for Dill to fly home from any aerodrome momentarily clear, and R.A.F. should send a whole squadron to escort him.

*Prime Minister to Secretary of State for War and C.I.G.S.*     25.V.40

Pray find out who was the officer responsible for sending the order to evacuate Calais yesterday, and by whom this very lukewarm telegram I saw this morning was drafted, in which mention is made of "for the sake of Allied solidarity". This is not the way to encourage men to fight to the end. Are you sure there is no streak of defeatist opinion in the General Staff?

* Semi-armour-piercing shell.

*Prime Minister to C.I.G.S.* 25.V.40

Something like this should be said to the Brigadier defending Calais: Defence of Calais to the utmost is of the highest importance to our country and our Army now. First, it occupies a large part of the enemy's armoured forces, and keeps them from attacking our line of communications. Secondly, it preserves a sally-port from which portions of the British Army may make their way home. Lord Gort has already sent troops to your aid, and the Navy will do all possible to keep you supplied. The eyes of the Empire are upon the defence of Calais, and His Majesty's Government are confident that you and your gallant regiment will perform an exploit worthy of the British name.

This message was sent to Brigadier Nicholson at about 2 p.m. on May 25.

The final decision not to relieve the garrison was taken on the evening of May 26. Till then the destroyers were held ready. Eden and Ironside were with me at the Admiralty. We three came out from dinner and at 9 p.m. did the deed. It involved Eden's own regiment, in which he had long served and fought in the previous struggle. One has to eat and drink in war, but I could not help feeling physically sick as we afterwards sat silent at the table.

Here was the message to the Brigadier:

Every hour you continue to exist is of the greatest help to the B.E.F. Government has therefore decided you must continue to fight. Have greatest possible admiration for your splendid stand. Evacuation will not (*repeat* not) take place, and craft required for above purpose are to return to Dover. *Verity* and *Windsor* to cover Commander Minesweeping and his retirement.

Calais was the crux. Many other causes might have prevented the deliverance of Dunkirk, but it is certain that the three days gained by the defence of Calais enabled the Gravelines waterline to be held, and that without this, even in spite of Hitler's vacillations and Rundstedt's orders, all would have been cut off and lost.

\* \* \* \* \*

Upon all this there now descended a simplifying catastrophe. The Germans, who had hitherto not pressed the Belgian front severely, on May 24 broke the Belgian line on either side of Courtrai, which is but thirty miles from Ostend and Dunkirk.

73

The King of the Belgians soon considered the situation hopeless, and prepared himself for capitulation.

By May 23 the Ist and IInd Corps of the B.E.F., withdrawn by stages from Belgium, were back again on the frontier defences north and east of Lille, which they had built for themselves during the winter. The German scythe-cut round our southern flank had reached the sea, and we had to shield ourselves from this. As the facts forced themselves upon Gort and his headquarters, troops had successively been sent to positions along the canal line La Bassée–Béthune–Aire–St. Omer–Watten. These, with elements of the French XVIth Corps, touched the sea at the Gravelines water-line. The British IIIrd Corps was responsible in the main for this curled-in flank facing south. There was no continuous line, but only a series of defended "stops" at the main crossings, some of which, like St. Omer and Watten, had already fallen to the enemy. The indispensable roads northwards from Cassel were threatened. Gort's reserve consisted only of the two British divisions, the 5th and 50th, which had, as we have seen, just been so narrowly extricated from their southerly counter-attack made at Arras in forlorn fulfilment of the Weygand plan. At this date the total frontage of the B.E.F. was about ninety miles, everywhere in close contact with the enemy.

To the south of the B.E.F. lay the First French Army, having two divisions in the frontier defences and the remainder, comprising eleven divisions in no good shape, cramped in the area north and east of Douai. This army was under attack from the south-east claw of the German encirclement. On our left the Belgian Army was being driven back from the Lys canal at many places, and with their retirement northwards a gap was developing north of Menin.

In the evening of the 25th Lord Gort took a vital decision. His orders still were to pursue the Weygand plan of a southerly attack towards Cambrai, in which the 5th and 50th Divisions, in conjunction with the French, were to be employed. The promised French attack northwards from the Somme showed no sign of reality. The last defenders of Boulogne had been evacuated. Calais still held out. Gort now abandoned the Weygand plan. There was in his view no longer hope of a march to the south and to the Somme. Moreover, at the same time the crumbling of the Belgian defence and the gap opening to the north created a new

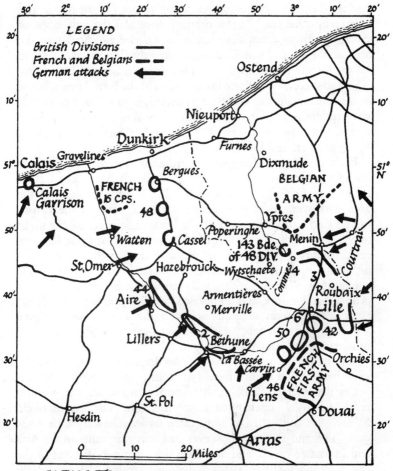

LEGEND
British Divisions ———
French and Belgians — — —
German attacks ◀━━

SITUATION AT NIGHTFALL 25 May

peril, dominating in itself. A captured order of the German Sixth Army showed that one corps was to march north-westwards towards Ypres and another corps westwards towards Wytschaete. How could the Belgians withstand this double thrust?

Confident in his military virtue, and convinced of the complete breakdown of all control, either by the British and French Governments or by the French Supreme Command, Gort resolved to abandon the attack to the southward, to plug the gap which a Belgian capitulation was about to open in the north, and to march to the sea. At this moment here was the only hope of saving anything from destruction or surrender. At 6 p.m. he ordered the 5th and 50th Divisions to join the IInd British Corps to till the impending Belgian gap. He informed General Blanchard, who had succeeded Billotte in command of the First Army Group, of his action; and this officer, acknowledging the force of events, gave orders at 11.30 p.m. for a withdrawal on the 26th to a line behind the Lys canal west of Lille, with a view to forming a bridgehead around Dunkirk.

Early on May 26 Gort and Blanchard drew up their plan for withdrawal to the coast. As the First French Army had farther to go, the first movements of the B.E.F. on the night of the 26th–27th were to be preparatory, and rearguards of the British Ist and IInd Corps remained on the frontier defences till the night of the 27th–28th. In all this Lord Gort had acted upon his own responsibility. But by now we at home, with a somewhat different angle of information, had already reached the same conclusions. On the 26th a telegram from the War Office approved his conduct, and authorised him "to operate towards the coast forthwith in conjunction with the French and Belgian armies". The emergency gathering on a vast scale of naval vessels of all kinds and sizes was already in full swing.

The reader must now look at the diagram which shows the areas held on the night of the 25th–26th by the British divisions.

On the western flank of the corridor to the sea the position remained largely unchanged during the 26th. The localities held by the 48th and 44th Divisions came under relatively little pressure. The 2nd Division however had heavy fighting on the Aire and La Bassée canals, and they held their ground. Farther to the east a strong German attack developed around Carvin, jointly defended by British and French troops. The situation was restored by the counter-attack of two battalions of the 50th Division, which were in bivouac close by. On the left of the British line the 5th Division, with the 143rd Brigade of the 48th Division under command, had travelled through the night, and at dawn took over the defence of the Ypres–Comines canal to close the gap which had opened between the British and Belgian armies. They were only just in time. Soon after they arrived the enemy attacked, and the fighting was heavy all day. Three battalions of the 1st Division in reserve were drawn in. The 50th Division, after bivouacking south of Lille, moved northwards to prolong the flank of the 5th Division around Ypres. The Belgian Army, heavily attacked throughout the day and with their right flank driven in, reported that they had no forces with which to regain touch with the British, and that they were unable to fall back to the line of the Yser canal in conformity with the British movement.

Meanwhile the organisation of the bridgeheads around Dunkirk was proceeding. The French were to hold from Gravelines to Bergues, and the British thence along the canal by Furnes to Nieuport and the sea. The various groups and parties of all arms which were arriving from both directions were woven into this line. Confirming the orders of the 26th, Lord Gort received from the War Office a telegram, dispatched at 1 p.m. on the 27th, telling him that his task henceforward was "to evacuate the maximum force possible". I had informed M. Reynaud the day before that the policy was to evacuate the British Expeditionary Force, and had requested him to issue corresponding orders. Such was the breakdown in communications that at 2 p.m. on the 27th the commander of the First French Army issued an order to his corps: "*La bataille sera livrée sans esprit de recul sur la position de la Lys.*"

Four British divisions and the whole of the First French Army were now in dire peril of being cut off around Lille. The two arms of the German encircling movement strove to close the pincers upon them. Although we had not in those days the admirable map rooms of more coherent periods, and although no control of the battle from London was possible, I had for three days past been harrowed by the position of the mass of Allied troops around Lille, including our four fine divisions. This however was one of those rare but decisive moments when mechanical transport exercises its rights. When Gort gave the order all these four divisions came back with surprising rapidity almost in a night. Meanwhile, by fierce battles on either side of the corridor, the rest of the British Army kept the path open to the sea. The pincer-claws, which were delayed by the 2nd Division, and checked for three days by the 5th Division, eventually met on the night of May 29 in a manner similar to the great Russian operation round Stalingrad in 1942. The trap had taken two and a half days to close, and in that time four British divisions and a great part of the First French Army, except the Vth Corps, which was lost, withdrew in good order through the gap, in spite of the French having only horse transport, and the main road to Dunkirk being already cut and the secondary roads filled with retiring troops, long trains of transport, and many thousands of refugees.

\* \* \* \* \*

The question about our ability to go on alone, which I had asked Mr. Chamberlain to examine with other Ministers ten days before, was now put formally by me to our military advisers. I drafted the reference purposely in terms which, while giving a lead, left freedom to the Chiefs of Staff to express their view, whatever it might be. I knew beforehand that they were absolutely determined; but it is wise to have written records of such decisions. I wished moreover to be able to assure Parliament that our resolve was backed by professional opinion. Here it is, with the answer:

1. We have reviewed our report on "British Strategy in a Certain Eventuality" in the light of the following terms of reference remitted to us by the Prime Minister.

"In the event of France being unable to continue in the war and becoming neutral, with the Germans holding their present position and the Belgian Army being forced to capitulate after assisting the British Expeditionary Force to reach the coast; in the event of terms being offered to Britain which would place her entirely at the mercy of Germany through disarmament, cession of naval bases in the Orkneys, etc.; what are the prospects of our continuing the war alone against Germany and probably Italy? Can the Navy and the Air Force hold out reasonable hopes of preventing serious invasion, and could the forces gathered in this Island cope with raids from the air involving detachments not greater than 10,000 men; it being observed that a prolongation of British resistance might be very dangerous for Germany, engaged in holding down the greater part of Europe?"

2. Our conclusions are contained in the following paragraphs:

3. While our Air Force is in being, our Navy and Air Force together should be able to prevent Germany carrying out a serious sea-borne invasion of this country.

4. Supposing Germany gained complete air superiority, we consider that the Navy could hold up an invasion for a time, but not for an indefinite period.

5. If, with our Navy unable to prevent it, and our Air Force gone, Germany attempted an invasion, our coast and beach defences could not prevent German tanks and infantry getting a firm footing on our

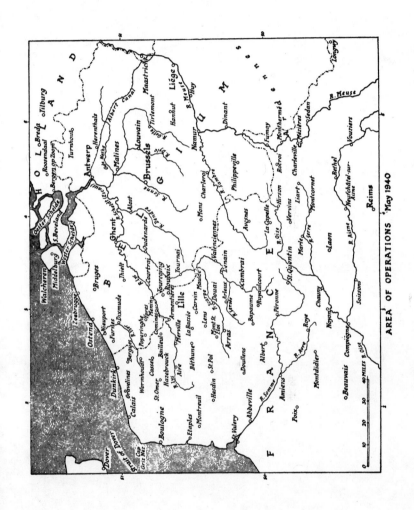

AREA OF OPERATIONS  May 1940

shores. In the circumstances envisaged above our land forces would be insufficient to deal with a serious invasion.

6. The crux of the matter is air superiority. Once Germany had attained this she might attempt to subjugate this country by air attack alone.

7. Germany could not gain complete air superiority unless she could knock out our Air Force, and the aircraft industries, some vital portions of which are concentrated at Coventry and Birmingham.

8. Air attacks on the aircraft factories would be made by day or by night. We consider that we should be able to inflict such casualties on the enemy by day as to prevent serious damage. Whatever we do however by way of defensive measures—and we are pressing on with these with all dispatch—we cannot be sure of protecting the large industrial centres, upon which our aircraft industries depend, from serious material damage by night attack. The enemy would not have to employ precision bombing to achieve this effect.

9. Whether the attacks succeed in eliminating the aircraft industry depends not only on the material damage by bombs, but on the moral effect on the workpeople and their determination to carry on in the face of wholesale havoc and destruction.

10. If therefore the enemy presses home night attacks on our aircraft industry, he is likely to achieve such material and moral damage within the industrial area concerned as to bring all work to a standstill.

11. It must be remembered that numerically the Germans have a superiority of four to one. Moreover, the German aircraft factories are well dispersed and relatively inaccessible.

12. On the other hand, so long as we have a counter-offensive bomber force we can carry out similar attacks on German industrial centres and by moral and material effect bring a proportion of them to a standstill.

13. To sum up, our conclusion is that *prima facie* Germany has most of the cards; but the real test is whether the morale of our fighting personnel and civil population will counterbalance the numerical and material advantages which Germany enjoys. We believe it will.

This report, which of course was written at the darkest moment before the Dunkirk Deliverance, was signed not only by the three Chiefs of Staff, Newall, Pound, and Ironside, but by the three Vice-Chiefs, Dill, Phillips, and Peirse. Reading it in after years, I must admit that it was grave and grim. But the War Cabinet and the few other Ministers who saw it were all of one mind. There was no discussion. Heart and soul we were together.

* * * * *

I now addressed myself to Lord Gort:

27.V.40

1. At this solemn moment I cannot help sending you my good wishes. No one can tell how it will go. But anything is better than being cooped up and starved out. I venture these few remarks. First, cannon ought to kill tanks, and they may as well be lost doing that as any other way. Second, I feel very anxious about Ostend till it is occupied by a brigade with artillery. Third, very likely the enemy tanks attacking Calais are tired, and, anyhow, busy on Calais. A column directed upon Calais while it is still holding out might have a good chance. Perhaps they will be less formidable when attacked themselves.

2. It is now necessary to tell the Belgians. I am sending following telegram to Keyes, but your personal contact with the King is desirable. Keyes will help. We are asking them to sacrifice themselves for us.

3. Presume [our] troops know they are cutting their way home to Blighty. Never was there such a spur for fighting. We shall give you all that the Navy and Air Force can do. Anthony Eden is with me now and joins his good wishes to mine.

*Enclosure*

*Prime Minister to Admiral Keyes*

Impart following to your friend [the King of the Belgians]. Presume he knows that British and French are fighting their way to coast between Gravelines and Ostend inclusive, and that we propose to give fullest support from Navy and Air Force during hazardous embarkation. What can we do for him? Certainly we cannot serve Belgium's cause by being hemmed in and starved out. Our only hope is victory, and England will never quit the war whatever happens till Hitler is beat or we cease to be a State. Trust you will make sure he leaves with you by aeroplane before too late. Should our operation prosper and we establish [an] effective bridgehead, we would try, if desired, to carry some Belgian divisions to France by sea. Vitally important Belgium should continue in war, and safety [of] King's person essential.

My telegram did not reach Admiral Keyes until after his return to England on the 28th. In consequence this particular message was not delivered to King Leopold. The fact is not however important because on the afternoon of the 27th between 5 and 6 o'clock Admiral Keyes spoke to me on the telephone. The following passage is taken from his report.

At about 5 p.m. on the 27th, when the King told me his Army had collapsed and he was asking for a cessation of hostilities, a cipher telegram was sent to Gort and to the War Office by wireless. The War Office received it at 5.54 p.m. I motored at once to La Panne and telephoned to the Prime Minister. The Prime Minister was not at all surprised, in view of the repeated warnings, but he told me that I must make every endeavour to persuade the King and Queen [Mother] to come to England with me, and dictated a message which he said I ought to have received that afternoon:

27.V.40

"Belgian Embassy here assumes from King's decision to remain that he regards the war as lost and contemplates separate peace.

"It is in order to dissociate itself from this that the constitutional Belgian Government has reassembled on foreign soil. Even if present Belgian Army has to lay down its arms there are 200,000 Belgians of military age in France, and greater resources than Belgium had in 1914, on which to fight back. By present decision the King is dividing the nation and delivering it into Hitler's protection. Please convey these considerations to the King, and impress upon him the disastrous consequences to the Allies and to Belgium of his present choice."

I gave King Leopold the Prime Minister's message, but he said that he had made up his mind that he must stay with his Army and people. . . .

*       *       *       *       *

At home I issued the following general injunction:

(Strictly Confidential)                                    28.V.40

In these dark days the Prime Minister would be grateful if all his colleagues in the Government, as well as important officials, would maintain a high morale in their circles; not minimising the gravity of events, but showing confidence in our ability and inflexible resolve to continue the war till we have broken the will of the enemy to bring all Europe under his domination.

No tolerance should be given to the idea that France will make a separate peace; but whatever may happen on the Continent, we cannot doubt our duty, and we shall certainly use all our power to defend the Island, the Empire, and our Cause.

During the morning of the 28th Lord Gort met General Blanchard again. I am indebted to General Pownall, Lord Gort's Chief of Staff, for this record made by him at the time:

Blanchard's enthusiasm at the Cassel meeting had evaporated when he visited us to-day. He had no constructive suggestions or plans. We read to him the telegram ordering us to proceed to the coast with a view to embarkation. He was horrified. And that was strange; for

81

what other reason did he think that he and Gort had been ordered to form bridgeheads? To what else could such a preliminary move lead? We pointed out that we had both received similar instructions regarding the bridgeheads. What had happened now was that we had got from our Government the next and logical step (which had no doubt been communicated to the French Government), whereas he had received as yet no such corresponding order. This pacified him somewhat, but by no means entirely. Then we said that we too, like him, wanted to keep the British and the First French Army together in this their last phase. Presumably therefore the First French Army would continue the retirement to-night, keeping aligned with us. Whereat he went completely off the deep end—it was impossible, he declared. We explained to him as clearly as the human tongue can explain the factors in the situation. The threat from the Germans on our north-eastern flank would probably not develop in strength for the next twenty-four hours (though when it did come it would be serious indeed). What was of immediate importance was the threat to our long south-western flank. There, as he well knew, advance-guards of German infantry divisions, supported by artillery, had made attacks yesterday at various points. Though the main points Wormhoudt, Cassel, Hazebrouck had held, there had been some penetration. The Germans might be relied upon to press these advantages, and we could be sure that the main bodies of the divisions would soon deploy and force themselves right across our line of withdrawal to the sea (a withdrawal which had been ordered for us, if not for him). There was therefore not a moment to be lost in getting back from the Lys, and we must get back to-night at least to the line Ypres-Poperinghe-Cassel. To wait till to-morrow night was to give *two* days to the Germans to get behind us, an act of madness. We thought it unlikely that we could get even 30 per cent. of our forces away even if we reached the sea; many indeed in forward positions would never reach it. But even if we could only save a small proportion of highly-trained officers and men it would be something useful to the continuance of the war. Everything possible must therefore be done, and the one thing that was possible, if only in part, was to get back some way to-night. . . .

Then came a liaison officer from General Prioux, now commanding the First Army. The liaison officer told Blanchard that Prioux had decided that he could *not* withdraw any farther to-night, and therefore intended to remain in the quadrangle of canals whose north-eastern corner is Armentières and south-western corner Béthune. This seemed to decide Blanchard against withdrawal. We begged him for the sake of the First Army and of the Allied cause to order Prioux to bring back at least some of his army in line with us. Not all of them could be so tired or so far away that it was impossible. For every man brought back there was at least *some* chance of embarkation, whereas every man who remained behind would certainly be eaten up. Why not *try* then? There was nothing to be gained by not trying: for those who did try there was at least *some* hope. But there was no shaking him. He declared that evacuation from the beach was impossible—no doubt the British Admiralty had arranged it for the B.E.F., but the French Marine would never be able to do it for French soldiers. It was therefore idle to try—the chance wasn't worth the effort involved; he agreed with Prioux.

He then asked, in terms, whether it was therefore Gort's intention to withdraw to-night to the line Ypres-Poperinghe-Cassel or not, knowing that in doing so Gort would be going *without* the French First Army. To which Gort replied that he *was* going. In the first place he had been ordered to re-embark, and to do so necessitated immediate withdrawal. To wait another twenty-four hours would mean that he would not be able to carry out his orders, for the troops would be cut off. In the second place, and apart from the formal aspect of obeying orders, it was madness to leave the troops forward in their present exposed positions. There they would certainly be overwhelmed very soon. For these reasons, therefore, and with great regret, it was necessary for the B.E.F. to withdraw even if the First French Army did not do so. . . .

*    *    *    *    *

In the early hours of the 28th the Belgian Army surrendered. Lord Gort received the formal intimation of this only one hour before the event, but the collapse had been foreseen three days earlier, and in one fashion or another the gap was plugged. I announced this event to the House in far more moderate terms than those M. Reynaud had thought it right to use.

The House will be aware that the King of the Belgians yesterday sent a plenipotentiary to the German Command asking for a suspension of arms on the Belgian front. The British and French Governments instructed their generals immediately to dissociate themselves from this procedure and to persevere in the operations in which they are now engaged. However, the German Command has agreed to the Belgian proposals and the Belgian Army ceased to resist the enemy's will at four o'clock this morning.

I have no intention of suggesting to the House that we should attempt at this moment to pass judgment upon the action of the King of the Belgians in his capacity as Commander-in-Chief of the Belgian

Army. This army has fought very bravely and has both suffered and inflicted heavy losses. The Belgian Government has dissociated itself from the action of the King, and, declaring itself to be the only legal Government of Belgium, has formally announced its resolve to continue the war at the side of the Allies.

Concern was expressed by the French Government that my reference to King Leopold's action was in sharp contrast to that of M. Reynaud. I thought it my duty, when speaking in the House on June 4, after a careful examination of the fuller facts then available, and in justice not only to our French Ally but also to the Belgian Government now in London, to state the truth in plain terms.

At the last moment, when Belgium was already invaded, King Leopold called upon us to come to his aid, and even at the last moment we came. He and his brave, efficient Army, nearly half a million strong, guarded our left flank and thus kept open our only line of retreat to the sea. Suddenly, without prior consultation, with the least possible notice, without the advice of his Ministers and upon his own personal act, he sent a plenipotentiary to the German Command, surrendered his Army, and exposed our whole flank and means of retreat.

The brave and efficient army of which I spoke had indeed conducted itself in accordance with its best traditions. They were overcome by an enemy whom it was beyond their power to resist for long. That they were defeated and ordered to surrender is no slur upon their honour or reputation.

\* \* \* \* \*

All this day of the 28th the escape of the British Army hung in the balance. On the front from Comines to Ypres and thence to the sea, facing east and attempting to fill the Belgian gap, General Brooke and his IInd Corps fought a magnificent battle. For two days past the 5th Division had held Comines against all attacks, but as the Belgians withdrew northwards, and then capitulated, the gap widened beyond repair. The protection of the flank of the B.E.F. was now their task. First the 50th Division came in to prolong the line; then the 4th and 3rd Divisions, newly withdrawn from east of Lille, hastened in motor transport to extend the wall of the vital corridor that led to Dunkirk. The German thrust between the British and Belgian Armies was not to be prevented, but its fatal consequence, an inward turn across the Yser,

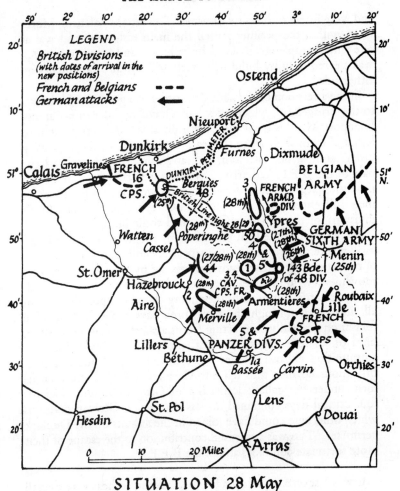

SITUATION 28 May

which would have brought the enemy on to the beaches behind our fighting troops, was foreseen and everywhere forestalled.

The Germans sustained a bloody repulse. Orders were given to the British artillery, both field and medium, to fire off all their ammunition at the enemy, and the tremendous fire did much to quell the German assault. All the time, only about four miles behind Brooke's struggling front, vast masses of transport and troops poured back into the developing bridgehead of Dunkirk, and were fitted with skilful improvisation into its defences. More-over, within the perimeter itself the main east-west road was at one time completely blocked by vehicles, and a one-way track was cleared only by bulldozers hurling them into the ditches on either side.

In the afternoon of the 28th Gort ordered a general withdrawal to the bridgehead, which now ran Gravelines–Bergues–Furnes–Nieuport. On this front the British divisions stood from right to left, and from Bergues to the sea by Nieuport, in the following order: 46th, 42nd, 1st, 50th, 3rd, and 4th. By the 29th a large part of the B.E.F. had arrived within the perimeter, and by this time the naval measures for evacuation were beginning to attain their full effect. On May 30 G.H.Q. reported that all British divisions, or the remains of them, had come in.

More than half the First French Army found their way to Dunkirk, where the great majority were safely embarked. But the line of retreat of at least five divisions was cut by the German pincers movement west of Lille. On the 28th they attempted to break out westwards, but in vain; the enemy closed in upon them from all sides. All through the next three days the French in Lille fought on gradually contracting fronts against increasing pressure, until on the evening of the 31st, short of food and with their ammunition exhausted, they were forced to surrender. About fifty thousand men thus fell into German hands. These French-men, under the gallant leadership of General Molinié, had for four critical days contained no less than seven German divisions which otherwise could have joined in the assaults on the Dunkirk perimeter. This was a splendid contribution to the escape of their more fortunate comrades and of the B.E.F.

\* \* \* \* \*

It was a severe experience for me, bearing so heavy an overall responsibility, to watch during these days in flickering glimpses this drama in which control was impossible, and intervention more likely to do harm than good. There is no doubt that by pressing in all loyalty the Weygand plan of retirement to the Somme as long as we did, our dangers, already so grave, were increased. But Gort's decision, in which we speedily concurred, to abandon the Weygand plan and march to the sea was executed by him and his staff with masterly skill, and will ever be regarded as a brilliant episode in British military annals.

# CHAPTER V

## THE DELIVERANCE OF DUNKIRK
### May 26 – June 4

*Service of Intercession and Prayer – "Hard and Heavy Tidings" – A Demonstration of Ministers – The Gathering of the Little Ships – Seven Hundred Vessels – Three Vital Factors – The Mosquito Armada – Bringing Off the French – Final Orders to Lord Gort – A Possible Consequence – Gort Transfers the Dunkirk Command to Alexander – My Third Visit to Paris, May 31 – General Spears and Marshal Pétain – The Evacuation Complete – My Statement to Parliament, June 4 – Significance of the Air Victory – Britain's Resolve.*

T HERE was a short service of Intercession and Prayer in Westminster Abbey on May 26. The English are loth to expose their feelings, but in my stall in the Choir I could feel the pent-up, passionate emotion, and also the fear of the congregation, not of death or wounds or material loss, but of defeat and the final ruin of Britain.

\*     \*     \*     \*     \*

It was Tuesday, May 28, and I did not again attend the House until that day week. There was no advantage to be gained by a further statement in the interval, nor did Members express a wish for one. But everyone realised that the fate of our Army and perhaps much else might well be decided before the week was out. "The House," I said, "should prepare itself for hard and heavy tidings. I have only to add that nothing which may happen in this battle can in any way relieve us of our duty to defend the world cause to which we have vowed ourselves; nor should it destroy our confidence in our power to make our way, as on former occasions in our history, through disaster and through grief to the ultimate defeat of our enemies." I had not seen many of my colleagues outside the War Cabinet, except individually, since the formation of the Government, and I thought it right to have a meeting in my room at the House of Commons of all Ministers of Cabinet rank other than the War Cabinet Members. We were perhaps twenty-five round the table. I described the course of events, and I showed them plainly where we were, and all that was in the balance. Then I said quite casually, and not treating it as a point of special significance:

"Of course, whatever happens at Dunkirk, we shall fight on."

There occurred a demonstration which, considering the character of the gathering—twenty-five experienced politicians and Parliament men, who represented all the different points of view, whether right or wrong, before the war—surprised me. Quite a number seemed to jump up from the table and come running to my chair, shouting and patting me on the back. There is no doubt that had I at this juncture faltered at all in the leading of the nation I should have been hurled out of office. I was sure that every Minister was ready to be killed quite soon, and have all his family and possessions destroyed, rather than give in. In this they represented the House of Commons and almost all the people. It fell to me in these coming days and months to express their sentiments on suitable occasions. This I was able to do because they were mine also. There was a white glow, overpowering, sublime, which ran through our Island from end to end.

\* \* \* \* \*

Accurate and excellent accounts have been written of the evacuation of the British and French armies from Dunkirk. Ever since the 20th the gathering of shipping and small craft had been proceeding under the control of Admiral Ramsay, who commanded at Dover. On the evening of the 26th (6.57 p.m.) an Admiralty signal put Operation "Dynamo" into play, and the first troops were brought home that night. After the loss of Boulogne and Calais only the remains of the port of Dunkirk and the open beaches next to the Belgian frontier were in our hands. At this time it was thought that the most we could rescue was about forty-five thousand men in two days. Early the next morning, May 27, emergency measures were taken to find additional small craft "for a special requirement". This was no less than the full evacuation of the British Expeditionary Force.

It was plain that large numbers of such craft would be required for work on the beaches, in addition to bigger ships which could load in Dunkirk harbour. On the suggestion of Mr. H. C. Riggs, of the Ministry of Shipping, the various boatyards, from Teddington to Brightlingsea, were searched by Admiralty officers, and yielded upwards of forty serviceable motor-boats or launches, which were assembled at Sheerness on the following day. At the same time lifeboats from liners in the London docks, tugs from the Thames, yachts, fishing-craft, lighters, barges, and pleasure-boats—anything that could be of use along the beaches—were called into service. By the night of the 27th a great tide of small vessels began to flow towards the sea, first to our Channel ports, and thence to the beaches of Dunkirk and the beloved Army.

Once the need for secrecy was relaxed the Admiralty did not hesitate to give full rein to the spontaneous movement which swept the seafaring population of our south and south-eastern shores. Everyone who had a boat of any kind, steam or sail, put out for Dunkirk, and the preparations, fortunately begun a week earlier, were now aided by the brilliant improvisation of volunteers on an amazing scale. The numbers arriving on the 29th were small, but they were the forerunners of nearly four hundred small craft which from the 31st were destined to play a vital part by ferrying from the beaches to the off-lying ships almost a hundred thousand men. In these days I missed the head of my Admiralty Map Room, Captain Pim, and one or two other familiar faces. They had got hold of a Dutch *schuit* which in four days brought off eight hundred soldiers. Altogether there came to the rescue of the Army under the ceaseless air bombardment of the enemy about eight hundred and sixty vessels, of which nearly seven hundred were British and the rest Allied.

Here is the official list, in which ships not engaged in embarking troops are omitted:

## BRITISH SHIPS

|  | Total engaged | Sunk | Damaged |
| --- | --- | --- | --- |
| A.A. Cruiser .. .. .. | 1 | — | 1 |
| Destroyers .. .. .. .. | 39 | 6 | 19 |
| Sloops, Corvettes, and Gunboats .. | 5 | 1 | 1 |

| | Total engaged | Sunk | Damaged |
|---|---|---|---|
| Minesweepers .. .. .. | 36 | 5 | 7 |
| Trawlers and Drifters .. .. | 77 | 17 | 6 |
| Special Service Vessels .. .. | 3 | 1 | — |
| Armed Boarding Vessels .. .. | 3 | 1 | 1 |
| Motor Torpedo-boats and Motor Anti-Submarine Boats .. | 4 | — | — |
| Dutch *Schuits* (British Naval Crews) | 40 | 4 | (Not recorded) |
| Yachts (Naval Crews) .. .. | 26 | 3 | (Not recorded) |
| Personnel Ships .. .. .. | 45 | 8 | 8 |
| Hospital Carriers .. .. | 8 | 1 | 5 |
| Naval Motor Boats .. .. | 12 | 6 | (Not recorded) |
| Tugs .. .. .. .. | 22 | 3 | (Not recorded) |
| *Other Small Craft .. .. | 372 | 170 | (Not recorded) |
| Total .. | 693 | 226 | |

ALLIED SHIPS

| | Total engaged | Sunk | Damaged |
|---|---|---|---|
| Warships (all types) .. .. | 49 | 8 | (Not recorded) |
| Other Ships and Craft .. .. | 119 | 9 | (Not recorded) |
| Total .. | 168 | 17 | |
| Grand Total .. | 861 | 243 | |

&ast; &ast; &ast; &ast; &ast;

Meanwhile ashore around Dunkirk the occupation of the perimeter was effected with precision. The troops arrived out of chaos and were formed in order along the defences, which even in two days had grown. Those men who were in best shape

* Omitting ships' lifeboats and some other privately owned small craft, of which no record is available.

90

THE DELIVERANCE OF DUNKIRK

turned about to form the line. Divisions like the 2nd and
5th, which had suffered most, were held in reserve on the
beaches and were then embarked early. In the first instance there
were to be three corps on the front, but by the 29th, with the
French taking a greater share in the defences, two sufficed. The
enemy had closely followed the withdrawal, and hard fighting
was incessant, especially on the flanks near Nieuport and Bergues.
As the evacuation went on the steady decrease in the number of
troops, both British and French, was accompanied by a corresponding contraction of the defence. On the beaches, among the
sand dunes, for three, four, or five days scores of thousands of
men dwelt under unrelenting air attack. Hitler's belief that the
German Air Force would render escape impossible, and that therefore he should keep his armoured formations for the final stroke
of the campaign, was a mistaken but not unreasonable view.

Three factors falsified his expectations. First, the incessant air-
bombing of the masses of troops along the seashore did them very
little harm. The bombs plunged into the soft sand, which muffled
their explosions. In the early stages, after a crashing air raid, the
troops were astonished to find that hardly anybody had been
killed or wounded. Everywhere there had been explosions, but
scarcely anyone was the worse. A rocky shore would have produced far more deadly results. Presently the soldiers regarded the
air attacks with contempt. They crouched in the sand dunes with
composure and growing hope. Before them lay the grey but not
unfriendly sea. Beyond, the rescuing ships and—Home.

The second factor which Hitler had not foreseen was the
slaughter of his airmen. British and German air quality was put
directly to the test. By intense effort Fighter Command maintained successive patrols over the scene, and fought the enemy at
long odds. Hour after hour they bit into the German fighter and
bomber squadrons, taking a heavy toll, scattering them and
driving them away. Day after day this went on, till the glorious
victory of the Royal Air Force was gained. Wherever German
aircraft were encountered, sometimes in forties and fifties, they
were instantly attacked, often by single squadrons or less, and shot
down in scores, which presently added up into hundreds. The
whole Metropolitan Air Force, our last sacred reserve, was used.
Sometimes the fighter pilots made four sorties a day. A clear
result was obtained. The superior enemy were beaten or killed,

and for all their bravery mastered, or even cowed. This was a decisive clash. Unhappily, the troops on the beaches saw very little of this epic conflict in the air, often miles away or above the clouds. They knew nothing of the loss inflicted on the enemy. All they felt was the bombs scourging the beaches, cast by the foes who had got through, but did not perhaps return. There was even a bitter anger in the Army against the Air Force, and some of the troops landing at Dover or at Thames ports in their ignorance insulted men in Air Force uniform. They should have clasped their hands; but how could they know? In Parliament I took pains to spread the truth.

But all the aid of the sand and all the prowess in the air would have been vain without the sea. The instructions given ten or twelve days before had, under the pressure and emotion of events, borne amazing fruit. Perfect discipline prevailed ashore and afloat. The sea was calm. To and fro between the shore and the ships plied the little boats, gathering the men from the beaches as they waded out or picking them from the water, with total indifference to the air bombardment, which often claimed its victims. Their numbers alone defied air attack. The Mosquito Armada as a whole was unsinkable. In the midst of our defeat glory came to the Island people, united and unconquerable; and the tale of the Dunkirk beaches will shine in whatever records are preserved of our affairs.

Notwithstanding the valiant work of the small craft it must not be forgotten that the heaviest burden fell on the ships plying from Dunkirk harbour, where two-thirds of the men were embarked. The destroyers played the predominant part, as the casualty list on pages 89–90 shows. Nor must the great part played by the personnel ships with their mercantile crews be overlooked.

\* \* \* \* \*

The progress of the evacuation was watched with anxious eyes and growing hope. On the evening of the 27th Lord Gort's position appeared critical to the naval authorities, and Captain Tennant, R.N., from the Admiralty, who had assumed the duties of Senior Naval Officer at Dunkirk, signalled for all available craft to be sent to the beaches immediately, as "evacuation to-morrow night is problematical". The picture presented was grim, even desperate. Extreme efforts were made to meet the call, and a

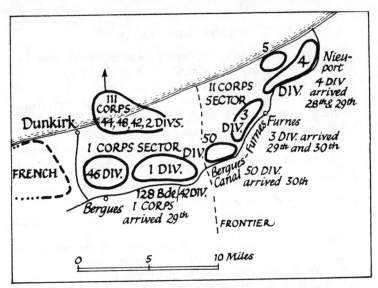

Diagram of DUNKIRK PERIMETER 29 & 30 May

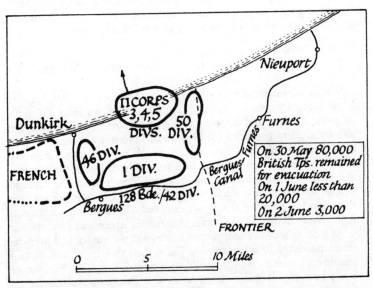

Diagram of DUNKIRK PERIMETER 31 May & 1 June

cruiser, eight destroyers, and twenty-six other vessels were sent. The 28th was a day of tension, which gradually eased as the position on land was stabilised with the powerful help of the Royal Air Force. The naval plans were carried through despite severe losses on the 29th, when three destroyers and twenty-one other vessels were sunk and many others damaged.

There was never any question of our leaving the French behind. Here was my order before any request or complaint from the French was received:

*Prime Minister to Secretary of State for War,*
*C.I.G.S., and General Ismay*    29.V.40
    (Original to C.I.G.S.)

It is essential that the French should share in such evacuations from Dunkirk as may be possible. Nor must they be dependent only upon their own shipping resources. Arrangements must be concerted at once with the French Missions in this country, or, if necessary, with the French Government, so that no reproaches, or as few as possible, may arise. It might perhaps be well if we evacuated the two French divisions from Dunkirk, and replaced them *pro tem.* with our own troops, thus simplifying the command. But let me have the best proposals possible, and advise me whether there is any action I should take.

*Prime Minister to General Spears (Paris)*    29.V.40
    Following for Reynaud, for communication to Weygand and Georges:

We have evacuated nearly 50,000 from Dunkirk and beaches, and hope another 30,000 to-night. Front may be beaten in at any time, or piers, beaches, and shipping rendered unusable by air attack, and also by artillery fire from the south-west. No one can tell how long present good flow will last, or how much we can save for future. We wish French troops to share in evacuation to fullest possible extent, and Admiralty have been instructed to aid French Marine as required. We do not know how many will be forced to capitulate, but we must share this loss together as best we can, and, above all, bear it without reproaches arising from inevitable confusion, stresses, and strains.

As soon as we have reorganised our evacuated troops, and prepared forces necessary to safeguard our life against threatened and perhaps imminent invasion, we shall build up a new B.E.F. from St. Nazaire. I am bringing Regulars from India and Palestine; Australians and Canadians are arriving soon. At present we are removing equipment south of Amiens beyond what is needed for five divisions. But this is only to get into order and meet impending shock, and we shall shortly send you new scheme for reinforcement of our troops in France. I send this in all comradeship. Do not hesitate to speak frankly to me.

*　*　*　*　*

On the 30th I held a meeting of the three Service Ministers and the Chiefs of Staff in the Admiralty War Room. We considered the events of the day on the Belgian coast. The total number of troops brought off had risen to 120,000, including only 6,000 French; 860 vessels of all kinds were at work. A message from Admiral Wake-Walker at Dunkirk said that, in spite of intense bombardment and air attack, 4,000 men had been embarked in the previous hour. He also thought that Dunkirk itself would probably be untenable by the next day. I emphasised the urgent need of getting off more French troops. To fail to do so might do irreparable harm to the relations between ourselves and our Ally. I also said that when the British strength was reduced to that of a corps we ought to tell Lord Gort to embark and return to England, leaving a Corps Commander in charge. The British Army would have to stick it out as long as possible so that the evacuation of the French could continue.

Knowing well the character of Lord Gort, I wrote out in my own hand the following order to him, which was sent officially by the War Office at 2 p.m. on the 30th:

Continue to defend the present perimeter to the utmost in order to cover maximum evacuation now proceeding well. Report every three hours through La Panne. If we can still communicate we shall send you an order to return to England with such officers as you may choose at the moment when we deem your command so reduced that it can be handed over to a Corps Commander. You should now nominate this Commander. If communications are broken you are to hand over and return as specified when your effective fighting force does not exceed the equivalent of three divisions. This is in accordance with correct military procedure, and no personal discretion is left you in the matter. On political grounds it would be a needless triumph to the enemy to capture you when only a small force remained under your orders. The Corps Commander chosen by you should be ordered to carry on the defence in conjunction with the French, and evacuation whether from Dunkirk or the beaches, but when in his judgment no further organised evacuation is possible and no further proportionate damage can be inflicted on the enemy he is authorised in consultation with the senior French Commander to capitulate formally to avoid useless slaughter.

It is possible that this last message influenced other great events and the fortunes of another valiant Commander. When I was at the White House at the end of December 1941 I learned from the President and Mr. Stimson of the approaching fate of General MacArthur and the American garrison at Corregidor. I thought it right to show them the way in which we had dealt with the position of a Commander-in-Chief whose force was reduced to a small fraction of his original command. The President and Mr. Stimson both read the telegram with profound attention, and I was struck by the impression it seemed to make upon them. A little later in the day Mr. Stimson came back and asked for a copy of it, which I immediately gave him. It may be (for I do not know) that this influenced them in the right decision which they took in ordering General MacArthur to hand over his command to one of his subordinate generals, and thus saved for all his future glorious services the great Commander who would otherwise have perished or passed the war as a Japanese captive. I should like to think this was true.

\* \* \* \* \*

On the 30th members of Lord Gort's staff in conference with Admiral Ramsay at Dover informed him that daylight on June 1 was the latest time up to which the eastern perimeter might be expected to hold. Evacuation was therefore pressed on with the utmost urgency to ensure, so far as possible, that a British rear-guard of no more than about four thousand men would then remain ashore. Later it was found that this number would be insufficient to defend the final covering positions, and it was decided to hold the British sector until midnight June 1-2, evacuation proceeding meanwhile on the basis of full equality between French and British forces.

Such was the situation when on the evening of May 31 Lord Gort in accordance with his orders handed over his command to Major-General Alexander and returned to England.

\* \* \* \* \*

To avoid misunderstandings by keeping personal contact it was necessary for me to fly to Paris on May 31 for a meeting of the Supreme War Council. With me in the plane came Mr. Attlee and Generals Dill and Ismay. I also took General Spears, who had flown over on the 30th with the latest news from Paris. This brilliant officer and Member of Parliament was a friend of mine from the First Great War. Liaison officer between the left of the French and the right of the British Armies, he had taken me round the Vimy Ridge in 1916, and had made me friends with General Fayolle, who commanded the 33rd French Corps. Speaking French with a perfect accent and bearing five wound-stripes on his sleeve, he was a personality at this moment fitted to our anxious relations. When Frenchmen and Englishmen are in trouble together and arguments break out, the Frenchman is often voluble and vehement, and the Englishman unresponsive or even rude. But Spears could say things to the high French personnel with an ease and force which I have never seen equalled.

This time we did not go to the Quai d'Orsay, but to M. Reynaud's room at the War Office in the Rue Saint-Dominique. Attlee and I found Reynaud and Marshal Pétain opposite to us as the only French Ministers. This was the first appearance of Pétain, now Vice-President of the Council, at any of our meetings. He wore plain clothes. Our Ambassador, Dill, Ismay, and Spears were with us, and Weygand and Darlan, Captain de Margerie, head of Reynaud's private office, and M. Baudouin, Secretary of the French War Cabinet, represented the French.

The first question was the position in Norway. I said that the British Government was of the considered opinion that the Narvik area should be evacuated at once. Our troops there, the destroyers involved, and a hundred anti-aircraft guns were badly wanted elsewhere. We therefore proposed an evacuation beginning on June 2. The British Navy would transport and repatriate the French forces, the King of Norway, and any Norwegian troops who wished to come. Reynaud said that the French Government agreed with this policy. The destroyers would be urgently required in the Mediterranean in the event of war with Italy. The 16,000 men would be very valuable on the line of the Aisne and the Somme. This matter was therefore settled.

I then turned to Dunkirk. The French seemed to have no more idea of what was happening to the Northern Armies than we had about the main French front. When I told them that 165,000 men, of whom 15,000 were French, had been taken off they were astonished. They naturally drew attention to the marked British preponderance. I explained that this was due largely to the fact that there had been many British administrative units in the back area who had been able to embark before fighting troops could be spared from the front. Moreover, the French up to the present had had no orders to evacuate. One of the chief reasons why I had come to Paris was to make sure that the same orders were given to the French troops as to the British. The three British divisions now holding the centre would cover the evacuation of all the Allied forces. That, and the sea transport, would be the British contribution to offset the heavy Allied losses which must now be faced. His Majesty's Government had felt it necessary in the dire circumstances to order Lord Gort to take off fighting men and leave the wounded behind. If present hopes were confirmed, 200,000 able-bodied troops might be got away. This would be almost a miracle. Four days ago I would not have wagered on more than 50,000 as a maximum. I dwelt upon our terrible losses in equipment. Reynaud paid a handsome tribute to the work of the British Navy and Air Force, for which I thanked him. We then spoke at some length upon what could be done to rebuild the British forces in France.

Meanwhile Admiral Darlan had drafted a telegram to Admiral Abrial at Dunkirk:

(1) A bridgehead shall be held round Dunkirk with the divisions under your command and those under British command.

(2) As soon as you are convinced that no troops outside the bridge-head can make their way to the points of embarkation the troops holding the bridgehead shall withdraw and embark, *the British forces embarking first.*

I intervened at once to say that the British would not embark first, but that the evacuation should proceed on equal terms between the British and the French—"*Bras dessus bras dessous.*" The British would form the rearguard. This was agreed.

The conversation next turned to Italy. I expressed the British view that if Italy came in we should strike at her at once in the most effective manner. Many Italians were opposed to war, and all should be made to realise its severity. I proposed that we should strike by air-bombing at the north-western industrial triangle enclosed by the three cities of Milan, Turin and Genoa. Reynaud agreed that the Allies must strike at once; and Admiral Darlan said he had a plan ready for the naval and aerial bombardment of

Italy's oil supplies, largely stored along the coast between the frontier and Naples. The necessary technical discussions were arranged.

I then mentioned my desire that more Ministers of the Administration I had just formed should become acquainted with their French opposite numbers as soon as possible. For instance, I should like Mr. Bevin, the Minister of Labour and trade union leader, to visit Paris. Mr. Bevin was showing great energy, and under his leadership the British working class was now giving up holidays and privileges to a far greater extent than in the last war. Reynaud cordially assented.

After some talk about Tangier and the importance of keeping Spain out of the war, I spoke on the general outlook. The Allies, I said, must maintain an unflinching front against all their enemies. ... The United States had been roused by recent events, and, even if they did not enter the war, would soon be prepared to give us powerful aid. An invasion of England, if it took place, would have a still more profound effect on the United States. England did not fear invasion, and would resist it most fiercely in every village and hamlet. It was only after her essential need of troops had been met that the balance of her armed forces could be put at the disposal of her French ally. . . . I was absolutely convinced we had only to carry on the fight to conquer. Even if one of us should be struck down, the other must not abandon the struggle. The British Government were prepared to wage war from the New World, if through some disaster England herself were laid waste. If Germany defeated either ally or both, she would give no mercy; we should be reduced to the status of vassals and slaves for ever. It would be better far that the civilisation of Western Europe with all its achievements should come to a tragic but splendid end than that the two great democracies should linger on, stripped of all that made life worth living.

Mr. Attlee then said that he entirely agreed with my view. "The British people now realise the danger with which they are faced, and know that in the event of a German victory everything they have built up will be destroyed. The Germans kill not only men, but ideas. Our people are resolved as never before in their history." Reynaud thanked us for what we had said. He was sure that the morale of the German people was not up to the level of the momentary triumph of their Army. If France could hold the Somme with the help of Britain and if American industry came in to make good the disparity in arms, then we could be sure of victory. He was most grateful, he said, for my renewed assurance that if one country went under the other would not abandon the struggle.

The formal meeting then ended.

After we rose from the table some of the principals talked together in the bay window in a somewhat different atmosphere. Chief among these was Marshal Pétain. Spears was with me, helping me out with my French and speaking himself. The young Frenchman, Captain de Margerie, had already spoken about fighting it out in Africa. But Marshal Pétain's attitude, detached and sombre, gave me the feeling that he would face a separate peace. The influence of his personality, his reputation, his serene acceptance of the march of adverse events, apart from any words he used, was almost overpowering to those under his spell. One of the Frenchmen, I cannot remember who, said in their polished way that a continuance of military reverses might in certain eventualities enforce a modification of foreign policy upon France. Here Spears rose to the occasion, and, addressing himself particularly to Marshal Pétain, said in perfect French: "I suppose you understand, M. le Maréchal, that that would mean blockade?" Someone else said: "That would perhaps be inevitable." But then Spears to Pétain's face: "That would not only mean blockade, but *bombardment* of all French ports in German hands." I was glad to have this said. I sang my usual song: we would fight on whatever happened or whoever fell out.

\* \* \* \* \*

Again we had a night of petty raids, and in the morning I departed. Here was the information that awaited me on my return:

*Prime Minister to General Weygand*                                    1.VI.40
Crisis in evacuation now reached. Five fighter squadrons, acting almost continuously, is the most we can do, but six ships, several filled with troops, sunk by bombing this morning. Artillery fire menacing only practicable channel. Enemy closing in on reduced bridgehead. By trying to hold on till to-morrow we may lose all. By going to-night much may certainly be saved, though much will be lost. Nothing like numbers of effective French troops you mention believed in bridge-

head now, and we doubt whether such large numbers remain in area. Situation cannot be fully judged by Admiral Abrial in the fortress, nor by you, nor by us here. We have therefore ordered General Alexander, commanding British sector of bridgehead, to judge, in consultation with Admiral Abrial, whether to try to stay over to-morrow or not. Trust you will agree.

May 31 and June 1 saw the climax though not the end at Dunkirk. On these two days over 132,000 men were safely landed in England, nearly one-third of them having been brought from the beaches in small craft under fierce air attack and shell-fire. On June 1 from early dawn onward the enemy bombers made their greatest efforts, often timed when our own fighters had withdrawn to refuel. These attacks took heavy toll of the crowded shipping, which suffered almost as much as in all the previous week. On this single day our losses by air attack, by mines, E-boats, or other misadventure were thirty-one ships sunk and eleven damaged. On land the enemy increased their pressure on the bridgehead, doing their utmost to break through. They were held at bay by the desperate resistance of the Allied rear-guards.

The final phase was carried through with much skill and precision. For the first time it became possible to plan ahead instead of being forced to rely on hourly improvisations. At dawn on June 2 about four thousand British with seven anti-aircraft guns and twelve anti-tank guns remained on the outskirts of Dunkirk with the still considerable French forces holding the contracting perimeter. Evacuation was now possible only in darkness, and Admiral Ramsay determined to make a massed descent on the harbour that night with all his available resources. Besides tugs and small craft, forty-four ships were sent that evening from England, including eleven destroyers and fourteen mine-sweepers. Forty French and Belgian vessels also participated. Before midnight the British rearguard was embarked.

This was not however the end of the Dunkirk story. We had been prepared to carry considerably greater numbers of French that night than had offered themselves. The result was that when our ships, many of them still empty, had to withdraw at dawn, great numbers of French troops, many still in contact with the enemy, remained ashore. One more effort had to be made. Despite the exhaustion of ships' companies after so many days without rest or respite, the call was answered. On June 4 26,175 Frenchmen were landed in England, over twenty-one thousand of them in British ships. Unfortunately several thousands remained, who continued the fight in the contracting bridgehead until the morning of the 4th, when the enemy was in the outskirts of the town and they had come to an end of their powers. They had fought gallantly for many days to cover the evacuation of their British and French comrades. They were to spend the next years in captivity. Let us remember that but for the endurance of the Dunkirk rearguard the re-creation of an army in Britain for home defence and final victory would have been gravely prejudiced.

Finally, at 2.23 p.m. on June 4 the Admiralty, in agreement with the French, announced that Operation "Dynamo" was now completed.

## BRITISH AND ALLIED TROOPS LANDED IN ENGLAND*

| Date | | From the Beaches | From Dunkirk Harbour | Total | Accumulated Total |
|---|---|---|---|---|---|
| May 27 | .. | Nil | 7,669 | 7,669 | 7,669 |
| 28 | .. | 5,930 | 11,874 | 17,804 | 25,473 |
| 29 | .. | 13,752 | 33,558 | 47,310 | 72,783 |
| 30 | .. | 29,512 | 24,311 | 53,823 | 126,606 |
| 31 | .. | 22,942 | 45,072 | 68,014 | 194,620 |
| June 1 | .. | 17,348 | 47,081 | 64,429 | 259,049 |
| 2 | .. | 6,695 | 19,561 | 26,256 | 285,305 |
| 3 | .. | 1,870 | 24,876 | 26,746 | 312,051 |
| 4 | .. | 622 | 25,553 | 26,175 | 338,226 |
| GRAND TOTAL | | 98,671 | 239,555 | 338,226 | |

\* \* \* \* \*

Parliament assembled on June 4, and it was my duty to lay the story fully before them both in public and later in secret session. The narrative requires only a few extracts from my speech, which is extant. It was imperative to explain not only to our own people but to the world that our resolve to fight on was based on serious

\* These figures are taken from a final analysis of the Admiralty records. The War Office figure for the total number of men landed in England is 336,427.

grounds, and was no mere despairing effort. It was also right to lay bare my own reasons for confidence.

*We must be very careful not to assign to this deliverance the attributes of a victory. Wars are not won by evacuations.* But there was a victory inside this deliverance, which should be noted. It was gained by the Air Force. Many of our soldiers coming back have not seen the Air Force at work; they saw only the bombers which escaped its protective attack. They underrate its achievements. I have heard much talk of this; that is why I go out of my way to say this. I will tell you about it.

This was a great trial of strength between the British and German Air Forces. Can you conceive a greater objective for the Germans in the air than to make evacuation from these beaches impossible, and to sink all these ships which were displayed, almost to the extent of thousands? Could there have been an objective of greater military importance and significance for the whole purpose of the war than this? They tried hard, and they were beaten back; they were frustrated in their task. We got the Army away; and they have paid fourfold for any losses which they have inflicted. . . . All of our types and all our pilots have been vindicated as superior to what they have at present to face.

When we consider how much greater would be our advantage in defending the air above this Island against an overseas attack, I must say that I find in these facts a sure basis upon which practical and reassuring thoughts may rest. I will pay my tribute to these young airmen. The great French Army was very largely, for the time being, cast back and disturbed by the onrush of a few thousands of armoured vehicles. May it not also be that the cause of civilisation itself will be defended by the skill and devotion of a few thousand airmen?

We are told that Herr Hitler has a plan for invading the British Isles. This has often been thought of before. When Napoleon lay at Boulogne for a year with his flat-bottomed boats and his Grand Army he was told by someone: "There are bitter weeds in England." There are certainly a great many more of them since the British Expeditionary Force returned.

The whole question of Home Defence against invasion is, of course, powerfully affected by the fact that we have for the time being in this Island incomparably stronger military forces than we have ever had at any moment in this war or the last. But this will not continue. We shall not be content with a defensive war. We have our duty to our Ally. We have to reconstitute and build up the British Expeditionary Force once again, under its gallant Commander-in-Chief, Lord Gort. All this is in train; but in the interval we must put our defences in this Island into such a high state of organisation that the fewest possible numbers will be required to give effective security and that the largest possible potential of offensive effort may be realised. On this we are now engaged.

I ended in a passage which was to prove, as will be seen, a timely and important factor in United States decisions.

"Even though large tracts of Europe and many old and famous States have fallen or may fall into the grip of the Gestapo and all the odious apparatus of Nazi rule, we shall not flag or fail. We shall go on to the end. We shall fight in France, we shall fight in the seas and oceans, we shall fight with growing confidence and growing strength in the air; we shall defend our Island, whatever the cost may be. We shall fight on the beaches, we shall fight on the landing-grounds, we shall fight in the fields and in the streets, we shall fight in the hills; we shall never surrender; and even if, which I do not for a moment believe, this Island or a large part of it were subjugated and starving, then our Empire beyond the seas, armed and guarded by the British Fleet, would carry on the struggle, until, in God's good time, the New World, with all its power and might, steps forth to the rescue and the liberation of the Old."

# CHAPTER VI

# THE RUSH FOR THE SPOILS

*Traditional British and Italian Friendship - Advantages to Italy and Mussolini of Neutrality - My Message to Mussolini on Becoming Prime Minister - His Hard Response - Reynaud's Visit to London of May 26 - France and Britain Invite President Roosevelt to Intervene - My Telegram Conveying the Cabinet Decision of May 28 - Preparations to Strike at Italy Should She Declare War - Italy and Yugoslavia - The Italian Declaration of War - The Attack on the Alpine Front Stopped by the French Army - Ciano's Letter to Me of December 23, 1943 - President Roosevelt's Denunciation of Italy - My Telegram to Him of June 11 - Anglo-Soviet Relations - Molotov's Congratulations upon German Victories - Sir Stafford Cripps Appointed Ambassador to Moscow - My Letter to Stalin of June 25, 1940 - The Soviet Share of the Spoil.*

THE friendship between the British and Italian peoples sprang from the days of Garibaldi and Cavour. Every stage in the liberation of Northern Italy from Austria and every step towards Italian unity and independence had commanded the sympathies of Victorian Liberalism. This had bred a warm and enduring response. The declaration in the original Treaty of Triple Alliance between Italy, Germany, and the Austro-Hungarian Empire stipulated that in no circumstances should Italy be drawn into war with Great Britain. British influence had powerfully contributed to the Italian accession to the Allied cause in the First World War. The rise of Mussolini and the establishment of Fascism as a counter to Bolshevism had in its early phases divided British opinion on party lines, but had not affected the broad foundations of goodwill between the peoples. We have seen that until Mussolini's designs against Abyssinia had raised grave issues he had ranged himself with Great Britain in opposition to Hitlerism and German ambitions. I have told in the previous volume the sad tale of how the Baldwin-Chamberlain policy about Abyssinia brought us the worst of both worlds, how we estranged the Italian dictator without breaking his power, and how the League of Nations was injured without Abyssinia being saved. We have also seen the earnest but futile efforts made by Mr. Chamberlain, Sir Samuel Hoare, and Lord Halifax to win back during the period of appeasement Mussolini's lost favour. And finally there was the growth of Mussolini's conviction that Britain's sun had set, and that Italy's future could, with German help, be founded on the ruins of the British Empire. This had been followed by the creation of the Berlin-Rome Axis, in accordance with which Italy might well have been expected to enter the war against Britain and France on its very first day.

It was certainly only common prudence for Mussolini to see how the war would go before committing himself and his country irrevocably. The process of waiting was by no means unprofitable. Italy was courted by both sides, and gained much consideration for her interests, many profitable contracts, and time to improve her armaments. Thus the twilight months had passed. It is an interesting speculation what the Italian fortunes would have been if this policy had been maintained. The United States with its large Italian vote might well have made it clear to Hitler that an attempt to rally Italy to his side by force of arms would raise the gravest issues. Peace, prosperity, and growing power would have been the prize of a persistent neutrality. Once Hitler was embroiled with Russia this happy state might have been almost indefinitely prolonged, with ever-growing benefits, and Mussolini might have stood forth in the peace or in the closing year of the war as the wisest statesman the sunny peninsula and its industrious and prolific people had known. This was a more agreeable situation than that which in fact awaited him.

At the time when I was Chancellor of the Exchequer under Mr. Baldwin in the years after 1924 I did what I could to preserve the traditional friendship between Italy and Britain. I made a debt settlement with Count Volpi which contrasted very favourably with the arrangements made with France. I received the warmest expressions of gratitude from the Duce, and with difficulty escaped the highest decoration. Moreover, in the conflict between Fascism and Bolshevism there was no doubt where my sympathies and convictions lay. On the two occasions in 1927 when I met Mussolini our personal relations had been intimate and easy. I would never have encouraged Britain to make a breach with him about Abyssinia or roused the League of Nations against him unless we were prepared to go to war in the last extreme. He, like Hitler, understood and in a way respected my campaign for British rearmament, though he was very glad British public opinion did not support my view.

In the crisis we had now reached of the disastrous Battle of France it was clearly my duty as Prime Minister to do my utmost to keep Italy out of the conflict, and though I did not indulge in vain hopes I at once used what resources and influence I might possess. Six days after becoming head of the Government I wrote at the Cabinet's desire the appeal to Mussolini which, together with his answer, was published two years later in very different circumstances.

*Prime Minister to Signor Mussolini*                                    16.v.40
Now that I have taken up my office as Prime Minister and Minister of Defence I look back to our meetings in Rome and feel a desire to speak words of goodwill to you as Chief of the Italian nation across what seems to be a swiftly-widening gulf. Is it too late to stop a river of blood from flowing between the British and Italian peoples? We can no doubt inflict grievous injuries upon one another and maul each other cruelly, and darken the Mediterranean with our strife. If you so decree, it must be so; but I declare that I have never been the enemy of Italian greatness, nor ever at heart the foe of the Italian lawgiver. It is idle to predict the course of the great battles now raging in Europe, but I am sure that whatever may happen on the Continent England will go on to the end, even quite alone, as we have done before, and I believe with some assurance that we shall be aided in increasing measure by the United States, and, indeed, by all the Americas.

I beg you to believe that it is in no spirit of weakness or of fear that I make this solemn appeal, which will remain on record. Down the ages above all other calls comes the cry that the joint heirs of Latin and Christian civilisation must not be ranged against one another in mortal strife. Hearken to it, I beseech you in all honour and respect, before the dread signal is given. It will never be given by us.

The response was hard. It had at least the merit of candour.

*Signor Mussolini to Prime Minister*                                    18.v.40
I reply to the message which you have sent me in order to tell you that you are certainly aware of grave reasons of an historical and

contingent character which have ranged our two countries in opposite camps. Without going back very far in time I remind you of the initiative taken in 1935 by your Government to organise at Geneva sanctions against Italy, engaged in securing for herself a small space in the African sun without causing the slightest injury to your interests and territories or those of others. I remind you also of the real and actual state of servitude in which Italy finds herself in her own sea. If it was to honour your signature that your Government declared war on Germany, you will understand that the same sense of honour and of respect for engagements assumed in the Italian-German Treaty guides Italian policy to-day and to-morrow in the face of any event whatsoever.

From this moment we could have no doubt of Mussolini's intention to enter the war at his most favourable opportunity. His resolve had in fact been made as soon as the defeat of the French armies was obvious. On May 13 he had told Ciano that he would declare war on France and Britain within a month. His official decision to declare war on any date suitable after June ‹ was imparted to the Italian Chiefs of Staff on May 29. At Hitler's request the date was postponed to June 10.

\* \* \* \* \*

On May 26, while the fate of the Northern Armies hung in the balance and no one could be sure that any would escape, Reynaud flew over to England to have a talk with us about this topic, which had not been absent from our minds. The Italian declaration of war must be expected at any moment. Thus France would burn upon another front, and a new foe would march hungrily upon her in the South. Could anything be done to buy off Mussolini? That was the question posed. I did not think there was the slightest chance, and every fact that the French Premier used as an argument for trying only made me surer there was no hope. However, Reynaud was under strong pressure at home, and we on our side wished to give full consideration to our Ally, whose one vital weapon, her Army, was breaking in her hand. M. Reynaud has published a full account of his visit, and especially of his conversations.\* Lord Halifax, Mr. Chamberlain, Mr. Attlee, and Mr. Eden were also at our meetings. Although there was no need to marshal the grave facts, M. Reynaud dwelt not obscurely upon the possible French withdrawal from the war.

\* Reynaud, *La France a sauvé l'Europe*, Vol. II, pp. 200 ff.

He himself would fight on, but there was always the possibility that he might soon be replaced by others of a different temper.

We had already on May 25 at the instance of the French Government made a joint request to President Roosevelt to intervene. In this message Britain and France authorised him to state that we understood Italy had territorial grievances against us in the Mediterranean, that we were disposed to consider at once any reasonable claims, that the Allies would admit Italy to the Peace Conference with a status equal to that of any belligerent, and that we would invite the President to see that any agreement reached now was carried out. The President acted accordingly; but his addresses were repulsed by the Italian dictator in the most abrupt manner. At our meeting with Reynaud we had already this answer before us. The French Premier now suggested more precise proposals. Obviously, if these were to remedy Italy's "state of servitude in her own sea", they must affect the status both of Gibraltar and Suez. France was prepared to make similar concessions about Tunis.

We were not able to show any favour to these ideas. This was not because it was wrong to examine them or because it did not seem worth while at this moment to pay a heavy price to keep Italy out of the war. My own feeling was that at the pitch in which our affairs lay we had nothing to offer which Mussolini could not take for himself or be given by Hitler if we were defeated. One cannot easily make a bargain at the last gasp. Once we started negotiating for the friendly mediation of the Duce we should destroy our power of fighting on. I found my colleagues very stiff and tough. All our minds ran much more on bombing Milan and Turin the moment Mussolini declared war, and seeing how he liked that. Reynaud, who did not at heart disagree, seemed convinced, or at least content. The most we could promise was to bring the matter before the Cabinet and send a definite answer the next day. Reynaud and I lunched alone together at the Admiralty. The following telegram, the greater part of which is my own wording, embodies the conclusions of the War Cabinet:

*Prime Minister to M. Reynaud*                                28.V.40

1. I have with my colleagues examined with the most careful and sympathetic attention the proposal for an approach by way of precise offer of concessions to Signor Mussolini that you have forwarded to

me to-day, fully realising the terrible situation with which we are both faced at this moment.

2. Since we last discussed this matter the new fact which has occurred, namely, the capitulation of the Belgian Army, has greatly changed our position for the worse, for it is evident that the chance of withdrawing the armies of Generals Blanchard and Gort from the Channel ports has become very problematical. The first effect of such a disaster must be to make it impossible at such a moment for Germany to put forward any terms likely to be acceptable, and neither we nor you would be prepared to give up our independence without fighting for it to the end.

3. In the formula prepared last Sunday by Lord Halifax it was suggested that if Signor Mussolini would co-operate with us in securing a settlement of all European questions which would safeguard our independence and form the basis of a just and durable peace for Europe we should be prepared to discuss his claims in the Mediterranean. You now propose to add certain specific offers, which I cannot suppose would have any chance of moving Signor Mussolini, and which once made could not be subsequently withdrawn, in order to induce him to undertake the *rôle* of mediator, which the formula discussed on Sunday contemplated.

4. I and my colleagues believe that Signor Mussolini has long had it in mind that he might eventually fill this *rôle*, no doubt counting upon substantial advantages for Italy in the process. But we are convinced that at this moment when Hitler is flushed with victory and certainly counts on early and complete collapse of Allied resistance it would be impossible for Signor Mussolini to put forward proposals for a conference with any success. I may remind you also that the President of the U.S.A. has received a wholly negative reply to the proposal which we jointly asked him to make, and that no response has been made to the approach which Lord Halifax made to the Italian Ambassador here last Saturday.

5. Therefore, without excluding the possibility of an approach to Signor Mussolini at some time, we cannot feel that this would be the right moment, and I am bound to add that in my opinion the effect on the morale of our people, which is now firm and resolute, would be extremely dangerous. You yourself can best judge what would be the effect in France.

6. You will ask, then, how is the situation to be improved? My reply is that by showing that after the loss of our two [Northern] armies and the support of our Belgian ally we still have stout hearts and confidence in ourselves we shall at once strengthen our hands in negotiations and draw the admiration and perhaps the material help of

the U.S.A. Moreover, we feel that as long as we stand together our undefeated Navy and our Air Force, which is daily destroying German fighters and bombers at a formidable rate, afford us the means of exercising in our common interest a continuous pressure upon Germany's internal life.

7. We have reason to believe that the Germans too are working to a time-table, and that their losses and the hardships imposed on them, together with the fear of our air raids, is undermining their courage. It would indeed be a tragedy if by too hasty an acceptance of defeat we threw away a chance that was almost within our grasp of securing an honourable issue from the struggle.

8. In my view, if we both stand out we may yet save ourselves from the fate of Denmark and Poland. Our success must depend first on our unity, then on our courage and endurance.

This did not prevent the French Government from making a few days later a direct offer of their own to Italy of territorial concessions, which Mussolini treated with disdain. "He was not interested," said Ciano to the French Ambassador on June 3, "in recovering any French territories by peaceful negotiation. He had decided to make war on France."* This was only what we had expected.

\*       \*       \*       \*       \*

I now gave daily a series of directives to make sure that if we were subjected to this odious attack by Mussolini we should be able to strike back at once.

*Prime Minister to General Ismay*                                        28.v.40
  1. Pray bring the following before the C.O.S. Committee:
What measures have been taken, in the event of Italy's going to war, to attack Italian forces in Abyssinia, sending rifles and money to the Abyssinian insurgents, and generally to disturb that country?

I understand General Smuts has sent a Union brigade to East Africa. Is it there yet? When will it be? What other arrangements are made? What is the strength of the Khartoum garrison, including troops in the Blue Nile Province? This is the opportunity for the Abyssinians to liberate themselves, with Allied help.

  2. If France is still our ally after an Italian declaration of war, it would appear extremely desirable that the combined Fleets, acting from opposite ends of the Mediterranean, should pursue an active offensive against Italy. It is important that at the outset collision should

* See Reynaud, Vol. II, p. 209.

take place both with the Italian Navy and Air Force, in order that we can see what their quality really is, and whether it has changed at all since the last war. The purely defensive strategy contemplated by Commander-in-Chief Mediterranean ought not to be accepted. Unless it is found that the fighting qualities of the Italians are high, it will be much better that the Fleet at Alexandria should sally forth and run some risks than that it should remain in a posture so markedly defensive. Risks must be run at this juncture in all theatres.

3. I presume that the Admiralty have a plan in the event of France becoming neutral.

*Prime Minister to General Ismay (and others)*                29.V.40
We must have eight battalions from Palestine home at the earliest moment. I regard the Mediterranean as closed to troopships. The choice is therefore between the Red Sea and the Persian Gulf. Let this alternative route [across the desert to the Gulf] be examined this afternoon, and Admiralty be consulted, and report to me on relative times and safety. The Australians can be left in Palestine for the moment, but the High Commissioner, like others, must conform to the supreme requirements of the State.

Admiralty should say whether it would be possible to pick these men up at the Cape in the big liners for extra speed.

*Prime Minister to First Lord of the Admiralty*                30.V.40
What measures have been taken to seize all Italian ships at the moment of war? How many are there in British ports, and what can be done about them on the seas or in foreign ports? Will you kindly pass this to the proper department immediately.

At the Supreme War Council in Paris on May 31, which has already been described, it was agreed that the Allies should undertake offensive operations against selected objectives in Italy at the earliest possible moment and that the French and British naval and air staffs should concert their plans. We had also agreed that in the event of Italian aggression against Greece, of which there were indications, we should make sure that Crete did not fall into enemy hands. I pursued the same theme in my minutes.

*Prime Minister to Secretary of State for Air and Chief of
Air Staff*                2.VI.40
It is of the utmost importance, in view of the [possible] raids on Lyons and Marseilles, that we should be able to strike back with our heavy bombers at Italy the moment she enters the war. I consider therefore that these squadrons should be flown to their aerodromes in

Southern France at the earliest moment when French permission can be obtained and when the servicing units are ready for their reception.

Pray let me know at our meeting to-night what you propose.

*Prime Minister to S. of S. for Air and C.A.S.*           6.VI.40

It is of the highest importance that we should strike at Italy the moment war breaks out, or an overbearing ultimatum is received. Please let me know the exact position of the servicing units which are on their way to the southern aerodromes in France.

An early Italian plan, favoured particularly by Ciano, had been that Italian action in Europe should be confined to the launching of an attack on Yugoslavia, thus consolidating Italy's power in Eastern Europe and strengthening her potential economic position. Mussolini himself was for a time won over to this idea. Graziani records that at the end of April the Duce told him: "We must bring Yugoslavia to her knees; we have need of raw materials, and it is in her mines that we must find them. In consequence my strategic directive is—defensive in the west (France) and offensive in the east (Yugoslavia). Prepare a study of the problem."\* Graziani claims that he advised strongly against committing the Italian armies, short as they were of equipment, particularly of artillery, to a repetition of the Isonzo campaign of 1915. There were also political arguments against the Yugoslav plan. The Germans were anxious at this moment to avoid disturbing Eastern Europe. They feared it would provoke British action in the Balkans and might inadvertently tempt Russia to further activity in the East. I was not unaware of this aspect of Italian policy.

*Prime Minister to Secretary of State for Foreign Affairs*      6.VI.40

I have hitherto argued against going to war with Italy because she attacked [*i.e.*, if she were to attack] Yugoslavia, and have wished to see whether it was a serious attack upon Yugoslavian independence or merely taking some naval bases in the Adriatic. However, this situation has changed. Italy is continually threatening to go to war with England and France, and not by "the back door". We are so near a break with Italy on grounds which have nothing to do with Yugoslavia that it would seem that our main aim might well be now to procure this Balkan mobilisation. Will you think this over?

\* \* \* \* \*

\* Graziani, *Ho Difeso la Patria*, p. 189.

In spite of the extreme efforts made by the United States, of which Mr. Hull has given an impressive account in his memoirs,* nothing could turn Mussolini from his course. Our preparations to meet the new assault and complication were well advanced when the moment came. On June 10 at 4.45 p.m. the Italian Minister for Foreign Affairs informed the British Ambassador that Italy would consider herself at war with the United Kingdom from midnight that day. A similar communication was made to the French Government. When Ciano delivered his note to the French Ambassador, M. François-Poncet remarked as he reached the door: "You too will find the Germans are hard masters." From his balcony in Rome Mussolini announced to well-organised crowds that Italy was at war with France and Britain. It was, as Ciano is said to have apologetically remarked later on, "a chance which comes only once in five thousand years." Such chances, though rare, are not necessarily good.

Forthwith the Italians attacked the French troops on the Alpine front and Great Britain reciprocally declared war on Italy. Five Italian ships detained at Gibraltar were seized, and orders were given to the Navy to intercept and bring into controlled ports all Italian vessels at sea. On the night of the 12th our bomber squadrons, after a long flight from England, which meant light loads, dropped their first bombs upon Turin and Milan. We looked forward however to a much heavier delivery as soon as we could use the French airfields at Marseilles.

It may be convenient at this point to dispose of the brief Franco-Italian campaign. The French could only muster three divisions, with fortress troops equivalent to three more, to meet invasion over the Alpine passes and along the Riviera coast by the western group of Italian armies. These comprised thirty-two divisions, under Prince Umberto. Moreover, strong German armour, rapidly descending the Rhone valley, soon began to traverse the French rear. Nevertheless the Italians were still confronted, and even pinned down, at every point on the new front by the French Alpine units, even after Paris had fallen and Lyons was in German hands. When on June 18 Hitler and Mussolini met at Munich the Duce had little cause to boast. A new Italian offensive was therefore launched on June 21. The French Alpine positions however

* *The Memoirs of Cordell Hull*, Vol. I, Chap. 56.

proved impregnable, and the major Italian effort towards Nice was halted in the outskirts of Mentone. But although the French Army on the south-eastern borders saved its honour, the German march to the south behind them made further fighting impossible, and the conclusion of the armistice with Germany was linked with a French request to Italy for the cessation of hostilities.

\* \* \* \* \*

My account of this Italian tragedy may fittingly be closed here by the letter which the unlucky Ciano wrote me shortly before his execution at the orders of his father-in-law.

VERONA, December 23, 1943

Signor Churchill.

You will not be surprised that as I approach the hour of my death I should turn to you whom I profoundly admire as the champion of a crusade, though you did at one time make an unjust statement against me.

I was never Mussolini's accomplice in that crime against our country and humanity, that of fighting side by side with the Germans. Indeed the opposite is the truth, and if last August I vanished from Rome it was because the Germans had convinced me that my children were in imminent danger. After they had pledged themselves to take me to Spain, they deported me and my family, against my will, to Bavaria. Now, I have been nearly three months in the prisons of Verona, abandoned to the barbarous treatment of the S.S. My end is near, and I have been told that in a few days my death will be decided, which to me will be no more nor less [than] a release from this daily martyrdom. And I prefer death to witnessing the shame and irreparable damage of an Italy which has been under Hun domination.

The crime which I am now about to expiate is that of having witnessed and been disgusted by the cold, cruel, and cynical preparation for this war by Hitler and the Germans. I was the only foreigner to see at close quarters this loathsome clique of bandits preparing to plunge the world into a bloody war. Now, in accordance with gangster rule, they are planning to suppress a dangerous witness. But they have miscalculated, for already a long time ago I put a diary of mine and various documents in a safe place, which will prove, more than I myself could, the crimes committed by those people with whom later that tragic and vile puppet Mussolini associated himself through his vanity and disregard of moral values.

I have made arrangements that as soon as possible after my death these documents, of the existence of which Sir Percy Loraine was aware at the time of his Mission in Rome, should be put at the disposal of the Allied Press.

Perhaps what I am offering you to-day is but little, but that and my life are all I can offer to the cause of liberty and justice, in the triumph of which I fanatically believe.

This testimony of mine should be brought to light so that the world may know, may hate and may remember, and that those who will have to judge the future should not be ignorant of the fact that the misfortune of Italy was not the fault of her people, but due to the shameful behaviour of one man.

Yours sincerely,

G. CIANO

\*     \*     \*     \*     \*

A speech from President Roosevelt had been announced for the night of the 10th. About midnight I listened to it with a group of officers in the Admiralty War Room, where I still worked. When he uttered the scathing words about Italy, "On this 10th day of June, 1940, the hand that held the dagger has struck it into the back of its neighbour," there was a deep growl of satisfaction. I wondered about the Italian vote in the approaching Presidential Election; but I knew that Roosevelt was a most experienced American party politician, although never afraid to run risks for the sake of his resolves. It was a magnificent speech, instinct with passion and carrying to us a message of hope. While the impression was strong upon me, and before going to bed, I expressed my gratitude.

*Former Naval Person to President Roosevelt*                    11.VI.40

We all listened to you last night, and were fortified by the grand scope of your declaration. Your statement that the material aid of the United States will be given to the Allies in their struggle is a strong encouragement in a dark but not unhopeful hour. Everything must be done to keep France in the fight and to prevent any idea of the fall of Paris, should it occur, becoming the occasion of any kind of parley. The hope with which you inspire them may give them the strength to persevere. They should continue to defend every yard of their soil and use the full fighting force of their Army. Hitler, thus baffled of quick results, will turn upon us, and we are preparing ourselves to resist his fury and defend our Island. Having saved the B.E.F., we do not lack troops at home, and as soon as divisions can be equipped on the much higher scale needed for Continental service they will be dispatched to France.

Our intention is to have a strong army fighting in France for the campaign of 1941. I have already cabled you about aeroplanes, including flying-boats, which are so needful to us in the impending struggle for the life of Great Britain. But even more pressing is the need for destroyers. The Italian outrage makes it necessary for us to cope with a much larger number of submarines, which may come out into the Atlantic and perhaps be based on Spanish ports. To this the only counter is destroyers. Nothing is so important as for us to have the thirty or forty old destroyers you have already had reconditioned. We can fit them very rapidly with our Asdics, and they will bridge the gap of six months before our war-time new construction comes into play. We will return them or their equivalents to you, without fail, at six months' notice if at any time you need them. The next six months are vital. If while we have to guard the East Coast against invasion a new heavy German-Italian submarine attack is launched against our commerce the strain may be beyond our resources and the ocean traffic by which we live may be strangled. Not a day should be lost. I send you my heartfelt thanks and those of my colleagues for all you are doing and seeking to do for what we may now indeed call the Common Cause.

\*    \*    \*    \*    \*

The rush for the spoils had begun. But Mussolini was not the only hungry animal seeking prey. To join the Jackal came the Bear.

I have recorded in the previous volume the course of Anglo-Soviet relations up till the outbreak of war, and the hostility, verging upon an actual breach with Britain and France, which arose during the Russian invasion of Finland. Germany and Russia now worked together as closely as their deep divergences of interest permitted. Hitler and Stalin had much in common as totalitarians, and their systems of government were akin. M. Molotov beamed on the German Ambassador, Count Schulenburg, on every important occasion, and was forward and fulsome in his approval of German policy and praise for Hitler's military measures. When the German assault had been made upon Norway he had said (April 9) that the Soviet Government understood the measures which were forced upon Germany. The English had certainly gone much too far. They had disregarded completely the rights of neutral nations. . . . *"We wish Germany complete success in her defensive measures."*\* Hitler had

\* *Nazi-Soviet Relations*, 1939-1941, p. 138.

taken pains to inform Stalin on the morning of May 10 of the onslaught he had begun upon France and the neutral Low Countries. "I called on Molotov," wrote Schulenburg. "He appreciated the news, and added that he understood that Germany had to protect herself against Anglo-French attack. He had no doubt of our success."[*]

Although these expressions of their opinion were of course unknown till after the war, we were under no illusions about the Russian attitude. We none the less pursued a patient policy of trying to re-establish relations of a confidential character with Russia, trusting to the march of events and to their fundamental antagonisms to Germany. It was thought wise to use the abilities of Sir Stafford Cripps as Ambassador to Moscow. He willingly accepted this bleak and unpromising task. We did not at that time realise sufficiently that Soviet Communists hate extreme Left Wing politicians even more than they do Tories or Liberals. The nearer a man is to Communism in sentiment the more obnoxious he is to the Soviets unless he joins the party. The Soviet Government agreed to receive Cripps as Ambassador, and explained this step to their Nazi confederates. "The Soviet Union", wrote Schulenburg to Berlin on May 29, "is interested in obtaining rubber and tin from England in exchange for lumber. There is no reason for apprehension concerning Cripps's mission, since there is no reason to doubt the loyal attitude of the Soviet Union towards us, and since the unchanged direction of Soviet policy towards England precludes damage to Germany or vital German interests. There are no indications of any kind here for belief that the latest German successes cause alarm or fear of Germany in the Soviet Government."[†]

The collapse of France and the destruction of the French armies and of all counter-poise in the West ought to have produced some reaction in Stalin's mind, but nothing seemed to warn the Soviet leaders of the gravity of their own peril. On June 18, when the French defeat was total, Schulenburg reported: "Molotov summoned me this evening to his office and expressed the warmest congratulations of the Soviet Government on *the splendid success* of the German Armed Forces."[‡] This was almost

* *Nazi-Soviet Relations*, 1939–1941, p. 142.
† *Ibid.*, p. 143.
‡ *Ibid.*, p. 154.

exactly a year from the date when these same Armed Forces, taking the Soviet Government by complete surprise, fell upon Russia in cataracts of fire and steel. We now know that only four months later in 1940 Hitler definitely decided upon a war of extermination against the Soviets, and began the long, vast, stealthy movement of these much-congratulated German armies to the East. No recollection of their miscalculation and former conduct ever prevented the Soviet Government and its Communist agents and associates all over the world from screaming for a Second Front, in which Britain, whom they had consigned to ruin and servitude, was to play a leading part.

However, we comprehended the future more truly than these cold-blooded calculators, and understood their dangers and their interests better than they did themselves. I now addressed myself for the first time to Stalin.

*Prime Minister to Monsieur Stalin*                                           25.VI.40

At this time, when the face of Europe is changing hourly, I should like to take the opportunity of your receiving His Majesty's new Ambassador to ask the latter to convey to you a message from myself.

Geographically our two countries lie at the opposite extremities of Europe, and from the point of view of systems of government it may be said that they stand for widely differing systems of political thought. But I trust that these facts need not prevent the relations between our two countries in the international sphere from being harmonious and mutually beneficial.

In the past—indeed in the recent past—our relations have, it must be acknowledged, been hampered by mutual suspicions; and last August the Soviet Government decided that the interests of the Soviet Union required that they should break off negotiations with us and enter into a close relation with Germany. Thus Germany became your friend almost at the same moment as she became our enemy.

But since then a new factor has arisen which I venture to think makes it desirable that both our countries should re-establish our previous contact, so that if necessary we may be able to consult together as regards those affairs in Europe which must necessarily interest us both. At the present moment the problem before all Europe—our two countries included—is how the States and peoples of Europe are going to react towards the prospect of Germany establishing a hegemony over the continent.

The fact that both our countries lie not in Europe but on her extremities puts them in a special position. We are better enabled than others less fortunately placed to resist Germany's hegemony, and, as you know, the British Government certainly intend to use their geographical position and their great resources to this end.

In fact, Great Britain's policy is concentrated on two objects—one, to save herself from German domination, which the Nazi Government wishes to impose, and the other, to free the rest of Europe from the domination which Germany is now in process of imposing on it.

The Soviet Union is alone in a position to judge whether Germany's present bid for the hegemony of Europe threatens the interests of the Soviet Union, and if so how best these interests can be safeguarded. But I have felt that the crisis through which Europe, and indeed the world, is passing is so grave as to warrant my laying before you frankly the position as it presents itself to the British Government. This, I hope, will ensure that in any discussion that the Soviet Government may have with Sir S. Cripps there should be no misunderstanding as to the policy of His Majesty's Government or of their readiness to discuss fully with the Soviet Government any of the vast problems created by Germany's present attempt to pursue in Europe a methodical process by successive stages of conquest and absorption.

There was of course no answer. I did not expect one. Sir Stafford Cripps reached Moscow safely, and even had an interview of a formal and frigid character with M. Stalin.

\*     \*     \*     \*     \*

Meanwhile the Soviet Government was busy collecting its spoils. On June 14, the day Paris fell, Moscow had sent an ultimatum to Lithuania accusing her and the other Baltic States of military conspiracy against the U.S.S.R. and demanding radical changes of government and military concessions. On June 15 Red Army troops invaded the country, and the President, Smetona, fled into East Prussia. Latvia and Estonia were exposed to the same treatment. Pro-Soviet Governments must be set up forthwith and Soviet garrisons admitted into these small countries. Resistance was out of the question. The President of Latvia was deported to Russia, and Mr. Vyshinsky arrived to nominate a Provisional Government to manage new elections. In Estonia the pattern was identical. On June 19 Zhdanov arrived in Tallinn to install a similar *régime*. On August 3–6 the pretence of pro-Soviet friendly and democratic Governments was swept away, and the Kremlin annexed the Baltic States to the Soviet Union.

The Russian ultimatum to Roumania was delivered to the

Roumanian Minister in Moscow at 10 p.m. on June 26. The cession of Bessarabia and the northern part of the province of Bukovina was demanded, and an immediate reply requested by the following day. Germany, though annoyed by this precipitate action of Russia, which threatened her economic interests in Roumania, was bound by the terms of the German-Soviet pact of August 1939, which recognised the exclusive political interest of Russia in these areas of South-east Europe. The German Government therefore counselled Roumania to yield. On June 27 Roumanian troops were withdrawn from the two provinces concerned, and the territories passed into Russian hands. The armed forces of the Soviet Union were now firmly planted on the shores of the Baltic and at the mouths of the Danube.

# CHAPTER VII

## BACK TO FRANCE

### June 4 – June 12

*High Morale of the Army – My First Thoughts and Directive, June 2, 1940 – The Lost Equipment – The President, General Marshall, and Mr. Stettinius – An Act of Faith – The Double Tensions of June – Reconstitution of the British Army – Its Fearful Lack of Modern Weapons – Decision to Send Our Only Two Well-armed Divisions to France – The Battle of France: Final Phase – Destruction of the 51st Highland Division, June 11–12 – "Auld Scotland Counts for Something Still" – My Fourth Visit to France: Briare – Weygand and Pétain – General Georges Summoned – My Discussion with Weygand – The French Prevent the Royal Air Force from Bombing Turin and Milan – The Germans Enter Paris – Renewed Conference Next Morning – Admiral Darlan's Promise – Farewell to G.Q.G. – Our Journey Home – My Report to the War Cabinet of the Conference.*

WHEN it was known how many men had been rescued from Dunkirk, a sense of deliverance spread in the Island and throughout the Empire. There was a feeling of intense relief, melting almost into triumph. The safe homecoming of a quarter of a million men, the flower of our Army, was a milestone in our pilgrimage through years of defeat. The achievement of the Southern Railway and the Movements Branch of the War Office, of the staffs at the ports in the Thames Estuary, and above all at Dover, where over 200,000 men were handled and rapidly distributed throughout the country, is worthy of the highest praise. The troops returned with nothing but rifles and bayonets and a few hundred machine-guns, and were forthwith sent to their homes for seven days' leave. Their joy at being once again united with their families did not overcome a stern desire to engage the enemy at the earliest moment. Those who had actually fought the Germans in the field had the belief that, given a fair chance, they could beat them. Their morale was high, and they rejoined their regiments and batteries with alacrity.

All the Ministers and departmental officers, permanent or newly-chosen, acted with confidence and vigour night and day, and there are many tales to be told besides this one. Personally I felt uplifted, and my mind drew easily and freely from the knowledge I had gathered in my life. I was exhilarated by the salvation of the Army. I present, for what they are worth, the directives to the departments and submissions to the War Cabinet which I issued day by day. Ismay carried them to the Chiefs of Staff, and Bridges to the War Cabinet and the departments. Mistakes were corrected and gaps filled. Amendments and improvements were often made, but in the main, to the degree perhaps of 90 per cent., action was taken, and with a speed and effectiveness which no Dictatorship could rival.

Here were my first thoughts at the moment when it became certain that the Army had escaped.

*Prime Minister to General Ismay* 2.VI.40
    Notes for C.O.S., etc., by the Minister of Defence.
    The successful evacuation of the B.E.F. has revolutionised the Home Defence position. As soon as the B.E.F. units can be re-formed on a Home Defence basis we have a mass of trained troops in the country which would require a raid to be executed on a prohibitively large scale. Even 200,000 men would not be beyond our compass. The difficulties of a descent and its risks and losses increase with every addition to the first 10,000. We must at once take a new view of the situation. Certain questions must be considered, chiefly by the War Office, but also by the Joint Staffs:
    1. What is the shortest time in which the B.E.F. can be given a new fighting value?
    2. Upon what scheme would they be organised? Will it be for service at Home in the first instance and only secondarily dispatch to France? On the whole, I prefer this.
    3. The B.E.F. in France must immediately be reconstituted, otherwise the French will not continue in the war. Even if Paris is lost, they must be adjured to continue a gigantic guerrilla. A scheme should be considered for a bridgehead and area of disembarkation in Brittany, where a large army can be developed. We must have plans worked

out which will show the French that there is a way through if they will only be steadfast.

4. As soon as the B.E.F. is reconstituted for Home Defence three divisions should be sent to join our two divisions south of the Somme, or wherever the French left may be by then. It is for consideration whether the Canadian Division should not go at once. Pray let me have a scheme.

5. Had we known a week ago what we now know about the Dunkirk evacuation, Narvik would have presented itself in a different light. Even now the question of maintaining a garrison there for some weeks on a self-contained basis should be reconsidered. I am deeply impressed with the vice and peril of chopping and changing. The letter of the Minister of Economic Warfare as well as the telegram of some days ago from the C.-in-C. must however receive one final weighing.

6. Ask Admiralty to supply a latest return of the state of the destroyer flotillas, showing what reinforcements have arrived or are expected within the month of June, and how many will come from repair.

7. It should now be possible to allow the eight Regular battalions in Palestine to be relieved by the eight native battalions from India *before* they are brought home, as brought home they must be, to constitute the cadres of the new B.E.F.

8. As soon as the Australians land, the big ships should be turned round and should carry eight or ten Territorial battalions to Bombay. They should bring back a second eight Regular battalions from India, and afterwards carry to India a second eight or ten Territorial battalions from England. It is for consideration how far the same principle should be applied to batteries in India.

9. Our losses in equipment must be expected to delay the fruition of our expansion of the B.E.F. from the twenty divisions formerly aimed at by $Z^* + 12$ months to no more than fifteen divisions by $Z + 18$; but we must have a project to put before the French. The essence of this should be the armoured division, the 51st, the Canadians, and two Territorial divisions under Lord Gort by mid-July, and the augmenting of this force by six divisions formed from the twenty-four Regular battalions in conjunction with Territorials, a second Canadian division, an Australian division, and two Territorial divisions by $Z + 18$. Perhaps we may even be able to improve on this.

10. It is of the highest urgency to have at least half a dozen brigade groups formed from the Regulars of the B.E.F. for Home Defence.

11. What air co-operation is arranged to cover the final evacuation to-night? It ought to be possible to reduce the pressure on the rear-guard at this critical moment.

* "Z" means the beginning of the war, September 3, 1939.

I close with a general observation. As I have personally felt less afraid of a German attempt at invasion than of the piercing of the French line on the Somme or Aisne and the fall of Paris, I have naturally believed the Germans would choose the latter. This probability is greatly increased by the fact that they will realise that the armed forces in Great Britain are now far stronger than they have ever been, and that their raiding parties would not have to meet half-trained formations, but the men whose mettle they have already tested, and from whom they have recoiled, not daring seriously to molest their departure. The next few days, before the B.E.F. or any substantial portion of it can be reorganised, must be considered as still critical.

★    ★    ★    ★    ★

There was of course a darker side to Dunkirk. We had lost the whole equipment of the Army to which all the first fruits of our factories had hitherto been given:

> 7,000 tons of ammunition.
> 90,000 rifles.
> 2,300 guns.
> 82,000 vehicles.
> 8,000 Bren guns.
> 400 anti-tank rifles.

Many months must elapse, even if the existing programmes were fulfilled without interruption by the enemy, before this loss could be repaired.

However, across the Atlantic in the United States strong emotions were already stirring in the breasts of its leading men. A precise and excellent account of these events is given by Mr. Stettinius,★ the worthy son of my old Munitions colleague of the First World War, one of our truest friends. It was at once realised that the bulk of the British Army had got away only with the loss of all their equipment. As early as June 1 the President sent out orders to the War and Navy Departments to report what weapons they could spare for Britain and France. At the head of the American Army as Chief of Staff was General Marshall, not only a soldier of proved quality, but a man of commanding vision. He instantly directed his Chief of Ordnance and his Assistant Chief of Staff to survey the entire list of the American reserve ordnance and munitions stocks. In forty-eight hours the

★ In *Lend-Lease—Weapon for Victory*, 1944.

answers were given, and on June 3 Marshall approved the lists. The first list comprised half a million .30 calibre rifles out of two million manufactured in 1917 and 1918 and stored in grease for more than twenty years. For these there were about 250 cartridges apiece. There were 900 "*soixante-quinze*" field guns, with a million rounds, 80,000 machine-guns, and various other items. In his excellent book Mr. Stettinius says: "Since every hour counted, it was decided that the Army should sell (for thirty-seven million dollars) everything on the list to one concern, which could in turn resell immediately to the British and French." The Chief of Ordnance, Major-General Wesson, was told to handle the matter, and immediately on June 3 all the American Army depots and arsenals started packing the material for shipment. By the end of the week more than six hundred heavily-loaded freight cars were rolling towards the Army docks at Raritan, New Jersey, up the river from Gravesend Bay. By June 11 a dozen British merchant ships moved into the bay and anchored, and loading from lighters began.

By these extraordinary measures the United States left themselves with the equipment for only 1,800,000 men, the minimum figure stipulated by the American Army's mobilisation plan. All this reads easily now, but at that time it was a supreme act of faith and leadership for the United States to deprive themselves of this very considerable mass of arms for the sake of a country which many deemed already beaten. They never had need to repent of it. As will presently be recounted, we ferried these precious weapons safely across the Atlantic during July, and they formed not only a material gain, but an important factor in all calculations made by friend or foe about invasion.

\* \* \* \* \*

Mr. Cordell Hull has a passage in his memoirs\* which is relevant at this point:

In response to Reynaud's almost pitiful pleas for backing, the President urged Mr. Churchill to send planes to France; but the Prime Minister refused. Bullitt [the United States Ambassador in Paris], outraged by this decision, communicated to the President and me on June 5 his fear that the British might be conserving their Air Force and Fleet so as to use them as bargaining points in negotiations with Hitler.

\* *The Memoirs of Cordell Hull*, Vol. I, pp. 774-5.

The President and I, however, thought differently. France was finished, but we were convinced that Britain, under Churchill's indomitable leadership, intended to fight on. There would be no negotiations between London and Berlin. Only the day before Bullitt's telegram Churchill had made his magnificent speech in the House of Commons. . . .

The President and I believed Mr. Churchill meant what he said. Had we had any doubt of Britain's determination to keep on fighting, we would not have taken the steps we did to get material aid to her. There would have been no logic in sending arms to Britain if we had thought that before they arrived there Churchill's Government would surrender to Germany.

★　　★　　★　　★　　★

The month of June was particularly trying to all of us, because of the dual and opposite stresses to which in our naked condition we were subjected by our duty to France on the one hand and the need to create an effective army at home and to fortify the Island on the other. The double tension of antagonistic but vital needs was most severe. Nevertheless we followed a firm and steady policy without undue excitement. First priority continued to be given to sending whatever trained and equipped troops we had in order to reconstitute the B.E.F. in France. After that our efforts were devoted to the defence of the Island—first, by re-forming and re-equipping the Regular Army; secondly, by fortifying the likely landing-places; thirdly, by arming and organising the population, so far as was possible; and of course by bringing home whatever forces could be gathered from the Empire. At this time the most imminent dangers seemed to be the landing of comparatively small but highly mobile German tank forces which would rip us up and disorganise our defence, and also parachute descents. In close contact with the new Secretary of State for War, Anthony Eden, I busied myself on all this.

The following scheme was devised by the Secretary of State and the War Office for reconstituting the Army in accordance with the directives which had been issued. Seven mobile brigade groups were already in existence. The divisions returned from Dunkirk were reconstituted, re-equipped as fast as possible, and took up their stations. In time the seven brigade groups were absorbed into the re-formed divisions. There were available fourteen Territorial divisions of high-quality men who had been nine months ardently training under war conditions and were partly equipped. One of these, the 52nd, was already fit for service overseas. There was a second armoured division and four. Army tank brigades in process of formation, but without tanks. There was the 1st Canadian Division, fully equipped.

It was not men that were lacking, but arms. Over 80,000 rifles were retrieved from the communications and bases south of the Seine, and by the middle of June every fighting man in the Regular forces had at least a personal weapon in his hand. We had very little field artillery, even for the Regular Army. Nearly all the new 25-pounders had been lost in France. There remained about five hundred 18-pounders, 4.5-inch and 6-inch howitzers. There were only 103 cruiser, 114 Infantry, and 252 light tanks. Fifty of the Infantry tanks were at home in a battalion of the Royal Tank Regiment, and the remainder were in training-schools. Never has a great nation been so naked before her foes.

\* \* \* \* \*

From the beginning I kept in the closest contact with my old friends now at the head of the Governments of Canada and South Africa.

*Prime Minister to Mr. Mackenzie King*        5.VI.40
British situation vastly improved by miraculous evacuation of B.E.F., which gives us an army in the Island more than capable, when re-equipped, of coping with any invading force likely to be landed. Also evacuation was a main trial of strength between British and German Air Forces. Germans have been unable to prevent evacuation, though largely superior in numbers, and have suffered at least three times our loss. For technical reasons, British Air Force would have many more advantages in defending the air above the Island than in operating overseas. Principal remaining danger is of course air [craft] factories, but if our air defence is so strong that enemy can only come on dark nights precision will not be easy. I therefore feel solid confidence in British ability to continue the war, defend the Island and the Empire, and maintain the blockade.

I do not know whether it will be possible to keep France in the war or not. I hope they will, even at the worst, maintain a gigantic guerrilla. We are reconstituting the B.E.F. out of other units.

We must be careful not to let Americans view too complacently prospect of a British collapse, out of which they would get the British Fleet and the guardianship of the British Empire, minus Great Britain.

If United States were in the war and England [were] conquered locally, it would be natural that events should follow the above course. But if America continued neutral, and we were overpowered, I cannot tell what policy might be adopted by a pro-German administration such as would undoubtedly be set up.

Although President is our best friend, no practical help has [reached us] from the United States as yet. We have not expected them to send military aid, but they have not even sent any worthy contribution in destroyers or planes, or by a visit of a squadron of their Fleet to Southern Irish ports. Any pressure which you can apply in this direction would be invaluable.

We are most deeply grateful to you for all your help and for [the four Canadian] destroyers, which have already gone into action against a U-boat. Kindest regards.

Smuts, far off in South Africa and without the latest information upon the specialised problems of Insular Air Defence, naturally viewed the tragedy of France according to orthodox principles. "Concentrate everything at the decisive point." I had the advantage of knowing the facts, and of the detailed advice of Air Chief Marshal Dowding, head of Fighter Command. If Smuts and I had been together for half an hour and I could have put the data before him we should have agreed, as we always did on large military issues.

*Prime Minister to General Smuts*                                  9.VI.40
We are of course doing all we can both from the air and by sending divisions as fast as they can be equipped to France. It would be wrong to send the bulk of our fighters to this battle, and when it was lost, as is probable, be left with no means of carrying on the war. I think we have a harder, longer, and more hopeful duty to perform. Advantages of resisting German air attack in this Island, where we can concentrate very powerful fighter strength, and hope to knock out four or five hostiles to one of ours, are far superior to fighting in France, where we are inevitably outnumbered and rarely exceed two to one ratio of destruction, and where our aircraft are often destroyed at exposed aerodromes. This battle does not turn on the score or so of fighter squadrons we could transport with their plant in the next month. Even if by using them up we held the enemy, Hitler could immediately throw his whole [air] strength against our undefended Island and destroy our means of future production by daylight attack. The classical principles of war which you mention are in this case modified by the actual quantitative data. *I see only one sure way through now.*

*to wit, that Hitler should attack this country, and in so doing break his air weapon.* If this happens he will be left to face the winter with Europe writhing under his heel, and probably with the United States against him after the Presidential election is over.

Am most grateful to you for cable. Please always give me your counsel, my old and valiant friend.

\* \* \* \* \*

Apart from our last twenty-five fighter squadrons, on which we were adamant, we regarded the duty of sending aid to the French Army as paramount. The movement of the 52nd Lowland Division to France, under previous orders, was due to begin on June 7. These orders were confirmed. The 3rd Division, under General Montgomery, was put first in equipment and assigned to France. The leading division of the Canadian Army, which had concentrated in England early in the year and was well armed, was directed, with the full assent of the Dominion Government, to Brest, to begin arriving there on June 11 for what might by this time already be deemed a forlorn hope. The two French light divisions evacuated from Norway were also sent home, together with all the French units and individuals we had carried away from Dunkirk.

That we should have sent our only two formed divisions, the 52nd Lowland Division and 1st Canadian Division, over to our failing French ally in this mortal crisis, when the whole fury of Germany must soon fall upon us, must be set to our credit against the very limited forces we had been able to put into France in the first eight months of war. Looking back on it, I wonder how, when we were resolved to continue the war to the death, and under the threat of invasion, and France was evidently falling, we had the nerve to strip ourselves of the remaining effective military formations we possessed. This was only possible because we understood the difficulties of the Channel crossing without the command of the sea or the air, or the necessary landing-craft.

\* \* \* \* \*

We had still in France, behind the Somme, the 51st Highland Division, which had been withdrawn from the Maginot Line and was in good condition, and the 52nd Lowland Division, which was arriving in Normandy. There was also our 1st (and only) Armoured Division, less the tank battalion and the support group, which had been sent to Calais. This however had lost heavily in attempts to cross the Somme as part of Weygand's plan. By June 1 it was reduced to one-third of its strength, and was sent back across the Seine to refit. At the same time a composite force known as "Beauman Force" was scraped together from the bases and lines of communication in France. It consisted of nine improvised infantry battalions, armed mainly with rifles, and very few anti-tank weapons. It had neither transport nor signals.

The Tenth French Army, with this British contingent, tried to hold the line of the Somme. The 51st Division alone had a front of sixteen miles, and the rest of the army was equally strained. On June 4, with a French division and French tanks, they attacked the German bridgehead at Abbeville, but without success.

On June 5 the final phase of the Battle of France began. The French front consisted of the Second, Third, and Fourth Groups of Armies. The Second defended the Rhine front and the Maginot Line; the Fourth stood along the Aisne, and the Third from the Aisne to the mouth of the Somme. This Third Army Group comprised the Sixth, Seventh, and Tenth Armies; and all the British forces in France formed part of the Tenth Army. All this immense line, in which there stood at this moment nearly one and a half million men, or perhaps sixty-five divisions, was now to be assaulted by one hundred and twenty-four German divisions, also formed in three Army Groups, viz.: Coastal Sector, Bock; Central Sector, Rundstedt; Eastern Sector, Leeb. These attacked on June 5, June 9, and June 15 respectively. On the night of June 5 we learned that a German offensive had been launched that morning on a seventy-mile front from Amiens to the Laon–Soissons road. This was war on the largest scale.

We have seen how the German armour had been hobbled and held back in the Dunkirk battle, in order to save it for the final phase in France. All this armour now rolled forward upon the weak and improvised or quivering French front between Paris and the sea. It is here only possible to record the battle on the coastal flank, in which we played a part. On June 7 the Germans renewed their attack, and two armoured divisions drove towards Rouen so as to split the Tenth French Army. The left French IXth Corps, including the Highland Division, two French infantry divisions, and two cavalry divisions, or what was left of them,

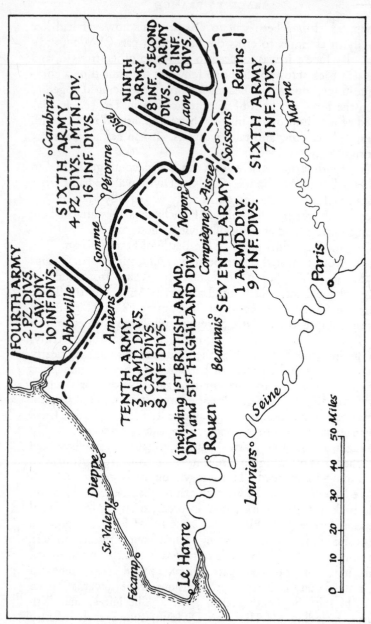

THE OPPOSING FORCES on the WESTERN FLANK 5 June 1940

FOURTH ARMY
2 PZ.DIVS.
1 CAV.DIV.
10 INF.DIVS.

SIXTH ARMY
4 PZ.DIVS. 1 MTN.DIV.
16 INF.DIVS.

NINTH
ARMY
8 INF.
DIVS.

SECOND
ARMY
8 INF.
DIVS.

TENTH ARMY
3 ARMD.DIVS.
3 CAV.DIVS.
8 INF.DIVS.
(including 1ST BRITISH ARMD.
DIV. and 51ST HIGHLAND Div)

SEVENTH ARMY
1 ARMD.DIV.
9 INF.DIVS.

SIXTH ARMY
7 INF.DIVS.

St. Valery
Dieppe
Fécamp
Le Havre
Abbeville
Amiens
Somme
Péronne
Cambrai
Oise
Laon
Reims
Soissons
Aisne
Noyon
Compiègne
Beauvais
Rouen
Louviers
Seine
Marne
Paris

0  10  20  30  40  50 Miles

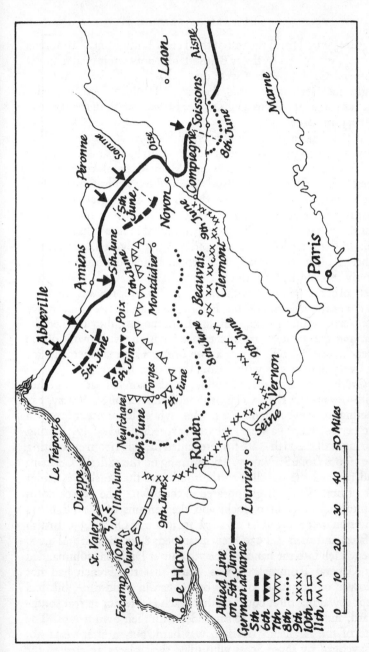

THE GERMAN ADVANCE 5 – 9 June

Le Tréport

Dieppe

St. Valery

Fécamp

Le Havre

Abbeville

Amiens

Péronne

Laon

Aisne

Soissons

Compiègne

Marne

Oise

Somme

Noyon

Beauvais

Clermont

Montdidier

Poix

Forges

Neufchatel

Rouen

Louviers

Vernon

Seine

Paris

5th June

5th June

5th June

5th June

6th June

7th June

7th June

7th June

8th June

8th June

8th June

8th June

9th June

9th June

9th June

10th June

11th June

Allied Line
on 5th June

German advance
5th
6th
7th
8th
9th
10th
11th

50 Miles

0  10  20  30  40  50

133

were separated from the rest of the Tenth Army front. "Beauman Force", supported by thirty British tanks, now attempted to cover Rouen. On June 8 they were driven back to the Seine, and that night the Germans entered the city. The 51st Division, with the remnants of the French IXth Corps, was cut off in the Rouen–Dieppe *cul-de-sac*.

We had been intensely concerned lest this division should be driven back to the Havre peninsula and thus be separated from the main armies, and its commander, Major-General Fortune, had been told to fall back if necessary in the direction of Rouen. This movement was forbidden by the already disintegrating French command. Repeated urgent representations were made by us, but they were of no avail. A dogged refusal to face facts led to the ruin of the French IXth Corps and our 51st Division. On June 9, when Rouen was already in German hands, our men had but newly reached Dieppe, thirty-five miles to the north. Only then were orders received to withdraw to Havre. A force was sent back to cover the movement, but before the main bodies could move the Germans interposed. Striking from the east, they reached the sea, and the greater part of the 51st Division, with many of the French, was cut off. It was a case of gross mismanagement, for this very danger was visible a full three days before.

On the 10th, after sharp fighting, the division fell back, together with the French IXth Corps, to the perimeter of St. Valéry, expecting to be evacuated by sea. Meanwhile all our other forces in the Havre peninsula were embarking speedily and safely. During the night of the 11th–12th fog prevented the ships from evacuating the troops from St. Valery. By morning on the 12th the Germans had reached the sea-cliffs to the south and the beach was under direct fire. White flags appeared in the town. The French corps capitulated at eight o'clock, and the remains of the Highland Division were forced to do so at 10.30 a.m. Only 1,350 British officers and men and 930 French escaped; 8,000 British and 4,000 French fell into the hands of the 7th Panzer Division, commanded by General Rommel. I was vexed that the French had not allowed our division to retire on Rouen in good time, but had kept it waiting till it could neither reach Havre nor retreat southward, and thus forced it to surrender with their own troops. The fate of the Highland Division was hard, but in after years not unavenged by those Scots who filled their places, re-created the division by merging it with the 9th Scottish, and marched across all the battlefields from Alamein to final victory beyond the Rhine.

Some lines* of Dr. Charles Murray's, written in the First World War, came into my mind, and it is fitting to print them here:

Half-mast the castle banner droops,
  The Laird's lament was played yestreen,
An' mony a widowed cottar wife
  Is greetin' at her shank aleen.
In Freedom's cause, for ane that fa's,
  We'll glean the glens an' send them three,
To clip the reivin' eagle's claws
  An' drook his feathers i' the sea.
    For gallant loons, in brochs an' toons,
      Are leavin' shop an' yaird an' mill,
    A' keen to show baith friend an' foe
      Auld Scotland counts for something still.

\*    \*    \*    \*    \*

About eleven o'clock on the morning of June 11 there was a message from Reynaud, who had also cabled to the President. The French tragedy had moved and slid downward. For several days past I had pressed for a meeting of the Supreme Council. We could no longer meet in Paris. We were not told what were the conditions there. Certainly the German spearheads were very close. I had had some difficulty in obtaining a rendezvous, but this was no time to stand on ceremony. We must know what the French were going to do. Reynaud now told me that he could receive us at Briare, near Orleans. The seat of government was moving from Paris to Tours. Grand-Quartier-Général was near Briare. He specified the airfield to which I should come. Nothing loth, I ordered the Flamingo to be ready at Hendon after luncheon, and, having obtained the approval of my colleagues at the morning Cabinet, we started about two o'clock. Before leaving I cabled to the President.

*Former Naval Person to President Roosevelt*          11.VI.40
  The French have sent for me again, which means that crisis has arrived. Am just off. Anything you can say or do to help them now may make the difference.

* From *Hamewith and Other Poems*, by courtesy of Messrs. Constable & Co.

We are also worried about Ireland. An American squadron at Berehaven would do no end of good, I am sure.

\*    \*    \*    \*    \*

This was my fourth journey to France; and since military conditions evidently predominated, I asked the Secretary of State for War, Mr. Eden, to come with me, as well as General Dill, now C.I.G.S., and of course Ismay. The German aircraft were now reaching far down into the Channel, and we had to make a still wider sweep. As before, the Flamingo had an escort of twelve Spitfires. After a couple of hours we alighted at a small landing-ground. There were a few Frenchmen about, and soon a colonel arrived in a motor-car. I displayed the smiling countenance and confident air which are thought suitable when things are very bad, but the Frenchman was dull and unresponsive. I realised immediately how very far things had fallen even since we were in Paris a week before. After an interval we were conducted to the château, where we found M. Reynaud, Marshal Pétain, General Weygand, the Air General Vuillemin, and some others, including the relatively junior General de Gaulle, who had just been appointed Under-Secretary for National Defence. Hard by on the railway was the Headquarters train, in which some of our party were accommodated. The château possessed but one telephone, in the lavatory. It was kept very busy, with long delays and endless shouted repetitions.

At seven o'clock we entered into conference. General Ismay kept a record. I merely reproduce my lasting impressions, which in no way disagree with it. There were no reproaches or recriminations. We were all up against brute facts. We British did not know where exactly the front line lay, and certainly there was anxiety about some dart by the German armour—even upon us. In effect, the discussion ran on the following lines: I urged the French Government to defend Paris. I emphasised the enormous absorbing power of the house-to-house defence of a great city upon an invading army. I recalled to Marshal Pétain the nights we had spent together in his train at Beauvais after the British Fifth Army disaster in 1918, and how he, as I put it, not mentioning Marshal Foch, had restored the situation. I also reminded him how Clemenceau had said: "I will fight in front of Paris, in Paris, and behind Paris." The Marshal replied very quietly and with dignity that in those days he had a mass of manœuvre of upwards of sixty divisions; now there was none. He mentioned that there were then sixty British divisions in the line. Making Paris into a ruin would not affect the final event.

Then General Weygand exposed the military position, so far as he knew it, in the fluid battle proceeding fifty or sixty miles away, and he paid a high tribute to the prowess of the French Army. He requested that every reinforcement should be sent— above all, that every British fighter air squadron should immediately be thrown into the battle. "Here," he said, "is the decisive point. Now is the decisive moment. It is therefore wrong to keep *any* squadrons back in England." But in accordance with the Cabinet decision, taken in the presence of Air Chief Marshal Dowding, whom I had brought specially to a Cabinet meeting, I replied: "This is not the decisive point and this is not the decisive moment. That moment will come when Hitler hurls his Luftwaffe against Great Britain. If we can keep command of the air, and if we can keep the seas open, as we certainly shall keep them open, we will win it all back for you."* Twenty-five fighter squadrons must be maintained at all costs for the defence of Britain and the Channel, and nothing would make us give up these. We intended to continue the war whatever happened, and we believed we could do so for an indefinite time, but to give up these squadrons would destroy our chance of life. At this stage I asked that General Georges, the Commander-in-Chief of the North-Western Front, who was in the neighbourhood, should be sent for, and this was accordingly done.

Presently General Georges arrived. After being apprised of what had passed, he confirmed the account of the French front which had been given by Weygand. I again urged my guerrilla plan. The German Army was not so strong as might appear at their points of impact. If all the French armies, every division and brigade, fought the troops on their front with the utmost vigour a general standstill might be achieved. I was answered by statements of the frightful conditions on the roads, crowded with refugees harried by unresisted machine-gun fire from the German aeroplanes, and of the wholesale flight of vast numbers of inhabitants and the increasing breakdown of the machinery of government and of military control. At one point General

* I am obliged to General Ismay for his recollection of these words.

Weygand mentioned that the French might have to ask for an armistice. Reynaud at once snapped at him: "That is a political affair." According to Ismay I said: "If it is thought best for France in her agony that her Army should capitulate, let there be no hesitation on our account, because whatever you may do we shall fight on for ever and ever and ever." When I said that the French Army, fighting on, wherever it might be, could hold or wear out a hundred German divisions, General Weygand replied: "Even if that were so, they would still have another hundred to invade and conquer you. What would you do then?" On this I said that I was not a military expert, but that my technical advisers were of the opinion that the best method of dealing with a German invasion of the Island of Britain was to drown as many as possible on the way over and knock the others on the head as they crawled ashore. Weygand answered with a sad smile: "At any rate I must admit you have a very good anti-tank obstacle." These were the last striking words I remember to have heard from him. In all this miserable discussion it must be borne in mind that I was haunted and undermined by the grief I felt that Britain, with her 48,000,000 population, had not been able to make a greater contribution to the land war against Germany, and that so far nine-tenths of the slaughter and ninety-nine-hundredths of the suffering had fallen upon France and upon France alone.

After another hour or so we got up and washed our hands while a meal was brought to the conference table. In this interval I talked to General Georges privately, and suggested first the continuance of fighting everywhere on the home front and a prolonged guerrilla in the mountainous regions, and secondly the move to Africa, which a week before I had regarded as "defeatist". My respected friend, who, although charged with much direct responsibility, had never had a free hand to lead the French armies, did not seem to think there was much hope in either of these.

I have written lightly of the happenings of these days, but here to all of us was real agony of mind and soul.

*    *    *    *    *

At about ten o'clock everyone took their places at the dinner. I sat on M. Reynaud's right and General de Gaulle was on my other side. There was soup, an omelette or something, coffee and light wine. Even at this point in our awful tribulation under the

German scourge we were quite friendly. But presently there was a jarring interlude. The reader will recall the importance I had attached to striking hard at Italy the moment she entered the war, and the arrangement that had been made with full French concurrence to move a force of British heavy bombers to the French airfields near Marseilles in order to attack Turin and Milan. All was now in readiness to strike. Scarcely had we sat down when Air Marshal Barratt, commanding the British Air Force in France, rang up Ismay on the telephone to say that the local authorities objected to the British bombers taking off, on the grounds that an attack on Italy would only bring reprisals upon the South of France, which the British were in no position to resist or prevent. Reynaud, Weygand, Eden, Dill, and I left the table, and, after some parleying, Reynaud agreed that orders should be sent to the French authorities concerned that the bombers were not to be stopped. But later that night Air Marshal Barratt reported that the French people near the airfields had dragged all kinds of country carts and lorries on to them, and that it had been impossible for the bombers to start on their mission.

Presently, when we left the dinner table and sat with some coffee and brandy, M. Reynaud told me that Marshal Pétain had informed him that it would be necessary for France to seek an armistice, and that he had written a paper upon the subject which he wished him to read. "He has not," said Reynaud, "handed it to me yet. He is still ashamed to do it." He ought also to have been ashamed to support even tacitly Weygand's demand for our last twenty-five squadrons of fighters, when he had made up his mind that all was lost and that France should give in. Thus we all went unhappily to bed in this disordered château or in the military train a few miles away. The Germans entered Paris on the 14th.

\* \* \* \* \*

Early in the morning we resumed our conference. Air Marshal Barratt was present. Reynaud renewed his appeal for five more squadrons of fighters to be based in France, and General Weygand said that he was badly in need of day bombers to make up for his lack of troops. I gave them an assurance that the whole question of increased air support for France would be examined carefully and sympathetically by the War Cabinet immediately I got back to London; but I again emphasised that it would be a vital mistake to denude the United Kingdom of its essential Home defences.*

Towards the end of this short meeting I put the following specific questions:

(i) Will not the mass of Paris and its suburbs present an obstacle dividing and delaying the enemy as in 1914, or like Madrid?

(ii) May this not enable a counter-stroke to be organised with British and French forces across the Lower Seine?

(iii) If the period of co-ordinated war ends, will that not mean an almost equal dispersion of the enemy forces? Would not a war of columns and [attacks] upon the enemy communications be possible? Are the enemy resources sufficient to hold down all the countries at present conquered as well as a large part of France, while they are fighting the French Army and Great Britain?

(iv) Is it not possible thus to prolong the resistance until the United States come in?

General Weygand, while agreeing with the conception of the counter-stroke on the Lower Seine, said that he had inadequate forces to implement it. He added that, in his judgment, the Germans had got plenty to spare to hold down all the countries at present conquered, as well as a large part of France. Reynaud added that the Germans had raised fifty-five divisions and had built four thousand to five thousand heavy tanks since the outbreak of war. This was of course an immense exaggeration of what they had built.

In conclusion, I expressed in the most formal manner my hope that if there was any change in the situation the French Government would let the British Government know at once, in order that we might come over and see them at any convenient spot, before they took any final decisions which would govern their action in the second phase of the war.

We then took leave of Pétain, Weygand, and the staff of G.Q.G., and this was the last we saw of them. Finally I took

* In this connection it is interesting to note a statement made by General Gamelin to the Conseil Supérieure de l'Air on March 15, 1938. "In the event of Britain coming to our aid, it is reasonable to hope that she would agree to reinforce our bomber arm by using our air bases. On the other hand, it is quite unlikely [peu probable] that she would agree to send to France the fighter units that are responsible for the defence of her own territory."

Admiral Darlan apart and spoke to him alone. "Darlan, you must never let them get the French Fleet." He promised solemnly that he would never do so.

\* \* \* \* \*

Lack of suitable petrol made it impossible for the twelve Spitfires to escort us. We had to choose between waiting till it cleared up or taking a chance in the Flamingo. We were assured that it would be cloudy all the way. It was urgently necessary to get back home. Accordingly we started alone, calling for an escort to meet us, if possible, over the Channel. As we approached the coast the skies cleared and presently became cloudless. Eight thousand feet below us on our right hand was Havre, burning. The smoke drifted away to the eastward. No new escort was to be seen. Presently I noticed some consultations going on with the captain, and immediately after we dived to a hundred feet or so above the calm sea, where aeroplanes are often invisible. What had happened? I learned later that they had seen two German aircraft below us firing at fishing-boats. We were lucky that their pilots did not look upwards. The new escort met us as we approached the English shore, and the faithful Flamingo alighted safely at Hendon.

\* \* \* \* \*

At five o'clock that evening I reported to the War Cabinet the results of my mission.

I described the condition of the French armies as it had been reported to the conference by General Weygand. For six days they had been fighting night and day, and they were now almost wholly exhausted. The enemy attack, launched by a hundred and twenty divisions, with supporting armour, had fallen on forty French divisions, which had been out-manœuvred and out-matched at every point. The enemy's armoured forces had caused great disorganisation among the headquarters of the higher formations, which were unwieldy and, when on the move, unable to exercise control over the lower formations. The French armies were now on the last line on which they could attempt to offer an organised resistance. This line had already been penetrated in two or three places; and if it collapsed General Weygand would not be responsible for carrying on the struggle.

General Weygand evidently saw no prospect of the French

going on fighting, and Marshal Pétain had quite made up his mind that peace must be made. He believed that France was being systematically destroyed by the Germans, and that it was his duty to save the rest of the country from this fate. I mentioned his memorandum to this effect, which he had shown to Reynaud but had not left with him. "There can be no doubt," I said, "that Pétain is a dangerous man at this juncture: he was always a defeatist, even in the last war." On the other hand, M. Reynaud had seemed quite determined to fight on, and General de Gaulle, who had attended the conference with him, was in favour of carrying on a guerrilla warfare. He was young and energetic and had made a very favourable impression on me. I thought it probable that if the present line collapsed Reynaud would turn to him to take command. Admiral Darlan also had declared that he would never surrender the French Navy to the enemy: in the last resort, he had said, he would send it over to Canada; but in this he might be overruled by the French politicians.

It was clear that France was near the end of organised resistance, and a chapter in the war was now closing. The French might by some means continue the struggle. There might even be two French Governments, one which made peace, and one which organised resistance from the French colonies, carrying on the war at sea through the French Fleet and in France through guerrillas. It was too early yet to tell. Though for a period we might still have to send some support to France, we must now concentrate our main efforts on the defence of our Island.

# CHAPTER VIII

## HOME DEFENCE

### June

*Intense British Effort – Imminent Dangers – The Question of "Commandos" – Local Defence Volunteers Renamed "Home Guard" – Lack of Means of Attacking Enemy Tanks – Major Jefferis' Experimental Establishment – The Sticky Bomb – Help for de Gaulle's Free French – Arrangements for Repatriation of other French Troops – Care of French Wounded – Freeing British Troops for Intensive Training – The Press and Air Raids – Danger of German Use of Captured European Factories – Questions Arising in the Middle East and India – Question of Arming the Jewish Colonists in Palestine – Progress of Our Plan of Defence – The Great Anti-Tank Obstacle and Other Measures.*

THE reader of these pages in future years should realise how dense and baffling is the veil of the Unknown. Now in the full light of the after-time it is easy to see where we were ignorant or too much alarmed, where we were careless or clumsy. Twice in two months we had been taken completely by surprise. The overrunning of Norway and the break-through at Sedan, with all that followed from these, proved the deadly power of the German initiative. What else had they got ready—prepared and organised to the last inch? Would they suddenly pounce out of the blue with new weapons, perfect planning, and overwhelming force upon our almost totally unequipped and disarmed Island at any one of a dozen or score of possible landing-places? Or would they go to Ireland? He would have been a very foolish man who allowed his reasoning, however clean-cut and seemingly sure, to blot out any possibility against which provision could be made.

143

"Depend upon it," said Dr. Johnson, "when a man knows he is going to be hanged in a fortnight, it concentrates his mind wonderfully." I was always sure we should win, but nevertheless I was highly geared up by the situation, and very thankful to be able to make my views effective. June 6 seems to have been for me an active and not barren day. My minutes, dictated as I lay in bed in the morning and pondered on the dark horizon, show the variety of topics upon which it was necessary to give directions.

First I called upon the Minister of Supply (Mr. Herbert Morrison) for an account of the progress of various devices connected with our rockets and sensitive fuzes for use against aircraft, on which some progress had been made, and upon the Minister of Aircraft Production (Lord Beaverbrook) for weekly reports on the design and production of automatic bomb sights and low altitude R.D.F. (Radio Direction Finding) and A.I. (Air Interception). I did this to direct the attention of these two new Ministers with their vast departments to those topics, in which I had already long been especially interested. I asked the Admiralty to transfer at least fifty trained and half-trained pilots temporarily to Fighter Command. Fifty-five actually took part in the great air battle. I called for a plan to be prepared to strike at Italy by air raids on Turin and Milan, should she enter the war against us. I asked the War Office for plans for forming a Dutch Brigade in accordance with the desires of the exiled Netherlands Government, and pressed the Foreign Secretary for the recognition of the Belgian Government, apart from the prisoner King, as the sole constitutional Belgian authority, and for the encouragement of mobilisation in Yugoslavia as a counter to Italian threats. I asked that the aerodromes at Bardufosse and Skaarnlands, which we had constructed in the Narvik area and were about to abandon, should be made unusable for as long as possible by means of delayed-action bombs buried in them. I remembered how effectively the Germans had by this method delayed our use in 1918 of the railways when they finally retreated. Alas! we had no bombs of long-delay in any numbers. I was worried about the many ships lying in Malta harbour in various conditions of repair in view of impending Italian hostility. I wrote a long minute to the Minister of Supply about timber felling and production at home. This was one of the most important methods of reducing the tonnage of our imports. Besides, we should get no more timber from Norway for a long time to come. Many of these minutes will be found in Appendix A.

I longed for more Regular troops with which to rebuild and expand the Army. Wars are not won by heroic militias.

*Prime Minister to Secretary of State for War*          6.VI.40

1. It is more than a fortnight since I was told that eight battalions could leave India and arrive in this country in forty-two days from the order being given. The order was given. Now it is not till June 6 (*i.e.*, to-day) that the first eight battalions leave India on their voyage round the Cape, arriving only July 25.

2. The Australians are coming in the big ships, but they seem to have wasted a week at Capetown, and are now only proceeding at eighteen knots, instead of the twenty I was assured were possible. It is hoped they will be here about the 15th. Is this so? At any rate, whenever they arrive the big ships should be immediately filled with Territorials—the more the better—preferably twelve battalions, and sent off to India at full speed. As soon as they arrive in India they should embark another eight Regular battalions for this country, making the voyage again at full speed. They should then take another batch of Territorials to India. Future transferences can be discussed later. . . . All I am asking now is that the big ships should go to and fro at full speed.

3. I am very sorry indeed to find the virtual deadlock which local objections have imposed upon the battalions from Palestine. It is quite natural that General Wavell should look at the situation only from his own viewpoint. Here we have to think of building up a good Army in order to make up, as far as possible, for the lamentable failure to support the French by an adequate B.E.F. during the first year of the war. Do you realise that in the first year of the late war we brought forty-seven divisions into action, and that these were divisions of twelve battalions plus one Pioneer battalion, not nine as now? We are indeed the victims of a feeble and weary departmentalism.

4. Owing to the saving of the B.E.F., I have been willing to wait for the relief of the eight battalions from Palestine by eight native Indian battalions, provided these latter were sent at once; but you give me no time-table for this. I have not yet received any report on whether it is possible to send these British battalions and their Indian relief via Basra and the Persian Gulf.

Perhaps you would very kindly let me have this in the first instance.

5. I am prepared also to consider as an alternative, or an immediate step, the sending home [*i.e.*, to Britain] of the rest of the Australian

corps. Perhaps you will let me have a note on this, showing especially dates at which the moves can be made.

6. You must not think I am ignoring the position in the Middle East. On the contrary, it seems to me that we should draw upon India much more largely, and that a ceaseless stream of Indian units should be passing into Palestine and Egypt via Bombay and [by] Karachi across the desert route. India is doing nothing worth speaking of at the present time. In the last war not only did we have all the [British] Regular troops out [of India] in the first nine months (many more than are there now), but also an Indian Corps fought by Christmas in France. Our weakness, slowness, lack of grip and drive are very apparent on the background of what was done twenty-five years ago. I really think that you, Lloyd, and Amery ought to be able to lift our affairs in the East and Middle East out of the catalepsy by which they are smitten.

\* \* \* \* \*

This was a time when all Britain worked and strove to the utmost limit and was united as never before. Men and women toiled at the lathes and machines in the factories till they fell exhausted on the floor and had to be dragged away and ordered home, while their places were occupied by newcomers ahead of time. The one desire of all the males and many women was to have a weapon. The Cabinet and Government were locked together by bonds the memory of which is still cherished by all. The sense of fear seemed entirely lacking in the people, and their representatives in Parliament were not unworthy of their mood. We had not suffered like France under the German flail. Nothing moves an Englishman so much as the threat of invasion, the reality unknown for a thousand years. Vast numbers of people were resolved to conquer or die. There was no need to rouse their spirit by oratory. They were glad to hear me express their sentiments and give them good reasons for what they meant to do, or try to do. The only possible divergence was from people who wished to do even more than was possible, and had the idea that frenzy might sharpen action.

Our decision to send our only two well-armed divisions back to France made it all the more necessary to take every possible measure to defend the Island against direct assault.

*Prime Minister to General Ismay*                                18.VI.40

I should like to be informed upon (1) the coastal watch and coastal batteries; (2) the gorging of the harbours and defended inlets (*i.e.*, the making of the landward defences); (3) the troops held in immediate support of the foregoing; (4) the mobile columns and brigade groups; (5) the General Reserve.

Someone should explain to me the state of these different forces, including the guns available in each area. I gave directions that the 8th Tank Regiment should be immediately equipped with Infantry and cruiser tanks until they have fifty-two new tanks, all well armoured and well gunned. What has been done with the output of this month and last month? Make sure it is not languishing in depots, but passes swiftly to troops. General Carr is responsible for this. Let him report.

What are the ideas of C.-in-C., H.F., about Storm Troops? We have always set our faces against this idea, but the Germans certainly gained in the last war by adopting it, and this time it has been a leading cause of their victory. There ought to be at least twenty thousand Storm Troops or "Leopards" [eventually called "Commandos"] drawn from existing units, ready to spring at the throat of any small landings or descents. These officers and men should be armed with the latest equipment, tommy guns, grenades, etc., and should be given great facilities in motor-cycles and armoured cars.

*   *   *   *   *

Mr. Eden's plan of raising Local Defence Volunteers, which he had proposed to the Cabinet on May 13, met with an immediate response in all parts of the country.

*Prime Minister to Secretary of State for War*     22.VI.40

Could I have a brief statement of the L.D.V. position, showing the progress achieved in raising and arming them, and whether they are designed for observation or for serious fighting? What is their relationship to the police, the Military Command, and the Regional Commissioners? From whom do they receive their orders, and to whom do they report? It would be a great comfort if this could be compressed on one or two sheets of paper.

I had always hankered for the name "Home Guard". I had indeed suggested it in October 1939.

*Prime Minister to Secretary of State for War*     26.VI.40

I don't think much of the name "Local Defence Volunteers" for your very large new force. The word "local" is uninspiring. Mr. Herbert Morrison suggested to me to-day the title "Civic Guard", but I think "Home Guard" would be better. Don't hesitate to change on account of already having made armlets, etc., if it is thought the title of Home Guard would be more compulsive.

*Prime Minister to Secretary of State for War*            27.VI.40

I hope you liked my suggestion of changing the name "Local Defence Volunteers", which is associated with Local Government and Local Option, to "Home Guard". I found everybody liked this in my tour yesterday.

The change was accordingly made, and the mighty organisation, which presently approached one and a half million men and gradually acquired good weapons, rolled forward.

*     *     *     *     *

In these days my principal fear was of German tanks coming ashore. Since my mind was attracted to landing tanks on their coasts, I naturally thought they might have the same idea. We had hardly any anti-tank guns or ammunition, or even ordinary field artillery. The plight to which we were reduced in dealing with this danger may be measured from the following incident. I visited our beaches in St. Margaret's Bay, near Dover. The Brigadier informed me that he had only three anti-tank guns in his brigade, covering four or five miles of this highly-menaced coastline. He declared that he had only six rounds of ammunition for each gun, and he asked me with a slight air of challenge whether he was justified in letting his men fire one single round for practice in order that they might at least know how the weapon worked. I replied that we could not afford practice rounds, and that fire should be held for the last moment at the closest range.

This was therefore no time to proceed by ordinary channels in devising expedients. In order to secure quick action, free from departmental processes, upon any bright idea or gadget, I decided to keep under my own hand as Minister of Defence the experimental establishment formed by Major Jefferis at Whitchurch. While engaged upon the fluvial mines in 1939 I had had useful contacts with this brilliant officer, whose ingenious, inventive mind proved, as will be seen, fruitful during the whole war. Lindemann was in close touch with him and me. I used their brains and my power. Major Jefferis and others connected with him were at work upon a bomb which could be thrown at a tank, perhaps from a window, and would stick upon it. The impact of a very high explosive in actual contact with a steel plate is particularly effective. We had the picture in mind that devoted soldiers or civilians would run close up to the tank and even thrust the bomb upon it, though its explosion cost them their lives. There were undoubtedly many who would have done it. I thought also that the bomb fixed on a rod might be fired with a reduced charge from rifles.

*Prime Minister to General Ismay*                    6.VI.40

It is of the utmost importance to find some projectile which can be fired from a rifle at a tank, like a rifle grenade, or from an anti-tank rifle, like a trench-mortar bomb. The "sticky" bomb seems to be useful for the first of these, but perhaps this is not so. Anyhow, concentrate attention upon finding something that can be fired from anti-tank rifles or from ordinary rifles.

I pressed the matter hard.

*Prime Minister to General Ismay*                    16.VI.40

Who is responsible for making the "sticky" bomb? I am told that a great sloth is being shown in pressing this forward. Ask General Carr to report to-day upon the position, and to let me have on one sheet of paper the back history of the subject from the moment when the question was first raised.

The matter is to be pressed forward from day to day, and I wish to receive a report every three days.

*Prime Minister to General Ismay*                    24.VI.40

I minuted some days ago about the "sticky" bombs. All preparations for manufacture should proceed in anticipation that the further trials will be successful. Let me have a time-table showing why it is that delay has crept into all the process, which is so urgent.

*Prime Minister to General Ismay*                    24.VI.40

I understand that the trials were not entirely successful and the bomb failed to stick on tanks which were covered with dust and mud. No doubt some more sticky mixture can be devised and Major Jefferis should persevere.

Any chortling by officials who have been slothful in pushing this bomb over the fact that at present it has not succeeded will be viewed with great disfavour by me.

In the end the "sticky" bomb was accepted as one of our best emergency weapons. We never had to use it at home; but in

Syria, where equally primitive conditions prevailed, it proved its value.

* * * * *

We had evidently to do our utmost to form French forces which might aid General de Gaulle in keeping the true personification of France alive.

*Prime Minister to First Lord of the Admiralty*
*and other Service Ministers*                                    27.VI.40

1. The French naval personnel at Aintree Camp, numbering 13,600, equally with the 5,530 military at Trentham Park, the 1,900 at Arrow Park, and the details at Blackpool, are to be immediately repatriated to French territory, *i.e.*, Morocco, in French ships now in our hands.

2. They should be told we will take them to French Africa because all French metropolitan ports are in German hands, and that the French Government will arrange for their future movements.

3. If however any wish to remain here to fight against Germany, they must immediately make this clear. Care must be taken that no officer or man is sent back into French jurisdiction against his will. The shipping is to be ready to-morrow. The troops should move under their own officers, and carry their personal arms, but as little ammunition as possible. Some arrangements should be made for their pay. The French material on board ships from Narvik will be taken over by us with the ammunition from the *Lombardy* and other ships as against expenses to which we are put.

4. Great care is to be taken of the French wounded. All who can be moved without danger should be sent back direct to France if possible. The French Government should be asked where they wish them delivered, and if at French metropolitan ports, should arrange with the Germans for their safe entry; otherwise Casablanca. All dangerous cases must be dealt with here.

5. Apart from any volunteers in the above groups of personnel who may wish to stay, there must be many individuals who have made their way here, hoping to continue to fight. These also should be given the option of returning to France, or serving in the French units under General de Gaulle, who should be told of our decisions, and given reasonable facilities to collect his people. I have abandoned the hope that he could address the formed bodies, as their morale has deteriorated too fast.

* * * * *

My desire that our own Army should regain its poise and fighting quality was at first hampered because so many troops were being absorbed in fortifying their own localities or sectors of the coast.

*Prime Minister to Secretary of State for War*                    25.VI.40

It is shocking that only fifty-seven thousand men [civilians] are being employed on all these [defence] works. Moreover, I fear that the troops are being used in large numbers on fortifications. At the present stage they should be drilling and training for at least eight hours a day, including one smart parade every morning. All the labour necessary should be found from civilian sources. I found it extremely difficult to see even a single battalion on parade in East Anglia during my visit. The fighting troops in the brigade groups should neither be used for guarding vulnerable points nor for making fortifications. Naturally a change like this cannot be made at once, but let me have your proposals for bringing it about as soon as possible.

\*     \*     \*     \*     \*

*Prime Minister to Minister of Information*                    26.VI.40

The Press and broadcast should be asked to handle air raids in a cool way and on a diminishing tone of public interest. The facts should be chronicled without undue prominence or headlines. The people should be accustomed to treat air raids as a matter of ordinary routine. Localities affected should not be mentioned with any precision. Photographs showing shattered houses should not be published unless there is something very peculiar about them, or to illustrate how well the Anderson shelters work. It must be clear that the vast majority of people are not at all affected by any single air raid, and would hardly sustain any evil impression if it were not thrust before them. Everyone should learn to take air raids and air-raid alarms as if they were no more than thunderstorms. Pray try to impress this upon the newspaper authorities, and persuade them to help. If there is difficulty in this, I would myself see the Newspaper Proprietors' Association, but I hope this will not be necessary. The Press should be complimented on their work so far in this matter.

*Prime Minister to Secretary of State for War*                    27.VI.40

Enclosed [dates of troop convoys from India] make me anxious to know how you propose to use these eight fine Regular battalions. Obviously they will be a reinforcement for your shock troops. One would suppose they might make the infantry of two divisions, with five good Territorial battalions added to each division, total eighteen. Should they not also yield up a certain number of officers and N.C.O.s to stiffen the Territorial battalions so attached? You would thus have six brigades of infantry quite soon. Alas, I fear the artillery must lag behind, but not, I trust, for long.

\* \* \* \* \*

As rumours grew of peace proposals and a message was sent to us from the Vatican through Berne I thought it right to send the following minute to the Foreign Secretary:

28.VI.40

I hope it will be made clear to the Nuncio that we do not desire to make any inquiries as to terms of peace with Hitler, and that all our agents are strictly forbidden to entertain any such suggestions.

But here is the record of a qualm:

*Prime Minister to Professor Lindemann* 29.VI.40

While we are hastening our preparations for air mastery, the Germans will be organising the whole industries of the captured countries for air production and other war production suitable [for use] against us. It is therefore a race. They will not be able to get the captured factories working immediately, and meanwhile we shall get round the invasion danger through the growth of our defences and Army strength. But what sort of relative outputs must be faced next year unless we are able to bomb the newly-acquired German plants? Germany also, being relieved from the need of keeping a gigantic army in constant contact with the French Army, must have spare capacity for the air and other methods of attacking us. Must we not expect this will be very great? How soon can it come into play? Hitherto I have been looking at the next three months because of the emergency, but what about 1941? It seems to me that only immense American supplies can be of use in turning the corner.

\* \* \* \* \*

As the month of June ground itself out, the sense of potential invasion at any moment grew upon us all.

*Prime Minister to General Ismay* 30.VI.40

The Admiralty charts of tides and state of the moon, Humber, Thames Estuary, Beachy Head, should be studied with a view to ascertaining on which days conditions will be most favourable to a seaborne landing. The Admiralty view is sought.

A landing or descent in Ireland was always a deep anxiety to the Chiefs of Staff. But our resources seemed to me too limited for serious troop movements.

*Prime Minister to General Ismay* 30.VI.40

It would be taking an undue risk to remove one of our only two thoroughly-equipped divisions out of Great Britain at this juncture. Moreover, it is doubtful whether the Irish situation will require the use of divisional formations complete with their technical vehicles as if for Continental war. The statement that it would take ten days to transport a division from this country to Ireland, even though every preparation can be made beforehand, is not satisfactory. Schemes should be prepared to enable two or three lightly-equipped brigades to move at short notice, and in not more than three days, into Northern Ireland. Duplicate transport should be sent on ahead. It would be a mistake to send any large force of artillery to Ireland. It is not at all likely that a naval descent will be effected there. Air-borne descents cannot carry much artillery. Finally, nothing that can happen in Ireland can be immediately decisive.

\* \* \* \* \*

In bringing home the troops from Palestine I had difficulties with both my old friends, the Secretary of State for India, Mr. Amery, and the Secretary of State for the Colonies, Lord Lloyd, who was a convinced anti-Zionist and pro-Arab. I wished to arm the Jewish colonists. Mr. Amery at the India Office had a different view from mine about the part which India should play. I wanted Indian troops at once to come into Palestine and the Middle East, whereas the Viceroy and the India Office were naturally inclined to a long-term plan of creating a great Indian Army based upon Indian munitions factories.

\* \* \* \* \*

*Prime Minister to Secretary of State for India* 22.VI.40

1. We have already very large masses of troops in India of which no use is being made for the general purposes of the war. The assistance of India this time is incomparably below that of 1914-18. . . . It seems to me very likely that the war will spread to the Middle East, and the climates of Iraq, Palestine, and Egypt are well suited to Indian troops. I recommend their organisation in brigade groups, each with a proportion of artillery on the new British model. I should hope that six or eight of these groups could be ready this winter. They should include some brigades of Gurkhas.

2. The process of liberating the Regular British battalions must continue, and I much regret that a fortnight's delay has become

inevitable in returning you the Territorial battalions in exchange. You should reassure the Viceroy that it is going forward.

\* \* \* \* \*

*Prime Minister to Secretary of State for the Colonies* 28.VI.40
The failure of the policy which you favour is proved by the very large numbers of sorely-needed troops you [we] have to keep in Palestine:

> 6 battalions of infantry,
> 9 regiments of Yeomanry,
> 8 battalions of Australian infantry,

the whole probably more than twenty thousand men. This is the price we have to pay for the anti-Jewish policy which has been persisted in for some years. Should the war go heavily into Egypt, all these troops will have to be withdrawn, and the position of the Jewish colonists will be one of the greatest danger. Indeed I am sure that we shall be told we cannot withdraw these troops, though they include some of our best, and are vitally needed elsewhere. If the Jews were properly armed, our forces would become available, and there would be no danger of the Jews attacking the Arabs, because they are entirely dependent upon us and upon our command of the seas. I think it is little less than a scandal that at a time when we are fighting for our lives these very large forces should be immobilised in support of a policy which commends itself only to a section of the Conservative Party.

I had hoped you would take a broad view of the Palestine situation, and would make it an earnest objective to set the British garrison free. I could certainly not associate myself with such an answer as you have drawn up for me. I do not at all admit that Arab feeling in the Near East and India would be prejudiced in the manner you suggest. Now that we have the Turks in such a friendly relationship the position is much more secure.

\* \* \* \* \*

For the first time in a hundred and twenty-five years a powerful enemy was now established across the narrow waters of the English Channel. Our re-formed Regular Army, and the larger but less well trained Territorials, had to be organised and deployed to create an elaborate system of defences, and to stand ready, if the invader came, to destroy him—for there could be no escape. It was for both sides "Kill or Cure". Already the Home Guard could be included in the general framework of defence. On June 25 General Ironside, Commander-in-Chief Home Forces, exposed his plans to the Chiefs of Staff. They were of course scrutinised with anxious care by the experts, and I examined them myself with no little attention. On the whole they stood approved. There were three main elements in this early outline of a great future plan: first, an entrenched "crust" on the probable invasion beaches of the coast, whose defenders should fight where they stood, supported by mobile reserves for immediate counter-attack; secondly, a line of anti-tank obstacles, manned by the Home Guard, running down the east centre of England and protecting London and the great industrial centres from inroads by armoured vehicles; thirdly, behind that line, the main reserves for major counter-offensive action.

Ceaseless additions and refinements to this first plan were effected as the weeks and months passed; but the general conception remained. All troops, if attacked, should stand firm, not in linear only but *in all-round defence*, whilst others moved rapidly to destroy the attackers, whether they came from sea or air. Men who had been cut off from immediate help would not have merely remained in position. Active measures were prepared to harass the enemy from behind; to interfere with his communications and to destroy material, as the Russians did with great results when the German tide flowed over their country a year later. Many people must have been bewildered by the innumerable activities all around them. They could understand the necessity for wiring and mining the beaches, the anti-tank obstacles at the defiles, the concrete pillboxes at the cross-roads, the intrusions into their houses to fill an attic with sandbags, on to their golfcourses or most fertile fields and gardens to burrow out some wide anti-tank ditch. All these inconveniences, and much more, they accepted in good part. But sometimes they must have wondered if there was a general scheme, or whether lesser individuals were not running amok in their energetic use of newly-granted powers of interference with the property of the citizen.

There was however a central plan, elaborate, co-ordinated, and all-embracing. As it grew it shaped itself thus: the overall Command was maintained at General Headquarters in London. All Great Britain and Northern Ireland were divided into seven Commands; these again into areas of Corps and Divisional Commands. Commands, Corps, and Divisions were each required to hold a proportion of their resources in mobile reserve, only the minimum being detailed to hold their own particular defences. Gradually there were built up in rear of the beaches zones of defence in each divisional area; behind these were similar "Corps Zones" and "Command Zones", the whole system amounting in depth to a hundred miles or more. And behind these was established the main anti-tank obstacle running across Southern England and northwards into Nottinghamshire. Above all was the final reserve directly under the Commander-in-Chief of the Home Forces. This it was our policy to keep as large and mobile as possible.

Within this general structure were many variations. Each of our ports on the east and south coasts was a special study. Direct frontal attack upon a defended port seemed an unlikely contingency, and all were made into strong points equally capable of defence from the landward or the seaward side. It astonishes me that when this principle of fortifying the gorges was so universally accepted and rigorously enforced by all military authorities at home no similar measures were adopted at Singapore by the succession of high officers employed there. But this is a later story. Obstacles were placed on many thousand square miles of Britain to impede the landing of air-borne troops. All our aerodromes, Radar stations, and fuel depots, of which even in the summer of 1940 there were three hundred and seventy-five, needed defence by special garrisons and by their own airmen. Many thousands of "vulnerable points", bridges, power-stations, depots, vital factories, and the like had to be guarded day and night from sabotage or sudden onset. Schemes were ready for the immediate demolition of resources helpful to the enemy if captured. The destruction of port facilities, the cratering of key roads, the paralysis of motor transport and of telephones and telegraph-stations, of rolling stock or permanent way, before they passed out of our control, were planned to the last detail. Yet despite all these wise and necessary precautions, in which the civilian departments gave unstinted help to the military, there was no question of a "scorched earth policy"; England was to be defended by its people, not destroyed.

# CHAPTER IX

# THE FRENCH AGONY

*F*UTURE generations may deem it noteworthy that the supreme question of whether we should fight on alone never found a place upon the War Cabinet agenda. It was taken for granted and as a matter of course by these men of all parties in the State, and we were much too busy to waste time upon such unreal, academic issues. We were united also in viewing the new phase with good confidence. It was decided to tell the Dominions the whole facts. I was invited to send a message in the same sense to President Roosevelt, and also to sustain the determination of the French Government and assure them of our utmost support.

*Former Naval Person to President Roosevelt*          12.VI.40

I spent last night and this morning at the French G.Q.G., where the situation was explained to me in the gravest terms by Generals Weygand and Georges. You have no doubt received full particulars from Mr. Bullitt. The practical point is what will happen when and if the French front breaks, Paris is taken, and General Weygand reports formally to his Government that France can no longer continue what he calls "co-ordinated war". The aged Marshal Pétain, who was none too good in April and July 1918, is, I fear, ready to lend his name and prestige to a treaty of peace for France. Reynaud, on the other hand, is for fighting on, and he has a young General de Gaulle, who believes much can be done. Admiral Darlan declares he will send the French Fleet to Canada. It would be disastrous if the two big modern ships fell into bad hands. It seems to me that there must be many elements in France who will wish to continue the struggle either in France or in the French colonies, or in both. This therefore is the moment for you to strengthen Reynaud the utmost you can, and try to tip the balance in favour of the best and longest possible French resistance. I venture to put this point before you, although I know you must understand it as well as I do.

*   *   *   *   *

On June 13 I made my last visit to France for four years almost to a day. The French Government had now withdrawn to Tours, and tension had mounted steadily. I took Edward Halifax and General Ismay with me, and Max Beaverbrook volunteered to come too. In trouble he is always buoyant. This time the weather was cloudless, and we sailed over in the midst of our Spitfire squadron, making however a rather wider sweep to the southward than before. Arrived over Tours, we found the airport had been heavily bombed the night before, but we and all our escort landed smoothly in spite of the craters. Immediately one sensed the increasing degeneration of affairs. No one came to meet us or seemed to expect us. We borrowed a service car from the station commander and motored into the city, making for the Prefecture, where it was said the French Government had their headquarters. No one of consequence was there, but Reynaud was reported to be motoring in from the country, and Mandel was also to arrive soon.

It being already nearly two o'clock, I insisted upon luncheon, and after some parleyings we drove through streets crowded with refugees' cars, most of them with a mattress on top and crammed with luggage. We found a café, which was closed, but after explanations we obtained a meal. During luncheon I was visited by M. Baudouin, whose influence had risen in these latter days. He began at once in his soft, silky manner about the hopelessness of the French resistance. If the United States would declare war on Germany it might be possible for France to continue. What did I think about this? I did not discuss the question further than to say that I hoped America would come in, and that we should certainly fight on. He afterwards, I was told, spread it about that I had agreed that France should surrender unless the United States came in.

We then returned to the Prefecture, where Mandel, Minister of the Interior, awaited us. This faithful former secretary of Clemenceau, and a bearer forward of his life's message, seemed in the best of spirits. He was energy and defiance personified. His luncheon, an attractive chicken, was uneaten on the tray before him. He was a ray of sunshine. He had a telephone in each hand, through which he was constantly giving orders and decisions. His ideas were simple: fight on to the end in France, in order to cover the largest possible movement into Africa. This was the last time I saw this valiant Frenchman. The restored French Republic rightly shot to death the hirelings who murdered him. His memory is honoured by his countrymen and their allies.

Presently M. Reynaud arrived. At first he seemed depressed. General Weygand had reported to him that the French armies were exhausted. The line was pierced in many places; refugees were pouring along all the roads through the country, and many of the troops were in disorder. The Generalissimo felt it was necessary to ask for an armistice while there were still enough French troops to keep order until peace could be made. Such was the military advice. He would send that day a further message to Mr. Roosevelt saying that the last hour had come and that the fate of the Allied cause lay in America's hand. Hence arose the alternative of armistice and peace.

M. Reynaud proceeded to say that the Council of Ministers had on the previous day instructed him to inquire what would be Britain's attitude should the worst come. He himself was well aware of the solemn pledge that no separate peace would be entered into by either ally. General Weygand and others pointed out that France had already sacrificed everything in the common cause. She had nothing left; but she had succeeded in greatly weakening the common foe. It would in those circumstances be a shock if Britain failed to concede that France was physically unable to carry on, if France was still expected to fight on and thus deliver up her people to the certainty of corruption and evil transformation at the hands of ruthless specialists in the art of bringing conquered peoples to heel. That then was the question which he had to put. Would Great Britain realise the hard facts with which France was faced?

The official British record reads as follows:

Mr. Churchill said that Great Britain realised how much France had suffered and was suffering. Her own turn would come, and she was ready. She grieved to find that her contribution to the land struggle was at present so small, owing to the reverses which had been met with as a result of applying an agreed strategy in the North. The British had not yet felt the German lash, but were aware of its force. They nevertheless had but one thought: to win the war and destroy Hitlerism. Everything was subordinate to that aim; no difficulties, no regrets, could stand in the way. He was well assured of British capacity for enduring and persisting, for striking back till the foe was beaten. They would therefore hope that France would carry on fighting south of Paris down to the sea, and if need be from North Africa. At all costs time must be gained. The period of waiting was not limitless: a pledge from the United States would make it quite short. The alternative course meant destruction for France quite as certainly. Hitler would abide by no pledges. If on the other hand France remained in the struggle, with her fine Navy, her great Empire, her Army still able to carry on guerrilla warfare on a gigantic scale, and if Germany failed to destroy England, which she must do or go under, if then Germany's might in the air was broken, then the whole hateful edifice of Nazidom would topple over. Given immediate help from America, perhaps even a declaration of war, victory was not so far off. At all events England would fight on. She had not and would not alter her resolve: no terms, no surrender. The alternatives for her were death or victory. That was his answer to M. Reynaud's question.

M. Reynaud replied that he had never doubted England's determination. He was however anxious to know how the British Government would react in a certain contingency. The French Government—the present one or another—might say: "We know you will carry on. We would also, if we saw any hope of a victory. But we see no sufficient hopes of an early victory. We cannot count on American help. *There is no light at the end of the tunnel.* We cannot abandon our people to indefinite German domination. We must come to terms. We have no choice. . . ." It was already too late to organise a redoubt in Brittany. Nowhere would a genuine French Government have a hope of escaping capture on French soil. . . . The question to Britain would therefore take the form: "Will you acknowledge that France has given her best, her youth and life-blood; that she can do no more; and that she is entitled, having nothing further to contribute to the common cause, to enter into a separate peace while maintaining the solidarity implicit in the solemn agreement entered into three months previously?"

Mr. Churchill said that in no case would Britain waste time and energy in reproaches and recriminations. That did not mean that she would consent to action contrary to the recent agreement. The first step ought to be M. Reynaud's further message putting the present position squarely to President Roosevelt. Let them await the answer before considering anything else. If England won the war France would be restored in her dignity and in her greatness.

All the same I thought the issue raised at this point was so serious that I asked to withdraw with my colleagues before answering it. So Lords Halifax and Beaverbrook and the rest of our party went out into a dripping but sunlit garden and talked things over for half an hour. On our return I re-stated our position. We could not agree to a separate peace however it might come. Our war aim remained the total defeat of Hitler, and we felt that we could still bring this about. We were therefore not in a position to release France from her obligation. Whatever happened, we would level no reproaches against France; but that was a different matter from consenting to release her from her pledge. I urged that the French should now send a new appeal to President Roosevelt, which we would support from London. M. Reynaud agreed to do this, and promised that the French would hold on until the result of his final appeal was known.

Before leaving I made one particular request to M. Reynaud. Over four hundred German pilots, the bulk of whom had been shot down by the R.A.F., were prisoners in France. Having regard to the situation, they should be handed over to our custody. M. Reynaud willingly gave this promise, but soon he had no power to keep it. These German pilots all became available for the Battle of Britain, and we had to shoot them down a second time.

\*    \*    \*    \*    \*

At the end of our talk M. Reynaud took us into the adjoining room, where MM. Herriot and Jeanneney, the Presidents of the Chamber and Senate respectively, were seated. Both these French patriots spoke with passionate emotion about fighting on to the death. As we went down the crowded passage into the courtyard I saw General de Gaulle standing stolid and expressionless at the doorway. Greeting him, I said in a low tone, in French: "*L'homme du destin.*" He remained impassive. In the courtyard there must have been more than a hundred leading Frenchmen in frightful misery. Clemenceau's son was brought up to me. I wrung his hand. The Spitfires were already in the air, and I slept sound on our swift and uneventful journey home. This was wise, for there was a long way to go before bed-time.

\* \* \* \* \*

After our departure from Tours at about half-past five M. Reynaud met his Cabinet again at Cangé. They were vexed that I and my colleagues had not come there to join them. We should have been very willing to do so, no matter how late we had to fly home. But we were never invited; nor did we know there was to be a French Cabinet meeting.

At Cangé the decision was taken to move the French Government to Bordeaux, and Reynaud sent off his telegram to Roosevelt with its desperate appeal for the entry on the scene at least of the American Fleet.

At 10.15 p.m. I made my new report to the Cabinet. My account was endorsed by my two companions. While we were still sitting Ambassador Kennedy arrived with President Roosevelt's reply to Reynaud's appeal of June 10.

*President Roosevelt to M. Reynaud*                                    13.VI.40
Your message of June 10 has moved me very deeply. As I have already stated to you and to Mr. Churchill, this Government is doing everything in its power to make available to the Allied Governments the material they so urgently require, and our efforts to do still more are being redoubled. This is so because of our faith in and our support of the ideals for which the Allies are fighting.

The magnificent resistance of the French and British Armies has profoundly impressed the American people.

I am, personally, particularly impressed by your declaration that France will continue to fight on behalf of Democracy, even if it means slow withdrawal, even to North Africa and the Atlantic. It is most important to remember that the French and British Fleets continue [in] mastery of the Atlantic and other oceans; also to remember that vital materials from the outside world are necessary to maintain all armies.

I am also greatly heartened by what Prime Minister Churchill said a few days ago about the continued resistance of the British Empire, and that determination would seem to apply equally to the great French Empire all over the world. Naval power in world affairs still carries the lessons of history, as Admiral Darlan well knows.

We all thought the President had gone a very long way. He had authorised Reynaud to publish his message of June 10, with all that that implied, and now he had sent this formidable answer. If, upon this, France decided to endure the further torture of the war, the United States would be deeply committed to enter it. At any rate, it contained two points which were tantamount to belligerence: first, a promise of all material aid, which implied active assistance; secondly, a call to go on fighting even if the Government were driven right out of France. I sent our thanks to the President immediately, and I also sought to commend the President's message to Reynaud in the most favourable terms. Perhaps these points were stressed unduly; but it was necessary to make the most of everything we had or could get.

*Former Naval Person to President Roosevelt*                    13.VI.40

Ambassador Kennedy will have told you about the British meeting to-day with the French at Tours, of which I showed him our record. I cannot exaggerate its critical character. They were very nearly gone. Weygand had advocated an armistice while he still has enough troops to prevent France from lapsing into anarchy. Reynaud asked us whether, in view of the sacrifices and sufferings of France, we would release her from the obligation about not making a separate peace. Although the fact that we have unavoidably been out of this terrible battle weighed with us, I did not hesitate in the name of the British Government to refuse consent to an armistice or separate peace. I urged that this issue should not be discussed until a further appeal had been made by Reynaud to you and the United States, which I undertook to second. Agreement was reached on this, and a much better mood prevailed for the moment in Reynaud and his Ministers.

Reynaud felt strongly that it would be beyond his power to encourage his people to fight on without hope of ultimate victory, and that that hope could only be kindled by American intervention up to the extreme limit open to you. As he put it, they wanted to see light at the end of the tunnel.

While we were flying back here your magnificent message was sent, and Ambassador Kennedy brought it to me on my arrival. The British Cabinet were profoundly impressed, and desire me to express their gratitude for it, but, Mr. President, I must tell you that it seems to me absolutely vital that this message should be published to-morrow, June 14, in order that it may play the decisive part in turning the course of world history. It will, I am sure, decide the French to deny Hitler a patched-up peace with France. He needs this peace in order to destroy us and take a long step forward to world mastery. All the far-reaching plans, strategic, economic, political, and moral, which your message expounds may be still-born if the French cut out now. Therefore I urge that the message should be published now. We realise fully that the moment Hitler finds he cannot dictate a Nazi peace in Paris he will turn his fury on to us. We shall do our best to withstand it, and if we succeed wide new doors are open upon the future and all will come out even at the end of the day.

To M. Reynaud I sent this message:

13.VI.40

On returning here we received a copy of President Roosevelt's answer to your appeal of June 10. Cabinet is united in considering this magnificent document as decisive in favour of the continued resistance of France in accordance with your own declaration of June 10 about fighting before Paris, behind Paris, in a province, or, if necessary, in Africa or across the Atlantic. The promise of redoubled material aid is coupled with definite advice and exhortation to France to continue the struggle even under the grievous conditions which you mentioned. If France on this message of President Roosevelt's continues in the field and in the war, we feel that the United States is committed beyond recall to take the only remaining step, namely, becoming a belligerent in form as she already has constituted herself in fact. Constitution of United States makes it impossible, as you foresaw, for the President to declare war himself, but if you act on his reply now received we sincerely believe that this must inevitably follow. We are asking the President to allow publication of the message, but, even if he does not agree to this for a day or two, it is on the record and can afford the basis for your action. I do beg you and your colleagues, whose resolution we so much admired to-day, not to miss this sovereign opportunity of bringing about the world-wide oceanic and economic coalition which must be fatal to Nazi domination. We see before us a definite plan of campaign, and the light which you spoke of shines at the end of the tunnel.

Finally, in accordance with the Cabinet's wishes, I sent a formal message of good cheer to the French Government, in which the note of an indissoluble union between our two countries was struck for the first time.

*Prime Minister to M. Reynaud*                      13.VI.40

In this solemn hour for the British and French nations and for the cause of Freedom and Democracy to which they have vowed themselves, His Majesty's Government desire to pay to the Government of the French Republic the tribute which is due to the heroic fortitude and constancy of the French armies in battle against enormous odds. Their effort is worthy of the most glorious traditions of France, and has inflicted deep and long-lasting injury upon the enemy's strength. Great Britain will continue to give the utmost aid in her power. We take this opportunity of proclaiming the indissoluble union of our two peoples and of our two Empires. We cannot measure the various forms of tribulation which will fall upon our peoples in the near future. We are sure that the ordeal by fire will only fuse them together into one unconquerable whole. We renew to the French Republic our pledge and resolve to continue the struggle at all costs in France, in this Island, upon the oceans and in the air, wherever it may lead us, using all our resources to the utmost limit and sharing together the burden of repairing the ravages of war. We shall never turn from the conflict until France stands safe and erect in all her grandeur, until the wronged and enslaved States and peoples have been liberated, and until civilisation is freed from the nightmare of Nazidom. That this day will dawn we are more sure than ever. It may dawn sooner than we now have the right to expect.

All these three messages were drafted by me before I went to bed after midnight on the 13th. They were written actually in the small hours of the 14th.

The next day arrived a telegram from the President explaining that he could not agree to the publication of his message to Reynaud. He himself, according to Mr. Kennedy, had wished to do so, but the State Department, while in full sympathy with him, saw the gravest dangers. The President thanked me for my account of the meeting at Tours and complimented the British and French Governments on the courage of their troops. He renewed the assurances about furnishing all possible material and supplies; but he then said he had told Ambassador Kennedy to inform me that his message of the 13th was in no sense intended to commit and did not commit the Government of the United States to military participation. There was no authority under the

American Constitution except Congress which could make any commitment of that nature. He bore particularly in mind the question of the French Fleet. Congress, at his desire, had appropriated fifty million dollars for the purpose of supplying food and clothing to civilian refugees in France. Finally he assured me that he appreciated the significance and weight of what I had set forth in my message.

This was a disappointing telegram.

Around our table we all fully understood the risks the President ran of being charged with exceeding his constitutional authority, and consequently of being defeated on this issue at the approaching election, on which our fate, and much more, depended. I was convinced that he would give up life itself, to say nothing of public office, for the cause of world freedom now in such awful peril. But what would have been the good of that? Across the Atlantic I could feel his suffering. In the White House the torment was of a different character from that of Bordeaux or London. But the degree of personal stress was not unequal.

In my reply I tried to arm the President with some arguments which he could use to others about the danger to the United States if Europe fell and Britain failed. This was no matter of sentiment, but of life and death.

*Former Naval Person to President Roosevelt* 14–15.VI.40

I am grateful to you for your telegram and I have reported its operative passages to Reynaud, to whom I had imparted a rather more sanguine view. He will, I am sure, be disappointed at non-publication. I understand all your difficulties with American public opinion and Congress, but events are moving downward at a pace where they will pass beyond the control of American public opinion when at last it is ripened. Have you considered what offers Hitler may choose to make to France? He may say: "Surrender the Fleet intact and I will leave you Alsace-Lorraine," or alternatively: "If you do not give me your ships I will destroy your towns." I am personally convinced that America will in the end go to all lengths, but this moment is supremely critical for France. A declaration that the United States will if necessary enter the war might save France. Failing that, in a few days French resistance may have crumpled and we shall be left alone.

Although the present Government and I personally would never fail to send the Fleet across the Atlantic if resistance was beaten down here, a point may be reached in the struggle where the present Ministers no longer have control of affairs and when very easy terms could be obtained for the British Island by their becoming a vassal state of the Hitler Empire. A pro-German Government would certainly be called into being to make peace, and might present to a shattered or a starving nation an almost irresistible case for entire submission to the Nazi will. The fate of the British Fleet, as I have already mentioned to you, would be decisive on the future of the United States, because if it were joined to the Fleets of Japan, France, and Italy and the great resources of German industry overwhelming sea-power would be in Hitler's hands. He might of course use it with a merciful moderation. On the other hand, he might not. This revolution in sea-power might happen very quickly, and certainly long before the United States would be able to prepare against it. If we go down you may have a United States of Europe under the Nazi command far more numerous, far stronger, far better armed than the New World.

I know well, Mr. President, that your eye will already have searched these depths, but I feel I have the right to place on record the vital manner in which American interests are at stake in our battle and that of France.

I am sending you through Ambassador Kennedy a paper on destroyer strength prepared by the Naval Staff for your information. If we have to keep, as we shall, the bulk of our destroyers on the East Coast to guard against invasion, how shall we be able to cope with a German-Italian attack on the food and trade by which we live? The sending of the thirty-five destroyers as I have already described will bridge the gap until our new construction comes in at the end of the year. Here is a definite practical and possibly decisive step which can be taken at once, and I urge most earnestly that you will weigh my words.

*     *     *     *     *

Meanwhile the situation on the French front went from bad to worse. The German operations north-west of Paris, in which our 51st Division had been lost, had brought the enemy, by June 9, to the lower reaches of the Seine and the Oise. On the southern banks the dispersed remnants of the Tenth and Seventh French Armies were hastily organising a defence; they had been riven asunder, and to close the gap the garrison of the capital, the so-called Armée de Paris, had been marched out and interposed.

Farther to the east, along the Aisne, the Sixth, Fourth, and Second Armies were in far better shape. They had had three weeks in which to establish themselves and to absorb such reinforcements as had been sent. During all the period of Dunkirk

and of the drive to Rouen they had been left comparatively
undisturbed, but their strength was small for the hundred miles
they had to hold, and the enemy had used the time to concentrate
against them a great mass of divisions to deliver the final blow.
On June 9 it fell. Despite a dogged resistance, for the French were
now fighting with great resolution, bridgeheads were established
south of the river from Soissons to Rethel, and in the next two
days these were expanded until the Marne was reached. German
Panzer divisions, which had played so decisive a part in the drive
down the coast, were brought across to join the new battle.
Eight of these, in two great thrusts, turned the French defeat
into a rout. The French armies, decimated and in confusion, were
quite unable to withstand this powerful assembly of superior
numbers, equipment, and technique. In four days, by June 16,
the enemy had reached Orleans and the Loire; while to the east
the other thrust had passed through Dijon and Besançon, almost
to the Swiss frontier.

West of Paris the remains of the Tenth Army, the equivalent
of no more than two divisions, had been pressed back south-
westwards from the Seine towards Alençon. The capital fell on
the 14th; its defending armies, the Seventh and the Armée de
Paris, were scattered; a great gap now separated the exiguous
French and British forces in the west from the rest and the remains
of the once proud Army of France.

And what of the Maginot Line, the shield of France, and its
defenders? Until June 14 no direct attack was made, and already
some of the active formations, leaving behind the garrison troops,
had started to join, if they could, the fast-withdrawing armies of
the centre. But it was too late. On that day the Maginot Line
was penetrated before Saarbrucken and across the Rhine by
Colmar; the retreating French were caught up in the battle and
unable to extricate themselves. Two days later the German penetration to Besançon had cut off their retreat. More than four
hundred thousand men were surrounded without hope of escape.
Many encircled garrisons held out desperately; they refused to
surrender until after the armistice, when French officers were
dispatched to give them the order. The last forts obeyed on
June 30, the commander protesting that his defences were still
intact at every point.

Thus the vast disorganised battle drew to its conclusion all along the French front. It remains only to recount the slender part which the British were able to play.

\* \* \* \* \*

General Brooke had won distinction in the retreat to Dunkirk, and especially by his battle in the gap opened by the Belgian surrender. We had therefore chosen him to command the British troops which remained in France and all reinforcements until they should reach sufficient numbers to require the presence of Lord Gort as an Army Commander. Brooke had now arrived in France, and on the 14th he met Generals Weygand and Georges. Weygand stated that the French forces were no longer capable of organised resistance or concerted action. The French Army was broken into four groups, of which its Tenth Army was the westernmost. Weygand also told him that the Allied Governments had agreed that a bridgehead should be created in the Brittany peninsula, to be held jointly by the French and British troops on a line running roughly north and south through Rennes. He ordered him to deploy his forces on a defensive line running through this town. Brooke pointed out that this line of defence was 150 kilometres long and required at least fifteen divisions. He was told that the instructions he was receiving must be regarded as an order.

It is true that on June 11 at Briare Reynaud and I had agreed to try to draw a kind of "Torres Vedras line" across the foot of the Brittany peninsula. Everything however was dissolving at the same time, and the plan, for what it was worth, never reached the domain of action. In itself the idea was sound, but there were no facts to clothe it with reality. Once the main French armies were broken or destroyed, this bridgehead, precious though it was, could not have been held for long against concentrated German attack. But even a few weeks' resistance here would have maintained contact with Britain and enabled large French withdrawals to Africa from other parts of the immense front, now torn to shreds. If the battle in France was to continue, it could be only in the Brest peninsula and in wooded or mountainous regions like the Vosges. The alternative for the French was surrender. Let none, therefore, mock at the conception of a bridgehead in Brittany. The Allied armies under Eisenhower, then an unknown American colonel, bought it back for us later at a high price.

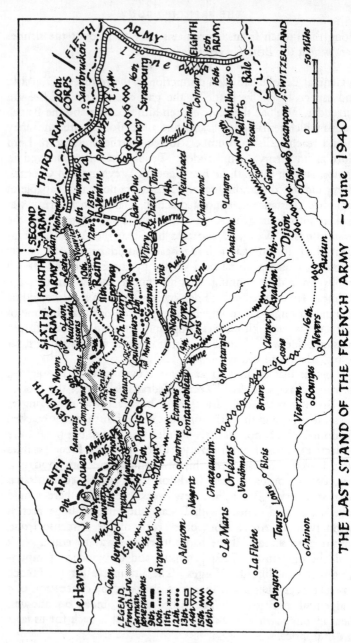

THE LAST STAND OF THE FRENCH ARMY — June 1940

General Brooke, after his talk with the French commanders, and having measured from his own headquarters a scene which was getting worse every hour, reported to the War Office and by telephone to Mr. Eden that the position was hopeless. All further reinforcements should be stopped, and the remainder of the British Expeditionary Force, now amounting to a hundred and fifty thousand men, should be re-embarked at once. On the night of June 14, as I was thought to be obdurate, he rang me up on a telephone line which by luck and effort was open, and pressed this view upon me. I could hear quite well, and after ten minutes I was convinced that he was right and we must go. Orders were given accordingly. He was released from French command. The back-loading of great quantities of stores, equipment, and men began. The leading elements of the Canadian Division which had landed got back into their ships, and the 52nd Division, which, apart from its 157th Brigade, had not yet been committed to action, retreated on Brest. No British troops operating under the Tenth French Army were withdrawn; but all else of ours took to the ships at Brest, Cherbourg, St. Malo, and St. Nazaire. On June 15 our troops were released from the orders of the Tenth French Army, and next day, when it carried out a further withdrawal to the south, they moved towards Cherbourg. The 157th Brigade, after heavy fighting, was extricated that night, and, retiring in their lorries, embarked during the night of the 17th–18th. On June 17 it was announced that the Pétain Government had asked for an armistice, ordering all French forces to cease fighting, without even communicating this information to our troops. General Brooke was consequently told to come away with all men he could embark and any equipment he could save.

We repeated now on a considerable scale, though with larger vessels, the Dunkirk evacuation. Over twenty thousand Polish troops who refused to capitulate cut their way to the sea and were carried by our ships to Britain. The Germans pursued our forces at all points. In the Cherbourg peninsula they were in contact with our rearguard ten miles south of the harbour on the morning of the 18th. The last ship left at 4 p.m., when the enemy, led by Rommel's 7th Panzer Division, were within three miles of the port. Very few of our men were taken prisoners.

In all there were evacuated from all French harbours 136,000 British troops and 310 guns; a total, with the Poles, of 156,000 men. This reflects great credit on General Brooke's embarkation staff, of whom the chief, General de Fonblanque, a British officer, died shortly afterwards as the result of his exertions.

At Brest and the western ports the evacuations were numerous. The German air attack on the transports was heavy. One frightful incident occurred on the 17th at St. Nazaire. The 20,000-ton liner *Lancastria*, with five thousand men on board, was bombed just as she was about to leave. Upwards of three thousand men perished. The rest were rescued under continued air attack by the devotion of the small craft. When this news came to me in the quiet Cabinet Room during the afternoon I forbade its publication, saying: "The newspapers have got quite enough disaster for to-day at least." I had intended to release the news a few days later, but events crowded upon us so black and so quickly that I forgot to lift the ban, and it was some time before the knowledge of this horror became public.

\* \* \* \* \*

To lessen the shock of the impending French surrender, it was necessary at this time to send a message to the Dominions Prime Ministers showing them that our resolve to continue the struggle although alone was not based upon mere obstinacy or desperation, and to convince them by practical and technical reasons, of which they might well be unaware, of the real strength of our position. I therefore dictated the following statement on the afternoon of June 16, a day already filled with much business.

*Prime Minister to the Prime Ministers of Canada, Australia, New Zealand, and South Africa* 16.VI.40

[Some sentences of introduction particular to each.]

I do not regard the situation as having passed beyond our strength. It is by no means certain that the French will not fight on in Africa and at sea, but, whatever they do, Hitler will have to break us in this Island or lose the war. Our principal danger is his concentrated air attack by bombing, coupled with parachute and air-borne landings and attempts to run an invading force across the sea. This danger has faced us ever since the beginning of the war, and the French could never have saved us from it, as he could always switch on to us. Undoubtedly it is aggravated by the conquests Hitler has made upon the European coast close to our shores. Nevertheless, in principle the danger is the same. I do not see why we should not be able to meet it. The Navy has never pretended to prevent a raid of five or ten thousand men, but we do not see how a force of, say, eighty to a hundred thousand could be transported across the sea, and still less maintained, in the teeth of superior sea-power. As long as our Air Force is in being it provides a powerful aid to the Fleet in preventing sea-borne landings and will take a very heavy toll of air-borne landings.

Although we have suffered heavy losses by assisting the French and during the Dunkirk evacuation, we have managed to husband our air fighter strength in spite of poignant appeals from France to throw it improvidently into the great land battle, which it could not have turned decisively. I am happy to tell you that it is now as strong as it has ever been, and that the flow of machines is coming forward far more rapidly than ever before; in fact, pilots have now become the limiting factor at the moment. Our fighter aircraft have been wont to inflict a loss of two or two and a half to one even when fighting under the adverse conditions in France. During the evacuation of Dunkirk, which was a sort of No-man's Land, we inflicted a loss of three or four to one, and often saw German formations turn away from a quarter of their numbers of our planes. But all air authorities agree that the advantage in defending this country against an oversea air attack will be still greater, because, first, we shall know pretty well by our various devices where they are coming, and because our squadrons lie close enough together to enable us to concentrate against the attackers and provide enough to attack both the bombers and the protecting fighters at the same time. All their shot-down machines will be total losses; many of ours and our pilots will fight again. Therefore I do not think it by any means impossible that we may so maul them that they will find daylight attacks too expensive.

The major danger will be from night attack on our aircraft factories, but this, again, is far less accurate than daylight attack, and we have many plans for minimising its effect. Of course their numbers are much greater than ours, but not so much greater as to deprive us of a good and reasonable prospect of wearing them out after some weeks or even months of air struggle. Meanwhile of course our bomber force will be striking continually at their key points, especially oil refineries and air factories, and at their congested and centralised war industry in the Ruhr. We hope our people will stand up to this bombardment as well as the enemy. It will, on both sides, be on an unprecedented scale. All our information goes to show that the Germans have not liked what they have got so far.

It must be remembered that, now that the B.E.F. is home and largely rearmed or rearming, if not upon a Continental scale, at any rate good enough for Home defence, we have far stronger military forces in this Island than we have ever had in the late war or in this war.

Therefore we hope that such numbers of the enemy as may be landed from the air or by sea-borne raid will be destroyed and be an example to those who try to follow. No doubt we must expect novel forms of attack and attempts to bring tanks across the sea. We are preparing ourselves to deal with these as far as we can foresee them. No one can predict or guarantee the course of a life-and-death struggle of this character, but we shall certainly enter upon it in good heart.

I have given you this full explanation to show you that there are solid reasons behind our resolve not to allow the fate of France, whatever it may be, to deter us from going on to the end. I personally believe that the spectacle of the fierce struggle and carnage in our Island will draw the United States into the war, and even if we should be beaten down through the superior numbers of the enemy's Air Force it will always be possible, as I indicated to the House of Commons in my last speech, to send our fleets across the oceans, where they will protect the Empire and enable it to continue the war and the blockade, I trust in conjunction with the United States, until the Hitler *régime* breaks under the strain. We shall let you know at every stage how you can help, being assured that you will do all in human power, as we, for our part, are entirely resolved to do.

I composed this in the Cabinet Room, and it was typed as I spoke. The door to the garden was wide open, and outside the sun shone warm and bright. Air Marshal Newall, the Chief of the Air Staff, sat on the terrace meanwhile, and when I had finished revising the draft I took it out to him in case there were any improvements or corrections to be made. He was evidently moved, and presently said he agreed with every word. I was comforted and fortified myself by putting my convictions upon record, and when I read the message over the final time before sending it off I felt a glow of sober confidence. This was certainly justified by what happened. All came true.

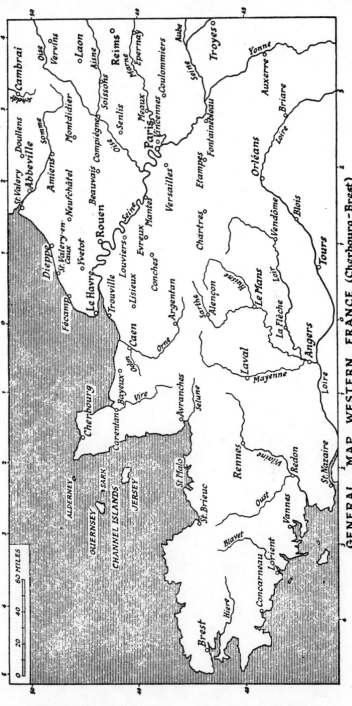

GENERAL MAP WESTERN FRANCE (Cherbourg–Brest)

# CHAPTER X

# THE BORDEAUX ARMISTICE

*The French Government Moves to Bordeaux – General Weygand's Attitude – Weygand and Reynaud – M. Chautemps' Insidious Proposal – The French Decision to Ask for Terms – British Insistence on the Safeguarding of the French Fleet – My Telegram to Reynaud of June 16 – A New Issue Arises – British Offer of Indissoluble Union with France – High Hopes of General de Gaulle that this would Strengthen M. Reynaud – M. Reynaud's Satisfaction – My Telegram of June 16 Suspended – Plan for Me to Visit Bordeaux by Cruiser with the Labour and Liberal Party Leaders Frustrated – Unfavourable Reception of the British Offer – Fall of the Reynaud Cabinet – Reynaud's Resignation – A Conversation with M. Monnet and General de Gaulle in Downing Street – Marshal Pétain Forms a French Government for an Armistice – My Message to Marshal Pétain and General Weygand, June 17 – My Broadcast of June 17 – General Spears Plans the Escape of General de Gaulle – Further Talk of Resistance in Africa – Mandel's Intentions – Admiral Darlan's Trap – Voyage of the "Massilia" – Mandel at Casablanca – Mr. Duff Cooper's Mission – Fate of the French Patriots – Jobbing Backwards – My Settled Conviction.*

W E must now quit the field of military disaster for the convulsions in the French Cabinet and the personages who surrounded it at Bordeaux.

It is not easy to establish the exact sequence of events. The British War Cabinet sat almost continuously, and messages were sent off from time to time as decisions were taken. As they took two or three hours to transmit in cipher, and probably another hour to deliver, the telephone was freely used by the officials of the Foreign Office to convey the substance to our Ambassador, and he also used the telephone frequently in reply. Therefore there are overlaps and short-circuits which are confusing. Events were moving at such a speed on both sides of the Channel that it would be misleading to present the tale as if it were an orderly flow of argument and decision.

M. Reynaud reached the new seat of government from Tours during the evening of the 14th. He received the British Ambassador about nine o'clock. Sir Ronald Campbell informed him that His Majesty's Government intended to insist on the terms of the agreement of March 28, binding both parties not to make any terms with the enemy. He also offered to provide all the necessary shipping in the event of the French Government resolving to move to North Africa. Both these statements were in accordance with the Ambassador's current instructions.

On the morning of the 15th Reynaud again received the Ambassador, and told him that he had definitely decided to divide the Government in half and to establish a centre of authority beyond the sea. Such a policy would obviously carry with it the removal of the French Fleet to ports beyond German power. Later that morning President Roosevelt's reply to Reynaud's appeal of June 13 was received. Although I had made the best of it in my telegram to the French Premier, I knew it was bound to disappoint him. Material aid, if Congress approved, was offered; but there was no question of any American entry into the war. France had no reason to expect such a declaration at this moment, and the President had not either the power to give it himself or to obtain it from Congress. There had been no meetings of the Council of Ministers since that at Cangé, near Tours, on the evening of the 13th. The Ministers having now all reached Bordeaux, the Council was summoned for the afternoon.

\* \* \* \* \*

General Weygand had been for some days convinced that all further resistance was vain. He therefore wished to force the French Government to ask for an armistice while the French Army still retained enough discipline and strength to maintain internal order on the morrow of defeat. He had a profound, lifelong dislike of the Parliamentary *régime* of the Third Republic. As an ardently religious Catholic, he saw in the ruin which had overwhelmed his country the chastisement of God for its abandonment of the Christian faith. He therefore used the power of his supreme military position far beyond the limits which his professional responsibilities, great as they were, justified or required. He confronted the Prime Minister with declarations that the French armies could fight no more, and that it was time to stop a horrible and useless massacre before general anarchy supervened.

Paul Reynaud, on the other hand, realised that the battle in France was over, but still hoped to carry on the war from Africa and the French Empire and with the French Fleet. None of the other States overrun by Hitler had withdrawn from the war. Physically in their own lands they were gripped, but from overseas their Governments had kept the flag flying and the national cause alive. Reynaud wished to follow their example, and with much more solid resources. He sought a solution on the lines of the Dutch capitulation. This, while it left the Army, whose chiefs had refused to fight any longer, free to lay down its arms wherever it was in contact with the enemy, nevertheless preserved to the State its sovereign right to continue the struggle by all the means in its power.

This issue was fought out between the Premier and the Generalissimo at a stormy interview before the Council meeting. Reynaud offered Weygand written authority from the Government to order the "Cease Fire". Weygand refused with indignation the suggestion of a military surrender. "He would never accept the casting of this shame upon the banners of the French Army." The Act of Surrender, which he deemed imperative, must be that of the Government and of the State, to which the army he commanded would dutifully conform. In so acting General Weygand, though a sincere and unselfish man, behaved wrongly. He asserted the right of a soldier to dominate the duly-constituted Government of the Republic, and thus to bring the whole resistance not only of France but of her Empire to an end contrary to the decision of his political and lawful chief.

Apart from these formalities and talk about the honour of the French Army there stood a practical point. An armistice formally entered into by the French Government would mean the end of the war for France. By negotiation part of the country might be left unoccupied and part of the Army free; whereas if the war were continued from overseas all who had not escaped from France would be controlled directly by the Germans, and millions of Frenchmen would be carried off to Germany as prisoners of war without the protection of any agreement. This was a substantial argument, but it belonged to the Government of the Republic and not to the Commander-in-Chief of the Army to decide upon it. Weygand's position that because the army under his orders would in his opinion fight no more the French Republic must give in and order its armed forces to obey an order which he was certainly willing to carry out finds no foundation in the law and practice of civilised States or in the professional honour of a soldier. In theory at least the Prime Minister had his remedy. He could have replied: "You are affronting the Constitution of the Republic. You are dismissed from this moment from your command. I will obtain the necessary sanction from the President."

Unfortunately M. Reynaud was not sufficiently sure of his position. Behind the presumptuous General loomed the illustrious Marshal Pétain, the centre of the band of defeatist Ministers whom Reynaud had so recently and so improvidently brought into the French Government and Council, and who were all resolved to stop the war. Behind these again crouched the sinister figure of Laval, who had installed himself at Bordeaux City Hall, surrounded by a clique of agitated Senators and Deputies. Laval's policy had the force and merit of simplicity. France must not only make peace with Germany; she must change sides, she must become the ally of the conqueror, and by her loyalty and services against the common foe across the Channel save her interests and her provinces and finish up on the victorious side. Evidently M. Reynaud, exhausted by the ordeals through which he had passed, had not the life or strength for so searching a personal ordeal, which would indeed have taxed the resources of an Oliver Cromwell or of a Clemenceau, of Stalin or of Hitler.

In the discussions on the afternoon of the 15th, at which the President of the Republic was present, Reynaud, having explained the situation to his colleagues, appealed to Marshal Pétain to persuade General Weygand to the Cabinet view. He could not have chosen a worse envoy. The Marshal left the room. There was an interval. After a while he returned with Weygand, whose position he now supported. At this serious juncture M. Chautemps, an important Minister, slid in an insidious proposal which wore the aspect of a compromise and was attractive to the waverers. He stated in the name of the Leftist elements of the Cabinet that Reynaud was right in affirming that an agreement with the enemy was impossible, but that it' would be prudent to make a gesture which would unite France. They should ask the Germans what the conditions of armistice would be, remaining entirely free to reject them. It was not of course possible to embark on this slippery slope and stop. The mere announcement that the French Government were asking the Germans on what terms an armistice would be granted was sufficient in itself to destroy what remained of the morale of the French Army. How could the soldier be ordered to cast away his life in obdurate resistance after so fatal a signal had been given? However, combined with the demonstration which they had witnessed from Pétain and Weygand, the Chautemps suggestion had a deadly effect on the majority. It was agreed to ask His Majesty's Government how they would view such a step, informing them at the same time that in no circumstances would the surrender of the Fleet be allowed. Reynaud now rose from the table and declared his intention to resign. But the President of the Republic restrained him, and declared that if Reynaud went he would go too. When the confused discussion was resumed no clear distinction was drawn between declining to surrender the French Fleet to the Germans and putting it out of German power by sailing it to ports outside France. It was agreed that the British Government should be asked to consent to the inquiry about the German terms. The message was immediately dispatched.

\* \* \* \* \*

The next morning Reynaud received the British Ambassador again, and was told that the British would accept the French request on the condition that the French Fleet was placed beyond German power—in fact, that it should be directed to British ports. These instructions had been telephoned to Campbell from London to save time. At eleven o'clock the distracted Council of Ministers met again, President Lebrun being present. The President of the Senate, M. Jeanneney, was brought in to endorse, both on his own behalf and on that of his colleague the President of the Chamber, M. Herriot, the proposal of the Premier to transfer the Government to North Africa. Up rose Marshal Pétain and read a letter, which it is believed had been written for him by another hand, resigning from the Cabinet. Having finished his speech, he prepared to leave the room. He was persuaded by the President of the Republic to remain on the condition that an answer would be given to him during the day. The Marshal had also complained of the delay in asking for an armistice. Reynaud replied that if one asked an ally to free one from an obligation it was customary to await the answer. The session then closed. After luncheon the Ambassador brought to Reynaud the textual answer of the British Government, of which he had already given the telephoned purport in his conversation of the morning.

\* \* \* \* \*

In these days the War Cabinet were in a state of unusual emotion. The fall and the fate of France dominated their minds. Our own plight, and what we should have to face and face alone, seemed to take a second place. Grief for our ally in her agony, and desire to do anything in human power to aid her, was the prevailing mood. There was also the overpowering importance of making sure of the French Fleet. It was in this spirit that a proposal for "an indissoluble union" between France and Britain was conceived.

I was not the prime mover. I first heard of a definite plan at a luncheon at the Carlton Club on the 15th, at which were present Lord Halifax, M. Corbin, Sir Robert Vansittart, and one or two others. It was evident that there had been considerable discussion beforehand. On the 14th Vansittart and Desmond Morton had met M. Monnet and M. Pleven (members of the French Economic Mission in London), and been joined by General de Gaulle, who had flown over to make arrangements for shipping to carry the French Government and as many French troops as possible to Africa. These gentlemen had evolved the outline of a declaration for a Franco-British Union with the object, apart from its general merits, of giving M. Reynaud some new fact of a vivid and stimulating nature with which to carry a majority of his Cabinet into the move to Africa and the continuance of the war. My first reaction was unfavourable. I asked a number of questions of a critical character, and was by no means convinced. However, at the end of our long Cabinet that afternoon the subject was raised. I was somewhat surprised to see the staid, stolid, experienced politicians of all parties engage themselves so passionately

180

in an immense design whose implications and consequences were not in any way thought out. I did not resist, but yielded easily to these generous surges which carried our resolves to a very high level of unselfish and undaunted action.

When the War Cabinet met the next morning we first addressed ourselves to the answer to be given to M. Reynaud's request sent the night before the formal release of France from her obligations under the Anglo-French Agreement. The Cabinet authorised the following reply, which at their request I went into the next room and drafted myself. It was dispatched from London at 12.35 p.m. on the 16th. It endorsed and repeated in a formal manner the telephoned instructions sent to Campbell early in the morning.

*Foreign Office to Sir R. Campbell*
Please give M. Reynaud the following message, which has been approved by the Cabinet:

*Mr. Churchill to M. Reynaud*             16 June, 1940, 12.35 p.m.
Our agreement forbidding separate negotiations, whether for armistice or peace, was made with the French Republic, and not with any particular French Administration or statesman. It therefore involves the honour of France. Nevertheless, *provided, but only provided, that the French Fleet is sailed forthwith for British harbours pending negotiations,* His Majesty's Government give their full consent to an inquiry by the French Government to ascertain the terms of an armistice for France. His Majesty's Government, being resolved to continue the war, wholly exclude themselves from all part in the above-mentioned inquiry concerning an armistice.

Early in the afternoon a second message in similar terms was sent by the Foreign Office to Sir Ronald Campbell (June 16, 3.10 p.m.).
Both messages were stiff, and embodied the main purpose of the War Cabinet at their morning meeting.

*Foreign Office to Sir R. Campbell*
You should inform M. Reynaud as follows:
We expect to be consulted as soon as any armistice terms are received. This is necessary not merely in virtue of treaty forbidding separate peace or armistice, but also in view of vital consequences of any armistice to ourselves, having regard especially to the fact that British troops are fighting with French Army. You should impress on

French Government that in stipulating for removal of French Fleet to British ports we have in mind French interests as well as our own, and are convinced that it will strengthen the hands of the French Government in any armistice discussions if they can show that the French Navy is out of reach of the German forces. As regards the French Air Force, we assume that every effort will be made to fly it to North Africa, unless indeed the French Government would prefer to send it to this country.

We count on the French Government doing all they can both before and during any armistice discussions to extricate the Polish, Belgian, and Czech troops at present in France, and to send them to North Africa. Arrangements are being made to receive Polish and Belgian Governments in this country.

\* \* \* \* \*

We reassembled at 3 p.m. that same afternoon. I recalled to the Cabinet that at the conclusion of our meeting the day before there had been some discussion on a proposal for the issue of some further declaration of closer union between France and Great Britain. I had seen General de Gaulle in the morning, and he had impressed on me that some dramatic move was essential to give M. Reynaud the support which he needed to keep his Government in the war, and suggested that a proclamation of the indissoluble union of the French and British peoples would serve the purpose. Both General de Gaulle and M. Corbin had been concerned at the sharpness of the decision reached by the War Cabinet that morning, and embodied in the telegrams already dispatched. I had heard that a new declaration had been drafted for consideration, and that General de Gaulle had telephoned to M. Reynaud. As a result it had seemed advisable to suspend action for the moment. A telegram had therefore been sent to Sir Ronald Campbell instructing him to suspend delivery accordingly.

The Foreign Secretary then said that after our morning meeting he had seen Sir Robert Vansittart, whom he had previously asked to draft some dramatic announcement which might strengthen M. Reynaud's hand. Vansittart had been in consultation with General de Gaulle, M. Monnet, M. Pleven, and Major Morton. Between them they had drafted a proclamation. General de Gaulle had impressed upon them the need for publishing the document as quickly as possible, and wished to take the draft back with him to France that night. De Gaulle had also suggested that I should go to meet M. Reynaud next day.

The draft statement was passed round, and everyone read it with deep attention. All the difficulties were immediately apparent, but in the end a Declaration of Union seemed to command general assent. I stated that my first instinct had been against the idea, but that in this crisis we must not let ourselves be accused of lack of imagination. Some dramatic announcement was clearly necessary to keep the French going. The proposal could not be lightly turned aside, and I was encouraged at finding so great a body of opinion in the War Cabinet favourable to it.

At 3.55 p.m. we were told that the French Council of Ministers would meet at 5 p.m. to decide whether further resistance was possible. Secondly, General de Gaulle had been informed by M. Reynaud on the telephone that if a favourable answer on the proposed proclamation of unity was received by 5 p.m. M. Reynaud felt he could hold the position. On this the War Cabinet approved the final draft proclamation of an Anglo-French Union, and authorised its dispatch to M. Reynaud by the hand of General de Gaulle. This was telephoned to M. Reynaud forthwith. The War Cabinet further invited me, Mr. Attlee, and Sir Archibald Sinclair, representing the three British parties, to meet M. Reynaud at the earliest moment to discuss the draft proclamation and related questions.

Here is the final draft:

## DECLARATION OF UNION

At this most fateful moment in the history of the modern world the Governments of the United Kingdom and the French Republic make this declaration of indissoluble union and unyielding resolution in their common defence of justice and freedom against subjection to a system which reduces mankind to a life of robots and slaves.

The two Governments declare that France and Great Britain shall no longer be two nations, but one Franco-British Union.

The constitution of the Union will provide for joint organs of defence, foreign, financial, and economic policies.

Every citizen of France will enjoy immediately citizenship of Great Britain; every British subject will become a citizen of France.

Both countries will share responsibility for the repair of the devastation of war, wherever it occurs in their territories, and the resources of both shall be equally, and as one, applied to that purpose.

During the war there shall be a single War Cabinet, and all the forces of Britain and France, whether on land, sea, or in the air, will be placed under its direction. It will govern from wherever it best can. The two Parliaments will be formally associated. The nations of the British Empire are already forming new armies. France will keep her available forces in the field, on the sea, and in the air. The Union appeals to the United States to fortify the economic resources of the Allies, and to bring her powerful material aid to the common cause.

The Union will concentrate its whole energy against the power of the enemy, no matter where the battle may be.

And thus we shall conquer.

Of all this Parliament was informed in due course. But the issue by then had ceased to count.

I did not, as has been seen, draft the statement myself. It was composed around the table, and I made my contribution to it. I then took it into the next room, where de Gaulle was waiting with Vansittart, Desmond Morton, and M. Corbin. The General read it with an air of unwonted enthusiasm, and, as soon as contact with Bordeaux could be obtained, began to telephone it to M. Reynaud. He hoped with us that this solemn pledge of union and brotherhood between the two nations and Empires would give the struggling French Premier the means to carry his Government to Africa with all possible forces and order the French Navy to sail for harbours outside impending German control.

\* \* \* \* \*

We must now pass to the other end of the wire. The British Ambassador delivered the two messages in answer to the French request to be released from their obligation of March 28. According to his account, M. Reynaud, who was in a dejected mood, did not take them well. He at once remarked that the withdrawal of the French Mediterranean Fleet to British ports would invite the immediate seizure of Tunis by Italy, and also create difficulties for the British Fleet. He had got no further than this when my message, telephoned by General de Gaulle, came through. "It acted," said the Ambassador, "like a tonic." Reynaud said that for a document like that he would fight to the last. In came at that moment M. Mandel and M. Marin. They obviously were equally relieved. M. Reynaud then left "with a light step" to read the document to the President of the Republic. He believed that, armed with this immense guarantee, he would be able to carry his Council with him on the policy of retiring to Africa and waging war. My telegram instructing the Ambassador to delay the presentation of the two stiff messages, or anyhow to suspend action upon them, arrived immediately after the Premier had gone. A messenger was therefore sent after him to say that the two earlier messages should be considered as "cancelled". "Suspended" would have been a better word. The War Cabinet had not altered its position in any respect. We felt however that it would be better to give the "Declaration of Union" its full chance under the most favourable conditions. If the French Council of Ministers were rallied by it, the greater would carry the less, and the removal of the Fleet from German power would follow automatically. If our offer did not find favour our rights and claims would revive in their full force. We could not tell what was going on inside the French Government, nor know that this was the last time we should ever be able to deal with M. Reynaud.

I had spoken to him on the telephone some time this day proposing that I should come out immediately to see him. In view of the uncertainty about what was happening or about to happen at Bordeaux, my colleagues in the War Cabinet wished me to go in a cruiser, and a rendezvous was duly arranged for the next day off the Brittany coast. I ought to have flown. But even so it would have been too late.

The following was sent from the Foreign Office:

*To Sir R. Campbell, Bordeaux*                     June 16, 6.45 p.m.
The P.M., accompanied by the Lord Privy Seal, Secretary of State for Air, and three Chiefs of Staff and certain others, arrives at Concarneau at twelve noon to-morrow, the 17th, in a cruiser for a meeting with M. Reynaud. General de Gaulle has been informed of the above, and has expressed the view that time and rendezvous would be convenient. We suggest the meeting be held on board as arousing less attention. H.M.S. *Berkeley* has been warned to be at the disposal of M. Reynaud and party if desired.

And also from the Foreign Secretary by telephone at 8 p.m., June 16:

Following is reason why you have been asked to suspend action on my last two telegrams.

After consultation with General de Gaulle, P.M. has decided to meet M. Reynaud to-morrow in Brittany to make a further attempt to dissuade the French Government from asking for an armistice. For this purpose, on the advice of General de Gaulle, he will offer to M. Reynaud to join in issuing forthwith a declaration announcing immediate constitution of closest Anglo-French Union in all spheres in order to carry on the war. Text of draft declaration as authorised by H.M.G. is contained in my immediately following telegram. You should read this text to M. Reynaud at once.

An outline of this proposed declaration has already been telephoned by General de Gaulle to M. Reynaud, who has replied that such a declaration by the two Governments would make all the difference to the decision of the French Government. General is returning to-night with copy.

Our War Cabinet sat until six o'clock on the 16th, and thereafter I set out on my mission. I took with me the Leaders of the Labour and Liberal Parties, the three Chiefs of Staff, and various important officers and officials. A special train was waiting at Waterloo. We could reach Southampton in two hours, and a night of steaming at thirty knots in the cruiser would bring us to the rendezvous by noon on the 17th. We had taken our seats in the train. My wife had come to see me off. There was an odd delay in starting. Evidently some hitch had occurred. Presently my private secretary arrived from Downing Street breathless with the following message from Campbell at Bordeaux:

Ministerial crisis has opened. . . . Hope to have news by midnight. Meanwhile meeting arranged for to-morrow impossible.

On this I returned to Downing Street with a heavy heart.

\* \* \* \* \*

The final scene in the Reynaud Cabinet was as follows.

The hopes which M. Reynaud had founded upon the Declaration of Union were soon dispelled. Rarely has so generous a proposal encountered such a hostile reception. The Premier read the document twice to the Council. He declared himself strongly for it, and added that he was arranging a meeting with me for the next day to discuss the details. But the agitated Ministers, some famous, some nobodies, torn by division and under the terrible hammer of defeat, were staggered. Some, we are told, had heard about it by a tapping of telephones. These were the defeatists. Most were wholly unprepared to receive such far-reaching themes. The overwhelming feeling of the Council was to reject the whole plan. Surprise and mistrust dominated the majority, and even the most friendly and resolute were baffled. The Council had met expecting to receive the answer to the French request, on which they had all agreed, that Britain should release France from her obligations of March 28, in order that the French might ask the Germans what their terms of armistice would be. It is possible, even probable, that if our formal answer had been laid before them the majority would have accepted our primary condition about sending their Fleet to Britain, or at least would have made some other suitable proposal and thus have freed them to open negotiations with the enemy, while reserving to themselves a final option of retirement to Africa if the German conditions were too severe. But now there was a classic example of "Order, counter-order, disorder".

Paul Reynaud was quite unable to overcome the unfavourable impression which the proposal of Anglo-French Union created. The defeatist section, led by Marshal Pétain, refused even to examine it. Violent charges were made. It was "a last-minute plan", "a surprise", "a scheme to put France in tutelage, or to carry off her colonial empire". It relegated France, so they said, to 'he position of a Dominion. Others complained that not even eq' ality of status was offered to the French, because Frenchmen were to receive only the citizenship of the British Empire instead of that of Great Britain, while the British were to be citizens of France. This suggestion is contradicted by the text.

Beyond these came other arguments. Weygand had convinced Pétain without much difficulty that England was lost. High French military authorities had advised: "In three weeks England will have her neck wrung like a chicken." To make a union with Great Britain was, according to Pétain, "fusion with a corpse". Ybarnegaray, who had been so stout in the previous war, exclaimed: "Better be a Nazi province. At least we know what that means." Senator Reibel, a personal friend of General Weygand's, declared that this scheme meant complete destruction for France, and anyhow definite subordination to England. In vain did Reynaud reply: "I prefer to collaborate with my allies rather than with my enemies." And

Mandel: "Would you rather be a German district than a British Dominion?" All was in vain.

We are assured that Reynaud's statement of our proposal was never put to a vote in the Council. It collapsed of itself. This was a personal and fatal reverse for the struggling Premier which marked the end of his influence and authority upon the Council. All further discussion turned upon the armistice and asking the Germans what terms they would give, and in this M. Chautemps was cool and steadfast. Our two telegrams about the Fleet were never presented to the Council. The demand that it should be sailed to British ports as a prelude to the negotiations with the Germans was never considered by the Reynaud Cabinet, which was now in complete decomposition. At about eight o'clock Reynaud, utterly exhausted by the physical and mental strain to which he had for so many days been subjected, sent his resignation to the President, and advised him to send for Marshal Pétain. This action must be judged precipitate. He still seems to have cherished the hope that he could keep his rendezvous with me the next day, and spoke of this to General Spears. "To-morrow there will be another Government, and you will no longer speak for anyone," said Spears.

According to Campbell (sent by telephone, June 16):

M. Reynaud, who had been so heartened this afternoon by P.M.'s magnificent message, told us later that forces in favour of ascertaining terms of armistice had become too strong for him. He had read the message twice to Council of Ministers and explained its import and the hope which it held out for the future. It had been of no avail.

We worked on him for half an hour, encouraging him to try to get rid of the evil influences among his colleagues. After seeing M. Mandel for a moment we then called for second time to-day on the President of Senate, M. Jeanneney, whose views (like those of President of Chamber) are sound, in hope of his being able to influence President of Republic to insist on M. Reynaud forming new Government.

We begged him to make it very clear to President that offer contained in P.M.'s message would not be extended to a Government which entered into negotiation with enemy.

An hour or so later M. Reynaud informed us that he was beaten and had handed in his resignation. Combination of Marshal Pétain and General Weygand (who were living in another world and imagined they could sit round a green table discussing armistice terms in the old manner) had proved too much for weak members of Government, on whom they worked by waving the spectre of revolution.

\* \* \* \* \*

On the afternoon of June 16 M. Monnet and General de Gaulle visited me in the Cabinet Room. The General in his capacity of Under-Secretary of State for National Defence had just ordered the French ship *Pasteur*, which was carrying weapons to Bordeaux from America, to proceed instead to a British port. Monnet was very active upon a plan to transfer all French contracts for munitions in America to Britain if France made a separate peace. He evidently expected this, and wished to save as much as possible from what seemed to him to be the wreck of the world. His whole attitude in this respect was most helpful. Then he turned to our sending all our remaining fighter air squadrons to share in the final battle in France, which was of course already over. I told him that there was no possibility of this being done. Even at this stage he used the usual arguments—"the decisive battle", "now or never", "if France falls all falls", and so forth. But I could not do anything to oblige him in this field. My two French visitors then got up and moved towards the door, Monnet leading. As they reached it, de Gaulle, who had hitherto scarcely uttered a single word, turned back, and, taking two or three paces towards me, said in English: "I think you are quite right." Under an impassive, unperturbable demeanour he seemed to me to have a remarkable capacity for feeling pain. I preserved the impression, in contact with this very tall, phlegmatic man: "Here is the Constable of France." He returned that afternoon in a British aeroplane, which I had placed at his disposal, to Bordeaux. But not for long.

\* \* \* \* \*

Forthwith Marshal Pétain formed a French Government with the main purpose of seeking an immediate armistice from Germany. Late on the night of June 16 the defeatist group of which he was the head was already so shaped and knit together that the process did not take long. M. Chautemps ("to ask for terms is not necessarily to accept them") was Vice-President of the Council. General Weygand, whose view was that all was over, held the Ministry of National Defence. Admiral Darlan was Minister of Marine, and M. Baudouin Minister for Foreign Affairs.

The only hitch apparently arose over M. Laval. The Marshal's first thought had been to offer him the post of Minister of Justice. Laval brushed this aside with disdain. He demanded the Ministry of Foreign Affairs, from which position alone he conceived it possible to carry out his plan of reversing the alliances of France, finishing up England, and joining as a minor partner the New Nazi Europe. Marshal Pétain surrendered at once to the vehemence of this formidable personality. M. Baudouin, who had already undertaken the Foreign Office, for which he knew himself to be utterly inadequate, was quite ready to give it up. But when he mentioned the fact to M. Charles-Roux, Permanent Under-Secretary to the Ministry of Foreign Affairs, the latter was indignant. He enlisted the support of Weygand. When Weygand entered the room and addressed the illustrious Marshal, Laval became so furious that both the military chiefs were overwhelmed. The permanent official however refused point-blank to serve under Laval. Confronted with this, the Marshal again subsided, and after a violent scene Laval departed in wrath and dudgeon.

This was a critical moment. When, four months later, on October 28, Laval eventually became Foreign Minister there was a new consciousness of military values. British resistance to Germany was by then a factor. Apparently the Island could not be entirely discounted. Anyhow, its neck had not been "wrung like a chicken's in three weeks". This was a new fact; and a fact at which the whole French nation rejoiced.

\*    \*    \*    \*    \*

Our telegram of the 16th had made our assent to inquiries about an armistice conditional upon the sailing of the French Fleet to British harbours. It had already been presented formally to Marshal Pétain. The War Cabinet approved, at my suggestion, a further message emphasising the point. But we were talking to the void.

On the 17th also I sent a personal message to Marshal Pétain and General Weygand, of which copies were to be furnished by our Ambassador to the French President and Admiral Darlan:

I wish to repeat to you my profound conviction that the illustrious Marshal Pétain and the famous General Weygand, our comrades in two great wars against the Germans, will not injure their ally by delivering over to the enemy the fine French Fleet. Such an act would scarify their names for a thousand years of history. Yet this result may easily come by frittering away these few precious hours when the Fleet can be sailed to safety in British or American ports, carrying with it the hope of the future and the honour of France.

In order that these appeals might not lack personal reinforcement on the spot, we sent the First Sea Lord, who believed himself to be in intimate personal and professional touch with Admiral Darlan, the First Lord, Mr. A. V. Alexander, and Lord Lloyd, Secretary of State for the Colonies, so long known as a friend of France. All these three laboured to make what contacts they could with the new Ministers during the 19th. They received many solemn assurances that the Fleet would never be allowed to fall into German hands. But no more French warships moved beyond the reach of the swiftly approaching German power.

\*    \*    \*    \*    \*

At the desire of the Cabinet I had broadcast the following statement on the evening of June 17:

The news from France is very bad, and I grieve for the gallant French people who have fallen into this terrible misfortune. Nothing will alter our feelings towards them or our faith that the genius of France will rise again. What has happened in France makes no difference to our actions and purpose. We have become the sole champions now in arms to defend the world cause. We shall do our best to be worthy of this high honour. We shall defend our Island home, and with the British Empire we shall fight on unconquerable until the curse of Hitler is lifted from the brows of mankind. We are sure that in the end all will come right.

\*    \*    \*    \*    \*

On the morning of the 17th I mentioned to my colleagues in the Cabinet a telephone conversation which I had had during the night with General Spears, who said he did not think he could perform any useful service in the new structure at Bordeaux. He spoke with some anxiety about the safety of General de Gaulle. Spears had apparently been warned that as things were shaping it might be well for de Gaulle to leave France. I readily assented to a good plan being made for this. So that very morning—the 17th—de Gaulle went to his office in Bordeaux, made a number of engagements for the afternoon, as a blind, and then drove to the airfield with his friend Spears to see him off. They shook hands and said good-bye, and as the plane began to move de Gaulle stepped in and slammed the door. The machine soared off into the air, while the French police and officials gaped. De Gaulle carried with him, in this small aeroplane, the honour of France.

That same evening he made his memorable broadcast to the French people. One passage should be quoted here:

France is not alone. She has a vast Empire behind her. She can unite with the British Empire, which holds the seas, and is continuing the struggle. She can utilise to the full, as England is doing, the vast industrial resources of the United States.

Other Frenchmen who wished to fight on were not so fortunate. When the Pétain Government was formed the plan of going to Africa to set up a centre of power outside German control was still open. It was discussed at a meeting of the Pétain Cabinet on June 18. The same evening President Lebrun, Pétain, and the Presidents of the Senate and the Chamber met together. There seems to have been general agreement at least to send a representative body to North Africa. Even the Marshal was not hostile. He himself intended to stay, but saw no reason why Chautemps, Vice-President of the Council, should not go and act in his name. When rumours of an impending exodus ran round agitated Bordeaux Weygand was hostile. Such a move, he thought, would wreck the "honourable" armistice negotiations which had already been begun by way of Madrid, on French initiative, on June 17. Laval was deeply alarmed. He feared that the setting up of an effective resistance administration outside France would frustrate the policy on which he was resolved. He set to work on the clusters of Deputies and Senators crowded into Bordeaux.

Darlan, as Minister of Marine, took a different view. To pack off all the principal critics of his conduct in a ship seemed at the moment to him a most convenient solution of many difficulties. Once aboard, all those who went would be in his power and there would be plenty of time for the Government to settle what to do. With the approval of the new Cabinet, he offered passages on the armed auxiliary cruiser *Massilia* to all political figures of influence who wished to go to Africa. The ship was to sail from the mouth of the Gironde on the 20th. Many however who had planned to go to Africa, including Jeanneney and Herriot, suspected a trap, and preferred to travel overland through Spain. The final party, apart from refugees, consisted of twenty-four Deputies and one Senator, and included Mandel, Campinchi, and Daladier, who had all been actively pressing for the move to Africa. On the afternoon of the 21st the *Massilia* sailed. On the 23rd the ship's radio announced that the Pétain Government had accepted and signed the armistice with Germany. Campinchi immediately tried to persuade the captain to set his course for England, but this officer no doubt had his instructions and met his former political chief of two days before with a bleak refusal. The unlucky band of patriots passed anxious hours, till on the evening of June 24 the *Massilia* anchored at Casablanca.

Mandel now acted with his usual decision. He had with Daladier drafted a proclamation setting up a resistance administration in North Africa with himself as Premier. He went on shore, and, after calling on the British Consul, established himself at the Hôtel Excelsior. He then attempted to send his proclamation out through the Havas Agency. When General Noguès read its text he was disturbed. He intercepted the message, and it was telegraphed not to the world but to Darlan and Pétain. They had now made up their minds to have no alternative and potentially rival Government outside German power. Mandel was arrested at his hotel and brought before the local court, but the magistrate, afterwards dismissed by Vichy, declared there was no case against him and set him free. He was however, by the orders of Governor-General Noguès, re-arrested and put back on the *Massilia*, which henceforth was detained in the harbour under strict control without its passengers having any communication with the shore.

Without of course knowing any of the facts here set forth, I was already concerned about the fate of Frenchmen who wished to fight on.

*Prime Minister to General Ismay*                    24.VI.40

It seems most important to establish now before the trap closes an organisation for enabling French officers and soldiers, as well as important technicians, who wish to fight, to make their way to various ports. A sort of "underground railway" as in the olden days of slavery should be established and a Scarlet Pimpernel organisation set up. I have no doubt there will be a steady flow of determined men, and we need all we can get for the defence of the French colonies. The Admiralty and Air Force must co-operate. General de Gaulle and his Committee would of course be the operative authority.

At our meeting of the War Cabinet late at night on June 25 we heard among other things that a ship with a large number of prominent French politicians on board had passed Rabat. We decided to establish contact with them at once. Mr. Duff Cooper, the Minister of Information, accompanied by Lord Gort, started for Rabat at dawn in a Sunderland flying-boat. They found the town in mourning. Flags were flying at half-mast, church bells were tolling, and a solemn service was taking place in the cathedral to bewail the defeat of France. All their attempts to get in touch with Mandel were prevented. The Deputy-Governor, named Morice, declared not only on the telephone, but in a personal interview which Duff Cooper demanded, that he had no choice but to obey the orders of his superiors. "If General Noguès tells me to shoot myself I will gladly obey. Unfortunately, the orders he has given me are more cruel." The former French Ministers and Deputies were in fact to be treated as escaped prisoners. Our mission had no choice but to return the way they came. A few days later (July 1) I gave instructions to the Admiralty to try to cut out the *Massilia* and rescue those on board. No plan could however be made, and for nearly three weeks she lay under the batteries of Casablanca, after which the whole party were brought back to France and disposed of as the Vichy Government thought convenient to themselves and agreeable to their German masters. Mandel began his long and painful internment which ended in his murder by German orders at the end of 1944. Thus perished the hope of setting up a strong representative French Government, either in Africa or in London.

*       *       *       *       *

Although vain, the process of trying to imagine what would have happened if some important event or decision had been different is often tempting and sometimes instructive. The manner of the fall of France was decided on June 16 by a dozen chances, each measured by a hair's-breadth. If Paul Reynaud had survived the 16th I should have been with him at noon on the 17th, accompanied by the most powerful delegation that has ever left our shores, armed with plenary powers in the name of the British nation. Certainly we should have confronted Pétain, Weygand, Chautemps, and the rest with our blunt proposition: "No release from the obligation of March 28 unless the French Fleet is sailed to British ports. On the other hand, we offer an indissoluble Anglo-French Union. Go to Africa and let us fight it out together." Surely we should have been aided by the President of the Republic, by the Presidents of the two French Chambers, and by all that resolute band who gathered behind Reynaud, Mandel, and de Gaulle. It seems to me probable that we should have uplifted and converted the defeatists round the table, or left them in a minority or even under arrest.

But let us pursue this ghostly speculation further. The French Government would have retired to North Africa. The Anglo-French Super-State or Working Committee, to which it would probably in practice have reduced itself, would have faced Hitler. The British and French Fleets from their harbours would have enjoyed complete mastery of the Mediterranean, and free passage through it for all troops and supplies. Whatever British air force could be spared from the defence of Britain, and what was left of the French Air Force, nourished by American production and based on the French North African airfields, would soon have become an offensive factor of the first importance. Malta, instead of being for so long a care and peril, would at once have taken its place as our most active naval base. Italy could have been attacked with heavy bombing from Africa far easier than from England. Her communications with the Italian armies in Libya and Tripolitania would have been effectively severed. Using no more fighter aircraft than we actually employed in the defence of Egypt and sending to the Mediterranean theatre no more troops than we actually sent, or held ready to send, we might well, with the remains of the French Army, have transferred the war from the East to the Central Mediterranean, and during 1941 the entire North African shore might have been cleared of Italian forces.

France would never have ceased to be one of the principal belligerent allies, and would have been spared the fearful schism which rent and still rends her people. Her homeland no doubt would have lain prostrate under the German rule, but that was only what actually happened after the Anglo-American descent in November 1942. Now that the whole story is before us no one can doubt that the armistice did not spare France a pang.

It is still more shadowy to guess what Hitler would have done. Would he have forced his way through Spain, with or without Spanish agreement, and, after assaulting and perhaps capturing Gibraltar, have invaded Tangier and Morocco? This was an area which deeply concerned the United States, and was ever prominent in President Roosevelt's mind. How could Hitler have made this major attack through Spain on Africa and yet have fought the Battle of Britain? He would have had to choose. If he chose Africa we, with the command of the sea and the French bases, could have moved both troops and air forces into Morocco and Algeria quicker than he, and in greater strength. We should certainly have welcomed in the autumn and winter of 1940 a vehement campaign in or from a friendly French North-West Africa.

Surveying the whole scene in the after-light, it seems unlikely that Hitler's main decision and the major events of the war, namely the Battle of Britain and the German surge to the East, would have been changed by the retirement of the French Government to North Africa. After the fall of Paris, when Hitler danced his jig of joy, he naturally dealt with very large propositions. Once France was prostrate he must if possible conquer or destroy Great Britain. His only other choice was Russia. A major operation through Spain into North-West Africa would have prejudiced both these tremendous adventures, or at least have prevented his attack on the Balkans. I have no doubt that it would have been better for all the Allies if the French Government had gone to North Africa; and that this would have remained true whether Hitler followed them and us thither or not.

One day when I was convalescing at Marrakesh in January 1944 General Georges came to luncheon. In the course of casual conversation I aired the fancy that perhaps the French Government's failure to go to Africa in June 1940 had all turned out for the best. At the Pétain trial in August 1945 the General thought it right to state this in evidence. I make no complaint, but my retrospective speculation on this occasion does not represent my considered opinion either during the war or now.

# CHAPTER XI

# ADMIRAL DARLAN AND THE
# FRENCH FLEET. ORAN

*Would Britain Surrender? – My Speech of June 18 – "Their Finest Hour" – Admiral Darlan's Opportunity – His Final Letter to Me – Armistice, Article 8 – A Dire Decision – Operation "Catapult": Zero Day, July 3 – Our Terms to the French – The Tragedy of Oran – My Report to Parliament, July 4 – World Impression on Elimination of French Navy.*

AFTER the collapse of France the question which arose in the minds of all our friends and foes was: "Will Britain surrender too?" So far as public statements count in the teeth of events, I had in the name of His Majesty's Government repeatedly declared our resolve to fight on alone. After Dunkirk on June 4 I had used the expression "if necessary for years, *if necessary alone*". This was not inserted without design, and the French Ambassador in London had been instructed the next day to inquire what I actually meant. He was told "exactly what was said". I could remind the House of my remark when I addressed it on June 18, the morrow of the Bordeaux collapse. I then gave "some indication of the solid practical grounds on which we based our inflexible resolve to continue the war". I was able to assure Parliament that our professional advisers of the three Services were confident that there were good and reasonable hopes of ultimate victory. I told them that I had received from all the four Dominions Prime Ministers messages in which they endorsed our decision to fight on and declared themselves ready to share our fortunes. "In casting up this dread balance-sheet and contemplating our dangers with a disillusioned eye I see great reasons for vigilance and exertion, but none whatever for panic or fear." I added: "During the first four years of the last war the Allies experienced nothing but disaster and disappointment. . . . We repeatedly asked ourselves the question 'How are we going to win?' and no one was ever able to answer it with much precision, until at the end, quite suddenly, quite unexpectedly, our terrible foe collapsed before us, and we were so glutted with victory that in our folly we threw it away.

"However matters may go in France or with the French Government or other French Governments, we in this Island and in the British Empire will never lose our sense of comradeship with the French people. . . . If final victory rewards our toils they shall share the gains—aye, and freedom shall be restored to all. We abate nothing of our just demands; not one jot or tittle do we recede. . . . Czechs, Poles, Norwegians, Dutch, Belgians, have joined their causes to our own. All these shall be restored." I ended:

"What General Weygand called the Battle of France is over. I expect that the Battle of Britain is about to begin. Upon this battle depends the survival of Christian civilisation. Upon it depends our own British life, and the long continuity of our institutions and our Empire. The whole fury and might of the enemy must very soon be turned on us. Hitler knows that he will have to break us in this Island or lose the war. If we can stand up to him, all Europe may be free and the life of the world may move forward into broad, sunlit uplands. But if we fail, then the whole world, including the United States, including all that we have known and cared for, will sink into the abyss of a new Dark Age, made more sinister, and perhaps more protracted, by the lights of perverted science. Let us therefore brace ourselves to our duties, and so bear ourselves that, if the British Empire and its Commonwealth last for a thousand years, men will still say: 'This was their finest hour.' "

All these often-quoted words were made good in the hour of victory. But now they were only words. Foreigners who do not understand the temper of the British race all over the globe when its blood is up might have supposed that they were only a bold front, set up as a good prelude for peace negotiations. Hitler's need to finish the war in the West was obvious. He was in a position to offer the most tempting terms. To those who like myself had studied his moves it did not seem impossible that he would consent to leave Britain and her Empire and Fleet intact and make a peace which would have secured him that free hand in the East of which Ribbentrop had talked to me in 1937, and which was his heart's main desire. So far we had not done him much harm. We had indeed only added our own defeat to his triumph over France. Can one wonder that astute calculators in many countries, ignorant as they mostly were of the problems of overseas invasion, and of the quality of our Air Force, and who dwelt under the overwhelming impression of German might and terror, were not convinced? Not every Government called into being by Democracy or by Despotism, and not every nation, while quite alone, and as it seemed abandoned, would have courted the horrors of invasion and disdained a fair chance of peace for which many plausible excuses could be presented. Rhetoric was no guarantee. Another administration might come into being. "The war-mongers have had their chance and failed." America had stood aloof. No one was under any obligation to Soviet Russia. Why should not Britain join the spectators who in Japan and in the United States, in Sweden, and in Spain, might watch with detached interest, or even relish, a mutually destructive struggle between the Nazi and Communist Empires? Future generations will find it hard to believe that the issues I have summarised here were never thought worth a place upon the Cabinet agenda, or even mentioned in our most private conclaves. Doubts could be swept away only by deeds. The deeds were to come.

\* \* \* \* \*

Meanwhile I telegraphed to Lord Lothian, who, at the desire of the United States naval authorities, had asked anxiously whether ammunition for the British Fleet and material for its repair ought not to be sent from England across the Atlantic:

22.VI.40

There is no warrant for such precautions at the present time.

I also sent the following telegrams to my Dominions friends.

To Mr. Mackenzie King:

24.VI.40

If you will read again my telegram of June 5 you will see that there is no question of trying to make a bargain with the United States about their entry into the war and our dispatch of the Fleet across the Atlantic should the Mother Country be defeated. On the contrary, I doubt very much the wisdom of dwelling upon the last contingency at the present time. I have good confidence in our ability to defend this Island, and I see no reason to make preparation for or give any countenance to the transfer of the British Fleet. I shall myself never enter into any peace negotiations with Hitler, but obviously I cannot bind a future Government, which, if we were deserted by the United States and beaten down here, might very easily be a kind of Quisling affair ready to accept German overlordship and protection. It would be a help if you would impress this danger upon the President, as I have done in my telegrams to him.

All good wishes, and we are very glad your grand Canadian Division is with us in our fight for Britain.

To Smuts I cabled again:

27.VI.40

Obviously, we have first to repulse any attack on Great Britain by invasion, and show ourselves able to maintain our development of air-power. This can only be settled by trial. *If Hitler fails to beat us here he will probably recoil eastwards. Indeed, he may do this even without trying invasion*, to find employment for his Army, and take the edge off the winter strain upon him.

I do not expect the winter strain will prove decisive, but to try to hold all Europe down in a starving condition with only Gestapo and military occupation *and no large theme appealing to the masses* is not an arrangement which can last long.

Development of our air-power, particularly in regions unaffected by bombing, should cause him ever-increasing difficulties, possibly decisive difficulties, in Germany, no matter what successes he has in Europe *or Asia*.

Our large Army now being created for Home Defence is being formed on the principle of attack, *and opportunity for large-scale offensive amphibious operations may come in 1940 and 1941*. We are still working on the 55-division basis here, but as our munitions supply expands and Empire resources are mobilised larger numbers may be possible. After all, we are now at last on interior lines. Hitler has vast hungry areas to defend, and we have the command of the seas. Choice of objectives in Western Europe is therefore wide.

I send you these personal notes in order to keep in closest contact with your thoughts, which ever weigh with me.

It was with good confidence that we entered upon the supreme test.

*Prime Minister to Lord Lothian* (*Washington*)                    28.VI.40
   No doubt I shall make some broadcast presently, but I don't think
words count for much now. Too much attention should not be paid
to eddies of United States opinion. Only force of events can govern
them. Up till April they were so sure the Allies would win that they
did not think help necessary. Now they are so sure we shall lose that
they do not think it possible. I feel good confidence we can repel
invasion and keep alive in the air. Anyhow, we are going to try.
Never cease to impress on President and others that if this country
were successfully invaded and largely occupied after heavy fighting
some Quisling Government would be formed to make peace on the
basis of our becoming a German Protectorate. In this case the British
Fleet would be the solid contribution with which this Peace Government would buy terms. Feeling in England against United States
would be similar to French bitterness against us now. We have really
not had any help worth speaking of from the United States so far.
[The rifles and field-guns did not arrive till the end of July. The
destroyers had been refused.] We know President is our best friend,
but it is no use trying to dance attendance upon Republican and
Democratic Conventions. What really matters is whether Hitler is
master of Britain in three months or not. I think not. But this is a
matter which cannot be argued beforehand. Your mood should be
bland and phlegmatic. No one is down-hearted here.

*        *        *        *        *

   In the closing days at Bordeaux Admiral Darlan became very
important. My contacts with him had been few and formal. I
respected him for the work he had done in re-creating the French
Navy, which after ten years of his professional control was more
efficient than at any time since the French Revolution. When in
December 1939 he had visited England we gave him an official
dinner at the Admiralty. In response to the toast, he began by
reminding us that his great-grandfather had been killed at the
Battle of Trafalgar. I therefore thought of him as one of those
good Frenchmen who hate England. Our Anglo-French naval
discussions in January had also shown how very jealous the
Admiral was of his professional position in relation to whoever
was the political Minister of Marine. This had become a positive
obsession, and, I believe, played a definite part in his action.
   For the rest, Darlan had been present at most of the conferences
which I have described, and as the end of the French resistance
approached he had repeatedly assured me that whatever hap-

pened the French Fleet should never fall into German hands. Now at Bordeaux came the fateful moment in the career of this ambitious, self-seeking, and capable Admiral. His authority over the Fleet was for all practical purposes absolute. He had only to order the ships to British, American, or French colonial harbours—some had already started—to be obeyed. In the morning of June 17, after the fall of M. Reynaud's Cabinet, he declared to General Georges that he was resolved to give the order. The next day Georges met him in the afternoon and asked him what had happened. Darlan replied that he had changed his mind. When asked why, he answered simply: "I am now Minister of Marine." This did not mean that he had changed his mind in order to become Minister of Marine, but that being Minister of Marine he had a different point of view.

How vain are human calculations of self-interest! Rarely has there been a more convincing example. Admiral Darlan had but to sail in any one of his ships to any port outside France to become the master of all French interests beyond German control. He would not have come, like General de Gaulle, with only an unconquerable heart and a few kindred spirits. He would have carried with him outside the German reach the fourth Navy in the world, whose officers and men were personally devoted to him. Acting thus, Darlan would have become the chief of the French Resistance with a mighty weapon in his hand. British and American dockyards and arsenals would have been at his disposal for the maintenance of his fleet. The French gold reserve in the United States would have assured him, once recognised, of ample resources. The whole French Empire would have rallied to him. Nothing could have prevented him from being the Liberator of France. The fame and power which he so ardently desired were in his grasp. Instead, he went forward through two years of worrying and ignominious office to a violent death, a dishonoured grave, and a name long to be execrated by the French Navy and the nation he had hitherto served so well.

*     *     *     *     *

There is a final note which should be struck at this point. In a letter which Darlan wrote to me on December 4, 1942, just three weeks before his assassination, he vehemently claimed that he had kept his word. As this letter states his case and should be on record, I print it here. It cannot be disputed that no French ship was ever manned by the Germans or used against us by them in the war. This was not entirely due to Admiral Darlan's measures; but he had certainly built up in the minds of the officers and men of the French Navy that at all costs their ships should be destroyed before being seized by the Germans, whom he disliked as much as he did the English.

*Admiral Darlan to Mr. Churchill**

Dear Mr. Prime Minister,                    ALGIERS, December 4, 1942

On June 12, 1940, at Briare, at the headquarters of General Weygand, you took me aside and said to me: "Darlan, I hope you will never surrender the Fleet." I answered you: "There is no question of doing so; it would be contrary to our naval traditions and honour." The First Lord of the Admiralty, Alexander, and the First Sea Lord, Pound, received the same reply on June 17, 1940, at Bordeaux, as did Lord Lloyd. If I did not consent to authorise the French Fleet to proceed to British ports, it was because I knew that such a decision would bring about the total occupation of Metropolitan France as well as North Africa.

I admit having been overcome by a great bitterness and a great resentment against England as the result of the painful events which touched me as a sailor; furthermore it seemed to me that you did not believe my word. One day Lord Halifax sent me word by M. Dupuy that in England my word was not doubted, but that it was believed that I should not be able to keep it. The voluntary destruction of the Fleet at Toulon has just proved that I was right, because even though I no longer commanded, the Fleet executed the orders which I had given and maintained, contrary to the wishes of the Laval Government. On the orders of my chief, the Marshal, I was obliged, from January 1941 to April 1942, to adopt a policy which would prevent France and its Empire from being occupied and crushed by the Axis Powers. This policy was by the force of events opposed to yours. What else could I do? At that time you were not able to help us, and any gesture towards you would have led to the most disastrous consequences for my country. If we had not assumed the obligation to defend the Empire by our own forces (I always refused German aid, even in Syria) the Axis would have come to Africa and our own Army would have been discarded; the First British Army undoubtedly would not be before Tunis to-day with French troops at its side to combat the Germans and Italians.

When the Allied Forces landed in Africa on November 8 I at first executed the orders I had received. Then as soon as this became

* Translated.

THE FALL OF FRANCE

impossible I ordered the cessation of the fighting in order to avoid unnecessary bloodshed and a fight which was contrary to the intimate sentiments of those engaged. Disavowed by Vichy and not wishing to resume the fight, I placed myself at the disposition of the American military authorities, only in that way being able to remain faithful to my oath. On November 11 I learned of the violation of the Armistice Convention by the Germans, the occupation of France, and the solemn protest of the Marshal. I then considered that I could resume my liberty of action, and that, remaining faithful to the person of the Marshal, I could follow that road which was most favourable to the welfare of the French Empire, that of the fight against the Axis. Supported by the high authorities of French Africa and by public opinion, and acting as the eventual substitute of the Chief of State, I formed the High Commissariat in Africa and ordered the French forces to fight at the side of the Allies. Since then French West Africa has recognised my authority. I should never have been able to accomplish this result if I had not acted under the aegis of the Marshal and if I were simply represented as a dissident. I have the conviction that all Frenchmen who now fight against Germany each in his own manner will finally achieve a general reconciliation, but I believe that for the moment they must continue their separate action. There is a certain resentment, notably in French West Africa, which is too active for me to obtain more, as you know. I follow my *rôle* without attacking anyone; I ask for reciprocity. For the moment the only thing that counts is to defeat the Axis; the French people when liberated will later choose their political *régime* and their leaders.

I thank you, Mr. Prime Minister, for having associated yourself with President Roosevelt in declaring that. like the United States, Great Britain wishes the integral re-establishment of French sovereignty as it existed in 1939. When my country has recovered its integrity and its liberty my only ambition will be to retire with the sentiment of having served it well.

Please accept, Mr. Prime Minister, the assurances of my highest consideration.
FRANÇOIS DARLAN, Admiral of the Fleet

\*    \*    \*    \*    \*

Those of us who were responsible at the summit in London understood the physical structure of our Island strength and were sure of the spirit of the nation. The confidence with which we faced the immediate future was not founded, as was commonly supposed abroad, upon audacious bluff or rhetorical appeal, but upon a sober consciousness and calculation of practical facts. When I spoke in the House of Commons I founded myself upon realities which I and others had carefully studied—some for many years. I will presently analyse in detail the Invasion problem as I and my expert advisers saw it in these memorable days. But first of all there was one step to take. It was obvious, and it was dire.

The addition of the French Navy to the German and Italian Fleets, with the menace of Japan measureless upon the horizon, confronted Great Britain with mortal dangers and gravely affected the safety of the United States. Article 8 of the Armistice prescribed that the French Fleet, except that part left free for safeguarding French colonial interests, "shall be collected in ports to be specified and there demobilised and disarmed under German or Italian control". It was therefore clear that the French war vessels would pass into that control while fully armed. It was true that in the same article the German Government solemnly declared that they had no intention of using them for their own purposes during the war. But who in his senses would trust the word of Hitler after his shameful record and the facts of the hour? Furthermore, the article excepted from this assurance "those units necessary for coast surveillance and mine-sweeping". The interpretation of this lay with the Germans. Finally, the Armistice could at any time be voided on any pretext of non-observance. There was in fact no security for us at all. At all costs, at all risks, in one way or another, we must make sure that the Navy of France did not fall into wrong hands, and then perhaps bring us and others to ruin.

The War Cabinet never hesitated. Those Ministers who, the week before, had given their whole hearts to France and offered common nationhood resolved that all necessary measures should be taken. This was a hateful decision, the most unnatural and painful in which I have ever been concerned. It recalled the episode of the seizure by the Royal Navy of the Danish fleet at Copenhagen in 1807; but now the French had been only yesterday our dear Allies, and our sympathy for the misery of France was sincere. On the other hand, the life of the State and the salvation of our cause were at stake. It was Greek tragedy. But no act was ever more necessary for the life of Britain and for all that depended upon it. I thought of Danton in 1793: "The coalesced Kings threaten us, and we hurl at their feet as a gage of battle the head of a King." The whole event was in this order of ideas.

*    *    *    *    *

The French Navy was disposed in the following manner: Two battleships, four light cruisers (or *contre-torpilleurs*), some submarines, including a very large one, the *Surcouf*, eight destroyers, and about two hundred smaller but valuable mine-sweeping and anti-submarine craft lay for the most part at Portsmouth and Plymouth. These were in our power. At Alexandria there were a French battleship, four French cruisers, three of them modern 8-inch-gun cruisers, and a number of smaller ships. These were covered by a strong British battle squadron. At Oran, at the other end of the Mediterranean, and at its adjacent military port of Mers-el-Kebir, were two of the finest vessels of the French fleet, the *Dunkerque* and the *Strasbourg*, modern battle-cruisers much superior to the *Scharnhorst* and *Gneisenau*, and built for the express purpose of being superior to them. These vessels in German hands on our trade routes would have been most disagreeable. With them were two French battleships, several light cruisers, and a number of destroyers, submarines, and other vessels. At Algiers were seven cruisers, of which four were 8-inch armed, and at Martinique an aircraft-carrier and two light cruisers. At Casablanca lay the *Jean Bart*, newly arrived from St. Nazaire, but without her guns. This was one of the key ships in the computation of world naval strength. She was unfinished, and could not be finished at Casablanca. She must not go elsewhere. The *Richelieu*, which was far nearer completion, had reached Dakar. She could steam, and her 15-inch guns could fire. There were many other French ships of minor importance in various ports. Finally, at Toulon a number of warships were beyond our reach. Operation "Catapult" comprised the simultaneous seizure, control, or effective disablement or destruction of all the accessible French Fleet.

*Prime Minister to General Ismay*                                      1.VII.40

1. The Admiralty are retaining *Nelson* and her four destroyers in home waters, and Operation "Catapult" should go forward, aiming at daybreak the 3rd.

2. During the night of 2nd-3rd all necessary measures should be taken at Portsmouth and Plymouth, at Alexandria, and if possible at Martinique, on the same lines as "Catapult". The reactions to these measures at Dakar and Casablanca must be considered, and every precaution taken to prevent the escape of valuable units.

On account of the pressure of events, I added also:

The Admiralty should endeavour to raise the flotillas in the narrow seas to a strength of forty destroyers, with additional cruiser support. An effort should be made to reach this strength during the next two or three days and hold it for the following fortnight, when the position can be reviewed. The losses in the Western Approaches must be accepted meanwhile. I should like also a daily return of the numbers of craft on patrol or available between Portsmouth and the Tyne.

\*     \*     \*     \*     \*

In the early morning of July 3 all the French vessels at Portsmouth and Plymouth were taken under British control. The action was sudden and necessarily a surprise. Overwhelming force was employed, and the whole transaction showed how easily the Germans could have taken possession of any French warships lying in ports which they controlled. In Britain the transfer, except in the *Surcouf*, was amicable, and the crews came willingly ashore. In the *Surcouf* two gallant British officers and a leading seaman were killed,\* and another seaman wounded. One French seaman also was killed, but many hundreds volunteered to join us. The *Surcouf*, after rendering distinguished service, perished on February 19, 1942, with all her gallant French crew.

\*     \*     \*     \*     \*

The deadly stroke was in the Western Mediterranean. Here, at Gibraltar, Vice-Admiral Somerville with "Force H", consisting of the battle-cruiser *Hood*, the battleships *Valiant* and *Resolution*, the aircraft-carrier *Ark Royal*, two cruisers, and eleven destroyers, received orders sent from the Admiralty at 2.25 a.m. on July 1:

Be prepared for "Catapult" July 3.

Among Somerville's officers was Captain Holland, a gallant and distinguished officer, lately Naval Attaché in Paris and with keen French sympathies, who was influential. In the early afternoon of July 1 the Vice-Admiral telegraphed:

After talk with Holland and others Vice-Admiral "Force H" is impressed with their view that the use of force should be avoided at all costs. Holland considers offensive action on our part would alienate all French wherever they are.

To this the Admiralty replied at 6.20 p.m.:

Firm intention of H.M.G. that if French will not accept any of your alternatives they are to be destroyed.

\* Commander D. V. Sprague, R.N., Lieutenant P. M. K. Griffiths, R.N., and Leading Seaman A. Webb, R.N.

Shortly after midnight (1.08 a.m., July 2) Admiral Somerville was sent the following carefully conceived text of the communication to be made to the French admiral:

His Majesty's Government have sent me to inform you as follows:

They agreed to the French Government approaching the German Government only on the condition that before an armistice was concluded the French Fleet should be sent to British ports to prevent it falling into the hands of the enemy. The Council of Ministers declared on 18th June that before capitulation on land the French Fleet would join up with the British Navy or sink itself.*

While the present French Government may consider that the terms of their armistices with Germany and Italy are reconcilable with these undertakings, H.M. Government find it impossible from our previous experiences to believe that Germany and Italy will not at any moment which suits them seize French warships and use them against Britain and her allies. The Italian armistice prescribes that French ships should return to metropolitan ports, and under the armistice France is required to yield up units for coast defence and minesweeping.

It is impossible for us, your comrades up to now, to allow your fine ships to fall into the power of the German or Italian enemy. We are determined to fight on to the end, and if we win, as we think we shall, we shall never forget that France was our Ally, that our interests are the same as hers, and that our common enemy is Germany. Should we conquer, we solemnly declare that we shall restore the greatness and territory of France. For this purpose, we must make sure that the best ships of the French Navy are not used against us by the common foe. In these circumstances His Majesty's Government have instructed me to demand that the French Fleet now at Mers-el-Kebir and Oran shall act in accordance with one of the following alternatives:

(a) Sail with us and continue to fight for victory against the Germans and Italians.

(b) Sail with reduced crews under our control to a British port. The reduced crews will be repatriated at the earliest moment.

If either of these courses is adopted by you, we will restore your ships to France at the conclusion of the war, or pay full compensation if they are damaged meanwhile.

(c) Alternatively, if you feel bound to stipulate that your ships should not be used against the Germans or Italians unless these break

---

* Misunderstanding was caused by this paragraph. As late as June 14 Admiral Darlan had favoured the idea of sending the French Fleet to British ports in certain eventualities, but by June 18 he had become Minister of Marine. Thereafter the new French Government under Marshal Pétain would not give the specific assurance demanded by the British. The second sentence in this paragraph therefore no longer represented the position of the French Government. In the crisis this last-minute change was not appreciated by the Admiralty officials concerned.

the Armistice, then sail them with us with reduced crews to some French port in the West Indies—Martinique, for instance—where they can be demilitarised to our satisfaction, or perhaps be entrusted to the United States and remain safe until the end of the war, the crews being repatriated.

If you refuse these fair offers, I must, with profound regret, require you to sink your ships within six hours.

Finally, failing the above, I have the orders of His Majesty's Government to use whatever force may be necessary to prevent your ships from falling into German or Italian hands.

In the evening of the 2nd I requested the Admiralty to send the Vice-Admiral the following message (dispatched 10.55 p.m.):

You are charged with one of the most disagreeable and difficult tasks that a British Admiral has ever been faced with, but we have complete confidence in you and rely on you to carry it out relentlessly.

The Admiral sailed at daylight and was off Oran at about nine-thirty. He sent Captain Holland himself in a destroyer to wait upon the French Admiral Gensoul. After being refused an interview Holland sent by messengers the document already quoted. Admiral Gensoul replied in writing, that in no case would the French warships be allowed to fall intact into German and Italian hands, and that force would be met with force.

All day negotiations continued. At 4.15 p.m. Captain Holland was at last permitted to board the Dunkerque, but the ensuing meeting with the French Admiral was frigid. Admiral Gensoul had meanwhile sent two messages to the French Admiralty, and at 3 p.m. the French Council of Ministers had met to consider the British terms. General Weygand was present at this meeting, and what transpired has now been recorded by his biographer. From this it seems that the third alternative, namely, the removal of the French Fleet to the West Indies, was never mentioned. He states, "... It would appear that Admiral Darlan, whether deliberately or not, or whether he was aware of them or not, I do not know, did not in fact inform us of all the details of the matter at the time. It now appears that the terms of the British ultimatum were less crude than we were led to believe, and suggested a third and far more acceptable alternative, namely, the departure of the Fleet for West Indian waters."* No explanation of this omission, if it were an omission, has so far been seen.

* The Rôle of General Weygand, by Jacques Weygand.

The distress of the British Admiral and his principal officers was evident to us from the signals which had passed. Nothing but the most direct orders compelled them to open fire on those who had been so lately their comrades. At the Admiralty also there was manifest emotion. But there was no weakening in the resolve of the War Cabinet. I sat all the afternoon in the Cabinet Room in frequent contact with my principal colleagues and the First Lord and First Sea Lord. A final signal was dispatched at 6.26 p.m.:

French ships must comply with our terms or sink themselves or be sunk by you before dark.

But the action had already begun. At 5.54 p.m. Admiral Somerville opened fire upon this powerful French fleet, which was also protected by its shore batteries. At 6.0 p.m. he reported that he was heavily engaged. The bombardment lasted for some ten minutes. The battleship *Bretagne* was blown up. The *Dunkerque* ran aground. The battleship *Provence* was beached. The *Strasbourg* escaped, and, though attacked by torpedo aircraft from the *Ark Royal*, reached Toulon, as did also the cruisers from Algiers.

At Alexandria, after protracted negotiations with Admiral Cunningham, the French Admiral Godefroy agreed to discharge his oil fuel, to remove important parts of his gun-mechanisms, and to repatriate some of his crews. At Dakar on July 8 an attack was made on the battleship *Richelieu* by the aircraft-carrier *Hermes*, and most gallantly by a motor-boat. The *Richelieu* was hit by an air torpedo and seriously damaged. The French aircraft-carrier and two light cruisers in the French West Indies were immobilised after long-drawn-out discussions under an agreement with the United States.

On July 4 I reported at length to the House of Commons what we had done. Although the battle-cruiser *Strasbourg* had escaped from Oran and the effective disablement of the *Richelieu* had not then been reported, the measures we had taken had removed the French Navy from major German calculations. I spoke for an hour or more that afternoon, and gave a detailed account of all these sombre events as they were known to me. I have nothing to add to the account which I then gave to Parliament and to the world. I thought it better, for the sake of proportion, to end upon a note which placed this mournful episode in true relation with the plight in which we stood. I therefore read to the House the admonition which I had, with Cabinet approval, circulated through the inner circles of the governing machine the day before.

On what may be the eve of an attempted invasion or battle for our native land, the Prime Minister desires to impress upon all persons holding responsible positions in the Government, in the Fighting Services, or in the Civil departments their duty to maintain a spirit of alert and confident energy. While every precaution must be taken that time and means afford, there are no grounds for supposing that more German troops can be landed in this country, either from the air or across the sea, than can be destroyed or captured by the strong forces at present under arms. The Royal Air Force is in excellent order and at the highest strength yet attained. The German Navy was never so weak nor the British Army at home so strong as now. The Prime Minister expects all His Majesty's servants in high places to set an example of steadiness and resolution. They should check and rebuke the expression of loose and ill-digested opinions in their circles, or by their subordinates. They should not hesitate to report, or if necessary remove, any persons, officers, or officials who are found to be consciously exercising a disturbing or depressing influence, and whose talk is calculated to spread alarm and despondency. Thus alone will they be worthy of the fighting men who, in the air, on the sea, and on land, have already met the enemy without any sense of being outmatched in martial qualities.

The House was very silent during the recital, but at the end there occurred a scene unique in my own experience. Everybody seemed to stand up all around, cheering, for what seemed a long time. Up till this moment the Conservative Party had treated me with some reserve, and it was from the Labour benches that I received the warmest welcome when I entered the House or rose on serious occasions. But now all joined in solemn stentorian accord.

The elimination of the French Navy as an important factor almost at a single stroke by violent action produced a profound impression in every country. Here was this Britain which so many had counted down and out, which strangers had supposed to be quivering on the brink of surrender to the mighty power arrayed against her, striking ruthlessly at her dearest friends of yesterday and securing for a while to herself the undisputed command of the sea. It was made plain that the British War Cabinet feared nothing and would stop at nothing. This was true.

\*     \*     \*     \*     \*

The Pétain Government had moved to Vichy on July 1, and proceeded to set itself up as the Government of Unoccupied France. On receiving the news of Oran they ordered retaliation by air upon Gibraltar, and a few bombs were dropped upon the harbour from their African stations. On July 5 they formally broke off relations with Great Britain. On July 11 President Lebrun gave place to Marshal Pétain, who was installed as Chief of the State by an enormous majority of 569 against 80, with 17 abstentions and many absentees.

The genius of France enabled her people to comprehend the whole significance of Oran, and in her agony to draw new hope and strength from this additional bitter pang. General de Gaulle, whom I did not consult beforehand, was magnificent in his demeanour, and France liberated and restored has ratified his conduct. I am indebted to M. Teitgen, a prominent member of the Resistance Movement, afterwards French Minister of Defence, for a tale which should be told. In a village near Toulon dwelt two peasant families, each of whom had lost their sailor son by British fire at Oran. A funeral service was arranged to which all their neighbours sought to go. Both families requested that the Union Jack should lie upon the coffins side by side with the Tricolour, and their wishes were respectfully observed. In this we may see how the comprehending spirit of simple folk touches the sublime.

\*     \*     \*     \*     \*

Immense relief spread through the high Government circles in the United States. The Atlantic Ocean seemed to regain its sheltering power, and a long vista of time opened out for the necessary preparations for the safety of the great Republic. Henceforth there was no more talk about Britain giving in. The only question was, would she be invaded and conquered? That was the issue which was now to be put to the proof.

CHAPTER XII

# THE APPARATUS OF COUNTER-
# ATTACK

1940

*My Own Reactions after Dunkirk – Minute to General Ismay of
June 4 – A Retrogression – My Old Plans of July 1917 – An Early
Idea of Tank Landing-craft – The Germ of the "Mulberry" Harbours
of 1944 – Directive to General Ismay on Counter-attack – "Com-
mandos" – Tank Landing-craft and Parachutists – My Minute of
July 7, 1940, Calling for Beach Landing-craft for Six or Seven
Hundred Tanks – Minute of August 5, 1940, on Programme of
Armoured Divisions – Overseas Transportation for Two Divisions at
a Time – Creation of the Combined Operations Command – Appointment of Sir Roger Keyes – The Joint Planning Committee is Placed
Directly Under the Minister of Defence – Progress of the Landing-
craft Construction in 1940 and 1941 – My Telegram to the President
of July 25, 1941 – My Consistent Purpose to Land Armour on
Beaches.*

MY first reaction to the "Miracle of Dunkirk" had been to
turn it to proper use by mounting a counter-offensive.
When so much was uncertain, the need to recover the
initiative glared forth. June 4 was much occupied for me by
the need to prepare and deliver the long and serious speech to
the House of Commons of which some account has been given,
but as soon as this was over I made haste to strike the note
which I thought should rule our minds and inspire our actions at
this moment.

We are greatly concerned—and it is certainly wise to be so—with the dangers of the Germans landing in England in spite of our possessing the command of the seas and having very strong defence by fighters in the air. Every creek, every beach, every harbour has become to us a source of anxiety. Besides this the parachutists may sweep over and take Liverpool or Ireland, and so forth. All this mood is very good if it engenders energy. But if it is so easy for the Germans to invade us, in spite of sea-power, some may feel inclined to ask the question, why should it be thought impossible for us to do anything of the same kind to them? The completely defensive habit of mind which has ruined the French must not be allowed to ruin all our initiative. It is of the highest consequence to keep the largest numbers of German forces all along the coasts of the countries they have conquered, and we should immediately set to work to organise raiding forces on these coasts where the populations are friendly. Such forces might be composed of self-contained, thoroughly-equipped units of say one thousand up to not more than ten thousand when combined. Surprise would be ensured by the fact that the destination would be concealed until the last moment. What we have seen at Dunkirk shows how quickly troops can be moved off (and I suppose on to) selected points if need be. How wonderful it would be if the Germans could be made to wonder where they were going to be struck next, instead of forcing us to try to wall in the Island and roof it over! An effort must be made to shake off the mental and moral prostration to the will and initiative of the enemy from which we suffer.

Ismay conveyed this to the Chiefs of Staff, and in principle it received their cordial approval and was reflected in many of the decisions which we took. Out of it gradually sprang a policy. My thought was at this time firmly fixed on tank warfare, not merely defensive but offensive. This required the construction of large numbers of tank-landing vessels, which henceforward became one of my constant cares. As all this was destined to become of major importance in the future I must now make a retrogression into a subject which had long ago lain in my mind and was now revived.

\* \* \* \* \*

I had always been fascinated by amphibious warfare, and the idea of using tanks to run ashore from specially-constructed landing-craft on beaches where they were not expected had long been in my mind. Ten days before I joined Mr. Lloyd George's

Government as Minister of Munitions on July 17, 1917, I had prepared, without expert assistance, a scheme for the capture of the two Frisian islands Borkum and Sylt. The object was to secure an overseas base for flotillas and cruisers and for such air forces as were available in those days, in order to force the naval fighting, in which we had a great numerical superiority, and by re-establishing close blockade relieve the pressure of the U-boat war, then at its height, against our Atlantic supply-line and the movement of the American armies to France. Mr. Lloyd George was impressed with the plan, and had it specially printed for the Admiralty and the War Cabinet.

It contained the following paragraph, 22c, which has never yet seen the light of day:

The landing of the troops upon the island [of Borkum or Sylt] under cover of the guns of the Fleet [should be] aided by gas and smoke from torpedo-proof transports by means of *bullet-proof lighters.* Approximately one hundred should be provided for landing a division. In addition a number—say fifty—*tank-landing lighters should be provided, each carrying a tank or tanks* [and] fitted for wire-cutting in its bow. By means of a drawbridge or shelving bow [the tanks] would land under [their] own power, and prevent the infantry from being held up by wire when attacking the gorges of the forts and batteries. This is a new feature, and removes one of the very great previous difficulties, namely, the rapid landing of [our] field artillery to cut wire.

And further, paragraph 27:

There is always the danger of the enemy getting wind of our intentions and reinforcing his garrisons with good troops beforehand, at any rate so far as Borkum, about which he must always be very sensitive, is concerned. On the other hand, *the landing could be effected under the shields of lighters, proof against machine-gun bullets,* and too numerous to be seriously affected by heavy gunfire [*i.e.,* the fire of heavy guns]; *and tanks employed in even larger numbers than are here suggested, especially the quick-moving tank and lighter varieties,* would operate in an area where no preparations could have been made to receive them. These may be thought new and important favourable considerations.

\* \* \* \* \*

In this paper also I had an alternative plan for making an artificial island in the shallow waters of Horn Reef (to the northward).

Para. 30. One of the methods suggested for investigation is as follows: *A number of flat-bottomed barges or caissons, made not of steel, but of concrete,* should be prepared in the Humber, at Harwich, and in the Wash, the Medway, and the Thames. These structures would be adapted to the depths in which they were to be sunk, according to a general plan. They would float when empty of water, and thus could be towed across to the site of the artificial island. On arrival at the buoys marking the island sea-cocks would be opened, and they would settle down on the bottom. They could subsequently be gradually filled with sand, as opportunity served, by suction dredgers. These structures would range in size from 50′ × 40′ × 20′ to 120′ × 80′ × 40′. *By this means a torpedo- and weather-proof harbour, like an atoll, would be created in the open sea, with regular pens for the destroyers and submarines, and alighting-platforms for aeroplanes.*

This project, if feasible, is capable of great elaboration, and it might be applied in various places. Concrete vessels can perhaps be made to carry a complete heavy gun turret, and these, on the admission of water to their outer chambers, would sit on the sea floor, like the Solent forts, at the desired points. Other sinkable structures could be made to contain store-rooms, oil-tanks, or living-chambers. It is not possible, without an expert inquiry, to do more here than indicate the possibilities, which embrace nothing less than the creation, transportation in pieces, assemblement and posing of an artificial island and destroyer base.

31. Such a scheme, if found mechanically sound, avoids the need of employing troops and all the risks of storming a fortified island. *It could be applied as a surprise, for although the construction of these concrete vessels would probably be known in Germany, the natural conclusion would be that they were intended for an attempt to block up the river-mouths, which indeed is an idea not to be excluded.* Thus, until the island or system of breakwaters actually began to grow the enemy would not penetrate the design.

A year's preparation would however be required.

For nearly a quarter of a century this paper had slumbered in the archives of the Committee of Imperial Defence. I did not print it in *The World Crisis,* of which it was to have been a chapter, for reasons of space, and because it was never put into effect. This was fortunate, because the ideas expressed were in this war more than ever vital; and the Germans certainly read my war books with attention. Indeed a staff study of the writings of any one in my position would be a matter of normal routine. The underlying conceptions of this old paper were deeply imprinted in my mind, and in the new emergency formed the foundation of action which, after a long interval, found memorable expression in the vast fleet of tank-landing craft of 1943 and in the "Mulberry" harbours of 1944.

\* \* \* \* \*

On this same not unfertile 6th of June, 1940, flushed with the sense of deliverance and the power to plan ahead, I began a long series of minutes in which the design and construction of tank-landing craft was ordered and steadily pressed.

*Prime Minister to General Ismay*                                    6.VI.40

Further to my minute of yesterday [dated June 4] about offensive action: when the Australians arrive it is a question whether they should not be organised in detachments of 250, equipped with grenades, trench-mortars, tommy-guns, armoured vehicles, and the like, capable of acting against an attack in this country, but also capable of landing on the friendly coasts now held by the enemy. We have got to get out of our minds the idea that the Channel ports and all the country between them are enemy territory. What arrangements are being made for good agents in Denmark, Holland, Belgium, and along the French coast? Enterprises must be prepared, with specially-trained troops of the hunter class, who can develop a reign of terror down these coasts, first of all on the "butcher and bolt" policy; but later on, or perhaps as soon as we are organised, we could surprise Calais or Boulogne, kill and capture the Hun garrison, and hold the place until all the preparations to reduce it by siege or heavy storm have been made, and then away. The passive resistance war, in which we have acquitted ourselves so well, must come to an end. I look to the Joint Chiefs of the Staff to propose me measures for a vigorous, enterprising and ceaseless offensive against the whole German-occupied coastline. *Tanks and A.F.V.s [Armoured Fighting Vehicles] must be made in flat-bottomed boats, out of which they can crawl ashore*, do a deep raid inland, cutting a vital communication, and then back, leaving a trail of German corpses behind them. It is probable that when the best troops go on to the attack of Paris only the ordinary German troops of the line will be left. The lives of these must be made an intense torment. The following measures should be taken:

1. *Proposals for organising the Striking Companies.*
2. *Proposals for transporting and landing tanks on the beach*, observing that we are supposed to have the command of the sea, while the enemy have not.

3. A proper system of espionage and intelligence along the whole coasts.

4. Deployment of parachute troops on a scale equal to five thousand.

5. Half a dozen of our 15-inch guns should be lined up [*i.e.*, with inner tubes] immediately to fire fifty or sixty miles, and should be mounted either on railway mountings or on steel and concrete platforms, so as to break up the fire of the German guns that will certainly in less than four months be firing across the Channel.

Action in many directions followed accordingly. The "Striking Companies" emerged under the name of "Commandos", ten of which were now raised from the Regular Army and the Royal Marines. The nucleus of this organisation had begun to take shape in the Norwegian campaign. An account will be given in its proper place of the cross-Channel heavy guns. I regret however that I allowed the scale I had proposed for British parachute troops to be reduced from five thousand to five hundred.

*  *  *  *  *

I recurred at intervals to the building of landing-craft, on which my mind constantly dwelt both as a peril to us and in the future as a project against the enemy. Development of small assault craft had been started before the outbreak of war, and a few had been employed at Narvik. Most of these had been lost either there or at Dunkirk. Now we required not only the small craft which could be lifted in the troop-carrying ships, but sea-going vessels capable themselves of transporting tanks and guns to the assault and landing them on to the beaches.

*Prime Minister to Minister of Supply*                           7.VII.40

What is being done about designing and planning vessels to transport tanks across the sea for a British attack on enemy countries? This might well be remitted as a study to Mr. Hopkins, former Chief Constructor of the Navy, who must have leisure now that Cultivator No. 6* is out of fashion. These must be able to move six or seven hundred vehicles in one voyage and land them on the beach, or, alternatively, take them off the beaches, as well, of course, as landing them on quays—if it be possible to combine the two.

*Prime Minister to General Ismay*                              5.VIII.40

I asked the other day for a forecast of the development of the armoured divisions which will be required in 1941—namely, five by

* A trench-cutting machine for attacking fortified lines.

the end of March and one additional every month until a total of ten is reached at the end of August 1941; and also for the composition of each division in armoured and ancillary vehicles of all kinds.

Pray let me know how far the War Office plans have proceeded, and whether the number of tanks ordered corresponds with a programme of these dimensions.

*Let me further have a report on the progress of the means of transportation overseas, which should be adequate to the movement at one moment of two armoured divisions.* Who is doing this—Admiralty or Ministry of Supply? I suggested that Mr. Hopkins might have some spare time available.

*Prime Minister to General Ismay*              9.VIII.40
Get me a further report about the designs and types of vessels to transport armoured vehicles by sea and land on [to] beaches.

In July I created a separate Combined Operations Command under the Chiefs of Staff for the study and exercise of this form of warfare, and Admiral of the Fleet Sir Roger Keyes became its chief. His close personal contact with me and with the Defence Office served to overcome any departmental difficulties arising from this unusual appointment.

*Prime Minister to General Ismay and Sir Edward Bridges*    17.VII.40
I have appointed Admiral of the Fleet Sir Roger Keyes as Director of Combined Operations. He should take over the duties and resources now assigned to General Bourne. General Bourne should be informed that, owing to the larger scope now to be given to these operations, it is essential to have an officer of higher rank in charge, and that the change in no way reflects upon him or those associated with him. Evidently he will have to co-operate effectively. I formed a high opinion of this officer's work as Adjutant-General Royal Marines, and in any case the Royal Marines must play a leading part in this organisation.

Pending any further arrangements, Sir Roger Keyes will form contact with the Service departments through General Ismay as representing the Minister of Defence.

\*　　\*　　\*　　\*　　\*

I have already explained how smoothly the office of Minister of Defence came into being and grew in authority. At the end of August I took the only formal step which I ever found necessary. Hitherto the Joint Planning Committee had worked under the Chiefs of Staff and looked to them as their immediate and official superiors. I felt it necessary to have this important though up till now not very effective body under my personal control. Therefore I asked the War Cabinet to give approval to this definite change in our war machine. This was readily accorded me by all my colleagues, and I gave the following instructions:

*Prime Minister to General Ismay and Sir Edward Bridges*     24.VIII.40

1. The Joint Planning Committee will from Monday next work directly under the orders of the Minister of Defence and will become a part of the Minister of Defence's office—formerly the C.I.D. Secretariat. Accommodation will be found for them at Richmond Terrace. They will retain their present positions in and contacts with the three Service departments. They will work out the details of such plans as are communicated to them by the Minister of Defence. They may initiate plans of their own after reference to General Ismay. They will, of course, be at the service of the Chiefs of Staff Committee for the elaboration of any matters sent to them.

2. All plans produced by the Joint Planning Committee or elaborated by them under instructions as above will be referred to the Chiefs of Staff Committee for their observations.

3. Thereafter should doubts and differences exist, or in important cases, all plans will be reviewed by the Defence Committee of the War Cabinet, which will consist of the Prime Minister, the Lord Privy Seal, and Lord Beaverbrook, and the three Service Ministers; the three Chiefs of Staff, with General Ismay, being in attendance.

4. The Prime Minister assumes the responsibility of keeping the War Cabinet informed of what is in hand; but the relation of the Chiefs of Staff to the War Cabinet is unaltered.

The Chiefs of Staff accepted this change without serious demur. Sir John Dill, however, wrote a minute to the Secretary of State for War on which I was able to reassure him.

*Prime Minister to Secretary of State for War*     31.VIII.40

There is no question of the Joint Planning Committee "submitting military advice" to me. They are merely to work out plans in accordance with directions which I shall give. The advice as to whether these plans or any variants of them should be adopted will rest as at present with the Chiefs of Staff. It is quite clear that the Chiefs of Staff also have their collective responsibility for advising the Cabinet as well as the Prime Minister or Minister of Defence. It has not been thought necessary to make any alteration in their constitutional position. Moreover, I propose to work with and through them as heretofore.

I have found it necessary to have direct access to and control of the Joint Planning Staffs because after a year of war I cannot recall a single plan initiated by the existing machinery. I feel sure that I can count upon you and the other two Service Ministers to help me in giving a vigorous and positive direction to the conduct of the war, and in overcoming the dead weight of inertia and delay which has so far led us to being forestalled on every occasion by the enemy.

It will of course be necessary from time to time to increase the number of the Joint Planning Staffs.

In practice the new procedure worked in an easy and agreeable manner, and I cannot recall any difficulties which arose.

\* \* \* \* \*

Henceforth intense energy was imparted to the development of all types of landing-craft, and a special department was formed in the Admiralty to deal with these matters. By October 1940 the trials of the first Landing-Craft Tank (L.C.T.) were in progress. Only about thirty of these were built, as they proved too small. An improved design followed, many of which were built in sections for more convenient transport by sea to the Middle East, where they began to arrive in the summer of 1941. These proved their worth, and as we gained experience the capabilities of later editions of these strange craft steadily improved. The Admiralty were greatly concerned at the inroads which this new form of specialised production might make into the resources of the shipbuilding industry. Fortunately it proved that the building of L.C.T. could be delegated to constructional engineering firms not engaged in shipbuilding, and thus the labour and plant of the larger shipyards need not be disturbed. This rendered possible the large-scale programme which we contemplated, but also placed a limit on the size of the craft.

The L.C.T. was suitable for cross-Channel raiding operations or for more extended work in the Mediterranean, but not for long voyages in the open sea. The need arose for a larger, more seaworthy vessel which besides transporting tanks and other vehicles on ocean voyages could also land them over beaches like the L.C.T. I gave directions for the design of such a vessel, which was first called an "Atlantic L.C.T.", but was soon re-named

"Landing Ship Tank" (L.S.T.). The building of these inevitably impinged on the resources of our hard-pressed shipyards. Thus, of the first design, nicknamed in the Admiralty the "Winette", only three were built; others were ordered in the United States and Canada, but were superseded by a later design. Meanwhile we converted three shallow-draft tankers to serve the same purpose, and these too rendered useful service later on.

By the end of 1940 we had a sound conception of the physical expression of amphibious warfare. The production of specialised craft and equipment of many kinds was gathering momentum, and the necessary formations to handle all this new material were being developed and trained under the Combined Operations Command. Special training centres for this purpose were established both at home and in the Middle East. All these ideas and their practical manifestation we presented to our American friends as they took shape. The results grew steadily across the years of struggle, and thus in good time they formed the apparatus which eventually played an indispensable part in our greatest plans and deeds. Our work in this field in these earlier years had such a profound effect on the future of the war that I must anticipate events by recording some of the material progress which we made later.

In the summer of 1941 the Chiefs of Staff pointed out that the programme of landing-craft construction was related only to small-scale operations and that our ultimate return to the Continent would demand a much greater effort than we could then afford. By this time the Admiralty had prepared a new design of the landing ship tank, and this was taken to the United States, where the details were jointly worked out. In February 1942 this vessel was put into production in America on a massive scale. It became the L.S.T.(2), which figured so prominently in all our later operations, making perhaps the greatest single contribution to the solution of the stubborn problem of landing heavy vehicles over beaches. Ultimately over a thousand of these were built.

Meanwhile the production of small craft of many types for use in a Continental assault was making steady progress on both sides of the Atlantic. All these required transport to the scene of action in the ships carrying the assaulting troops. Thus an immense conversion programme was initiated to fit British and American troopships to carry these craft as well as great quantities of other specialised equipment. These ships became known as Landing Ships Infantry (L.S.I.). Some were commissioned into the Royal Navy; others preserved their mercantile status, and their masters and crews served them with distinction in all our offensive operations. Such ships could ill be spared from the convoys carrying the endless stream of reinforcements to the Middle East and elsewhere, yet this sacrifice had to be made. Many other ancillary types of ship for use in the assault also came into being at this time. In 1940 and 1941 our efforts in this field were limited by the demands of the U-boat struggle. Not more than seven thousand men could be spared for landing-craft production up to the end of 1940, nor was this number greatly exceeded in the following year. However, by 1944 no less than seventy thousand men in Britain alone were dedicated to this stupendous task, besides much larger numbers in the United States.

* * * * *

As all our work in this sphere had a powerful bearing on the future of the war, I print at this point a telegram which I sent to the President in 1941:

25.VII.41

We have been considering here our war plans, not only for the fighting of 1942, but also for 1943. After providing for the security of essential bases, it is necessary to plan on the largest scale the forces needed for victory. In broad outline we must aim first at intensifying the blockade and propaganda. Then we must subject Germany and Italy to a ceaseless and ever-growing air bombardment. These measures may themselves produce an internal convulsion or collapse. *But plans ought also to be made for coming to the aid of the conquered populations by landing armies of liberation when opportunity is ripe. For this purpose it will be necessary not only to have great numbers of tanks, but also of vessels capable of carrying them and landing them direct on to beaches.* It ought not to be difficult for you to make the necessary adaptation in some of the vast numbers of merchant vessels you are building so as to fit them for tank-landing ships.

And a little later:

*Prime Minister to First Sea Lord*                                    8.IX.41

My idea was not that the President should build Winettes as such, apart from any already arranged for, but that, out of the great number of merchant vessels being constructed in the United States for 1942, he would fit out a certain number with brows and side-ports to enable tanks to be landed from them on beaches, or into tank landing-craft which would take them to the beaches.

Please help me to explain this point to him, showing what kind of alteration would be required in the American merchant ships now projected.

In view of the many accounts which are extant and multiplying of my supposed aversion from any kind of large-scale opposed-landing, such as took place in Normandy in 1944, it may be convenient if I make it clear that from the very beginning I provided a great deal of the impulse and authority for creating the immense apparatus and armada for the landing of armour on beaches, without which it is now universally recognised that all such major operations would have been impossible. I shall unfold this theme step by step in these volumes by means of documents written by me at the time, which show a true and consistent purpose on my part in harmony with the physical facts, and a close correspondence with what was actually done.

# CHAPTER XIII

## AT BAY

### July 1940

*Can Britain Survive? – Anxiety in the United States – Resolute Demeanour of the British Nation – The Relief of Simplicity – Hitler's "Peace Offer", July 19 – Our Response – German Diplomatic Approaches Rejected – The King of Sweden's Démarche – I Visit the Threatened Coasts – General Montgomery and the 3rd Division at Brighton – The Importance of Buses – General Brooke Succeeds Ironside in Command of the Home Army – My Contacts with General Brooke – Some Directives and Minutes of July – The Defence of London – Conditions in the Threatened Coastal Zones – Statistics on the Growth and Equipment of the Army – Lindemann's Diagrams – The Canadian Second Division Retrieved from Iceland – Need to Prevent Enemy Concentration of Shipping in the Channel – Arrival of the American Rifles – Special Precautions – The French "Seventy-fives" – The Growth of the German Channel Batteries – Our Counter-measures – My Visit to Admiral Ramsay at Dover – Progress of Our Batteries Coaxed and Urged – The Monitor "Erebus"–The Defence of the Kentish Promontory – British Heavy-Gun Concentration, September – Our Rising Strength – An Ordeal Averted.*

I N these summer days of 1940 after the fall of France we were all alone. None of the British Dominions or India or the Colonies could send decisive aid, or send what they had in time. The victorious, enormous German armies, thoroughly equipped and with large reserves of captured weapons and arsenals behind them, were gathering for the final stroke. Italy, with numerous and imposing forces, had declared war upon us, and eagerly sought our destruction in the Mediterranean and in Egypt. In the Far East Japan glared inscrutably, and pointedly requested the closing of the Burma Road against supplies for China. Soviet

Russia was bound to Nazi Germany by her pact, and lent important aid to Hitler in raw materials. Spain, which had already occupied the International Zone of Tangier, might turn against us at any moment and demand Gibraltar, or invite the Germans to help her attack it, or mount batteries to hamper passage through the Straits. The France of Pétain and Bordeaux, soon moved to Vichy, might any day be forced to declare war upon us. What was left at Toulon of the French Fleet seemed to be in German power. Certainly we had no lack of foes.

After Oran it became clear to all countries that the British Government and nation were resolved to fight on to the last. But even if there were no moral weakness in Britain, how could the appalling physical facts be overcome? Our armies at home were known to be almost unarmed except for rifles. There were in fact hardly five hundred field-guns of any sort and hardly two hundred medium or heavy tanks in the whole country. Months must pass before our factories could make good even the munitions lost at Dunkirk. Can one wonder that the world at large was convinced that our hour of doom had struck?

Deep alarm spread through the United States, and indeed through all the surviving free countries. Americans gravely asked themselves whether it was right to cast away any of their own severely-limited resources to indulge a generous though hopeless sentiment. Ought they not to strain every nerve and nurse every weapon to remedy their own unpreparedness? It needed a very sure judgment to rise above these cogent, matter-of-fact arguments. The gratitude of the British nation is due to the noble President and his great officers and high advisers for never, even in the advent of the Third Term Presidential Election, losing their confidence in our fortunes or our will.

The buoyant and imperturbable temper of Britain, which I had the honour to express, may well have turned the scale. Here was this people, who in the years before the war had gone to the extreme bounds of pacifism and improvidence, who had indulged in the sport of party politics, and who, though so weakly armed, had advanced light-heartedly into the centre of European affairs, now confronted with the reckoning alike of their virtuous impulses and neglectful arrangements. They were not even dismayed. They defied the conquerors of Europe. They seemed willing to have their Island reduced to a shambles rather than give in. This would make a fine page in history. But there were other tales of this kind. Athens had been conquered by Sparta. The Carthaginians made a forlorn resistance to Rome. Not seldom in the annals of the past—and how much more often in tragedies never recorded or long-forgotten—had brave, proud, easy-going states, and even entire races, been wiped out, so that only their name or even no mention of them remains.

Few British and very few foreigners understood the peculiar technical advantages of our insular position; nor was it generally known how even in the irresolute years before the war the essentials of sea and latterly air defence had been maintained. It was nearly a thousand years since Britain had seen the fires of a foreign camp on English soil. At the summit of British resistance everyone remained calm, content to set their lives upon the cast. That this was our mood was gradually recognised by friends and foes throughout the whole world. What was there behind the mood? That could be settled only by brute force.

\*       \*       \*       \*       \*

There was also another aspect. One of our greatest dangers during June lay in having our last reserves drawn away from us into a wasting, futile French resistance in France, and the strength of our air forces gradually worn down by their flights or transference to the Continent. If Hitler had been gifted with supernatural wisdom he would have slowed down the attack on the French front, making perhaps a pause of three or four weeks after Dunkirk on the line of the Seine, and meanwhile developing his preparations to invade England. Thus he would have had a deadly option, and could have tortured us with the hooks of either deserting France in her agony or squandering the last resources for our future existence. The more we urged the French to fight on, the greater was our obligation to aid them, and the more difficult it would have become to make any preparations for defence in England, and above all to keep in reserve the twenty-five squadrons of fighter aircraft on which all depended. On this point we should never have given way, but the refusal would have been bitterly resented by our struggling Ally, and would have poisoned all our relations. It was even with an actual sense of relief that some of our high commanders addressed themselves to our new and grimly simplified problem. As the commissionaire at one of the Service clubs in London said to a rather downcast member: "Anyhow, sir, we're in the Final, and it's to be played on the Home Ground."

\*     \*     \*     \*     \*

The strength of our position was not, even at this date, under-rated by the German High Command. Ciano tells how, when he visited Hitler in Berlin on July 7, 1940, he had a long conversation with General von Keitel. Keitel, like Hitler, spoke to him about the attack on England. He repeated that up to the present nothing definite had been decided. He regarded the landing as possible, but considered it an "extremely difficult operation, which must be approached with the utmost caution, in view of the fact that the intelligence available on the military preparedness of the island and on the coastal defences is meagre and not very reliable".\* What would appear to be easy and also essential was a major air attack upon the airfields, factories, and the principal communication centres in Great Britain. It was necessary however to bear in mind that the British Air Force was extremely efficient. Keitel calculated that the British had about fifteen hundred machines ready for defence and counter-attack. He admitted that recently the offensive action of the British Air Force had been greatly intensified. Bombing missions were carried out with noteworthy accuracy, and the groups of aircraft which appeared numbered up to eighty machines at a time. There was however in England a great shortage of pilots, and those who were now attacking the German cities could not be replaced by the new pilots, who were completely untrained. Keitel also insisted upon the necessity of striking at Gibraltar in order to disrupt the British imperial system. Neither Keitel nor Hitler made any reference to the duration of the war. Only Himmler said incidentally that the war ought to be finished by the beginning of October.

Such was Ciano's report. He also offered Hitler, at "the earnest wish of the Duce", an army of ten divisions and an air component of thirty squadrons to take part in the invasion. The army was politely declined. Some of the air squadrons came, but, as will be presently related, fared ill.

\*     \*     \*     \*     \*

\* Ciano, *Diplomatic Papers*, p. 378.

On July 19 Hitler delivered his triumphant speech in the Reichstag, in which, after predicting that I would shortly take refuge in Canada, he made what has been called his Peace Offer. The operative sentences were:

In this hour I feel it to be my duty before my own conscience to appeal once more to reason and common sense in Great Britain as much as elsewhere. I consider myself in a position to make this appeal, since I am not a vanquished foe begging favours, but the victor, speaking in the name of reason. I can see no reason why this war need go on. I am grieved to think of the sacrifices it must claim. . . . Possibly Mr. Churchill will brush aside this statement of mine by saying it is merely born of fear and doubt of final victory. In that case I shall have relieved my conscience in regard to the things to come.

This gesture was accompanied during the following days by diplomatic representations through Sweden, the United States, and at the Vatican. Naturally Hitler would be very glad, after having subjugated Europe to his will, to bring the war to an end by procuring British acceptance of what he had done. It was in fact an offer not of peace but of readiness to accept the surrender by Britain of all she had entered the war to maintain. As the German Chargé d'Affaires in Washington had attempted some communication with our Ambassador there, I sent the following telegram:

20.VII.40

I do not know whether Lord Halifax is in town to-day, but Lord Lothian should be told on no account to make any reply to the German Chargé d'Affaires' message.

My first thought, however, was a solemn, formal debate in both Houses of Parliament. I therefore wrote at the same time to Mr. Chamberlain and Mr. Attlee.

20.VII.40

It might be worth while meeting Hitler's speech by resolutions in both Houses. These resolutions should be proposed by private Peers and Members. On the other hand, the occasion will add to our burdens. What do you say?

My colleagues thought that this would be making too much of the matter, upon which we were all of one mind. It was decided instead that the Foreign Secretary should dismiss Hitler's gesture in a broadcast. On the night of the 22nd he "brushed aside" Hitler's "summons to capitulate to his will". He contrasted Hitler's picture of Europe with the picture of the Europe for which we were fighting, and declared that "we shall not stop fighting until Freedom is secure". In fact however the rejection of any idea of a parley had already been given in the British Press and by the B.B.C., without any prompting from His Majesty's Government, as soon as Hitler's speech was heard over the radio.

Ciano, in his account of another meeting with Hitler on July 20, observes:

The reaction of the English Press to yesterday's speech has been such as to allow of no prospect of an understanding. Hitler is therefore preparing to strike the military blow at England. He stresses that Germany's strategic position, as well as her sphere of influence and of economic control, are such as to have already greatly weakened the possibilities of resistance by Great Britain, which will collapse under the first blows. The air attack already began some days ago, and is continually growing in intensity. The reaction of the anti-aircraft defences and of the British fighters is not seriously hindering the German air attack. The decisive offensive operation is now being studied, since the fullest preparations have been made.*

Ciano also records in his diaries that "Late in the evening of the 19th, when the first cold British reaction to the speech arrived, a sense of ill-concealed disappointment spread among the Germans." Hitler "would like an understanding with Great Britain. He knows that war with the British will be hard and bloody, and knows also that people everywhere are averse from bloodshed". Mussolini, on the other hand, "fears that the English may find in Hitler's much too cunning speech a pretext to begin negotiations". "That", remarks Ciano, "would be sad for Mussolini, because now more than ever he wants war."† He need not have fretted himself. He was not to be denied all the war he wanted.

There was no doubt continuous German diplomatic activity behind the scenes, and when on August 3 the King of Sweden thought fit to address us on the subject I suggested to the Foreign Secretary the following reply, which formed the basis of the official answer:

* Ciano, *Diplomatic Papers*, p. 381.
† *Ciano's Diaries*, pp. 277–8.

(*Begins.*) On October 12, 1939, His Majesty's Government defined at length their position towards German peace offers in maturely considered statements to Parliament. Since then a number of new hideous crimes have been committed by Nazi Germany against the smaller States upon her borders. Norway has been overrun, and is now occupied by a German invading army. Denmark has been seized and pillaged. Belgium and Holland, after all their efforts to placate Herr Hitler, and in spite of all the assurances given to them by the German Government that their neutrality would be respected, have been conquered and subjugated. In Holland particularly acts of long-prepared treachery and brutality culminated in the massacre of Rotterdam, where many thousands of Dutchmen were slaughtered and an important part of the city destroyed.

These horrible events have darkened the pages of European history with an indelible stain. His Majesty's Government see in them not the slightest cause to recede in any way from their principles and resolves as set forth in October 1939. On the contrary, their intention to prosecute the war against Germany by every means in their power until Hitlerism is finally broken and the world relieved from the curse which a wicked man has brought upon it has been strengthened to such a point that they would rather all perish in the common ruin than fail or falter in their duty. They firmly believe however that with the help of God they will not lack the means to discharge their task. This task may be long; but it will always be possible for Germany to ask for an armistice, as she did in 1918, or to publish her proposals for peace. Before however any such requests or proposals could even be considered it would be necessary that effective guarantees by deeds, not words, should be forthcoming from Germany which would ensure the restoration of the free and independent life of Czechoslovakia, Poland, Norway, Denmark, Holland, Belgium, and above all France, as well as the effectual security of Great Britain and the British Empire in a general peace. (*Ends.*)

I added:

The ideas set forth in the Foreign Office memo. appear to me to err in trying to be too clever, and to enter into refinements of policy unsuited to the tragic simplicity and grandeur of the times and the issues at stake. At this moment, when we have had no sort of success, the slightest opening will be misjudged. Indeed, a firm reply of the kind I have outlined is the only chance of extorting from Germany any offers which are not fantastic.

On the same day I issued the following statement to the Press:

3.VIII.40

The Prime Minister wishes it to be known that the possibility of German attempts at invasion has by no means passed away. The fact that the Germans are now putting about rumours that they do not intend an invasion should be regarded with a double dose of the suspicion which attaches to all their utterances. Our sense of growing strength and preparedness must not lead to the slightest relaxation of vigilance or moral alertness.

*　　*　　*　　*　　*

At the end of June the Chiefs of Staff through General Ismay had suggested to me at the Cabinet that I should visit the threatened sectors of the east and south coasts. Accordingly I devoted a day or two every week to this agreeable task, sleeping when necessary in my train, where I had every facility for carrying on my regular work and was in constant contact with Whitehall. I inspected the Tyne and the Humber and many possible landing-places. The Canadian Division, soon to be reinforced to a corps by the division sent to Iceland, did an exercise for me in Kent. I examined the landward defences of Harwich and Dover. One of my earliest visits was to the 3rd Division, commanded by General Montgomery, an officer whom I had not met before. My wife came with me. The 3rd Division was stationed near Brighton. It had been given the highest priority in re-equipment, and had been about to sail for France when the French resistance ended. General Montgomery's headquarters were near Steyning, and he showed me a small exercise of which the central feature was a flanking movement of Bren-gun carriers, of which he could at that moment muster only seven or eight. After this we drove together along the coast through Shoreham and Hove till we came to the familiar Brighton front, of which I had so many schoolboy memories. We dined in the Royal Albion Hotel, which stands opposite the end of the pier. The hotel was entirely empty, a great deal of evacuation having taken place; but there were still a number of people airing themselves on the beaches or the parade. I was amused to see a platoon of the Grenadier Guards making a sandbag machine-gun post in one of the kiosks of the pier, like those where in my childhood I had often admired the antics of the performing fleas. It was lovely weather. I had very good talks with the General, and enjoyed my outing thoroughly. However:

(*Action this Day*)
*Prime Minister to Secretary of State for War* 3.VII.40

I was disturbed to find the 3rd Division spread along thirty miles of coast, instead of being, as I had imagined, held back concentrated in reserve, ready to move against any serious head of invasion. But much more astonishing was the fact that the infantry of this division, which is otherwise fully mobile, are not provided with the buses necessary to move them to the point of action.* This provision of buses, waiting always ready and close at hand, is essential to all mobile units, and to none more than the 3rd Division while spread about the coast.

I heard the same complaint from Portsmouth that the troops there had not got their transport ready and close at hand. Considering the great masses of transport, both buses and lorries, which there are in this country, and the large numbers of drivers brought back from the B.E.F., it should be possible to remedy these deficiencies at once. I hope, at any rate, that the G.O.C. 3rd Division will be told to-day to take up, as he would like to do, the large number of buses which are even now plying for pleasure traffic up and down the sea front at Brighton.

\* \* \* \* \*

In mid-July the Secretary of State for War recommended that General Brooke should replace General Ironside in command of our Home Forces. On July 19, in the course of my continuous inspection of the invasion sectors, I visited the Southern Command. Some sort of tactical exercise was presented to me in which no fewer than twelve tanks(!) were able to participate. All the afternoon I drove with General Brooke, who commanded this front. His record stood high. Not only had he fought the decisive flank-battle near Ypres during the retirement to Dunkirk, but he had acquitted himself with singular firmness and dexterity, in circumstances of unimaginable difficulty and confusion, when in command of the new forces we had sent to France during the first three weeks of June. I also had a personal link with Alan Brooke through his two gallant brothers—the friends of my early military life.†

* This was an old device which I had used for the Marine Brigade of the Royal Naval Division when we landed on the French coast in September 1914. We took fifty of them from the London streets, and the Admiralty carried them across in a night.

† His brother Victor was a subaltern in the 9th Lancers when I joined the 4th Hussars, and I formed a warm friendship with him in 1895 and 1896. His horse reared up and fell over backwards, breaking his pelvis, and he was sorely stricken for the rest of his life. However, he continued to be able to serve and ride, and perished gloriously from sheer

These connections and memories did not decide my opinion on the grave matters of selection; but they formed a personal foundation upon which my unbroken war-time association with Alan Brooke was maintained and ripened. We were four hours together in the motor-car on this July afternoon, 1940, and we seemed to be in agreement on the methods of Home Defence. After the necessary consultations with others, I approved the Secretary of State for War's proposal to place Brooke in command of the Home Forces in succession to General Ironside. Ironside accepted his retirement with the soldierly dignity which on all occasions characterised his actions.

During the invasion menace for a year and a half Brooke organised and commanded the Home Forces, and thereafter when he had become C.I.G.S. we continued together for three and a half years until victory was won. I shall presently narrate the benefits which I derived from his advice in the decisive changes of command in Egypt and the Middle East in August 1942, and also the heavy disappointment which I had to inflict upon him about the command of the cross-Channel invasion Operation "Overlord" in 1944. His long tenure as chairman of the Chiefs of Staff Committee during the greater part of the war and his work as C.I.G.S. enabled him to render services of the highest order, not only to the British Empire, but also to the Allied Cause. These volumes will record occasional differences between us, but also an overwhelming measure of agreement, and will witness to a friendship which I cherish.

\*     \*     \*     \*     \*

Meanwhile we all faced in ever-increasing detail and tenacity the possibility of Invasion. Some of my minutes illustrate this process.

exhaustion whilst acting as liaison officer with the French Cavalry Corps in the retreat from Mons in 1914.

General Brooke had another brother, Ronnie. He was older than Victor and several years older than me. In the years 1895–98 he was thought to be a rising star in the British Army. Not only did he serve with distinction in all the campaigns which occurred, but he shone at the Staff College among his contemporaries. In the Boer War he was Adjutant of the South African Light Horse, and I for some months during the relief of Ladysmith was Assistant Adjutant, the regiment having six squadrons. Together we went through the fighting at Spion Kop, Vaal Krantz, and the Tugela. I learned much about tactics from him. Together we galloped into Ladysmith on the night of its liberation. Later on, in 1903, although I was only a youthful Member of Parliament, I was able to help him to the Somaliland campaign, in which he added to his high reputation. He was stricken down by arthritis at an early age, and could only command a reserve brigade at home during the First World War. Our friendship continued till his premature death in 1930.

(*Action this Day*)
*Prime Minister to Secretary of State for Air and C.A.S.*    3.VII.40

I hear from every side of the need for throwing your main emphasis on bombing the ships and barges in all the ports under German control.

*Prime Minister to General Ismay*    2.VII.40

See the letter [on the defence of London] from Mr. Wedgwood, M.P., which is interesting and characteristic. What is the position about London? I have a very clear view that we should fight every inch of it, and that it would *devour* quite a large invading army.

*Prime Minister to Mr. Wedgwood*    5.VII.40

Many thanks for your letters. I am hoping to get a great many more rifles very soon, and to continue the process of arming the Home Guard (L.D.V.). You may rest assured that we should fight every street of London and its suburbs. It would *devour* an invading army, assuming one ever got so far. We hope however to drown the bulk of them in the salt sea.

It is curious that the German army commander charged with the invasion plan used this same word "devour" about London, and determined to avoid it.

*Prime Minister to General Ismay*    4.VII.40

What is being done to encourage and assist the people living in threatened sea-ports to make suitable shelters for themselves in which they could remain during an invasion? Active measures must be taken forthwith. Officers or representatives of the local authority should go round explaining to families that if they decide not to leave in accordance with our general advice, they should remain in the cellars, and arrangements should be made to prop up the building overhead. They should be assisted in this both with advice and materials. Their gas-masks should be inspected. All this must be put actively in operation from to-day. The process will stimulate voluntary evacuation, and at the same time make reasonable provision for those who remain.

*Prime Minister to General Ismay*    5.VII.40

Clear instructions should now be issued about the people living in the threatened coastal zones: (1) They should be encouraged as much as possible to depart voluntarily, both by the pressure of a potential compulsory order hanging over them, and also by local (not national) propaganda through their Regional Commissioners or local bodies. Those who wish to stay, or can find nowhere to go on their own, should be told that if invasion impact occurs in their town or village on the coast they will not be able to leave till the battle is over. They should therefore be encouraged and helped to put their cellars in order so that they have fairly safe places to go to. They should be supplied with whatever form of Anderson shelter is now available (I hear there are new forms not involving steel). Only those who are trustworthy should be allowed to stay. All doubtful elements should be removed.

Pray have precise proposals formulated upon these lines for my approval.

*Prime Minister to Professor Lindemann* 7.VII.40
(Copy to General Ismay)

I want my "S" Branch to make a chart of all the thirty divisions, showing their progress towards complete equipment. Each division would be represented by a square divided into sections: officers and men, rifles, Bren guns, Bren-gun carriers, anti-tank rifles, anti-tank guns, field artillery, medium ditto (if any), transport sufficient to secure mobility of all three brigades simultaneously, etc. As and when a proportion of these subsidiary squares is completed a chart can be painted red. I should like to see this chart every week. A similar diagram can be prepared for the Home Guard. In this case it is only necessary to show rifles and uniforms.

*Prime Minister to Secretary of State for War* 7.VII.40

You shared my astonishment yesterday at the statement made to us by General McNaughton that the whole of the 2nd Canadian Division was destined for Iceland. It would surely be a very great mistake to allow these fine troops to be employed in so distant a theatre. Apparently the first three battalions have already gone there. No one was told anything about this. We require two Canadian divisions to work as a corps as soon as possible.

I am well aware of the arguments about training, etc., but they did not convince me. We ought to have another thorough re-examination of this point. Surely it should be possible to send second-line Territorial troops to Iceland, where they should fortify themselves at the key points, and then to have, say, one very high-class battalion of the "Gubbins" type in order to strike at any landing. I should be most grateful if you would deal with this.

*Prime Minister to First Lord and First Sea Lord* 7.VII.40

1. I cannot understand how we can tolerate the movement at sea along the French coast of any vessels without attacking them. It is not sufficient surely to use the air only. Destroyers should be sent under air escort. Are we really to resign ourselves to the Germans building up a large armada under our noses in the Channel, and conducting vessels through the Straits of Dover with impunity? This is the

beginning of a new and very dangerous threat which must be countered.

2. I should be glad of a report not only on the points mentioned above, but also on the state of our minefield there, and how it is to be improved. Is it true the mines have become defective after ten months? If so, several new rows should be laid. Why should not an effort be made to lay a minefield by night in the French passage, and lie in wait for any craft sent to sweep a channel through it? We really must not be put off from asserting our sea-power by the fact that the Germans are holding the French coast. If German guns open upon us a heavy ship should be sent to bombard them under proper air protection.

\*     \*     \*     \*     \*

During this month of July American weapons in considerable quantities were safely brought across the Atlantic. This seemed to me so vital that I issued reiterated injunctions for care in their transportation and reception.

*Prime Minister to Secretary of State for War*        7.VII.40

I have asked the Admiralty to make very special arrangements for bringing in your rifle convoys. They are sending four destroyers far out to meet them, and all should arrive during the 9th. You can ascertain the hour from the Admiralty. I was so glad to hear that you were making all preparations for the unloading, reception, and distribution of these rifles. At least 100,000 ought to reach the troops that very night, or in the small hours of the following morning. Special trains should be used to distribute them and the ammunition according to a plan worked out beforehand exactly, and directed from the landing-port by some high officer thoroughly acquainted with it. It would seem likely that you would emphasise early distribution to the coastal districts, so that all the Home Guard in the danger areas should be the first served. Perhaps you would be good enough to let me know beforehand what you decide.

*Prime Minister to General Ismay*        8.VII.40

Have any steps been taken to load the later portions of American ammunition, rifles, and guns upon faster ships than was the case last time? What are the ships in which the latest consignments are being packed, and what are their speeds? Will you kindly ascertain this from the Admiralty.

*Prime Minister to First Lord*        27.VII.40

The great consignments of rifles and guns, together with their ammunition, which are now approaching this country are entirely on a different level from anything else we have transported across the ocean except the Canadian Division itself. Do not forget that 200,000 rifles mean 200,000 men, as the men are waiting for the rifles. The convoys approaching on July 31 are unique, and a special effort should be made to ensure their safe arrival. The loss of these rifles and field-guns would be a disaster of the first order.

When the ships from America approached our shores with their priceless arms special trains were waiting in all the ports to receive their cargoes. The Home Guard in every county, in every town, in every village, sat up all through the night to receive them. Men and women worked night and day making them fit for use. By the end of July we were an armed nation, so far as parachute or air-borne landings were concerned. We had become a "hornets' nest". Anyhow, if we had to go down fighting (which I did not anticipate) a lot of our men and some women had weapons in their hands. The arrival of the first instalment of the half-million .300 rifles for the Home Guard (albeit with only about fifty cartridges apiece, of which we dared only issue ten, and no factories yet set in motion) enabled us to transfer three hundred thousand .303 British-type rifles to the rapidly-expanding formations of the Regular Army.

At the "seventy-fives", with their thousand rounds apiece, some fastidious experts presently turned their noses up. There were no limbers and no immediate means of procuring more ammunition. Mixed calibres complicate operations. But I would have none of this, and during all 1940 and 1941 these nine hundred "seventy-fives" were a great addition to our military strength for Home Defence. Arrangements were devised and men were drilled to run them up on planks into lorries for movement. When you are fighting for existence any cannon is better than no cannon at all, and the French "seventy-five", although out-dated by the British 25-pounder and the German field-gun howitzer, was still a splendid weapon.

\* \* \* \* \*

We had watched with attention the growth of the German heavy batteries along the Channel coast during August and September. By far the strongest concentration of this artillery was around Calais and Cape Gris-Nez, with the apparent purpose not only of forbidding the Straits to our warships but also of commanding the shortest route across them. We now know

# STATE OF READINESS 13ᵀᴴ July 1940
## INFANTRY DIVISIONS

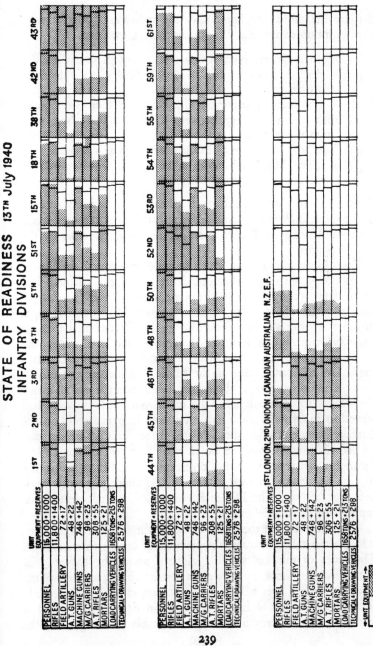

| UNIT | EQUIPMENT + RESERVES |
|---|---|
| PERSONNEL | 15,000 + 1000 |
| RIFLES | 11,800 + 1400 |
| FIELD ARTILLERY | 72 + 17 |
| A.T. GUNS | 48 + 22 |
| MACHINE GUNS | 746 + 142 |
| M/G CARRIERS | 96 + 23 |
| A.T. RIFLES | 308 + 55 |
| MORTARS | 125 + 21 |
| LOAD CARRYING VEHICLES | 1658 TONS + 213 TONS |
| TECHNICAL & DRAWING VEHICLES | 2576 + 298 |

1ST  2ND  3RD  4TH  5TH  51ST  15TH  18TH  38TH  42ND  43RD

| UNIT | EQUIPMENT + RESERVES |
|---|---|
| PERSONNEL | 15,000 + 1000 |
| RIFLES | 11,800 + 1400 |
| FIELD ARTILLERY | 72 + 17 |
| A.T. GUNS | 48 + 22 |
| MACHINE GUNS | 746 + 142 |
| M/G CARRIERS | 96 + 23 |
| A.T. RIFLES | 308 + 55 |
| MORTARS | 125 + 21 |
| LOAD CARRYING VEHICLES | 1658 TONS + 213 TONS |
| TECHNICAL & DRAWING VEHICLES | 2576 + 298 |

44TH  45TH  46TH  48TH  50TH  52ND  53RD  54TH  55TH  59TH  61ST

ISTLONDON. 2ND LONDON. 1. CANADIAN AUSTRALIAN  N.Z.E.F.

| UNIT | EQUIPMENT + RESERVES |
|---|---|
| PERSONNEL | 15,000 + 1000 |
| RIFLES | 11,800 + 1400 |
| FIELD ARTILLERY | 72 + 17 |
| A.T. GUNS | 48 + 22 |
| MACHINE GUNS | 746 + 142 |
| M/G CARRIERS | 96 + 23 |
| A.T. RIFLES | 308 + 55 |
| MORTARS | 125 + 21 |
| LOAD CARRYING VEHICLES | 1658 TONS + 213 TONS |
| TECHNICAL & DRAWING VEHICLES | 2576 + 298 |

← UNIT EQUIPMENT →
← RESERVES →

Ɪ  *Levels on 13th July 1940*

239

that by the middle of September the following batteries were already mounted and ready for use in this region alone:

(a) Siegfried battery, south of Gris-Nez, with four 38-cm. guns.
(b) Friedrich-August battery, north of Boulogne, with three 30.5-cm. guns.
(c) Grosser Kurfuerst battery, at Gris-Nez, with four 28-cm. guns.
(d) Prinz Heinrich battery, between Calais and Blanc-Nez, with two 28-cm. guns.
(e) Oldenburg battery, east of Calais, with two 24-cm. guns.
(f) M.1, M.2, M.3, M.4 batteries, in the sector of Gris-Nez—Calais, with a total of fourteen 17-cm. guns.

Besides this no fewer than thirty-five heavy and medium batteries of the German Army, as well as seven batteries of captured guns, were sited along the French coast for defensive purposes by the end of August.

The orders which I had given in June for arming the Dover promontory with guns that could fire across the Channel had borne fruit, though not on the same scale. I took a personal interest in the whole of this business. I visited Dover several times in these anxious summer months. In the Citadel of the Castle large underground galleries and chambers had been cut in the chalk, and there was a wide balcony from which on clear days the shores of France, now in the hands of the enemy, could be seen. Admiral Ramsay, who commanded, was a friend of mine. He was the son of a colonel of the 4th Hussars under whom I had served in my youth, and I had often seen him as a child on the Barrack Square at Aldershot. When three years before the war he had resigned his position as Chief of Staff to the Home Fleet through a difference with its Commander-in-Chief, it was to me that he had come to seek advice. I had long talks with him, and together with the Dover Fortress Commander visited our rapidly-improving defences.

I carefully studied there and at home the Intelligence reports, which almost daily showed the progress of the German batteries. The series of minutes which I dictated about the Dover guns during August show my very great desire to break up some of the heaviest battery sites before their guns could reply. I certainly thought this ought to have been done in August, as we had at least three of the very heaviest guns capable of firing across the

Channel. Later on the Germans became too strong for us to court a duel.

*Prime Minister to General Ismay*                                      3.VIII.40

1. The 14-inch gun I ordered to be mounted at Dover should be ready in ample time to deal with this new German battery. It certainly should not fire until all the guns are in position. The plan for the shoot may however now be made, and I should like to know what arrangements for spotting aircraft, protected by fighters in strength, will be prepared for that joyous occasion. Also, when the two guns, 13.5's on railway mountings, will be ready, and whether they can reach the target mentioned. Several other camouflaged guns should be put up at various points, with arrangements to make suitable flashes, smoke and dust. Let me know what arrangements can be devised. I presume work on the railway extensions for the 13.5's is already in hand. Please report.

2. The movement of the German warships southward to Kiel creates a somewhat different situation from that dealt with in C.-in-C. Home Fleet's appreciation asked for some time ago about an invasion across the narrow waters supported by heavy ships. The Admiralty should be asked whether C.-in-C.'s attention should not be drawn to the altered dispositions of the enemy, in case he has anything further to say.

*Prime Minister to First Lord*                                      8.VIII.40

I am impressed by the speed and efficiency with which the emplacement for the 14-inch gun at Dover has been prepared and the gun itself mounted. Will you tell all those who have helped in this achievement how much I appreciate the sterling effort they have made.

The enemy batteries first opened fire on August 22, engaging a convoy without effect and later firing on Dover. They were replied to by one of our 14-inch guns which was now in action. Thenceforward there were artillery duels at irregular intervals. Dover was engaged six times in September, the heaviest day being September 9, when over one hundred and fifty shells were fired. Very little damage was done to convoys.

*Prime Minister to First Lord and First Sea Lord*                   25.VIII.40

I shall be much obliged if you will make proposals for a shoot by *Erebus** against the German batteries at Gris-Nez. I was very glad to

---

* H.M.S. *Erebus* was a monitor of the first war mounting two 15-inch guns. After being refitted she went to Scapa for target practice in August. Delay arose in her working up practices through defects and bad weather and she did not reach Dover until late in September. It was therefore not until the night of September 29–30 that she carried out a bombardment of Calais.

hear you thought this practicable. It is most desirable. There is no reason why it should wait for the railway guns, though of course if they were ready they could follow on with the 14-inch at daybreak. We ought to smash these batteries. I hope we have not got to wait for the next moon for *Erebus*, and I shall be glad to know what are the moon conditions which you deem favourable.

*Prime Minister to General Ismay and C.O.S. Committee*     27.VIII.40

It would not seem unreasonable that the enemy should attempt gradually to master the Dover promontory and command the Channel at its narrowest point. This would be a natural preliminary to invasion. It would give occasion for continued fighting with our Air Force in the hope of exhausting them by numbers. It would tend to drive our warships from all the Channel bases. The concentration of many batteries on the French coast must be expected. What are we doing in defence of the Dover promontory by heavy artillery? Ten weeks ago I asked for heavy guns. One has been mounted. Two railway guns are expected. Now we are told these will be very inaccurate on account of super-charging. We ought to have a good many more heavy guns lined up inside to smaller calibre with stiffer rifling and a range of at least fifty miles, and firing at twenty-five or thirty miles would then become more accurate. I do not understand why I have not yet received proposals on this subject. We must insist upon maintaining superior artillery positions on the Dover promontory, no matter what form of attack they are exposed to. We have to fight for the command of the Straits by artillery, to destroy the enemy's batteries, and to multiply and fortify our own.

I have sent on other papers a request for a surprise attack by *Erebus*, which should be able to destroy the batteries at Gris-Nez. She has an armoured deck against air bombing. What is being done about this? When is she going into action? The Air Ministry should of course co-operate. The operation would take an offensive turn. We should require spotting aircraft by day. It may be that the first squadrons of Hurricanes fitted with Merlin 20 would be the best for this. If *Erebus* is attacked from the air she should be strongly defended, and action sought with the enemy Air Force.

Pray let me have your plans.

*Prime Minister to General Ismay, for C.O.S. Committee*     30.VIII.40

Further to my previous minute on defence of the Kentish promontory, we must expect that very powerful batteries in great numbers will be rapidly brought into being on the French coast. It would be a natural thought for the Germans to try to dominate the Straits by artillery. At present we are ahead of them with our 14-inch and two

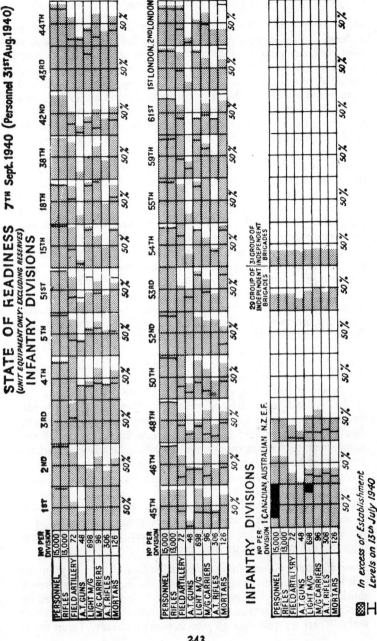

## STATE OF READINESS 7ᵀᴴ Sept.1940 (Personnel 31ˢᵀ Aug.1940)

*(UNIT EQUIPMENT ONLY: EXCLUDING RESERVES)*

### INFANTRY DIVISIONS

| Nᵒ PER DIVISION | |
|---|---|
| PERSONNEL | 15,000 |
| RIFLES | 13,000 |
| FIELD ARTILLERY | 72 |
| A.T. GUNS | 48 |
| LIGHT M/G | 698 |
| M/G CARRIERS | 96 |
| A.T. RIFLES | 306 |
| MORTARS | 126 |

1ST  2ND  3RD  4TH  5TH  15TH  51ST  18TH  38TH  42ND  43RD  44TH

| Nᵒ PER DIVISION | |
|---|---|
| PERSONNEL | 15,000 |
| RIFLES | 13,000 |
| FIELD ARTILLERY | 72 |
| A.T. GUNS | 48 |
| LIGHT M/G | 698 |
| M/G CARRIERS | 96 |
| A.T. RIFLES | 306 |
| MORTARS | 126 |

45TH  46TH  48TH  50TH  52ND  53RD  54TH  55TH  59TH  61ST  1ST LONDON, 2ND LONDON

### INFANTRY DIVISIONS

| Nᵒ PER DIVISION | |
|---|---|
| PERSONNEL | 15,000 |
| RIFLES | 13,000 |
| FIELD ARTILLERY | 72 |
| A.T. GUNS | 48 |
| LIGHT M/G | 698 |
| M/G CARRIERS | 96 |
| A.T. RIFLES | 306 |
| MORTARS | 126 |

1 CANADIAN, AUSTRALIAN, N.Z.E.F.   29 GROUP OF INDEPENDENT BRIGADES | 31 GROUP OF INDEPENDENT BRIGADES

In excess of Establishment

Levels on 13ᵗʰ July 1940

243

13.5 railway guns. The Admiral at Dover should be furnished in addition, as soon as possible, with a large number of the most modern 6-inch or 8-inch guns. I understand the Admiralty is considering taking guns from [H.M.S.] *Newcastle* or *Glasgow*, which are under long repair. A record evolution should be made of getting one or two of these turrets mounted. Report to me about this and dates. There is a 9.2 Army experimental gun and mounting, and surely we have some 12-inch on railway mountings. If our ships cannot use the Straits the enemy must not be able to. Even if guns cannot fire on to the French shore they are none the less very valuable.

Some of our heavy artillery—the 18-inch howitzer and the 9.2's—should be planted in positions whence they could deny the ports and landings to the enemy, and, as C.I.G.S. mentioned, support the counter-attack which would be launched against any attempted bridgehead. Much of this mass of artillery I saved from the last war has done nothing, and has been under reconditioning for a whole year. Let me have a good programme for using it to support counter-strokes and deny landings, both north and south of the Thames. Farther north I have seen already some very good heavy batteries.

I should like also to be informed of the real [actual] lines of defence drawn up between Dover and London and Harwich and London. Now that the coast is finished there is no reason why we should not develop these lines, which in no way detract from the principle of vehement counter-attack.

But the most urgent matter is one or two modern 6-inch to shoot all German craft up to 35,000 yards.

I am also endeavouring to obtain from United States at least a pair of their 16-inch coast defence weapons. These fire 45,000 yards, throwing a ton and a quarter, without being super-charged. They should therefore be very accurate. General Strong, United States Army, mentioned this to me as a promising line. He thought, without committing his Government, that the United States Army might be prepared to take a couple of these guns and their carriages away from some of their twin batteries.

Let me know all details about these guns. It ought to be possible to make the concrete foundation in three months, and I expect it would take as long to get these guns over here. There are very few ships that can carry them on their decks.

*Prime Minister to General Ismay and First Sea Lord*      31.VIII.40

It becomes particularly urgent to attack the batteries on the French shore. Yesterday's photographs show guns being actually hoisted up into position, and it will be wise to fire on them before they are able to reply. There are quite enough guns in position already. I trust therefore *Erebus* will not be delayed, as every day our task will become harder.

It seems most necessary to damage and delay the development of the hostile batteries in view of the fact that we are so far behindhand with our own.

At the beginning of September our heavy gun strength towards the sea was:

*Pre-War Coast Defence*

| | |
|---|---|
| 9.2-inch | two |
| 6-inch | six |

*Recent Additions*

| | |
|---|---|
| 14-inch (Naval) | one |
| 9.2-inch | two (railway mountings) |
| 6-inch (Naval) | two |
| 4-inch (Naval) | two |

These were soon to be further reinforced by two 13.5-inch guns from the old battleship *Iron Duke*, which were being erected on railway mountings, and a battery of four 5.5-inch guns from H.M.S. *Hood*. Many of these additional guns were manned by the Royal Navy and Royal Marines.

Although still inferior in numbers to the enemy we thus had a powerful fire concentration.

In addition, one of the 18-inch howitzers I had saved after the first war and twelve 12-inch howitzers were installed for engaging enemy landings. All these were mobile and would have brought a terrible fire on any landing area.

\*    \*    \*    \*    \*

As the months of July and August passed without any disaster we settled ourselves down with increasing assurance that we could make a long and hard fight. Our gains of strength were borne in upon us from day to day. The entire population laboured to the last limit of its strength, and felt rewarded when they fell asleep after their toil or vigil by a growing sense that we should have time and that we should win. All the beaches now bristled with defences of various kinds. The whole country was organised in defensive localities. The factories poured out their weapons. By the end of August we had over two hundred and fifty new tanks! The fruits of the American "Act of Faith" had been gathered. The whole trained professional British Army and its Territorial comrades drilled and exercised from morn till night, and longed to meet the foe. The Home Guard overtopped the million mark, and when rifles were lacking grasped lustily the shotgun, the sporting rifle, the private pistol, or, when there was no firearm, the pike and the club. No Fifth Column existed in Britain, though a few spies were carefully rounded up and examined. What few Communists there were lay low. Everyone else gave all they had to give.

When Ribbentrop visited Rome in September he said to Ciano: "The English territorial defence is non-existent. A single German division will suffice to bring about a complete collapse." This merely shows his ignorance. I have often wondered however what would have happened if two hundred thousand German storm troops had actually established themselves ashore. The massacre would have been on both sides grim and great. There would have been neither mercy nor quarter. They would have used Terror, and we were prepared to go all lengths. I intended to use the slogan "You can always take one with you". I even calculated that the horrors of such a scene would in the last resort turn the scale in the United States. But none of these emotions was put to the proof. Far out on the grey waters of the North Sea and the Channel coursed and patrolled the faithful, eager flotillas peering through the night. High in the air soared the fighter pilots, or waited serene at a moment's notice around their excellent machines. This was a time when it was equally good to live or die.

# CHAPTER XIV

# THE INVASION PROBLEM

*Former Studies of Invasion - The New Air-Power - My Statement
to Parliament of June 18 - The First Rumours - My Note of June 28
- My Minute on "Invasion" of July 10 - Importance of Mobile
Reserves - Two Thousand Miles of British Coastline - The First Sea
Lord's Memorandum - Distribution of Potential Attack - I Double
His Estimate for Safety - My Minute of August 5, 1940 - My Sug-
gested Distribution of Our Army - Coincidence of Chiefs of Staff
View - Our Emphasis on East Coast - The Germans Choose the
South Coast - We Turn Our Front - Change in Our Dispositions
between August and September - Persisting Dangers from Across the
North Sea - Tension in July and August.*

AFTER Dunkirk, and still more when three weeks later the
French Government capitulated, the questions whether
Hitler would, or secondly could, invade and conquer our
Island rose, as we have seen, in all British minds. I was no novice
at this problem. As First Lord I had for three years before the
First Great War taken part in all the discussions of the Committee of Imperial Defence upon the point. On behalf of the
Admiralty I had always argued that at least two divisions out of
our Expeditionary Force of six should be kept at home until the
Territorial Army and other war-time forces became militarily
effective. As Admiral "Tug" Wilson put it: "The Navy cannot
play international football without a goalkeeper." However,
when at the outbreak of that war we found ourselves with the
Navy fully mobilised, the Grand Fleet safe beyond hostile ken at
sea, all surprises, treacheries, and accidents left behind us, we
felt ourselves able at the Admiralty to be better than our word.
At the extraordinary meeting of Ministers and high military

authorities which Mr. Asquith summoned to the Cabinet Room on August 5, 1914, I declared formally, with the full agreement of the First Sea Lord (Prince Louis of Battenberg), that the Navy would guarantee the protection of the Island against invasion or serious raid even if all the Regular troops were immediately sent to the great battle impending in France. So far as we were concerned, the whole Army could go. In the course of the first six weeks all the six divisions went.

Sea-power, when properly understood, is a wonderful thing. The passage of an army across salt water in the face of superior fleets and flotillas is an almost impossible feat. Steam had added enormously to the power of the Navy to defend Great Britain. In Napoleon's day the same wind which would carry his flat-bottomed boats across the Channel from Boulogne would drive away our blockading squadrons. But everything that had happened since then had magnified the power of the superior navy to destroy the invaders in transit. Every complication which modern apparatus had added to armies made their voyage more cumbrous and perilous, and the difficulties of their maintenance when landed probably insuperable. At that former crisis in our Island fortunes we possessed superior and, as it proved, ample sea-power. The enemy was unable to gain a major sea battle against us. He could not face our cruiser forces. In flotillas and light craft we outnumbered him tenfold. Against this must be set the incalculable chances of weather, particularly fog. But even if this were adverse and a descent were effected at one or more points the problem of maintaining a hostile line of communications and of nourishing any lodgments remained unsolved. Such was the position in the First Great War.

But now there was the air. What effect had this sovereign development produced upon the invasion problem? Evidently if the enemy could dominate the narrow seas, on both sides of the Straits of Dover, by superior air-power, the losses of our flotillas would be very heavy and might eventually be fatal. No one would wish, except on a supreme occasion, to bring heavy battleships or large cruisers into waters commanded by the German bombers. We did not in fact station any capital ships south of the Forth or east of Plymouth. But from Harwich, the Nore, Dover, Portsmouth, and Portland we maintained a tireless, vigilant patrol of light fighting vessels which steadily increased in number. By September they exceeded eight hundred, which only a hostile air-power could destroy, and then only by degrees.

But who had the power in the air? In the Battle of France we had fought the Germans against odds of two and three to one and inflicted losses in similar proportions. Over Dunkirk, where we had to maintain continuous patrol to cover the escape of the Army, we had fought at four or five to one with success and profit. Over our own waters and exposed coasts and counties Air Chief Marshal Dowding contemplated profitable fighting at seven or eight to one. The strength of the German Air Force at this time, taken as a whole, so far as we knew—and we were well informed —apart from particular concentrations, was about three to one Although these were heavy odds at which to fight the brave and efficient German foe, I rested upon the conclusion that in our own air, over our own country and its waters, we could beat the German Air Force. And if this were true our naval power would continue to rule the seas and oceans, and would destroy all enemies who set their course towards us.

There was of course a third potential factor. Had the Germans with their renowned thoroughness and foresight secretly prepared a vast armada of special landing-craft, which needed no harbours or quays, but could land tanks, cannon, and motor vehicles anywhere on the beaches, and which thereafter could supply the landed troops? As has been shown, such ideas had risen in my mind long ago in 1917, and were now being actually developed as the result of my directions. We had however no reason to believe that anything of this kind existed in Germany, though it is always best when counting the cost not to exclude the worst. It took us four years of intense effort and experiment and immense material aid from the United States to provide such equipment on a scale equal to the Normandy landing. Much less would have sufficed the Germans at this moment. But they had only a few Siebel ferries.

Thus the invasion of England in the summer and autumn of 1940 required from Germany local naval superiority and air superiority and immense special fleets and landing-craft. But it was we who had the naval superiority; it was we who conquered the mastery in the air; and finally we believed, as we now know rightly, that they had not built or conceived any special craft. These were the foundations of my thought about invasion in 1940, from which I gave from day to day the instructions and directives which these chapters contain.

\* \* \* \* \*

I laid the broad outlines plainly before Parliament on June 18:

The Navy has never pretended to be able to prevent raids by bodies of five or ten thousand men flung suddenly across and thrown ashore at several points on the coast some dark night or foggy morning. The efficacy of sea-power, especially under modern conditions, depends upon the invading force being of large size. It has to be of large size, in view of our military strength, to be of any use. If it is of large size, then the Navy have something they can find and meet and, as it were, bite on. Now we must remember that even five divisions, however lightly equipped, would require 200 to 250 ships, and with modern air reconnaissance and photography it would not be easy to collect such an armada, marshal it, and conduct it across the sea without any powerful naval forces to escort it; and there would be very great possibilities, to put it mildly, that this armada would be intercepted long before it reached the coast, and all the men drowned in the sea, or, at the worst, blown to pieces with their equipment while trying to land.

\* \* \* \* \*

As early as the end of June some reports indicated that the enemy's plans might include the Channel, and I immediately called for inquiry.

*Prime Minister to General Ismay* 27.VI.40
It seems difficult to believe that any large force of transports could be brought to the Channel ports without our being aware of it, or that any system of mining would prevent our sweepers from clearing a way for attack on such transports on passage. However, it would be well if the Chiefs of Staff gave their attention to this rumour.

Anyhow, the possibility of a cross-Channel invasion, improbable though it was at that time, had to be most closely examined. I was not entirely satisfied with the military dispositions. It was imperative that the Army should know the exact task assigned to it, and above all should not fritter away strength in a sedentary dispersion along the threatened coasts or exhaust the national resources by manning unduly all the coasts. Therefore I wrote:

*Prime Minister to General Ismay*                                    28.VI.40

### NOTE BY THE PRIME MINISTER TO C.O.S. COMMITTEE

1. See papers by Vice-Chiefs of Staff and further papers by C.O.S. Committee.

2. It is prudent to block off likely sections of the beaches with a good defence and to make secure all creeks and harbours on the east coast. The south coast is less immediately dangerous. No serious invasion is possible without a harbour with its quays, etc. No one can tell, should the Navy fail, on what part of the east coast the impact will fall. Perhaps there will be several lodgments. Once these are made all troops employed on other parts of the coastal crust will be as useless as those in the Maginot Line. Although fighting on the beaches is favourable to the defence, this advantage cannot be purchased by trying to guard all the beaches. The process must be selective. But if time permits defended sectors may be widened and improved.

3. Every effort should be made to man coast defences with sedentary troops, well sprinkled with experienced late-war officers. The safety of the country depends [however] on having a large number (now only nine, but should soon be fifteen) of "Leopard" brigade groups which can be directed swiftly, *i.e.*, within four hours, to the points of lodgment. Difficulties of landing on beaches are serious, even when the invader has reached them; but difficulties of nourishing a lodgment when exposed to heavy attack by land, air, and sea are far greater. All therefore depends on rapid, resolute engagement of any landed forces which may slip through the sea control. This should not be beyond our means provided the field troops are not consumed in beach defences, and are kept in a high condition of mobility, crouched and ready to spring.

4. In the unhappy event of the enemy capturing a port, larger formations with artillery will be necessary. There should be four or five good divisions held in general reserve to deal with such an improbable misfortune. The scale of lodgment to be anticipated should be not more than ten thousand men landed at three points simultaneously—say thirty thousand in all; the scale of air attack not more than fifteen thousand landed simultaneously at two or three points in all. The enemy will not have strength to repeat such descents often. It is very doubtful whether air-borne troops can be landed in force by night; by day they should be an easy prey [to our Air Force].

5. The tank story is somewhat different, and it is right to minimise by local cannon and obstacles the landing-places of tanks. The Admiralty should report upon the size, character, and speed of potential tank-carrying barges or floats, whether they will be self-propelled or towed and by what craft. As they can hardly go above seven miles

an hour they should be detected in summer-time after they have started, and even in fog or haze the R.D.F. stations should give warning while they are still several hours from land. The destroyers issuing from the sally-ports must strike at these with gusto. The arrangement of stops and blocks held by local sedentary forces should be steadily developed, and anti-tank squads formed. Our own tank reserve must engage the surviving invader tanks, and no doubt it is held in a position which allows swift railing [transport by rail] to the attacked area.

6. Parachutists, Fifth-Columnists, and enemy motor-cyclists who may penetrate or appear in disguise in unexpected places must be left to the Home Guard, reinforced by special squads. Much thought must be given to the [enemy] trick of wearing British uniform.

7. In general I find myself in agreement with the Commander-in-Chief's plan, but all possible field troops must be saved from the beaches and gathered into the "Leopard" brigades and other immediate mobile supports. Emphasis should be laid upon the main reserve. The battle will be won or lost, not on the beaches, but by the mobile brigades and the main reserve. Until the Air Force is worn down by prolonged air fighting and destruction of aircraft supply the power of the Navy remains decisive against any serious invasion.

8. The above observations apply only to the immediate summer months. We must be much better equipped and stronger before the autumn.

In July there was growing talk and anxiety on the subject both inside the British Government and at large. In spite of ceaseless reconnaissance and all the advantages of air photography, no evidence had yet reached us of large assemblies of transport in the Baltic or in the Rhine or Scheldt harbours, and we were sure that no movement either of shipping or self-propelled barges through the Straits into the Channel had taken place. Nevertheless preparation to resist invasion was the supreme task before us all, and intense thought was devoted to it throughout our war circle and Home Command.

## INVASION

### NOTE BY THE PRIME MINISTER

*Prime Minister to C.-in-C. Home Forces, C.I.G.S., and General Ismay*                                                                 10.VII.40

1. I find it very difficult to visualise the kind of invasion all along the coast by troops carried in small craft, and even in boats. I have not seen any serious evidence of large masses of this class of craft being

assembled, and, except in very narrow waters, it would be a most hazardous and even suicidal operation to commit a large army to the accidents of the sea in the teeth of our very numerous armed patrolling forces. The Admiralty have over a thousand armed patrolling vessels, of which two or three hundred are always at sea, the whole being well manned by competent seafaring men. A surprise crossing should be impossible, and in the broader parts of the North Sea the invaders should be an easy prey, as part of their voyage would be made by daylight. Behind these patrolling craft are the flotillas of destroyers, of which forty destroyers are now disposed between the Humber and Portsmouth, the bulk being in the narrowest waters. The greater part of these are at sea every night, and rest in the day. They would therefore probably encounter the enemy vessels in transit during the night, but also could reach any landing-point or points on the front mentioned in two or three hours. They could immediately break up the landing-craft, interrupt the landing, and fire upon the landed troops, who, however lightly equipped, would have to have some proportion of ammunition and equipment carried on to the beaches from their boats. The flotillas would however need strong air support from our fighter aircraft during their intervention from dawn onwards. The provision of the air fighter escort for our destroyers after daybreak is essential to their most powerful intervention on the beaches.

2. You should see the Commander-in-Chief's (Home Fleet) reply to the question put to him by the desire of the Cabinet, i.e., what happens if the enemy cover the passage of their invading army with their heavy warships? The answer is that, as far as we know at present, they have no heavy ships not under long repair, except those at Trondheim,* which are closely watched by our very largely superior forces. When the Nelson and Barham are worked up after refit in a few days' time (the 13th and 16th), it would be easily possible to make two forces of British heavy ships, either of which would be sufficiently strong; thus the danger of a northern outbreak could be contained, and at the same time a dart to the south by the Trondheim ships could be rapidly countered. Moreover, the cruisers in the Thames and Humber are themselves strong enough, with the flotillas, to attack effectively any light cruisers with which the enemy could cover an invasion. I feel therefore that it will be very difficult for the enemy to place large well-equipped bodies of troops on the east coast of England, whether in formed bodies or flung piecemeal on the beaches as they get across. Even greater difficulties would attend expeditions in larger vessels seeking to break out to the northward. It may further be added that at

* Actually the Scharnhorst and Gneisenau, which had been at Trondheim, had both been torpedoed and were out of action.

present there are no signs of any assemblies of ships or small craft sufficient to cause anxiety, except perhaps in Baltic ports. Frequent reconnaissance by the air and the constant watching by our submarines should give timely warning, and our minefields are an additional obstruction.

3. Even more unlikely is it that the south coast would be attacked. We know that no great mass of shipping exists in the French ports, and that the numbers of small boats there are not great. The Dover barrage is being replenished and extended to the French shore. This measure is of the utmost consequence, and the Admiralty are being asked to press it forward constantly and rapidly. *They do not think that any important vessels, warships or transports, have come through the Straits of Dover.* Therefore I find it difficult to believe that the south coast is in serious danger at the present time. Of course a small raid might be made upon Ireland from Brest. But this also would be dangerous to the raiders while at sea.

4. The main danger is from the Dutch and German harbours, which bear principally upon the coast from Dover to the Wash. As the nights lengthen this danger zone will extend northwards, but then, again, the weather becomes adverse and the "fishing-boat invasion" far more difficult. Moreover, with cloud the enemy air support may be lacking at the moment of his impact.

5. I hope therefore, relying on the above reasoning, which should be checked with the Admiralty, that you will be able *to bring an ever larger proportion of your formed divisions back from the coast into support or reserve, so that their training may proceed in the highest forms of offensive warfare and counter-attack,* and that the coast, as it becomes fortified, will be increasingly confided to troops other than those of the formed divisions, and also to the Home Guard. I am sure you will be in agreement with this view in principle, and the only question open would be the speed of the transformation. Here too I hope we shall be agreed that the utmost speed shall rule.

6. Air-borne attack is not dealt with in this note. It does not alter its conclusions.

\* \* \* \* \*

It will be noted that my advisers and I deemed the east coast more likely to be attacked during July and August than the south coast. There was in fact no chance of either assault during these two months. As will presently be described, the German plan was to invade across the Channel with medium ships (4,000 to 5,000 tons) and small craft, and we now know that they never had any hope or intention of moving an army from the Baltic and North Sea ports in large transports; still less did they make any plans for an invasion from the Biscay ports. This does not mean that in choosing the south coast as their target they were thinking rightly and we wrongly. The east coast invasion was by far the more formidable if the enemy had had the means to attempt it. There could of course be no south coast invasion unless or until the necessary shipping had passed southwards through the Straits of Dover and had been assembled in the French Channel ports. Of this, during July, there was no sign.

We had none the less to prepare against all variants, and yet at the same time avoid the dispersion of our mobile forces, and to gather reserves. This nice and difficult problem could only be solved in relation to the news and events from week to week. The British coastline, indented with innumerable inlets, is over two thousand miles in circumference, without including Ireland. The only way of defending so vast a perimeter, any part or parts of which may be simultaneously or successively attacked, is by lines of observation and resistance around the coast or frontiers with the object of delaying an enemy, and meanwhile creating the largest possible reserves of highly-trained mobile troops so disposed as to be able to reach any point assailed in the shortest time for strong counter-attack. When in the last phases of the war Hitler found himself encircled and confronted with a similar problem he made, as we shall see, the gravest possible mistakes in handling it. He had created a spider's web of communications, *but he forgot the spider*. With the example of the unsound French dispositions for which such a fatal penalty had just been exacted fresh in our memories, we did not forget the "mass of manœuvre"; and I ceaselessly inculcated this policy to the utmost extent that our growing resources would allow.

The views in my paper of July 10 were in general harmony with Admiralty thought, and two days later Admiral Pound sent me a full and careful statement which he and the Naval Staff had drawn up in pursuance of it. Naturally and properly, the dangers we had to meet were forcefully stated.

But in summing up Admiral Pound said: "*It appears probable that a total of some hundred thousand men might reach these shores without being intercepted by naval forces . . . but the maintenance of their line of supply, unless the German Air Force had overcome both our Air Force and our Navy, seems practically impossible.*

... If the enemy undertook this operation he would do so in the hope that he could make a quick rush on London, living on the country as he went, and force the Government to capitulate." The First Sea Lord divided the hundred-thousand maximum figure both as to enemy ports of departure and the possible impact on our coasts as in the following table:

| | |
|---|---:|
| South coast from Bay of Biscay ports | 20,000 |
| South coast from Channel ports | 5,000 |
| East coast from Dutch and Belgian ports | 12,000 |
| East coast from German ports | 50,000 |
| Shetlands, Iceland, and coast of Scotland from Norwegian ports | 10,000 |
| | 97,000 |

I was content with this estimate. As the enemy could not bring heavy weapons with them, and would speedily have the supply lines of any lodgments cut, the invading strength seemed even in July to be well within the capacity of our rapidly-improving Army. I remitted the two documents to the Staffs and Home Command.

### MINUTE BY THE PRIME MINISTER
15.VII.40

The Chiefs of Staff and Home Defence should consider these papers. The First Sea Lord's memorandum may be taken as a working basis, and although I personally believe that the Admiralty will in fact be better than their word, and that the invaders' losses in transit would further reduce the scale of attack, yet the preparations of the land forces should be such as to make assurance doubly sure. *Indeed, for the land forces the scale of attack might well be doubled, namely, 200,000 men distributed as suggested [i.e., in the proportion suggested] by the First Sea Lord.* Our Home Army is already at a strength when it should be able to deal with such an invasion, and its strength is rapidly increasing.

I should be very glad if our plans to meet invasion on shore could be reviewed on this basis, so that the Cabinet may be informed of any modifications. It should be borne in mind that although the heaviest attack would seem likely to fall in the north, *yet the sovereign importance of London and the narrowness of the seas in this quarter make the south the theatre where the greatest precautions must be taken.*

There was general acceptance of this basis, and for the next few weeks we proceeded upon it. Upon the action to be taken

by our main Fleet in the narrow waters precise orders were issued
with which I was in full agreement. On July 20, after considerable discussion with Admiral Forbes, the Commander-in-Chief
the following decisions were promulgated by the Admiralty:

(i) Their Lordships do not expect our heavy ships to go south to
break up an expedition landing on our coast in the absence of
any reports indicating the presence of enemy heavy ships.

(ii) If enemy heavy ships support an expedition, accepting the risks
involved in an approach to our coast in the southern part of the
North Sea, then it is essential that our heavy ships should move
south against them, also accepting risks.

In order to reach more definite conclusions about the varying
probabilities and scales of attack on our extended coastline, so as
to avoid undue spreading of our forces, I sent the Chiefs of Staff
a further minute early in August.

## DEFENCE AGAINST INVASION

Minute by the Prime Minister and Minister of Defence

5.VIII.40

Bearing in mind the immense cost in war energy and the disadvantages of attempting to defend the whole coast of Great Britain, and
the dangers of being unduly committed to systems of passive defence
I should be glad if the following notes could be considered:

1. Our first line of defence against invasion must be as ever the
enemy's ports. Air reconnaissance, submarine watching, and other
means of obtaining information should be followed by resolute attacks
with all our forces available and suitable upon any concentrations of
enemy shipping.

2. Our second line of defence is the vigilant patrolling of the sea
to intercept any invading expedition, and to destroy it in transit.

3. Our third line is the counter-attack upon the enemy when he
makes any landfall, and particularly while he is engaged in the act
of landing. This attack, which has long been ready from the sea,
must be reinforced by air action; and both sea and air attacks must
be continued so that it becomes impossible for the invader to nourish
his lodgments.

4. The land defences and the Home Army are maintained primarily
for the purpose of making the enemy come in such large numbers as to
afford a proper target to the sea and air forces above mentioned, and

to make hostile preparations and movements noticeable to air and other forms of reconnaissance.

5. However, should the enemy succeed in landing at various points, he should be made to suffer as much as possible by local resistance on the beaches, combined with the aforesaid attack from the sea and the air. This forces him to use up his ammunition, and confines him to a limited area. The defence of any part of the coast must be measured, not by the forces on the coast, but by the number of hours within which strong counter-attacks by mobile troops can be brought to bear upon the landing-places. Such attacks should be hurled with the utmost speed and fury upon the enemy at his weakest moment, which is not, as is sometimes suggested, when actually getting out of his boats, but when sprawled upon the shore with his communications cut and his supplies running short. It ought to be possible to concentrate ten thousand men fully equipped within six hours, and twenty thousand men within twelve hours, upon any point where a serious lodgment has been effected. The withholding of the reserves until the full gravity of the attack is known is a nice problem for the Home Command.

6. It must be admitted that the task of the Navy and Air Force in preventing invasion becomes more difficult in the Narrow Seas, namely, from the Wash to Dover. This sector of the coast front is also nearest to the supreme enemy objective, London. The sector from Dover to Land's End is far less menaced, because the Navy and Air Force must make sure that no mass of shipping, still less protecting warships, can be passed into the French Channel ports. At present the scale of attack on this wide front is estimated by the Admiralty at no more than five thousand men.* Doubling this for greater security, it should be possible to make good arrangements for speedy counter-attack in superior numbers, and at the same time to achieve large economies of force on this southern sector, in which the beach troops should be at their minimum and the mobile reserves at their maximum. These mobile reserves must be available to move to the south-eastern sectors at short notice. *Evidently this situation can be judged only from week to week.*

7. When we come to the west coast of Britain, a new set of conditions rules. The enemy must commit himself to the broad seas, and there will be plenty of time, if his approach is detected, to attack him with cruisers and flotillas. The Admiralty dispositions should conform to this need. The enemy has at present no warships to escort him. Should we, for instance, care to send twelve thousand men unescorted in merchant ships to land on the Norwegian coast, or in the Skagerrak

---

* Here I omitted to mention the 20,000 which might come from the distant Biscay ports; but, as will be seen, my proposed disposition of our forces guarded against this potential but, as we now know, non-existent danger.

and Kattegat, in face of superior sea-power and air-power? It would be thought madness.

8. However, to make assurance triply sure, the Admiralty should pursue their plan of laying a strong minefield from Cornwall to Ireland, covering the Bristol Channel and the Irish Sea from southward attack. This minefield is all the more necessary now that by the adoption of the northabout route for commerce we have transferred a large part of our patrolling craft from the [South-] Western Approaches, which have become permanently more empty and unwatched.

9. The establishment of this minefield will simplify and mitigate all questions of local defence north of its point of contact with Cornwall. We must consider this sector from Cornwall to the Mull of Kintyre as the least vulnerable to sea-borne invasion. Here the works of defence should be confined to guarding by a few guns or land torpedo-tubes the principal harbours, and giving a moderate scale of protection to their gorges.* It is not admissible to lavish our limited resources upon this sector.

10. North of the Mull of Kintyre to Scapa Flow, the Shetlands, and the Faroes, all lies in the orbit of the main Fleet. The voyage of an expedition from the Norwegian coast would be very hazardous, and its arrival anywhere right round to Cromarty Firth would not raise immediately decisive issues. The enemy, who is now crouched, would then be sprawled. His advance would lie in difficult and sparsely-inhabited country. He could be contained until sufficient forces were brought to bear, and his communications immediately cut from the sea. This would make his position all the more difficult, as the distances to any important objective are much longer and he would require considerable wheeled transport. It would be impossible to fortify all landing-points in this sector, and it would be a waste of energy to attempt to do so. A much longer period may be allowed for counter-attack than in the south-east, opposite London.

11. From Cromarty Firth to the Wash is the second most important sector, ranking next after the Wash to Dover. Here however all the harbours and inlets are defended, both from the sea and from the rear, and it should be possible to counter-attack in superior force within twenty-four hours. The Tyne must be regarded as the second major objective after London, for here (and to a lesser extent at the Tees) grievous damage could be done by an invader or large-scale raider in a short time. On the other hand, the sea and air conditions are more favourable to us than to the southward.

12. The combined Staffs should endeavour to assign to all these sectors their relative scales of vulnerability and defence, both in the

* I.e., their approaches from the rear.

number of men employed in the local defence of beaches and of harbours, and also in the number of days or hours within which heavy counter-attacks should be possible. As an indication of these relative scales of attack and defence, I set down for consideration the following:

| | |
|---|---|
| Cromarty Firth to Wash inclusive* .. .. .. .. | 3 |
| Wash to Dover promontory .. .. .. .. .. | 5 |
| Dover promontory to Land's End, and round to start of minefield .. .. .. .. .. .. .. | $1\frac{1}{2}$ |
| Start of minefield to the Mull of Kintyre .. .. .. | $\frac{1}{4}$ |
| Mull of Kintyre northabout to Cromarty Firth .. .. | $\frac{1}{2}$ |

The Chiefs of Staff Committee, after another review of all our information, replied to this paper by a report from Colonel Hollis, who acted as their secretary.

### DEFENCE AGAINST INVASION

*Prime Minister* 13.VIII.40

1. The Chiefs of Staff have examined, in consultation with the Commander-in-Chief Home Forces, your minute [of August 5], and find themselves in complete agreement with the principles enunciated in paragraphs 1 to 5.

2. The Commander-in-Chief assures us that the paramount importance of immediate counter-attack upon the enemy, should he obtain a temporary footing on these shores, has been impressed on all ranks, and that it is his policy to bring back divisions into reserve as soon as they are adequately trained and equipped for offensive operations.

3. The Chiefs of Staff also agree with your assessment of the relative scales of vulnerability to sea-borne attack of the various sectors of the coast. Indeed, it is remarkable how closely the present distribution of Home Defence divisions corresponds with your figures in paragraph 12. This is worked out as follows:

4. Your theoretical scales of defence are:

| | |
|---|---|
| Cromarty Firth to Wash .. .. .. .. | 3 |
| Wash to Dover .. .. .. .. .. | 5 |
| Dover to North Cornwall .. .. .. .. | $1\frac{1}{2}$ |
| North Cornwall to Mull of Kintyre .. .. | $\frac{1}{4}$ |
| Mull of Kintyre to Cromarty Firth .. .. | $\frac{1}{2}$ |
| Total .. .. .. .. .. | $10\frac{1}{4}$ |

* These are, of course, *proportions*, not divisional formations.

5. A force of ten divisions, if distributed in the above proportions, would give three divisions on Sector Forth-Wash, five divisions on Sector Wash-Dover, and so on. There are in fact twenty-six divisions in this Island, and, if your figures are multiplied by 2.6 and compared with actual distribution of these twenty-six divisions, the following picture results:

| Sector | Distribution in accordance with Prime Minister's assessment of vulnerability | Actual distribution of divisions |
|---|---|---|
| Cromarty-Wash .. .. .. | $7\frac{1}{2}$ | $8\frac{1}{2}$ |
| Wash–Dover .. .. .. | $12\frac{1}{2}$ | 7–10 |
| Dover–North Cornwall .. .. | $4\frac{1}{4}$ | 5–8 |
| North Cornwall–Mull of Kintyre | $\frac{1}{2}$ | 2 |
| Mull of Kintyre–Cromarty .. | $1\frac{1}{4}$ | $\frac{1}{2}$ |

6. The similarity between the two sets of figures is even closer than appears at first sight, by reason of the fact that the reserve divisions located immediately north and north-west of London are available for deployment in either the Sector Wash–Dover or the Sector Dover–Portsmouth, and therefore the number of "available" divisions for these two sectors is variable. A total of fifteen divisions is available on the combined sectors against your suggested requirements of sixteen and three-quarters.

7. The Chiefs of Staff point out that your figures are based on scales of sea-borne attack, whereas the actual distribution takes into account the threat from air-borne attack as well. Thus although we may seem at present to be slightly over-insured along the south coast, the reason for this is that our defences there can be brought under the enemy fighter "umbrella" and can be subjected to assault across the Channel at comparatively short range.

\* \* \* \* \*

Even while these documents were being considered and printed the situation had begun to change in a decisive manner. Our excellent Intelligence confirmed that the operation "Sea Lion" had been definitely ordered by Hitler and was in active preparation. It seemed certain that the man was going to try. Moreover, the front to be attacked was altogether different from *or additional to* the east coast, on which the Chiefs of Staff, the Admiralty and I, in full agreement, still laid the major emphasis.

But thereafter came a rapid transformation. A large number of self-propelled barges and motor-boats began to pass by night through the Straits of Dover, creeping along the French coast and gradually assembling in all the French Channel ports from Calais to Brest. Our daily photographs showed this movement with precision. It had not been found possible to re-lay our minefields close to the French shore. We immediately began to attack the vessels in transit with our small craft, and Bomber Command was concentrated upon the new set of invasion ports now opening upon us. At the same time a great deal of information came to hand about the assembly of a German Army or Armies of Invasion along this stretch of the hostile coast, of movement on the railways, and of large concentrations in the Pas de Calais and Normandy. Later on two mountain divisions with mules, evidently meant to scale the Folkestone cliffs, were reported near Boulogne. Meanwhile large numbers of powerful long-range batteries all along the French Channel coast came into existence.

In response to the new menace we began to shift our weight from one leg to the other and to improve all our facilities for moving our increasingly large mobile reserves towards the southern front. About the end of the first week of August General Brooke, now Commander-in-Chief Home Forces, pointed out that the threat of invasion was developing on the south coast as much as on the east. All the time our forces were increasing in numbers, efficiency, mobility, and equipment.

\* \* \* \* \*

The change in our dispositions between August and September was as follows:

|  | AUGUST | SEPTEMBER |
|---|---|---|
| Wash–Thames | 7 divisions | 4 divisions plus 1 armoured brigade |
| South Coast | 5 divisions | 9 divisions plus 2 armoured brigades |
| Reserve for either sector | 3 divisions | 3 divisions plus 2 armoured divisions, plus 1 division (equivalent) London District |
| Total available South Coast | 8 divisions | 13 divisions plus 3 armoured divisions |

Thus in the last half of September we were able to bring into action on the south coast front, including Dover, sixteen divisions of high quality, of which three were armoured divisions or their equivalent in brigades, all of which were additional to the local coastal defence and could come into action with great speed against any invasion landing. This provided us with a punch or series of punches which General Brooke was well poised to deliver as might be required; and no one more capable.

*     *     *     *     *

All this while we could not feel any assurance that the inlets and river-mouths from Calais to Terschelling and Heligoland, with all that swarm of islands off the Dutch and German coasts (the "Riddle of the Sands" of the previous war), might not conceal other large hostile forces with small or moderate-sized ships. An attack from Harwich right round to Portsmouth, Portland, or even Plymouth, centring upon the Kent promontory, seemed to impend. We had nothing but negative evidence that a third wave of invasion harmonised with the others might not be launched from the Baltic through the Skaggerak in large ships. This was indeed essential to a German success, because in no other way could heavy weapons reach the landed armies or large depots of supply be established in and around store-ships stranded near the east coast beaches.

We now entered upon a period of extreme tension and vigilance. We had of course all this time to maintain heavy forces north of the Wash, right up to Cromarty; and arrangements were perfected to draw from these should the assault declare itself decidedly in the south. The abundant intricate railway system of the Island and our continued mastery of our home air would have enabled us to move with certainty another four or five divisions to reinforce the southern defence if it were necessary on the fourth, fifth, and sixth days after the enemy's full effort had been exposed.

A very careful study was made of the moon and the tides. We thought that the enemy would like to cross by night and land at dawn; and we now know the German Army Command felt like this too. They would also be glad of a half-moonlight on the way over, so as to keep their order and make their true landfall. Measuring it all with precision, the Admiralty thought the most favourable conditions for the enemy would arise between the 15th and 30th of September. Here also we now find that we were in agreement with our foes. We had little doubt of our ability to destroy anything that got ashore on the Dover promontory or on the sector of coast from Dover to Portsmouth, or even Portland. As all our thoughts at the summit moved together in harmonious and detailed agreement, one could not help liking the picture which presented itself with growing definition. Here perhaps was the chance of striking a blow at the mighty enemy which would resound throughout the world. One could not help being inwardly excited alike by the atmosphere and the evidence of Hitler's intention which streamed in upon us. There were indeed some who on purely technical grounds, and for the sake of the effect the total defeat and destruction of his expedition would have on the general war, were quite content to see him try.

In July and August we had asserted air mastery over Great Britain, and were especially powerful and dominant over the Home Counties of the south-east. The Canadian Army Corps stood most conveniently posted between London and Dover. Their bayonets were sharp and their hearts were high. Proud would they have been to strike the decisive blow for Britain and Freedom. Similar passions burned in all breasts. Vast intricate systems of fortifications, defended localities, anti-tank obstacles, block-houses, pill-boxes, and the like laced the whole area. The coastline bristled with defences and batteries, and at the cost of heavier losses through reduced escorts in the Atlantic, and also by new construction coming into commission, the flotillas grew substantially in numbers and quality. We had brought the battleship *Revenge*, and the old target-ship and dummy-battleship *Centurion*, and a cruiser to Plymouth. The Home Fleet was at its maximum strength and could operate without much risk to the Humber and even to the Wash. In all respects therefore we were fully prepared.

Finally, we were already not far from the equinoctial gales customary in October. Evidently September was the month for Hitler to strike if he dared, and the tides and the moon-phase were favourable in the middle of that month.

\* \* \* \* \*

There was some talk in Parliament after the danger had passed away of the "invasion scare". Certainly those who knew most were the least scared. Apart from mastery of the air and command of the sea, we had as large (if not so well equipped) an army, fresh and ardent, as that which Germany assembled in Normandy four years later to oppose our return to the Continent. In that case, although we landed a million men in the first month, with vast apparatus, and with every other condition favourable, the battle was long and severe, and nearly three months were required to enlarge the area originally seized and break out into the open field. But these were values only to be tested and known in the future.

$$* \quad * \quad * \quad * \quad *$$

It is time to go over to the other camp and set forth the enemy's preparations and plans as we now know them.

# CHAPTER XV

# OPERATION "SEA LION"

*Plan of the German Admiralty – Their Conditions Met by the Conquest of France and the Low Countries – Heads of the Services' Meeting with the Fuehrer of July 21 – Hitler Comprehends the Difficulties but Gives the Order – Controversy between the German Navy and Army Staffs – Raeder and Halder at Variance – The Compromise Plan Agreed – Further Misgivings of the German Admiralty – Both German Navy and Army Chiefs Cast the Burden on Goering and the Air – Goering Accepts – Hitler Postpones D-Day – British Counter-activities – The "Cromwell" Order of September 7 – A Healthy Tonic – German Ignorance of Amphibious Warfare – Service Disunion – The Germans Stake All on the Air Battle.*

$S$OON after war broke out on September 3, 1939, the German Admiralty, as we have learned from their captured archives, began their Staff study of the invasion of Britain. Unlike us, they had no doubt that the only way was across the narrow waters of the English Channel. They never considered any other alternative. If we had known this it would have been an important relief. An invasion across the Channel came upon our best-defended coast, the old sea front against France, where all the ports were fortified and our main flotilla bases and in later times most of our airfields and air-control stations for the defence of London were established. There was no part of the Island where we could come into action more quickly or in such great strength with all three Services. Admiral Raeder was anxious not to be found wanting should the demand to invade Britain be made upon the German Navy. At the same time he asked for a lot of conditions. The first of these was the entire control of the French, Belgian, and Dutch coasts, harbours, and river-mouths. Therefore the project slumbered during the Twilight War.

Suddenly all these conditions were surprisingly fulfilled, and it must have been with some misgivings but also satisfaction that on the morrow of Dunkirk and the French surrender he could present himself to the Fuehrer with a plan.  On May 21 and again on June 20 he spoke to Hitler on the subject, not with a view to proposing an invasion, but in order to make sure that if it were ordered the planning in detail should not be rushed. Hitler was sceptical, saying that "he fully appreciated the exceptional difficulties of such an undertaking".  He also nursed the hope that England would sue for peace.  It was not until the last week in June that the Supreme Headquarters turned to this idea, nor till July 2 that the first directive was issued for planning the invasion of Britain as a possible event.  "The Fuehrer has decided that under certain conditions—the most important of which is achieving air superiority—a landing in England may take place."  On July 16 Hitler issued his directive: "Since England in spite of her militarily hopeless position shows no sign of coming to terms, I have decided to prepare a landing operation against England, and if necessary to carry it out. . . . The preparations for the entire operation must be completed by mid-August." Active measures in every direction were already in progress.

*　　*　　*　　*　　*

The German Navy plan, of which it is clear I had received an inkling in June, was essentially mechanical.  Under the cover of heavy-gun batteries firing from Gris-Nez towards Dover, and a very strong artillery protection along the French coast in the Straits, they proposed to make a narrow corridor across the Channel on the shortest convenient line and to wall this in by minefields on either side, with outlying U-boat protection. Through this the Army was to be ferried over and supplied in a large number of successive waves.  There the Navy stopped, and on this the German Army chiefs were left to address themselves to the problem.

Considering that we could, with our overwhelming naval superiority, tear these minefields to pieces with small craft under superior air-power and also destroy the dozen or score of U-boats concentrated to protect them, this was at the outset a bleak proposition.  Nevertheless, after the fall of France anyone could see that the only hope of avoiding a long war, with all that it might entail, was to bring Britain to her knees. The German Navy itself had been, as we have recorded, knocked about in a most serious manner in the fighting off Norway; and in their crippled condition they could not offer more than minor support to the Army. Still, they had their plan, and no one could say that they had been caught unawares by good fortune.

The German Army Command had from the first regarded the invasion of England with considerable qualms. They had made no plans or preparations for it; and there had been no training. As the weeks of prodigious, delirious victory succeeded one another they were emboldened. The responsibility for the safe crossing was not departmentally theirs, and, once landed in strength, they felt that the task was within their power. Indeed, already in August Admiral Raeder felt it necessary to draw their attention to the dangers of the passage, during which perhaps the whole of the Army forces employed might be lost. Once the responsibility for putting the Army across was definitely thrust upon the Navy, the German Admiralty became consistently pessimistic.

On July 21 the heads of the three Services met the Fuehrer. He informed them that the decisive stage of the war had already been reached, but that England had not yet recognised it and still hoped for a turn of fate. He spoke of the support of England by the United States and of a possible change in German political relations with Soviet Russia. The execution of "Sea Lion", he said, must be regarded as the most effective means of bringing about a rapid conclusion of the war. After his long talks with Admiral Raeder, Hitler had begun to realise what the crossing of the Channel, with its tides and currents, and all the mysteries of the sea, involved. He described "Sea Lion" as "an exceptionally bold and daring undertaking". "Even if the way is short, this is not just a river crossing, but the crossing of a sea which is dominated by the enemy. This is not a case of a single-crossing operation, as in Norway; operational surprise cannot be expected; a defensively-prepared *and utterly determined enemy* faces us and dominates the sea area which we must use. For the Army operation forty divisions will be required. The most difficult part will be the material reinforcements and stores. We cannot count on supplies of any kind being available to us in England." The prerequisites were complete mastery of the air, the operational use of powerful artillery in the Dover straits, and protection by minefields. "The time of year," he said, "is an important factor, since the weather in the North Sea and in the Channel during the second half of September is very bad, and the fogs begin in the middle of October. The main operation must therefore be completed by September 15, for after that date co-operation between the Luftwaffe and the heavy weapons becomes too unreliable. But as air co-operation is decisive it must be regarded as the principal factor in fixing the date."

A vehement controversy, conducted with no little asperity, arose in the German Staffs about the width of the front and the number of points to be attacked. The Army demanded a series of landings along the whole English southern coast from Dover to Lyme Regis, west of Portland. They also desired an ancillary landing north of Dover at Ramsgate. The German Naval Staff now stated that the most suitable area for the safe crossing of the English Channel was between the North Foreland and the western end of the Isle of Wight. On this the Army Staff developed a plan for a landing of 100,000 men, followed almost immediately by 160,000 more at various points from Dover westward to Lyme Bay. Colonel-General Halder, Chief of the Army Staff, declared that it was necessary to land at least four divisions in the Brighton area. He also required landings in the area Deal-Ramsgate; at least thirteen divisions must be deployed, as far as possible simultaneously, at points along the whole front. In addition, the Luftwaffe demanded shipping to transport fifty-two A.A. batteries with the first wave.

The Chief of the Naval Staff however made it clear that nothing like so large or rapid a movement was possible. He could not physically undertake to escort a landing fleet across the whole width of the area mentioned. All he had meant was that within these limits the Army should pick the best place. The Navy had not enough strength, even with air supremacy, to protect more than one passage at a time, and they thought the narrowest parts of the Straits of Dover the least difficult. To carry the whole of the 160,000 men of the second wave and their equipment in a single operation would require two million tons of shipping. Even if this fantastic requirement could have been met, such quantities of shipping could not have been accommodated in the area of embarkation. Only the first échelons could be thrown across for the formation of narrow bridgeheads, and at least two days would be needed to land the second échelons of these divisions, to say nothing of the second six divisions which were thought indispensable. He further pointed out that a broad-front landing would mean three to five and a half hours' difference in the times of high water at the various points selected. Either therefore unfavourable tide conditions must be accepted at some places, or simultaneous landings renounced. This objection must have been very difficult to answer.

Much valuable time had been consumed in these exchanges of memoranda. It was not until August 7 that the first verbal discussion took place between General Halder and the Chief of the Naval Staff. At this meeting Halder said: "I utterly reject the Navy's proposals. From the Army view-point I regard it as complete suicide. I might just as well put the troops that have been landed straight through the sausage-machine." The Naval Chief of Staff rejoined that he must equally reject the landing on a broad front, as that would lead only to a sacrifice of the troops on the passage over. In the end a compromise decision was given by Hitler which satisfied neither the Army nor the Navy. A Supreme Command Directive, issued on August 27, decided that "the Army operations must allow for the facts regarding available shipping space and security of the crossing and disembarkation". All landings in the Deal-Ramsgate area were abandoned, but the front was extended from Folkestone to Bognor. Thus it was nearly the end of August before even this measure of agreement was reached; and of course everything was subject to victory being gained in the air battle, which had now been raging for six weeks.

On the basis of the frontage at last fixed the final plan was made. The military command was entrusted to Rundstedt, but shortage of shipping reduced his force to thirteen divisions with twelve in reserve. The Sixteenth Army, from ports between Rotterdam and Boulogne. were to land in the neighbourhood of Hythe, Rye, Hastings, and Eastbourne, the Ninth Army, from ports between Boulogne and Havre, attacking between Brighton and Worthing. Dover was to be captured from the landward side; then both armies would advance to the covering line of Canterbury-Ashford-Mayfield-Arundel. In all. eleven divisions were to be landed in the first waves. A week after the landing it was hoped, optimistically, to advance yet farther, to Gravesend,

Reigate, Petersfield, Portsmouth. In reserve lay the Sixth Army, with divisions ready to reinforce, or, if circumstances allowed, to extend the frontage of attack to Weymouth. It would have been easy to increase these three armies, once the bridgeheads were gained, "because", says General Halder, "no military forces were facing the Germans on the Continent". There was indeed no lack of fierce and well-armed troops, but they required shipping and safe conveyance.

On the Naval Staff fell the heaviest initial task. Germany had about 1,200,000 tons of seagoing shipping available to meet all her needs. To embark the invasion force would require more than half this amount, and would involve great economic disturbance. By the beginning of September the Naval Staff were able to report that the following had been requisitioned:

> 168 transports (of 700,000 tons)
> 1,910 barges
> 419 tugs and trawlers
> 1,600 motor-boats

All this Armada had to be manned, and brought to the assembly ports by sea and canal. Meanwhile since early July we had made a succession of attacks on the shipping in Wilhelmshaven, Kiel, Cuxhaven, Bremen, and Emden; and raids were made on small craft and barges in French ports and Belgian canals. When on September 1 the great southward flow of invasion shipping began it was watched, reported, and violently assailed by the Royal Air Force along the whole front from Antwerp to Havre. The German Naval Staff recorded: "The enemy's continuous fighting defence off the coast, his concentration of bombers on the 'Sea Lion' embarkation ports, and his coastal reconnaissance activities indicate that he is now expecting an immediate landing."

And again: "The English bombers however and the mine-laying forces of the British Air Force . . . are still at full operational strength, and it must be confirmed that the activity of the British forces has undoubtedly been successful even if no decisive hindrance has yet been caused to German transport movement."

Yet, despite delays and damage, the German Navy completed the first part of its task. The 10 per cent. margin for accidents and losses it had provided was fully expended. What survived

however did not fall short of the minimum it had planned to
have for the first stage.

Both Navy and Army now cast their burden on the German
Air Force. All this plan of the corridor, with its balustrades of
minefields to be laid and maintained under the German Air Force
canopy against the overwhelming superiority of the British
flotillas and small craft, depended upon the defeat of the British
Air Force and the complete mastery of the air by Germany over
the Channel and South-East England, and not only over the
crossing but over the landing-points. Both the older services
passed the buck to Reichsmarschall Goering.

Goering was by no means unwilling to accept this responsibility, because he believed that the German Air Force, with
its large numerical superiority, would, after some weeks of hard
fighting, beat down the British air defence, destroy their airfields
in Kent and Sussex, and establish a complete domination of the
Channel. But apart from this he felt assured that the bombing
of England, and particularly of London, would reduce the
decadent, peace-loving British to a condition in which they would
sue for peace, more especially if the threat of invasion grew
steadily upon their horizon. The German Admiralty were by no
means convinced; indeed their misgivings were profound. They
considered "Sea Lion" should be launched only in the last resort,
and in July they had recommended the postponement of the
operation till the spring of 1941, unless *the unrestricted air attack
and the unlimited U-boat warfare* should "cause the enemy to
negotiate with the Fuehrer on his own terms". But Feldmarschall
Keitel and General Jodl were glad to find the Air Supreme
Commander so confident.

These were great days for Nazi Germany. Hitler had danced
his jig of joy before enforcing the humiliation of the French
Armistice at Compiègne. The German Army marched triumph-
antly through the Arc de Triomphe and down the Champs
Elysées. What was there they could not do? Why hesitate to
play out a winning hand? Thus each of the three services involved
in the operation "Sea Lion" worked upon the hopeful factors in
their own theme and left the ugly side to their companions.

As the days passed doubts and delays appeared and multiplied.
Hitler's order of July 16 had laid down that all preparations were
to be completed by the middle of August. All three services

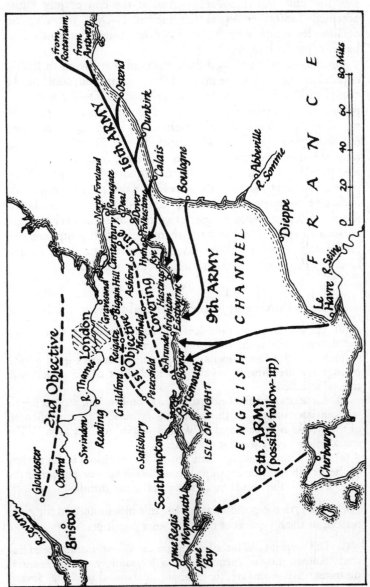

Sketch Map of GERMAN INVASION PLAN

saw that this was impossible. And at the end of July Hitler accepted September 15 as the earliest D-Day, reserving his decision for action until the results of the projected intensified air battle could be known.

On August 30 the Naval Staff reported that owing to British counter-action against the invasion fleet preparations could not be completed by September 15. At their request D-Day was postponed to September 21, with a proviso of ten days' previous warning. This meant that the preliminary order had to be issued on September 11. On September 10 the Naval Staff again reported their various difficulties from the weather, which is always tiresome, and from British counter-bombing. They pointed out that although the necessary naval preparations could in fact be completed by the 21st, the stipulated operational condition of undisputed air superiority over the Channel had not been achieved. On the 11th therefore Hitler postponed the preliminary order by three days, thus setting back the earliest D-Day to the 24th; on the 14th he further put it off.

\* \* \* \* \*

On the 14th Admiral Raeder expressed the view that:

(*a*) The present air situation does not provide conditions for carrying out the operation, as the risk is still too great.

(*b*) If the "Sea Lion" operation fails, this will mean a great gain in prestige for the British; and the powerful effect of our attacks will thus be annulled.

(*c*) Air attacks on England, particularly on London, must continue without interruption. If the weather is favourable an intensification of the attacks is to be aimed at, without regard to "Sea Lion". The attacks must have a decisive outcome.

(*d*) "Sea Lion" however must not yet be cancelled, as the anxiety of the British must be kept up; if cancellation became known to the outside world, this would be a great relief to the British.

On the 17th the postponement became indefinite, and for good reason, in their view as in ours. Raeder continues:

(i) The preparations for a landing on the Channel coast are extensively known to the enemy, who is increasingly taking countermeasures. Symptoms are, for example, operational use of his aircraft for attacks and reconnaissances over the German operational harbours, frequent appearance of destroyers off the south coast of England, in the Straits of Dover and on the Franco-Belgian coast, stationing of his patrol vessels off the north coast of France, Churchill's last speech, etc.

(ii) The main units of the Home Fleet are being held in readiness to repel the landing, though the majority of the units are still in western bases.

(iii) Already a large number of destroyers (over thirty) have been located by air reconnaissance in the southern and south-eastern harbours.

(iv) All available information indicates that the enemy's naval forces are solely occupied with this theatre of operations.

\*      \*      \*      \*      \*

During August the corpses of about forty German soldiers were washed up at scattered points along the coast between the Isle of Wight and Cornwall. The Germans had been practising embarkations in the barges along the French coast. Some of these barges put out to sea in order to escape British bombing and were sunk, either by bombing or bad weather. This was the source of a widespread rumour that the Germans had attempted an invasion and had suffered very heavy losses either by drowning or by being burnt in patches of sea covered with flaming oil. We took no steps to contradict such tales, which spread freely through the occupied countries in a wildly exaggerated form and gave much encouragement to the oppressed populations. In Brussels, for instance, a shop exhibited men's bathing-suits marked "For Channel swimming".

On September 7 the information before us showed that the westerly and southerly movement of barges and small ships to ports between Ostend and Havre was in progress, and as these assembly harbours were under heavy British air attack it was not likely the ships would be brought to them until shortly before the actual attempt. The striking strength of the German Air Force between Amsterdam and Brest had been increased by the transfer of one hundred and sixty bomber aircraft from Norway; and short-range dive-bomber units were observed on the forward airfields in the Pas de Calais area. Four Germans captured a few days earlier after landing from a rowing-boat on the south-east coast had confessed to being spies, and said that they were to be ready at any time during the next fortnight to report the movement of British reserve formations in the area Ipswich-London-Reading-Oxford. Moon and tide conditions between the 8th and

10th of September were favourable for invasion on the south-east coast. On this the Chiefs of Staff concluded that the possibility of invasion had become imminent and that the defence forces should stand by at immediate notice.

There was however at that time no machinery at General Headquarters, Home Forces, by which the existing eight hours' notice for readiness could be brought to "readiness for immediate action" by intermediate stages. The code-word "Cromwell", which meant "invasion imminent", was therefore issued by Home Forces at 8 p.m., September 7, to the Eastern and Southern Commands, implying action stations for the forward coastal divisions. It was also sent to all formations in the London area and to the IVth and VIIth Corps in G.H.Q. Reserve. It was repeated for information to all other commands in the United Kingdom. On this, in some parts of the country, the Home Guard commanders, acting on their own initiative, called out the Home Guard by ringing the church bells. This led to rumours of enemy parachutist landings, and also that German E-boats were approaching the coast. Neither I nor the Chiefs of Staff were aware that the decisive code-word "Cromwell" had been used, and the next morning instructions were given to devise intermediate stages by which vigilance could be increased on future occasions without declaring an invasion imminent. Even on receipt of the code-word "Cromwell" the Home Guard were not to be called out except for special tasks; and also church bells were to be rung only by order of a Home Guard who had himself seen as many as twenty-five parachutists landing, and not because other bells had been heard or for any other reason. As may be imagined, this incident caused a great deal of talk and stir, but no mention of it was made in the newspapers or in Parliament. It served as a useful tonic and rehearsal for all concerned.

*  *  *  *  *

Having traced the German invasion preparations steadily mounting to a climax, we have seen how the early mood of triumph changed gradually to one of doubt and finally to complete loss of confidence in the outcome. Confidence was in fact already destroyed in 1940, and, despite the revival of the project in 1941, it never again held the imagination of the German leaders as it had done in the halcyon days following the fall of France.

During the fateful months of July and August we see the Naval Commander, Raeder, endeavouring to teach his military and air colleagues about the grave difficulties attending large-scale amphibious war. He realised his own weakness and the lack of time for adequate preparation, and sought to impose limits on the grandiose plans advanced by Halder for landing immense forces simultaneously over a wide front. Meanwhile Goering with soaring ambition was determined to achieve spectacular victory with his Air Force alone and was disinclined to play the humbler *rôle* of working to a combined plan for the systematic reduction of opposing sea and air forces in the invasion area.

It is apparent from the records that the German High Command were very far from being a co-ordinated team working together with a common purpose and with a proper understanding of each other's capabilities and limitations. Each wished to be the brightest star in the firmament. Friction was apparent from the outset, and so long as Halder could thrust responsibility on to Raeder he did little to bring his own plans into line with practical possibilities. Intervention by the Fuehrer was necessary, but seems to have done little to improve the relations between the services. In Germany the prestige of the Army was paramount and the military leaders regarded their naval colleagues with some condescension. It is impossible to resist the conclusion that the German Army was reluctant to place itself in the hands of its sister service in a major operation. When questioned after the war about these plans, General Jodl impatiently remarked, "Our arrangements were much the same as those of Julius Cæsar". Here speaks the authentic German soldier in relation to the sea affair, having little conception of the problems involved in landing and deploying large military forces on a defended coast exposed to all the hazards of the sea.

In Britain, whatever our shortcomings, we understood the sea affair very thoroughly. For centuries it has been in our blood, and its traditions stir not only our sailors but the whole race. It was this above all things which enabled us to regard the menace of invasion with a steady gaze. The system of control of operations by the three Chiefs of Staff concerted under a Minister of Defence produced a standard of team-work, mutual understanding, and ready co-operation unrivalled in the past. When in course of time our opportunity came to undertake great invasions from the sea it was upon a foundation of solid achievement in preparation for the task and with a full understanding of the technical needs of such vast and hazardous undertakings. Had the Germans possessed in 1940 well-trained amphibious forces equipped with all the apparatus of modern amphibious war their task would still have been a forlorn hope in the face of our sea- and air-power. In fact they had neither the tools nor the training.

\* \* \* \* \*

We have seen how our many anxieties and self-questionings led to a steady increase in the confidence with which from the beginning we had viewed the invasion project. On the other hand, the more the German High Command and the Fuehrer looked at the venture the less they liked it. We could not of course know each other's moods and valuations; but with every week from the middle of July to the middle of September the unknown identity of views upon the problem between the German and British Admiralties, between the German Supreme Command and the British Chiefs of Staff, and also between the Fuehrer and the author of this book, became more definitely pronounced. If we could have agreed equally well about other matters there need have been no war. It was of course common ground between us that all depended upon the battle in the air. The question was how this would end between the combatants; and in addition the Germans wondered whether the British people would stand up to the air bombardment, the effect of which in these days was greatly exaggerated, or whether they would crumple and force His Majesty's Government to capitulate. About this Reichsmarschall Goering had high hopes, and we had no fears.

# BOOK II

# ALONE

BOOK II

ATOMS

# CHAPTER XVI

# THE BATTLE OF BRITAIN

*The Decisive Struggle – Hitler's Dilemma – Three Phases – Advantages of Fighting in One's Own Air – "Sea Lion" and the Air Assault – The German Raid against Tyneside – Massacre of the Heinkels – Lord Beaverbrook's Hour – Mr. Ernest Bevin and Labour – Cabinet Solidarity – Checking German Losses – First Attacks on London – Uneasiness of the German Naval Staff – My Broadcast of September 11 – The Hard Strain from August 24 to September 6 – The Articulation of Fighter Command Endangered – A Quarter of Our Pilots Killed or Disabled in a Fortnight – Goering's Mistake of Turning on London Too Soon – A Breathing-space – September 15 the Culminating Date – With No. 11 Group – Air Vice-Marshal Park – The Group Operations Room – The Attack Begins – All Reserves Employed – A Cardinal Victory – Hitler Postpones "Sea Lion", September 17 – After-light on Claims and Losses – Honour for All.*

OUR fate now depended upon victory in the air. The German leaders had recognised that all their plans for the invasion of Britain depended on winning air supremacy above the Channel and the chosen landing-places on our south coast. The preparation of the embarkation ports, the assembly of the transports, the mine-sweeping of the passages, and the laying of the new minefields were impossible without protection from British air attack. For the actual crossing and landings complete mastery of the air over the transports and the beaches was the decisive condition. The result therefore turned upon the destruction of the Royal Air Force and the system of airfields between London and the sea. We now know that Hitler said to Admiral Raeder on July 31: "If after eight days of intensive air war the Luftwaffe has not achieved considerable destruction of the enemy's Air Force, harbours, and naval forces, the operation will have to be put off till May 1941." This was the battle that had now to be fought.

I did not myself at all shrink mentally from the impending trial of strength. I had told Parliament on June 4: "The great French Army was very largely, for the time being, cast back and disturbed by the onrush of a few thousand armoured vehicles. May it not also be that the cause of civilisation itself will be defended by the skill and devotion of a few thousand airmen?" And to Smuts, on June 9: "I see only one sure way through now—to wit, that Hitler should attack this country, and in so doing break his air weapon." The occasion had now arrived.

Admirable accounts have been written of the struggle between the British and German Air Forces which constitutes the Battle of Britain. In Air Chief Marshal Dowding's dispatch and the Air Ministry pamphlet No. 156 the essential facts are fully recorded as they were known to us in 1941 and 1943. We have now also access to the views of the German High Command and to their inner reactions in the various phases. It appears that the German losses in some of the principal combats were a good deal less than we thought at the time, and that reports on both sides were materially exaggerated. But the main features and the outline of this famous conflict, upon which the life of Britain and the freedom of the world depended, are not in dispute.

The German Air Force had been engaged to the utmost limit in the Battle of France, and, like the German Navy after the Norway campaign, they required a period of weeks or months for recovery. This pause was convenient to us too, for all but three of our fighter squadrons had at one time or another been engaged in the Continental operations. Hitler could not conceive that Britain would not accept a peace offer after the collapse of France. Like Marshal Pétain, Weygand, and many of the French generals and politicians, he did not understand the separate, aloof resources of an Island State, and like these Frenchmen he misjudged our will-power. We had travelled a long way and learned a lot since Munich. During the month of June he had addressed himself to the new situation as it gradually dawned upon him, and meanwhile the German Air Force recuperated and redeployed for their next task. There could be no doubt what this would be. Either Hitler must invade and conquer England, or he must face an indefinite prolongation of the war, with all its incalculable hazards and complications. There was always the possibility that victory over Britain in the air would bring about the end of the British resistance, and that actual invasion, even if it became practicable, would also become unnecessary, except for the occupying of a defeated country.

During June and early July the German Air Force revived and regrouped its formations and established itself on all the French and Belgian airfields from which the assault had to be launched, and by reconnaissance and tentative forays sought to measure the character and scale of the opposition which would be encountered. It was not until July 10 that the first heavy onslaught began, and this date is usually taken as the opening of the battle. Two other dates of supreme consequence stand out, August 15 and September 15. There were also three successive but overlapping phases in the German attack. First, from July 10 to August 18, the harrying of British convoys in the Channel and of our southern ports from Dover to Plymouth, whereby our Air Force should be tested, drawn into battle, and depleted; whereby also damage should be done to those seaside towns marked as objectives for the forthcoming invasion. In the second phase, August 24 to September 27, a way to London was to be forced by the elimination of the Royal Air Force and its installations, leading to the violent and continuous bombing of the capital. This would also cut communications with the threatened shores. But in Goering's view there was good reason to believe that a greater prize was here in sight, no less than throwing the world's largest city into confusion and paralysis, the cowing of the Government and the people, and their consequent submission to the German will. Their Navy and Army Staffs devoutly hoped that Goering was right. As the situation developed they saw that the R.A.F. was not being eliminated, and meanwhile their own urgent needs for the "Sea Lion" adventure were neglected for the sake of destruction in London. And then, when all were disappointed, when invasion was indefinitely postponed for lack of the vital need, air supremacy, there followed the third and last phase. The hope of daylight victory had faded, the Royal Air Force remained vexatiously alive, and Goering in October resigned himself to the indiscriminate bombing of London and the centres of industrial production.

\* \* \* \* \*

In the quality of the fighter aircraft there was little to choose. The Germans' were faster, with a better rate of climb; ours more manœuvrable, better armed. Their airmen, well aware of their great numbers, were also the proud victors of Poland, Norway, the Low Countries, France; ours had supreme confidence in themselves as individuals and that determination which the British race displays in fullest measure when in supreme adversity. One important strategical advantage the Germans enjoyed and skilfully used: their forces were deployed on many and widely-spread bases, whence they could concentrate upon us in great strengths and with feints and deceptions as to the true points of attack. But the enemy may have underrated the adverse conditions of fighting above and across the Channel compared with those which had prevailed in France and Belgium. That they regarded them as serious is shown by the efforts they made to organise an efficient Sea Rescue Service. German transport planes, marked with the Red Cross, began to appear in some numbers over the Channel in July and August whenever there was an air fight. We did not recognise this means of rescuing enemy pilots who had been shot down in action, in order that they might come and bomb our civil population again. We rescued them ourselves whenever it was possible, and made them prisoners of war. But all German air ambulances were forced or shot down by our fighters on definite orders approved by the War Cabinet. The German crews and doctors on these machines professed astonishment at being treated in this way, and protested that it was contrary to the Geneva Convention. There was no mention of such a contingency in the Geneva Convention, which had not contemplated this form of warfare. The Germans were not in a strong position to complain, in view of all the treaties, laws of war, and solemn agreements which they had violated without compunction whenever it suited them. They soon abandoned the experiment, and the work of sea rescue for both sides was carried out by our small craft, on which of course the Germans fired on every occasion.

\* \* \* \* \*

By August the Luftwaffe had gathered 2,669 operational air-craft, comprising 1,015 bombers, 346 dive-bombers, 933 fighters, and 375 heavy fighters. The Fuehrer's Directive No. 17 authorised the intensified air war against England on August 5. Goering never set much store by "Sea Lion"; his heart was in the "absolute" air war. His consequent distortion of the arrangements disturbed the German Naval Staff. The destruction of the Royal Air Force and our aircraft industry was to them but a means to an end: when this was accomplished the air war should be turned against the enemy's warships and shipping. They regretted the lower priority assigned by Goering to the naval targets, and they were irked by the delays. On August 6 they reported to the Supreme Command that the preparations for German mine-laying in the Channel area could not proceed because of the constant British threat from the air. On August 10 the Naval Staff's War Diary records:

Preparations for "Sea Lion", particularly mine-clearance, are being affected by the inactivity of the Luftwaffe, which is at present prevented from operating by the bad weather, and, for reasons not known to the Naval Staff, the Luftwaffe has missed opportunities afforded by the recent very favourable weather. . . .

The continuous heavy air fighting of July and early August had been directed upon the Kent promontory and the Channel coast. Goering and his skilled advisers formed the opinion that they must have drawn nearly all our fighter squadrons into this southern struggle. They therefore decided to make a daylight raid on the manufacturing cities north of the Wash. The distance was too great for their first-class fighters, the Me. 109's. They would have to risk their bombers with only escorts from the Me. 110's, which, though they had the range, had nothing like the quality, which was what mattered now. This was nevertheless a reasonable step for them to take, and the risk was well run.

Accordingly, on August 15, about a hundred bombers, with an escort of forty Me. 110's, were launched against Tyneside. At the same time a raid of more than eight hundred planes was sent to pin down our forces in the South, where it was thought they were already all gathered. But now the dispositions which Dowding had made of the Fighter Command were signally vin-

dicated. The danger had been foreseen. Seven Hurricane or Spitfire squadrons had been withdrawn from the intense struggle in
the South to rest in and at the same time to guard the North.
They had suffered severely, but were none the less deeply grieved
to leave the battle. The pilots respectfully represented that they
were not at all tired. Now came an unexpected consolation.
These squadrons were able to welcome the assailants as they
crossed the coast. Thirty German planes were shot down, most
of them heavy bombers (Heinkel 111's, with four trained men in
each crew), for a British loss of only two pilots injured. The
foresight of Air Marshal Dowding in his direction of Fighter
Command deserves high praise, but even more remarkable had
been the restraint and the exact measurement of formidable
stresses which had reserved a fighter force in the North through
all these long weeks of mortal conflict in the South. We must
regard the generalship here shown as an example of genius in the
art of war. Never again was a daylight raid attempted outside
the range of the highest-class fighter protection. Henceforth
everything north of the Wash was safe by day.

August 15 was the largest air battle of this period of the war;
five major actions were fought, on a front of five hundred miles.
It was indeed a crucial day. In the South all our twenty-two
squadrons were engaged, many twice, some three times, and the
German losses, added to those in the North, were seventy-six to
our thirty-four. This was a recognisable disaster to the German
Air Force.

It must have been with anxious minds that the German Air
Chiefs measured the consequences of this defeat, which boded
ill for the future. The German Air Force however had still as
their target the Port of London, all that immense line of docks
with their masses of shipping, and the largest city in the world,
which did not require much accuracy to hit.

*     *     *     *     *

During these weeks of intense struggle and ceaseless anxiety
Lord Beaverbrook rendered signal service. At all costs the fighter
squadrons must be replenished with trustworthy machines. This
was no time for red tape and circumlocution, although these have
their place in a well-ordered, placid system. All his remarkable
qualities fitted the need. His personal buoyancy and vigour

were a tonic. I was glad to be able sometimes to lean on him. He did not fail. This was his hour. His personal force and genius, combined with so much persuasion and contrivance, swept aside many obstacles. Everything in the supply pipe-line was drawn forward to the battle. New or repaired aeroplanes streamed to the delighted squadrons in numbers they had never known before. All the services of maintenance and repair were driven to an intense degree. I felt so much his value that on August 2, with the King's approval, I invited him to join the War Cabinet. At this time also his eldest son, Max Aitken, gained high distinction and at least six victories as a fighter pilot.

Another Minister I consorted with at this time was Ernest Bevin, Minister of Labour and National Service, with the whole man-power of the nation to manage and animate. All the workers in the munitions factories were ready to take his direction. In September he too joined the War Cabinet. The trade unionists cast their slowly-framed, jealously-guarded rules and privileges upon the altar where wealth, rank, privilege, and property had already been laid. I was much in harmony with both Beaverbrook and Bevin in the white-hot weeks. Afterwards they quarrelled, which was a pity, and caused much friction. But at this climax we were all together. I cannot speak too highly of the loyalty of Mr. Chamberlain, or of the resolution and efficiency of all my Cabinet colleagues. Let me give them my salute.

$$* \quad * \quad * \quad * \quad *$$

I was most anxious to form a true estimate of the German losses. With all strictness and sincerity, it is impossible for pilots fighting often far above the clouds to be sure how many enemy machines they have shot down, or how many times the same machine has been claimed by others.

*Prime Minister to General Ismay*                                    17.VIII.40

Lord Beaverbrook told me that in Thursday's action upwards of eighty German machines had been picked up on our soil. Is this so? If not, how many?

I asked C.-in-C. Fighter Command if he could discriminate in this action between the fighting over the land and over the sea. This would afford a good means of establishing for our own satisfaction the results which are claimed.

*Prime Minister to C.A.S.*                                             17.VIII.40

While our eyes are concentrated on the results of the air fighting over this country, we must not overlook the serious losses occurring in the Bomber Command. Seven heavy bombers [lost] last night and also twenty-one aircraft now destroyed on the ground—the bulk at Tangmere—total twenty-eight. These twenty-eight, added to the twenty-two fighters, make our loss fifty on the day, and very much alters the picture presented by the German loss of seventy-five. In fact, on the day we have lost two to three.

Let me know the types of machines destroyed on the ground.

*Prime Minister to Secretary of State for Air*                        21.VIII.40

The important thing is to bring the German aircraft down and to win the battle, and the rate at which American correspondents and the American public are convinced that we are winning, and that our figures are true, stands at a much lower level. They will find out quite soon enough when the German air attack is plainly shown to be repulsed. It would be a pity to tease the Fighter Command at the present time, when the battle is going on from hour to hour and when continuous decisions have to be taken about air-raid warnings, etc. I confess I should be more inclined to let the facts speak for themselves. There is something rather obnoxious in bringing correspondents down to air squadrons in order that they may assure the American public that the fighter pilots are not bragging and lying about their figures. We can, I think, afford to be a bit cool and calm about all this.

I should like you to see on other papers an inquiry I have been making of my own in order to check up on the particular day when M.A.P. [Ministry of Aircraft Production] said they picked up no fewer than eighty German machines brought down over the land alone. This gives us a very good line for our own purposes. I must say I am a little impatient about the American scepticism. The event is what will decide all.

*     *     *     *     *

On August 20 I could report to Parliament:

The enemy is of course far more numerous than we are. But our new production already largely exceeds his, and the American production is only just beginning to flow in. Our bomber and fighter strengths now, after all this fighting, are larger than they have ever been. We believe that we should be able to continue the air struggle indefinitely and as long as the enemy pleases, and the longer it continues the more rapid will be our approach, first towards that parity, and then into that superiority in the air upon which in large measure the decision of the war depends.

Up till the end of August Goering did not take an unfavourable view of the air conflict. He and his circle believed that the English ground organisation and aircraft industry and the fighting strength of the R.A.F. had already been severely damaged. They estimated that since August 8 we had lost 1,115 aircraft, against the German losses of 467. But of course each side takes a hopeful view, and it is in the interest of their leaders that they should. There was a spell of fine weather in September, and the Luftwaffe hoped for decisive results. Heavy attacks fell upon our aerodrome installations round London, and on the night of the 6th sixty-eight aircraft attacked London, followed on the 7th by the first large-scale attack of about three hundred. On this and succeeding days, during which our anti-aircraft guns were doubled in numbers, very hard and continuous air fighting took place over the capital, and the Luftwaffe were still confident through their over-estimation of our losses. But we now know that the German Naval Staff, in anxious regard for their own interests and responsibilities, wrote in their diary on September 10:

There is no sign of the defeat of the enemy's Air Force over Southern England and in the Channel area, and this is vital to a further judgment of the situation. The preliminary attacks by the Luftwaffe have indeed achieved a noticeable weakening of the enemy's fighter defence, so that considerable German fighter superiority can be assumed over the English area. However . . . we have not yet attained the operational conditions which the Naval Staff stipulated to the Supreme Command as being essential for the enterprise, namely, undisputed air supremacy in the Channel area and the elimination of the enemy's air activity in the assembly area of the German naval forces and ancillary shipping. . . . It would be in conformity with the time-table preparations for "Sea Lion" if the Luftwaffe now concentrated less on London and more on Portsmouth and Dover, as well as on the naval ports in and near the operational area. . . .

As by this time Hitler had been persuaded by Goering that the major attack on London would be decisive, the Naval Staff did not venture to appeal to the Supreme Command; but their uneasiness continued, and on the 12th they reached this sombre conclusion:

The air war is being conducted as an "absolute air war", without regard to the present requirements of the naval war, and outside the framework of Operation "Sea Lion". In its present form the air war cannot assist preparations for "Sea Lion", which are predominantly in the hands of the Navy. In particular one cannot discern any effort on the part of the Luftwaffe to engage the units of the British Fleet, which are now able to operate almost unmolested in the Channel, and this will prove extremely dangerous to the transportation. Thus the main safeguard against British naval forces would have to be the minefields, which, as repeatedly explained to the Supreme Command, cannot be regarded as reliable protection for shipping.

The fact remains that up to now the intensified air war has not contributed towards the landing operation; hence, for operational and military reasons the execution of the landing cannot yet be considered.

\* \* \* \* \*

I stated in a broadcast on September 11:

Whenever the weather is favourable waves of German bombers, protected by fighters, often three or four hundred at a time, surge over this Island, especially the promontory of Kent, in the hope of attacking military and other objectives by daylight. However, they are met by our fighter squadrons and nearly always broken up; and their losses average three to one in machines and six to one in pilots.

This effort of the Germans to secure daylight mastery of the air over England is of course the crux of the whole war. So far it has failed conspicuously. It has cost them very dear, and we have felt stronger, and actually are relatively a good deal stronger, than when the hard fighting began in July. There is no doubt that Herr Hitler is using up his fighter force at a very high rate, and that if he goes on for many more weeks he will wear down and ruin this vital part of his Air Force. That will give us a great advantage.

On the other hand, for him to try to invade this country without having secured mastery in the air would be a very hazardous undertaking. Nevertheless, all his preparations for invasion on a great scale are steadily going forward. Several hundreds of self-propelled barges are moving down the coasts of Europe, from the German and Dutch harbours to the ports of Northern France, from Dunkirk to Brest, and beyond Brest to the French harbours in the Bay of Biscay.

Besides this, convoys of merchant ships in tens and dozens are being moved through the Straits of Dover into the Channel, dodging along from port to port under the protection of the new batteries which the Germans have built on the French shore. There are now considerable gatherings of shipping in the German, Dutch, Belgian, and French harbours, all the way from Hamburg to Brest. Finally, there are some preparations made of ships to carry an invading force from the Norwegian harbours.

Behind these clusters of ships or barges there stand large numbers of German troops, awaiting the order to go on board and set out on their very dangerous and uncertain voyage across the seas. We cannot tell when they will try to come; we cannot be sure that in fact they will try at all; but no one should blind himself to the fact that a heavy full-scale invasion of this Island is being prepared with all the usual German thoroughness and method, and that it may be launched now—upon England, upon Scotland, or upon Ireland, or upon all three.

If this invasion is going to be tried at all, it does not seem that it can be long delayed. The weather may break at any time. Besides this, it is difficult for the enemy to keep these gatherings of ships waiting about indefinitely while they are bombed every night by our bombers, and very often shelled by our warships which are waiting for them outside.

Therefore we must regard the next week or so as a very important period in our history. It ranks with the days when the Spanish Armada was approaching the Channel, and Drake was finishing his game of bowls; or when Nelson stood between us and Napoleon's Grand Army at Boulogne. We have read all about this in the history books; but what is happening now is on a far greater scale and of far more consequence to the life and future of the world and its civilisation than those brave old days.

$$\ast \quad \ast \quad \ast \quad \ast \quad \ast$$

In the fighting between August 24 and September 6 the scales had tilted against Fighter Command. During these crucial days the Germans had continuously applied powerful forces against the airfields of South and South-East England. Their object was to break down the day fighter defence of the capital, which they were impatient to attack. Far more important to us than the protection of London from terror-bombing was the functioning and articulation of these airfields and the squadrons working from them. In the life-and-death struggle of the two Air Forces this was a decisive phase. We never thought of the struggle in terms of the defence of London or any other place, but only who won in the air. There was much anxiety at Fighter Headquarters at Stanmore, and particularly at the headquarters of No. 11 Fighter Group at Uxbridge. Extensive damage had been done to five of the Group's forward airfields, and also to the six Sector Stations. Manston and Lympne on the Kentish coast were on several occasions and for days unfit for operating fighter aircraft. Biggin Hill Sector Station, to the south of London, was so severely damaged that for a week only one fighter squadron could operate from it. If the enemy had persisted in heavy attacks against the adjacent sectors and damaged their operations rooms or telephone communications the whole intricate organisation of Fighter Command might have been broken down. This would have meant not merely the maltreatment of London, but the loss to us of the perfected control of our own air in the decisive area. As will be seen in the minutes printed in Appendix A, I was led to visit several of these stations, particularly Manston (August 28), and Biggin Hill, which is quite near my home. They were getting terribly knocked about, and their runways were ruined by craters. It was therefore with a sense of relief that Fighter Command felt the German attack turn on to London on September 7, and concluded that the enemy had changed his plan. Goering should certainly have persevered against the air-fields, on whose organisation and combination the whole fighting power of our Air Force at this moment depended. By departing from the classical principles of war, as well as from the hitherto accepted dictates of humanity, he made a foolish mistake.

This same period (August 24–September 6) had seriously drained the strength of Fighter Command as a whole. The Command had lost in this fortnight 103 pilots killed and 128 seriously wounded, while 466 Spitfires and Hurricanes had been destroyed or seriously damaged. Out of a total pilot strength of about a thousand nearly a quarter had been lost. Their places could only be filled by 260 new, ardent, but inexperienced pilots drawn from training units, in many cases before their full courses were complete. The night attacks on London for ten days after September 7 struck at the London docks and railway centres, and killed and wounded many civilians, but they were in effect for us a breathing-space of which we had the utmost need.

During this period I usually managed to take two afternoons a week in the areas under attack in Kent or Sussex in order to see for myself what was happening. For this purpose I used my train, which was now most conveniently fitted and carried a bed, a bath, an office, a connectible telephone, and an effective staff. I was thus able to work continuously, apart from sleeping, and with almost all the facilities available at Downing Street.

\* \* \* \* \*

THE BATTLE OF BRITAIN

We must take September 15 as the culminating date. On this day the Luftwaffe, after two heavy attacks on the 14th, made its greatest concentrated effort in a resumed daylight attack on London.

It was one of the decisive battles of the war, and, like the Battle of Waterloo, it was on a Sunday. I was at Chequers. I had already on several occasions visited the headquarters of No. 11 Fighter Group in order to witness the conduct of an air battle, when not much had happened. However, the weather on this day seemed suitable to the enemy, and accordingly I drove over to Uxbridge and arrived at the Group Headquarters. No. 11 Group comprised no fewer than twenty-five squadrons covering the whole of Essex, Kent, Sussex, and Hampshire, and all the approaches across them to London. Air Vice-Marshal Park had for six months commanded this group, on which our fate largely depended. From the beginning of Dunkirk all the daylight actions in the South of England had already been conducted by him, and all his arrangements and apparatus had been brought to the highest perfection. My wife and I were taken down to the bomb-proof Operations Room, fifty feet below ground. All the ascendancy of the Hurricanes and Spitfires would have been fruitless but for this system of underground control centres and telephone cables, which had been devised and built before the war by the Air Ministry under Dowding's advice and impulse. Lasting credit is due to all concerned. In the South of England there were at this time No. 11 Group H.Q. and six subordinate Fighter Station Centres. All these were, as has been described, under heavy stress. The Supreme Command was exercised from the Fighter Headquarters at Stanmore, but the actual handling of the direction of the squadrons was wisely left to No. 11 Group, which controlled the units through its Fighter Stations located in each county.

The Group Operations Room was like a small theatre, about sixty feet across, and with two storeys. We took our seats in the Dress Circle. Below us was the large-scale map-table, around which perhaps twenty highly-trained young men and women, with their telephone assistants, were assembled. Opposite to us, covering the entire wall, where the theatre curtain would be, was a gigantic blackboard divided into six columns with electric bulbs, for the six fighter stations, each of their squadrons having a sub-column of its own, and also divided by lateral lines. Thus the lowest row of bulbs showed as they were lighted the squadrons which were "Standing By" at two minutes' notice, the next row those at "Readiness", five minutes, then at "Available", twenty minutes, then those which had taken off, the next row those which had reported having seen the enemy, the next—with red lights—those which were in action, and the top row those which were returning home. On the left-hand side, in a kind of glass stage-box, were the four or five officers whose duty it was to weigh and measure the information received from our Observer Corps, which at this time numbered upwards of fifty thousand men, women, and youths. Radar was still in its infancy, but it gave warning of raids approaching our coast, and the observers, with field-glasses and portable telephones, were our main source of information about raiders flying overland. Thousands of messages were therefore received during an action. Several roomfuls of experienced people in other parts of the underground headquarters sifted them with great rapidity, and transmitted the results from minute to minute directly to the plotters seated around the table on the floor and to the officer supervising from the glass stage-box.

On the right hand was another glass stage-box containing Army officers who reported the action of our anti-aircraft batteries, of which at this time in the Command there were two hundred. At night it was of vital importance to stop these batteries firing over certain areas in which our fighters would be closing with the enemy. I was not unacquainted with the general outlines of this system, having had it explained to me a year before the war by Dowding when I visited him at Stanmore. It had been shaped and refined in constant action, and all was now fused together into a most elaborate instrument of war, the like of which existed nowhere in the world.

"I don't know," said Park, as we went down, "whether anything will happen to-day. At present all is quiet." However, after a quarter of an hour the raid-plotters began to move about. An attack of "40 plus" was reported to be coming from the German stations in the Dieppe area. The bulbs along the bottom of the wall display-panel began to glow as various squadrons came to "Stand By". Then in quick succession "20 plus", "40 plus" signals were received, and in another ten minutes it was evident that a serious battle impended. On both sides the air began to fill.

One after another signals came in, "40 plus", "60 plus"; there was even an "80 plus". On the floor-table below us the movement of all the waves of attack was marked by pushing discs forward from minute to minute along different lines of approach, while on the blackboard facing us the rising lights showed our fighter squadrons getting into the air, till there were only four or five left at "Readiness". These air battles, on which so much depended, lasted little more than an hour from the first encounter. The enemy had ample strength to send out new waves of attack, and our squadrons, having gone all out to gain the upper air, would have to refuel after seventy or eighty minutes, or land to rearm after a five-minute engagement. If at this moment of refuelling or rearming the enemy were able to arrive with fresh unchallenged squadrons some of our fighters could be destroyed on the ground. It was therefore one of our principal objects to direct our squadrons so as not to have too many on the ground refuelling or rearming simultaneously during daylight.

Presently the red bulbs showed that the majority of our squadrons were engaged. A subdued hum arose from the floor, where the busy plotters pushed their discs to and fro in accordance with the swiftly-changing situation. Air Vice-Marshal Park gave general directions for the disposition of his fighter force, which were translated into detailed orders to each Fighter Station by a youngish officer in the centre of the Dress Circle, at whose side I sat. Some years after I asked his name. He was Lord Willoughby de Broke. (I met him next in 1947, when the Jockey Club, of which he was a Steward, invited me to see the Derby. He was surprised that I remembered the occasion.) He now gave the orders for the individual squadrons to ascend and patrol as the result of the final information which appeared on the map-table. The Air Marshal himself walked up and down behind, watching with vigilant eye every move in the game, supervising his junior executive hand, and only occasionally intervening with some decisive order, usually to reinforce a threatened area. In a little while all our squadrons were fighting, and some had already begun to return for fuel. All were in the air. The lower line of bulbs was out. There was not one squadron left in reserve. At this moment Park spoke to Dowding at Stanmore, asking for three squadrons from No. 12 Group to be put at his disposal in case of another major attack while his squadrons were rearming and refuelling. This was done. They were specially needed to cover London and our fighter aerodromes, because No. 11 Group had already shot their bolt.

The young officer, to whom this seemed a matter of routine, continued to give his orders, in accordance with the general directions of his Group Commander, in a calm, low monotone, and the three reinforcing squadrons were soon absorbed. I became conscious of the anxiety of the Commander, who now stood still behind his subordinate's chair. Hitherto I had watched in silence. I now asked: "What other reserves have we?" "There are none," said Air Vice-Marshal Park. In an account which he wrote about it afterwards he said that at this I "looked grave". Well I might. What losses should we not suffer if our refuelling planes were caught on the ground by further raids of "40 plus" or "50 plus"! The odds were great; our margins small; the stakes infinite.

Another five minutes passed, and most of our squadrons had now descended to refuel. In many cases our resources could not give them overhead protection. Then it appeared that the enemy were going home. The shifting of the discs on the table below showed a continuous eastward movement of German bombers and fighters. No new attack appeared. In another ten minutes the action was ended. We climbed again the stairways which led to the surface, and almost as we emerged the "All Clear" sounded.

"We are very glad, sir, you have seen this," said Park. "Of course, during the last twenty minutes we were so choked with information that we couldn't handle it. This shows you the limitation of our present resources. They have been strained far beyond their limits to-day." I asked whether any results had come to hand, and remarked that the attack appeared to have been repelled satisfactorily. Park replied that he was not satisfied that we had intercepted as many raiders as he had hoped we should. It was evident that the enemy had everywhere pierced our defences. Many scores of German bombers, with their fighter escort, had been reported over London. About a dozen had been brought down while I was below, but no picture of the results of the battle or of the damage or losses could be obtained.

It was 4.30 p.m. before I got back to Chequers, and I immediately went to bed for my afternoon sleep. I must have been tired by the drama of No. 11 Group, for I did not wake till eight. When I rang, John Martin, my Principal Private Secretary, came in with the evening budget of news from all over the world. It was repellent. This had gone wrong here; that had been delayed there; an unsatisfactory answer had been received from so-and-so; there had been bad sinkings in the Atlantic. "However," said Martin, as he finished this account, "all is redeemed by the air. We have shot down one hundred and eighty-three for a loss of under forty."

\* \* \* \* \*

Although post-war information has shown that the enemy's losses on this day were only fifty-six, September 15 was the crux of the Battle of Britain. That same night our Bomber Command attacked in strength the shipping in the ports from Boulogne to Antwerp. At Antwerp particularly heavy losses were inflicted. On September 17, as we now know, the Fuehrer decided to postpone "Sea Lion" indefinitely. It was not till October 12 that the invasion was formally called off till the following spring. In July 1941 it was postponed again by Hitler till the spring of 1942, "by which time the Russian campaign will be completed". This was a vain but an important imagining. On February 13, 1942, Admiral Raeder had his final interview on "Sea Lion" and got Hitler to agree to a complete "stand-down". Thus perished Operation "Sea Lion". And September 15 may stand as the date of its demise.

\* \* \* \* \*

The German Naval Staff were in hearty accord with all the postponements; indeed they instigated them. The Army leaders made no complaint. On the 17th I said in Parliament: "The process of waiting keyed up to concert pitch day after day is apt in time to lose its charm of novelty. Sunday's action was the most brilliant and fruitful of any fought up to that date by the fighters of the Royal Air Force. . . . We may await the decision of this long air battle with sober but increasing confidence." An impartial observer, Brigadier-General Strong, Assistant Chief of the United States War Plans Division and Head of the American Military Mission which had been sent to London to observe the results of the Luftwaffe attacks, arrived back in New York on the

19th, and reported that the Luftwaffe had made no serious inroad on the strength of the R.A.F., that the military damage done by air bombardment had been comparatively small, and that British claims of German aircraft losses were "on the conservative side".

Yet the Battle of London was still to be fought out. Although invasion had been called off, it was not till September 27 that Goering gave up hope that his method of winning the war might succeed. In October, though London received its full share, the German effort was spread by day and night in frequent small-scale attacks on many places. Concentration of effort gave way to dispersion; the battle of attrition began. Attrition! But whose?

\*　　\*　　\*　　\*　　\*

In cold blood, with the knowledge of the after-time, we may study the actual losses of the British and German Air Forces in what may well be deemed one of the decisive battles of the world. From the tables which follow our hopes and fears may be contrasted with what happened.

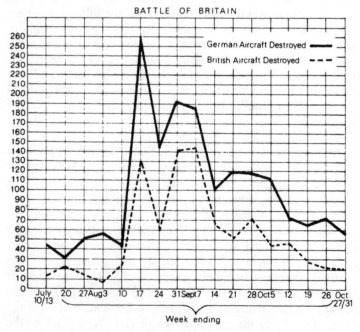

BATTLE OF BRITAIN

## AIRCRAFT LOSSES

| | British Fighters LOST by R.A.F. (complete write-off or missing) | Enemy Aircraft actually DESTROYED (according to German records) | Enemy Aircraft CLAIMED by us (Fighter Command, A.A., Balloons, etc.) |
|---|---|---|---|
| **WEEKLY TOTALS:** | | | |
| July 10–13 .. | 15 | 45 | 63 |
| Week to July 20 .. | 22 | 31 | 49 |
| ,, ,, ,, 27 .. | 14 | 51 | 58 |
| ,, ,, Aug. 3 .. | 8 | 56 | 39 |
| ,, ,, ,, 10 .. | 25 | 44 | 64 |
| ,, ,, ,, 17 .. | 134 | 261 | 496 |
| ,, ,, ,, 24 .. | 59 | 145 | 251 |
| ,, ,, ,, 31 .. | 141 | 193 | 316 |
| ,, ,, Sept. 7 .. | 144 | 187 | 375 |
| ,, ,, ,, 14 .. | 67 | 102 | 182 |
| ,, ,, ,, 21 .. | 52 | 120 | 268 |
| ,, ,, ,, 28 .. | 72 | 118 | 230 |
| ,, ,. Oct. 5 .. | 44 | 112 | 100 |
| ,, ,, ,, 12 .. | 47 | 73 | 66 |
| ,, ,, ,, 19 .. | 29 | 67 | 38 |
| ,, ,, ,, 26 .. | 21 | 72 | 43 |
| Oct. 27–31 .. .. | 21 | 56 | 60 |
| | | | |
| **MONTHLY TOTALS:** | | | |
| July (from July 10) | 58 | 164 | 203 |
| August .. .. | 360 | 662 | 1,133 |
| September .. .. | 361 | 582 | 1,108 |
| October .. .. | 136 | 325 | 254 |
| | — | — | — |
| **TOTALS** .. .. | 915 | 1,733 | 2,698 |

Further tables will be found in Appendix C.

No doubt we were always over-sanguine in our estimates of enemy scalps. In the upshot we got two to one of the German assailants, instead of three to one, as we believed and declared. But this was enough. The Royal Air Force, far from being destroyed, was triumphant. A strong flow of fresh pilots was provided. The aircraft factories, upon which not only our immediate need but our power to wage a long war depended, were mauled but not paralysed. The workers, skilled and unskilled, men and women alike, stood to their lathes and manned the workshops under fire as if they were batteries in action—which indeed they were. At the Ministry of Supply Herbert Morrison spurred all in his wide sphere. "Go to it," he adjured, and to it they went. Skilful and ever-ready support was given to the air-fighting by the Anti-Aircraft Command under General Pile. Their main contribution came later. The Observer Corps, devoted and tireless, were hourly at their posts. The carefully-wrought organisation of Fighter Command, without which all might have been in vain, proved equal to months of continuous strain. All played their part.

At the summit the stamina and valour of our fighter pilots remained unconquerable and supreme. Thus Britain was saved. Well might I say in the House of Commons: "Never in the field of human conflict was so much owed by so many to so few."

# CHAPTER XVII

# THE BLITZ

*Successive Phases of the German Attack – Goering Assumes Command of the Air Battle – His Attempt to Conquer London – Hitler's Boast – Fifty-seven Nights' Bombardment (September 7–November 3) – General Pile's Barrage – Some Personal Notes – Downing Street and the Annexe – Mr. Chamberlain's Fortitude after His Major Operation – He Consents to Leave London – His Stoical Death – A Dinner at Number 10 – My Lucky Inspiration – The Bomb in the Treasury Courtyard – Burning Pall Mall – Destruction of the Carlton Club – Courage of the People – The Ramsgate Hotel and the War Damage Insurance Scheme – The Tubes as Air-Raid Shelters – Our Expectation that London would be Reduced to Rubble – Rules for the Public Departments – The "Alert" and the "Alarm" – The "Banshee Howlings" – Improving the Anderson Shelters – The Cabinet Advances its Meal-times – The Mood of Parliament – I Persuade Members to Act with Prudence – Their Good Fortune.*

THE German air assault on Britain is a tale of divided counsels, conflicting purposes, and never fully accomplished plans. Three or four times in these months the enemy abandoned a method of attack which was causing us severe stress, and turned to something new. But all these stages overlapped one another, and cannot be readily distinguished by precise dates. Each one merged into the next. The early operations sought to engage our air forces in battle over the Channel and the south coast; next the struggle was continued over our southern counties, principally Kent and Sussex, the enemy aiming to destroy our air-power organisation; then nearer to and over London; then London became the supreme target; and finally, when London triumphed, there was a renewed dispersion to the provincial cities and to our sole Atlantic life-line by the Mersey and the Clyde.

We have seen how very hard they had run us in the attack on

the south coast airfields in the last week of August and the first week of September. But on September 7 Goering publicly assumed command of the air battle, and turned from daylight to night attack and from the fighter airfields of Kent and Sussex to the vast built-up areas of London. Minor raids by daylight were frequent, indeed constant, and one great daylight attack was still to come; but in the main the whole character of the German offensive was altered. For fifty-seven nights the bombing of London was unceasing. This constituted an ordeal for the world's largest city, the results of which no one could measure beforehand. Never before was so wide an expanse of houses subjected to such bombardment or so many families required to face its problems and its terrors.

The sporadic raiding of London towards the end of August was promptly answered by us in a retaliatory attack on Berlin. Because of the distance we had to travel, this could only be on a very small scale compared with attacks on London from near-by French and Belgian airfields. The War Cabinet were much in the mood to hit back, to raise the stakes, and to defy the enemy. I was sure they were right, and believed that nothing impressed or disturbed Hitler so much as his realisation of British wrath and will-power. In his heart he was one of our admirers. He took of course full advantage of our reprisal on Berlin, and publicly announced the previously-settled German policy of reducing London and other British cities to chaos and ruin. "If they attack our cities," he declared on September 4, "we will simply erase theirs." He tried his best.

The first German aim had been the destruction of our air-power; the second was to break the spirit of the Londoner, or at least render uninhabitable the world's largest city. In these new purposes the enemy did not succeed. The victory of the Royal Air Force had been gained by the skill and daring of our pilots, by the excellence of our machines, and by their wonderful organisation. Other virtues not less splendid, not less indispensable to the life of Britain, were now to be displayed by millions of ordinary humble people, who proved to the world the strength of a community nursed in freedom.

\*　　\*　　\*　　\*　　\*

From September 7 to November 3 an average of two hundred

German bombers attacked London every night. The various preliminary raids which had been made upon our provincial cities in the previous three weeks had led to a considerable dispersion of our anti-aircraft artillery, and when London first became the main target there were but ninety-two guns in position. It was thought better to leave the air free for our night-fighters, working under No. 11 Group. Of these there were six squadrons of Blenheims and Defiants. Night-fighting was in its infancy, and very few casualties were inflicted on the enemy. Our batteries therefore remained silent for three nights in succession. Their own technique was at this time woefully imperfect. Nevertheless, in view of the weakness of our night-fighters and of their unsolved problems it was decided that the anti-aircraft gunners should be given a free hand to fire at unseen targets, using any methods of control they liked. In forty-eight hours General Pile, commanding the Air Defence Artillery, had more than doubled the number of guns in the capital by withdrawals from the provincial cities. Our own aircraft were kept out of the way, and the batteries were given their chance.

For three nights Londoners had sat in their houses or inadequate shelters enduring what seemed to be an utterly unresisted attack. Suddenly, on September 10, the whole barrage opened, accompanied by a blaze of searchlights. This roaring cannonade did not do much harm to the enemy, but gave enormous satisfaction to the population. Everyone was cheered by the feeling that we were hitting back. From that time onwards the batteries fired regularly, and of course practice, ingenuity, and grinding need steadily improved the shooting. A slowly increasing toll was taken of the German raiders. Upon occasions the batteries were silent and the night-fighters, whose methods were also progressing, came on the scene. The night raids were accompanied by more or less continuous daylight attacks by small groups or even single enemy planes, and the sirens often sounded at brief intervals throughout the whole twenty-four hours. To this curious existence the seven million inhabitants of London accustomed themselves.

\*     \*     \*     \*     \*

In the hope that it may lighten the hard course of this narrative I record a few personal notes about the "Blitz", well knowing how many thousands have far more exciting tales to tell.

When the bombardment first began the idea was to treat it with disdain. In the West End everybody went about their business and pleasure and dined and slept as they usually did. The theatres were full, and the darkened streets were crowded with casual traffic. All this was perhaps a healthy reaction from the frightful squawk which the defeatist elements in Paris had put up on the occasion when they were first seriously raided in May. I remember dining in a small company when very lively and continuous raids were going on. The large windows of Stornoway House opened upon the Green Park, which flickered with the flashes of the guns and was occasionally lit by the glare of an exploding bomb. I felt that we were taking unnecessary risks. After dinner we went to the Imperial Chemicals building overlooking the Embankment. From these high stone balconies there was a splendid view of the river. At least a dozen fires were burning on the south side, and while we were there several heavy bombs fell, one near enough for my friends to pull me back behind a substantial stone pillar. This certainly confirmed my opinion that we should have to accept many restrictions upon the ordinary amenities of life.

The group of Government buildings around Whitehall were repeatedly hit. Downing Street consists of houses two hundred and fifty years old, shaky and lightly built by the profiteering contractor whose name they bear. At the time of the Munich alarm shelters had been constructed for the occupants of No. 10 and No. 11, and the rooms on the garden level had had their ceilings propped up with a wooden under-ceiling and strong timbers. It was believed that this would support the ruins if the building was blown or shaken down; but of course neither these rooms nor the shelters were effective against a direct hit. During the last fortnight of September preparations were made to transfer my Ministerial headquarters to the more modern and solid Government offices looking over St. James's Park by Storey's Gate. These quarters we called "the Annexe". Below them were the War Room and a certain amount of bomb-proof sleeping accommodation. The bombs at this time were of course smaller than those of the later phases. Still, in the interval before the new apartments were ready life at Downing Street was exciting. One might as well have been at a battalion headquarters in the line.

* * * * *

304

In these months we held our evening Cabinets in the War Room in the Annexe basement. To get there from Downing Street it was necessary to walk through the Foreign Office quadrangle and then clamber through the working parties who were pouring in the concrete to make the War Room and basement offices safer. I did not realise what a trial this was to Mr. Chamberlain, with all the consequences of his major operation upon him. Nothing deterred him, and he was never more spick and span or cool and determined than at the last Cabinets which he attended.

One evening in late September 1940 I looked out of the Downing Street front door and saw workmen piling sandbags in front of the low basement windows of the Foreign Office opposite. I asked what they were doing. I was told that after his operation Mr. Neville Chamberlain had to have special periodical treatment, and that it was embarrassing to carry this out in the shelter of No. 11, where at least twenty people were gathered during the constant raids, so a small private place was being prepared over there for him. Every day he kept all his appointments, reserved, efficient, faultlessly attired. But here was the background. It was too much. I used my authority. I walked through the passage between No. 10 and No. 11 and found Mrs. Chamberlain. I said: "He ought not to be here in this condition. You must take him away till he is well again. I will send all the telegrams to him each day." She went off to see her husband. In an hour she sent me word. "He will do what you wish. We are leaving tonight." I never saw him again. In less than two months he was no more. I am sure he wanted to die in harness. This was not to be.

       \*     \*     \*     \*     \*

Another evening (October 17) stands out in my mind. We were dining in the garden-room of No. 10 when the usual night raid began. My companions were Archie Sinclair, Oliver Lyttelton, and Moore-Brabazon. The steel shutters had been closed. Several loud explosions occurred around us at no great distance, and presently a bomb fell, perhaps a hundred yards away, on the Horse Guards Parade, making a great deal of noise. Suddenly I had a providential impulse. The kitchen at No. 10 Downing Street is lofty and spacious, and looks out through a large plate-glass window about twenty-five feet high. The butler and parlourmaid continued to serve the dinner with complete detachment, but I became acutely aware of this big window, behind which Mrs. Landemare, the cook, and the kitchen-maid, never turning a hair, were at work. I got up abruptly, went into the kitchen, told the butler to put the dinner on the hot plate in the dining-room, and ordered the cook and the other servants into the shelter, such as it was. I had been seated again at table only about three minutes when a really very loud crash, close at hand, and a violent shock showed that the house had been struck. My detective came into the room and said much damage had been done. The kitchen, the pantry, and the offices on the Treasury side were shattered.

We went into the kitchen to view the scene. The devastation was complete. The bomb had fallen fifty yards away on the Treasury, and the blast had smitten the large, tidy kitchen, with all its bright saucepans and crockery, into a heap of black dust and rubble. The big plate-glass window had been hurled in fragments and splinters across the room, and would of course have cut its occupants, if there had been any, to pieces. But my fortunate inspiration, which I might so easily have neglected, had come in the nick of time. The underground Treasury shelter across the court had been blown to pieces by a direct hit, and the four civil servants who were doing Home Guard night-duty there were killed. All however were buried under tons of brick rubble, and we did not know who was missing.

As the raid continued and seemed to grow in intensity we put on our tin hats and went out to view the scene from the top of the Annexe buildings. Before doing so, however, I could not resist taking Mrs. Landemare and the others from the shelter to see their kitchen. They were upset at the sight of the wreck, but principally on account of the general untidiness!

Archie and I went up to the cupola of the Annexe building. The night was clear and there was a wide view of London. It seemed that the greater part of Pall Mall was in flames. At least five fierce fires were burning there, and others in St. James's Street and Piccadilly. Farther back over the river in the opposite direction there were many conflagrations. But Pall Mall was the vivid flame-picture. Gradually the attack died down, and presently the "All Clear" sounded, leaving only the blazing fires. We went downstairs to my new apartments on the first floor of the Annexe, and there found Captain David Margesson, the Chief

Whip, who was accustomed to live at the Carlton Club. He told us the club had been blown to bits, and indeed we had thought, by the situation of the fires, that it must have been hit. He was in the club with about two hundred and fifty members and staff. It had been struck by a heavy bomb. The whole of the façade and the massive coping on the Pall Mall side had fallen into the street, obliterating his motor-car, which was parked near the front door. The smoking-room had been full of members, and the whole ceiling had come down upon them. When I looked at the ruins next day it seemed incredible that most of them should not have been killed. However, by what seemed a miracle, they had all crawled out of the dust, smoke, and rubble, and though many were injured not a single life was lost. When in due course these facts came to the notice of the Cabinet our Labour colleagues facetiously remarked: "The devil looks after his own." Mr. Quintin Hogg had carried his father, a former Lord Chancellor, on his shoulders from the wreck, as Æneas had borne Pater Anchises from the ruins of Troy. Margesson had nowhere to sleep, and we found him blankets and a bed in the basement of the Annexe. Altogether it was a lurid evening, and considering the damage to buildings it was remarkable that there were not more than five hundred people killed and about a couple of thousand injured.

\* \* \* \* \*

One day after luncheon the Chancellor of the Exchequer, Kingsley Wood, came to see me on business at No. 10, and we heard a very heavy explosion take place across the river in South London. I took him to see what had happened. The bomb had fallen in Peckham. It was a very big one—probably a land-mine. It had completely destroyed or gutted twenty or thirty small three-storey houses and cleared a considerable open space in this very poor district. Already little pathetic Union Jacks had been stuck up amid the ruins. When my car was recognised the people came running from all quarters, and a crowd of more than a thousand was soon gathered. All these folk were in a high state of enthusiasm. They crowded round us, cheering and manifesting every sign of lively affection, wanting to touch and stroke my clothes. One would have thought I had brought them some fine substantial benefit which would improve their lot in life. I was completely undermined, and wept. Ismay, who was with me, records that he heard an old woman say: "You see, he really cares. He's crying." They were tears not of sorrow but of wonder and admiration. "But see, look here," they said, and drew me to the centre of the ruins. There was an enormous crater, perhaps forty yards across and twenty feet deep. Cocked up at an angle on the very edge was an Anderson shelter, and we were greeted at its twisted doorway by a youngish man, his wife, and three children, quite unharmed but obviously shell-jarred. They had been there at the moment of the explosion. They could give no account of their experiences. But there they were, and proud of it. Their neighbours regarded them as enviable curiosities. When we got back into the car a harsher mood swept over this haggard crowd. "Give it 'em back," they cried, and "Let *them* have it too." I undertook forthwith to see that their wishes were carried out; and this promise was certainly kept. The debt was repaid tenfold, twentyfold, in the frightful routine bombardment of German cities, which grew in intensity as our air-power developed, as the bombs became far heavier and the explosives more powerful. Certainly the enemy got it all back in good measure, pressed down and running over. Alas for poor humanity!

*　*　*　*　*

Another time I visited Ramsgate. An air raid came upon us, and I was conducted into their big tunnel, where quite large numbers of people lived permanently. When we came out, after a quarter of an hour, we looked at the still-smoking damage. A small hotel had been hit. Nobody had been hurt, but the place had been reduced to a litter of crockery, utensils, and splintered furniture. The proprietor, his wife, and the cooks and waitresses were in tears. Where was their home? Where was their livelihood? Here is a privilege of power. I formed an immediate resolve. On the way back in my train I dictated a letter to the Chancellor of the Exchequer laying down the principle that all damage from the fire of the enemy must be a charge upon the State and compensation be paid in full and at once. Thus the burden would not fall alone on those whose homes or business premises were hit, but would be borne evenly on the shoulders of the nation. Kingsley Wood was naturally a little worried by the indefinite character of this obligation. But I pressed hard, and an insurance scheme was devised in a fortnight which afterwards played a substantial part in our affairs. In explaining this to Parliament on September 5 I said:

It is very painful to me to see, as I have seen in my journeys about the country, a small British house or business smashed by the enemy's fire, and to see that without feeling assured that we are doing our best to spread the burden so that we all stand in together. Damage by enemy action stands on a different footing from any other kind of loss or damage, because the nation undertakes the task of defending the lives and property of its subjects and taxpayers against assaults from outside. Unless public opinion and the judgment of the House were prepared to separate damage resulting from the fire of the enemy from all other forms of war loss, and unless the House was prepared to draw the distinction very sharply between war damage by bomb and shell and the other forms of loss which are incurred, we could not attempt to deal with this matter; otherwise we should be opening up a field to which there would be no bounds. If however we were able to embark upon such a project as would give complete insurance, at any rate up to a certain minimum figure, for every one against war damage by shell or bomb, I think it would be a very solid mark of the confidence which after some experience we are justified in feeling about the way in which we are going to come through this war.

The Treasury went through various emotions about this insurance scheme. First they thought it was going to be their ruin; but when, after May 1941, the air raids ceased for over three years they began to make a great deal of money, and considered the plan provident and statesmanlike. However, later on in the war, when the "doodle-bugs" and rockets began, the accounts swung the other way, and eight hundred and ninety millions have in fact already been paid out. I am very glad it is so.

\*　　\*　　\*　　\*　　\*

Our outlook at this time was that London, except for its strong modern buildings, would be gradually and soon reduced to a rubble-heap. I was deeply anxious about the life of the people of London, the greater part of whom stayed, slept, and took a chance where they were. The brick and concrete shelters were multiplying rapidly. The Tubes offered accommodation for a good many. There were several large shelters, some of which held as many as seven thousand people, who camped there in confidence night after night, little knowing what the effect of a direct hit would have been upon them. I asked that brick traverses should be built in these as fast as possible. About the Tubes there was an argument which was ultimately resolved by a compromise.

*Prime Minister to Sir Edward Bridges, Home Secretary*
*and Minister of Transport*                                    21.IX.40

1. When I asked at the Cabinet the other day why the Tubes could not be used to some extent, even at the expense of transport facilities, as air-raid shelters, I was assured that this was most undesirable, and that the whole matter had been reviewed before that conclusion was reached. I now see that the Aldwych Tube is to be used as a shelter. Pray let me have more information about this, and what has happened to supersede the former decisive arguments.

2. I still remain in favour of a widespread utilisation of the Tubes, by which I mean not only the stations but the railway lines, and I should like a short report on one sheet of paper showing the numbers that could be accommodated on various sections and the structural changes that would be required to fit these sections for their new use. Is it true, for instance, that 750,000 people could be accommodated in the Aldwych section alone? We may well have to balance the relative demands of transport and shelter.

3. I am awaiting the report of the Home Secretary on the forward policy of—

    (*a*) Making more shelters.
    (*b*) Strengthening existing basements.
    (*c*) Making empty basements and premises available.
    (*d*) Most important. Assigning fixed places by tickets to a large proportion of the people, thus keeping them where we want them, and avoiding crowding.

In this new phase of warfare it became important to extract the optimum of work not only from the factories but even more from the departments in London which were under frequent bombardment during both the day and night. At first, whenever the sirens gave the alarm, all the occupants of a score of Ministries were promptly collected and led down to the basements, for what these were worth. Pride, even, was being taken in the efficiency and thoroughness with which this evolution was performed. In many cases it was only half a dozen aeroplanes which approached—sometimes only one. Often they did not arrive. A petty raid might bring to a standstill for over an hour the whole executive and administrative machine in London.

I therefore proposed the stage "Alert", operative on the siren warning, as distinct from the "Alarm", which should be enforced only when the spotters on the roof, or Jim Crows, as they came to be called, reported "Imminent danger", which meant that the enemy were actually overhead or very near. Schemes were worked out accordingly. In order to enforce rigorous compliance, while we lived under these repeated daylight attacks, I called for a weekly return of the number of hours spent by the staff of each department in the shelters.

*Prime Minister to Sir Edward Bridges and General Ismay*      17.IX.40
    Please report by to-morrow night the number of hours on September 16 that the principal offices in London were in their dug-outs and out of action through air alarm.
    General Ismay should find out how the Air Ministry and Fighter Command view the idea that no red warning should be given when only two or three aircraft are approaching London.

*Prime Minister to Sir Horace Wilson and Sir Edward Bridges*      19.IX.40
    Let me have a further return [of time lost in Government departments owing to air-raid warnings] for the 17th and 18th, and henceforward daily, from all Ministries, including the Service departments. These returns will be circulated to heads of all departments at the same time as they are sent to me. Thus it will be possible to see who are doing best. If all returns are not received on any day from some departments, those that are should nevertheless be circulated.

\*     \*     \*     \*     \*

    This put everybody on their mettle. Eight of these returns were actually furnished. It was amusing to see that the fighting departments were for some time in the worst position. Offended and spurred by this implied reproach, they very quickly took their proper place. The loss of hours in all departments was reduced to a fraction. Presently our fighters made daylight attack too costly to the enemy, and this phase passed away. In spite of the almost continuous Alerts and Alarms which were sounded, hardly a single Government department was hit during daylight when it was full of people, nor any loss of life sustained. But how much time might have been wasted in the functioning of the war machine if the civil and military staffs had shown any weakness, or been guided up the wrong alley!
    As early as September 1, before the heavy night attacks began, I had addressed the Home Secretary and others.

## AIR-RAID WARNINGS AND PRECAUTIONS

1. The present system of air-raid warnings was designed to cope with occasional large mass raids on definite targets, not with waves coming over several times a day, and still less with sporadic bombers roaming about at nights. We cannot allow large parts of the country to be immobilised for hours every day and to be distracted every night. The enemy must not be permitted to prejudice our war effort by stopping work in the factories which he has been unable to destroy.

2. There should be instituted therefore a new system of warnings:

The Alert.
The Alarm.
The All Clear.

The Alert should not interrupt the normal life of the area. People not engaged on national work could, if they desired, take refuge or put their children in a place of safety. But in general they should learn, and they do learn, to adapt themselves to their dangers and take only such precautions as are compatible with their duties and imposed by their temperament.

3. The air-raid services should be run on an increased nucleus staff, and not all be called out every time as on a present red warning. The look-out system should be developed in all factories where war work is proceeding, and should be put into effect when the Alert is given; the look-outs would have full authority to give local factory or office alarms. The signal for the Alert might be given during the day by the hoisting of a display of yellow flags by a sufficient number of specially-charged air-raid wardens. At night flickering yellow (or perhaps red) lamps could be employed. The use of electric street lighting should be studied, and the possibility of sounding special signals on the telephone.

4. The Alarm is a direct order to "Take cover" and for the full manning of all A.R.P. positions. This will very likely synchronise with or precede by only a brief interval the actual attack. The routine in each case must be subject to local conditions.

The signal for the Alarm would be the siren. It would probably be unnecessary to supplement this by light or telephone signals.

5. The All Clear could be sounded as at present. It would end the Alarm period. If the Alert continued, the flags would remain hoisted; if the enemy had definitely turned back, the Alert flags and lights would be removed.

The use of the Alert and Alarm signals might vary in different parts of the country. In areas subject to frequent attack, such as East Kent, South and South-East London, south East Anglia, Birmingham,

Derby, Liverpool, Bristol, and some other places, the Alert would be a commonplace. The Alarm would denote actual attack. This would also apply to the Whitehall district. In other parts of the country a somewhat less sparing use of the Alarm might be justified in order to keep the air-raid services from deteriorating.

6. In Government offices in London no one should be forced to take cover until actual firing has begun and the siren ordering the Alarm under the new conditions has been sounded. No one is to stop work merely because London is under Alert conditions.

\* \* \* \* \*

I had to give way about the sirens, or "Banshee howlings", as I described them to Parliament.

*Prime Minister to Home Secretary and others concerned* 14.IX.40

I promised the House that new regulations about air-raid warnings, sirens, whistles, Jim Crow, etc., should be considered within the past week. However, the intensification of raiding has made it inexpedient to abolish the sirens at this moment. I shall be glad however to have a short statement prepared of what is the practice which has in fact developed during the last week.

\* \* \* \* \*

One felt keenly for all the poor people, most of them in their little homes, with nothing over their heads.

*Prime Minister to Home Secretary* 3.IX.40

In spite of the shortage of materials, a great effort should be made to help people to drain their Anderson shelters, which reflect so much credit on your name, and to make floors for them against the winter rain. Bricks on edge placed loosely together without mortar, covered with a piece of linoleum, would be quite good, but there must be a drain and a sump. I am prepared to help you in a comprehensive scheme to tackle this. Instruction can be given on the broadcast, and of course the Regional Commissioners and local authorities should be used. Let me have a plan.

*Prime Minister to General Ismay and Private Office* 11.IX.40

Please call for reports on whether any serious effects are being produced by the air attack on—

    (*a*) food supplies and distribution;
    (*b*) numbers of homeless, and provision therefor;
    (*c*) exhaustion of Fire Brigade personnel;
    (*d*) sewage in London area;
    (*e*) gas and electricity;

(f) water supplies in London area.

(g) General Ismay to find out what is the practical effect of the bombing on Woolwich production. See also my report from the Minister of Supply.

*Prime Minister to Sir Edward Bridges*                    12.IX.40

Will you kindly convey to the Cabinet and Ministers the suggestion which I make that our hours should be somewhat advanced. Luncheon should be at one o'clock, and Cabinet times moved forward by half an hour. In principle it will be convenient if we aim at an earlier dinner-hour, say 7.15 p.m. Darkness falls earlier, and for the next few weeks severe bombing may be expected once the protection of the fighter aircraft is withdrawn. It would be a good thing if staffs and servants could be under shelter as early as possible, and Ministers are requested to arrange to work in places of reasonable security during the night raids, and especially to find places for sleeping where they will not be disturbed by anything but a direct hit.

I propose to ask Parliament when it meets at the usual time on Tuesday to meet in these occasional sittings at 11 a.m. and separate at 4 or 5 p.m. This will allow Members to reach their homes, and I hope their shelters, by daylight. We must adapt ourselves to these conditions, which will probably be accentuated. Indeed, it is likely we shall have to move our office hours forward by another half-hour as the days shorten.

\*     \*     \*     \*     \*

Parliament also required guidance about the conduct of its work in these dangerous days. Members felt that it was their duty to set an example. This was right, but it might have been pushed too far; I had to reason with the Commons to make them observe ordinary prudence and conform to the peculiar conditions of the time. I convinced them in Secret Session of the need to take necessary and well-considered precautions. They agreed that their days and hours of sitting should not be advertised, and to suspend their debates when the Jim Crow reported to the Speaker "Imminent danger". Then they all trooped down dutifully to the crowded, ineffectual shelters that had been provided. It will always add to the renown of the British Parliament that its Members continued to sit and discharge their duties through all this period. The Commons are very touchy in such matters, and it would have been easy to misjudge their mood. When one Chamber was damaged they moved to another, and I did my utmost to persuade them to follow wise advice with good grace. Their migrations will be recorded in due course. In short, everyone behaved with sense and dignity. It was also lucky that when the Chamber was blown to pieces a few months later it was by night and not by day, when empty and not when full. With our mastery of the daylight raids there came considerable relief in personal convenience. But during the first few months I was never free from anxiety about the safety of the Members. After all, a free sovereign Parliament, fairly chosen by universal suffrage, able to turn out the Government any day, but proud to uphold it in the darkest days, was one of the points which were in dispute with the enemy. Parliament won.

I doubt whether any of the Dictators had as much effective power throughout his whole nation as the British War Cabinet. When we expressed our desires we were sustained by the people's representatives, and cheerfully obeyed by all. Yet at no time was the right of criticism impaired. Nearly always the critics respected the national interest. When on occasions they challenged us the Houses voted them down by overwhelming majorities, and this, in contrast with totalitarian methods, without the slightest coercion, intervention, or use of the police or Secret Service. It was a proud thought that Parliamentary Democracy, or whatever our British public life can be called, can endure, surmount, and survive all trials. Even the threat of annihilation did not daunt our Members, but this fortunately did not come to pass.

# CHAPTER XVIII

# "LONDON CAN TAKE IT"

*Grim and Gay – Passion in the United States – The London Drains – Danger of Epidemics – Broken Windows – The Delayed-Action Bombs – Minutes Thereupon – The U.X.B. Detachments – The Peril Mastered – Heavy Parachute Mines – The Question of Reprisals – Later German Experiences Compared with Ours – Need of Security for the Central Government – "Paddock" Rehearsal – Herbert Morrison Succeeds John Anderson as Home Secretary – The Incendiary Attacks Begin – The National Fire Service – Civil Defence, a Fourth Arm of the Crown – Power of London to Take Punishment – Permanent Arrangements for Safeguarding the War Machine – I Am Placed in Safety in Piccadilly Underground – Return to the Annexe – Another Change of the German Plan – The Provincial Cities – Coventry – Birmingham – Attacks on the Ports – Great Burning of the City of London, December 29, 1940 – The King at Buckingham Palace – His Majesty's Mastery of Business – A Thought for the Future.*

*T*HESE were the times when the English, and particularly the Londoners, who had the place of honour, were seen at their best. Grim and gay, dogged and serviceable, with the confidence of an unconquered people in their bones, they adapted themselves to this strange new life, with all its terrors, with all its jolts and jars. One evening when I was leaving for an inspection on the East Coast, on my way to King's Cross the sirens sounded, the streets began to empty, except for long queues of very tired, pale people, waiting for the last bus that would run. An autumn mist and drizzle shrouded the scene. The air was cold and raw. Night and the enemy were approaching. I felt, with a spasm of mental pain, a deep sense of the strain and suffering that was being borne throughout the world's largest capital city. How

316

long would it go on? How much more would they have to bear? What were the limits of their vitality? What effects would their exhaustion have upon our productive war-making power?*

Away across the Atlantic the prolonged bombardment of London, and later of other cities and sea-ports, aroused a wave of sympathy in the United States, stronger than any ever felt before or since in the English-speaking world. Passion flamed in American hearts, and in none more than in the heart of President Roosevelt. The temperature rose steadily in the United States. I could feel the glow of millions of men and women eager to share the suffering, burning to strike a blow. As many Americans as could get passages came, bringing whatever gifts they could, and their respect, reverence, deep love, and comradeship were very inspiring. However, this was only September, and we had many months before us of this curious existence.

Under the pressure of the bombardment the shelters and defences grew continually. I was worried principally on three counts. The first was the drains. When you had six or seven million people living in a great built-up area the smashing of their sewers and water supply seemed to me a very great danger. Could we keep the sewage system working or would there be a pestilence? What would happen if the drains got into the water supply? Actually, early in October the main sewage outfall was destroyed and we had to let all our sewage flow into the Thames, which stank, first of sewage and afterwards of the floods of chemicals we poured into it. But all was mastered. Secondly, I feared that the long nights for millions in the crowded street-shelters—only blast-proof at that—would produce epidemics of influenza, diphtheria, the common cold, and what not. But it appeared that Nature had already provided against this danger. Man is a gregarious animal, and apparently the mischievous microbes he exhales fight and neutralise each other. They go out and devour each other, and Man walks off unharmed. If this is not scientifically correct, it ought to be. The fact remains that during this rough winter the health of the Londoners was actually above the average. Moreover, the power of enduring suffering

* I was coming in one night to the Annexe when there was a lot of noise and something cracked off not far away, and I saw in the obscurity seven or eight men of the Home Guard gathered about the doorway on some patrol or duty. We exchanged greetings, and a big man said from among them: "It's a grand life, if we don't weaken."

in the ordinary people of every country, when their spirit is roused, seemed to have no bounds.

My third fear was a glass famine. Sometimes whole streets had every window-frame smashed by the blast of a single bomb. In a series of minutes I inquired anxiously about this, and proposed to stop all export of glass forthwith. I was however re-assured by facts and figures, and this danger also never came to pass.

*       *       *       *       *

In the middle of September a new and damaging form of attack was used against us. Large numbers of delayed-action bombs were now widely and plentifully cast upon us and became an awkward problem. Long stretches of railway-line, important junctions, the approaches to vital factories, airfields, main thoroughfares, had scores of times to be blocked off and denied to us in our need. These bombs had to be dug out and exploded or rendered harmless. This was a task of the utmost peril, especially at the beginning, when the means and methods had all to be learned by a series of decisive experiences. I have already recounted in Volume I the drama of dismantling the magnetic mine, but this form of self-devotion now became commonplace while remaining sub-lime. I had always taken an interest in the delayed-action fuze, which had first impressed itself on me in 1918, when the Germans had used it on a large scale to deny us the use of the railways by which we planned to advance into Germany. I had urged its use by us both in Norway and in the Kiel Canal. There is no doubt that it is a most effective agent in warfare, on account of the prolonged uncertainty which it creates. We were now to taste it ourselves. A special organisation to deal with it was set up under General King, a highly capable officer, whom I interviewed my-self at Chequers. He handed over the work shortly afterwards to General Taylor. In a series of minutes I tried to stimulate the work.

*Prime Minister to Secretary of State for War*                13.IX.40

As I telephoned to you last night, it appears to be of high importance to cope with the U.X.B. [unexploded bombs] in London, and especially on the railways. The congestion in the marshalling-yards is becoming acute, mainly from this cause. It would be well to bring in clearance parties both from the north and the west, and also to expand as rapidly as possible General King's organisation. It must be planned on large enough lines to cope with this nuisance, which may soon wear a graver aspect.

*Prime Minister to Minister of Supply*                    21.IX.40

The rapid disposal of unexploded bombs is of the highest importance. Any failure to grapple with this problem may have serious results on the production of aircraft and other vital war material. The work of the Bomb Disposal Squads must be facilitated by the provision of every kind of up-to-date equipment. The paper which I have received from the Secretary of State for War shows the experiments on foot and the equipment being planned. Priority 1 (a) should be allotted to the production of the equipment required, and to any further requirements which may come to light.

*Prime Minister to Secretary of State for War*                    14.IX.40

I hear that there is a special type of auger manufactured in the United States which is capable of boring in the space of less than an hour a hole of such a size and depth as would take two to three days to dig manually.

You should, I think, consider ordering a number of these appliances for the use of the bomb-disposal squads. The essence of this business is to reach the bomb and deal with it with the least possible delay.

These augers may perhaps be expensive, but they will pay for themselves many times over by the saving they will effect in life and property. Besides, I consider that we owe it to these brave men to provide them with the very best technical equipment.

*Prime Minister to Secretary of State for War*                    28.IX.40

I am told that there is good evidence to show that the system of dealing with time-bombs by trepanning* is proving very successful. In view of the serious and growing trouble that is being caused by these bombs, I should like to be assured that this method is being used on a large enough scale. Will you please let me have a report on the extent to which trepanning is being used.

Special companies were formed in every city, town, and district. Volunteers pressed forward for the deadly game. Teams were formed which had good or bad luck. Some survived this phase of our ordeal. Others ran twenty, thirty, or even forty courses before they met their fate. The Unexploded Bomb detachments presented themselves wherever I went on my tours. Somehow or other their faces seemed different from those of ordinary men, however brave and faithful. They were gaunt, they were haggard, their faces had a bluish look, with bright gleaming eyes and exceptional compression of the lips; withal a

* Trepanning consisted of making a hole in the bomb casing in order to deal with the explosive contents.

perfect demeanour. In writing about our hard times we are apt to overuse the word "grim". It should have been reserved for the U.X.B. Disposal Squads.*

One squad I remember which may be taken as symbolic of many others. It consisted of three people—the Earl of Suffolk, his lady private secretary, and his rather aged chauffeur. They called themselves "the Holy Trinity". Their prowess and continued existence got around among all who knew. Thirty-four unexploded bombs did they tackle with urbane and smiling efficiency. But the thirty-fifth claimed its forfeit. Up went the Earl of Suffolk in his Holy Trinity. But we may be sure that, as for Mr. Valiant-for-truth, "all the trumpets sounded for them on the other side".

Very quickly, but at heavy sacrifice of our noblest, the devotion of the U.X.B. detachments mastered the peril. In a month I could write:

*Prime Minister to General Ismay*                                9.X.40
We have not heard much lately about the delayed-action bomb, which threatened to give so much trouble at the beginning of September. I have a sort of feeling that things are easier in this respect. Let me have a report showing how many have been cast upon us lately, and how many have been handled successfully or remain a nuisance.

Is the easement which we feel due to the enemy's not throwing them, or to our improved methods of handling?

The reply was reassuring.

*       *       *       *       *

About the same time the enemy began to drop by parachute numbers of naval mines of a weight and explosive power never carried by aircraft before. Many formidable explosions took place. To this there was no defence except reprisal. The abandonment by the Germans of all pretence of confining the air war to

---

* It seems incongruous to record a joke in such sombre scenes. But in war the soldier's harsh laugh is often a measure of inward compressed emotions. The party were digging out a bomb, and their prize man had gone down the pit to perform the delicate act of disconnection. Suddenly he shouted to be drawn up. Forward went his mates and pulled him out. They seized him by the shoulders and, dragging him along, all rushed off together for the fifty or sixty yards which were supposed to give a chance. They flung themselves on the ground. But nothing happened. The prize man was seriously upset. He was blanched and breathless. They looked at him inquiringly. "My God," he said, "there was a *rat*!"

military objectives had also raised this question of retaliation. I was for it, but I encountered many conscientious scruples.

*Prime Minister to V.C.A.S.*                                                6.IX.40

I never suggested any departure from our main policy, but I believe that moral advantage would be gained in Germany at the present time if on two or three nights in a month a number of minor, unexpected, widespread attacks were made upon the smaller German centres. You must remember that these people are never told the truth, and that wherever the Air Force has not been they are probably told that the German defences are impregnable. Many factors have to be taken into consideration, and some of them are those which are not entirely technical. I hope therefore you will consider my wish and make me proposals for giving effect to it as opportunity serves.

Among those who demurred was my friend Admiral Tom Phillips, Vice-Chief of the Naval Staff.

*Prime Minister to General Ismay, for C.O.S. Committee*          19.IX.40
(Admiral Phillips to see)

1. It was not solely on moral grounds that we decided against retaliation upon Germany. It pays us better to concentrate upon limited high-class military objectives. Moreover, in the indiscriminate warfare the enemy's lack of skill in navigation, etc., does not tell against him so much.

2. However, the dropping of large mines by parachute proclaims the enemy's entire abandonment of all pretence of aiming at military objectives. At five thousand feet he cannot have the slightest idea what he is going to hit. This therefore proves the "act of terror" intention against the civil population. We must consider whether his morale would stand up to this as well as ours. Here is a simple war thought.

3. My inclination is to say that we will drop a heavy parachute mine on German cities for every one he drops on ours; and it might be an intriguing idea to mention a list of cities that would be black-listed for this purpose. I do not think they would like it, and there is no reason why they should not have a period of suspense.

4. The time and character of the announcement is a political decision. Meanwhile I wish to know when the tackle could be ready. Let care be taken to make a forthcoming response to this. Let officers be set to propose the best method on a substantial scale in the shortest time. It would be better to act by parachute mines upon a number of German towns not hitherto touched, but if we have to use 1,000-lb. air-bombs which we have because otherwise the delay would be too long, let the case be stated.

5. I wish to know by Saturday night what is the worst form of proportionate retaliation, *i.e.*, *equal* retaliation, that we can inflict upon ordinary German cities for what they are now doing to us by means of the parachute mine. To-day we were informed that thirty-six had been dropped, but by to-morrow it may be one hundred. Well, let it be a hundred, and make the best plan possible on that scale for action within, say, a week or ten days. If we have to wait longer so be it, but make sure there is no obstruction.

6. Pending the above information I agree that we should not make a wail or a whine about what has happened. Let me have practical propositions by Saturday night.

A month later I was still pressing for retaliation; but one objection after another, moral and technical, obstructed it.

*Prime Minister to Secretary of State for Air and C.A.S.*     16.X.40

I see it reported that last night a large number of land mines were dropped here, many of which have not yet gone off, and that great harm was done.

Let me have your proposals forthwith for effective retaliation upon Germany.

I am informed that it is quite possible to carry similar mines or large bombs to Germany, and that the squadrons wish to use them, but that the Air Ministry are refusing permission. I trust that due consideration will be given to my views and wishes. It is now about three weeks since I began pressing for similar treatment of German military objectives to that which they are meting out to us. Who is responsible for paralysing action?

It is difficult to compare the ordeal of the Londoners in the winter of 1940–41 with that of the Germans in the last three years of the war. In this latter phase the bombs were much more powerful and the raids far more intense. On the other hand, long preparation and German thoroughness had enabled a complete system of bomb-proof shelters to be built, into which all were forced to go by iron routine. When eventually we got into Germany we found cities completely wrecked, but strong buildings standing up above the ground, and spacious subterranean galleries where the inhabitants slept night after night, although their houses and property were being destroyed above. In many cases only the rubble-heaps were stirred. But in London, although the attack was less overpowering, the security arrangements were far less developed. Apart from the Tubes there were no really safe places. There were very few basements or cellars which could withstand a direct hit. Virtually the whole mass of the London population lived and slept in their homes or in their Anderson shelters under the fire of the enemy, taking their chance with British phlegm after a hard day's work. Not one in a thousand had any protection except against blast and splinters. But there was as little psychological weakening as there was physical pestilence. Of course, if the bombs of 1943 had been applied to the London of 1940 we should have passed into conditions which might have pulverised all human organisation. However, everything happens in its turn and in its relation, and no one has a right to say that London, which was certainly unconquered, was not also unconquerable.

Little or nothing had been done before the war or during the passive period to provide bomb-proof strongholds from which the central government could be carried on. Elaborate plans had been made to move the seat of government from London. Complete branches of many departments had already been moved to Harrogate, Bath, Cheltenham, and elsewhere. Accommodation had been requisitioned over a wide area, providing for all Ministers and important functionaries in the event of an evacuation of London. But now under the bombardment the desire and resolve of the Government and of Parliament to remain in London was unmistakable, and I shared this feeling to the full. I, like others, had often pictured the destruction becoming so overpowering that a general move and dispersal would have to be made. But under the impact of the event all our reactions were in the contrary sense.

*Prime Minister to Sir Edward Bridges, General Ismay or Colonel Jacob, and Private Office*                    14.IX.40

1. I have not at any time contemplated wholesale movement from London of black or yellow Civil Servants.* Anything of this nature is so detrimental that it could only be forced upon us by Central London becoming practically uninhabitable. Moreover, new resorts of Civil Servants would soon be identified and harassed, and there is more shelter in London than anywhere else.

2. The movement of the high control from the Whitehall area to "Paddock" or other citadels stands on a different footing. We must

---

* These were the official categories. "Yellow" civil servants were those performing less essential tasks and who could therefore be evacuated earlier than "black" ones. The latter would remain in London as long as conditions made it possible to carry on.

make sure that the centre of Government functions harmoniously and vigorously. This would not be possible under conditions of almost continuous air raids. A movement to "Paddock" by échelons of the War Cabinet, War Cabinet Secretariat, Chiefs of Staff Committee, and Home Forces G.H.Q. must now be planned, and may even begin in some minor respects. War Cabinet Ministers should visit their quarters in "Paddock" and be ready to move there at short notice. They should be encouraged to sleep there if they want quiet nights. Secrecy cannot be expected, but publicity must be forbidden.

We must expect that the Whitehall-Westminster area will be the subject of intensive air attack any time now. The German method is to make the disruption of the Central Government a vital prelude to any major assault upon the country. They have done this everywhere. They will certainly do it here, where the landscape can be so easily recognised, and the river and its high buildings afford a sure guide, both by day and night. We must forestall this disruption of the Central Government.

3. It is not necessary to move the Admiralty yet. They are well provided for. The Air Ministry should begin to get from one leg to the other. The War Office and Home Forces must have all their preparations made.

4. Pray concert forthwith all the necessary measures for moving not more than two or three hundred principal persons and their immediate assistants to the new quarters, and show how it should be done step by step. Let me have this by Sunday night, in order that I may put a well-thought-out scheme before the Cabinet on Monday. On Monday the Cabinet will meet either in the Cabinet Room or in the Central War Room, in accordance with the rules already prescribed.

\* \* \* \* \*

On the line of sticking it out in London it was necessary to construct all kinds of strongholds under or above ground from which the Executive, with its thousands of officials, could carry out their duties. A citadel for the War Cabinet had already been prepared near Hampstead, with offices and bedrooms, and wire and fortified telephone communication. This was called "Paddock". On September 29 I prescribed a dress rehearsal, so that everybody should know what to do if it got too hot. "I think it important that 'Paddock' should be broken in. Thursday next therefore the Cabinet will meet there. At the same time, other departments should be encouraged to try a preliminary move of a skeleton staff. If possible, lunch should be provided for the Cabinet and those attending it." We held a Cabinet meeting at "Paddock" far from the light of day, and each Minister was requested to inspect and satisfy himself about his sleeping and working apartments. We celebrated this occasion by a vivacious luncheon, and then returned to Whitehall. This was the only time "Paddock" was ever used by Ministers. Over the War Room and offices in the basement of the Annexe we floated in six feet of steel and concrete, and made elaborate arrangements for ventilation, water supply, and above all telephones. As these offices were far below the level of the Thames, only two hundred yards away, care had to be taken that those in them were not trapped by an inrush of water.

\* \* \* \* \*

October came in raw and rough. But it seemed that London was adapting itself to the new peculiar conditions of existence or death. In some directions even there was an easement. Transport into and out of the Whitehall area became an outstanding problem, with the frequently-repeated daily raids, the rush hour, and the breakdowns on the railways. I cast about for some solution.

*Prime Minister to Sir Horace Wilson*                    12.X.40
About a fortnight ago I directed that the talk about four days a week for Civil Servants should stop, because I feared the effect in the factories of such an announcement. I am however now coming round to the idea of a five-day week, sleeping in for four nights (and where possible feeding in), and three nights and two days away at home. This of course would apply only to people who work in London and live in the suburbs. I see such queues at the bus stops, and no doubt it is going to become increasingly difficult to get in and out of London quickly. Each department should work out a scheme to suit their own and their staff's convenience. The same amount of work must be crowded into the five days as is now done. Efforts should also be made to stagger the hours of arrival and departure, so as to get as many away as possible before the rush hour and spread the traffic over the day.

Let me have your views on this, together with proposals for action in a circular to departments.

Nothing came of this plan, which broke down under detailed examination.

\* \* \* \* \*

The retirement of Mr. Chamberlain, enforced by grave illness, led to important Ministerial changes. Mr. Herbert Morrison had been an efficient and vigorous Minister of Supply, and Sir John Anderson had faced the Blitz of London with firm and competent management. By the early days of October the continuous attack on the largest city in the world was so severe and raised so many problems of a social and political character in its vast harassed population that I thought it would be a help to have a long-trained Parliamentarian at the Home Office, which was now also the Ministry of Home Security. London was bearing the brunt. Herbert Morrison was a Londoner, versed in every aspect of Metropolitan administration. He had unrivalled experience of London government, having been leader of the County Council, and in many ways the principal figure in its affairs. At the same time I needed John Anderson, whose work at the Home Office had been excellent, as Lord President of the Council in the wider sphere of the Home Affairs Committee, to which an immense mass of business was referred, with great relief to the Cabinet. This also lightened my own burden and enabled me to concentrate upon the military conduct of the war, in which my colleagues seemed increasingly disposed to give me latitude.

I therefore invited these two high Ministers to change their offices. It was no bed of roses which I offered Herbert Morrison. These pages certainly cannot attempt to describe the problems of London government, when often night after night ten or twenty thousand people were made homeless, and when nothing but the ceaseless vigil of the citizens as Fire Guards on the roofs prevented uncontrollable conflagrations; when hospitals, filled with mutilated men and women, were themselves struck by the enemy's bombs; when hundreds of thousands of weary people crowded together in unsafe and insanitary shelters; when communications by road and rail were ceaselessly broken down; when drains were smashed and light, power, and gas paralysed; and when nevertheless the whole fighting, toiling life of London had to go forward, and nearly a million people be moved in and out for their work every night and morning. We did not know how long it would last. We had no reason to suppose that it would not go on getting worse. When I made the proposal to Mr. Morrison he knew too much about it to treat it lightly. He asked for a few hours' consideration; but in a short time he returned and said he would be proud to shoulder the job. I highly approved his manly decision.

In Mr. Chamberlain's day a Civil Defence Committee of the Cabinet had already been set up. This met regularly every morning to review the whole situation. In order to make sure that the new Home Secretary was armed with all the powers of State I also held a weekly meeting, usually on Fridays, of all authorities concerned. The topics discussed were often far from pleasant.

\* \* \* \* \*

Quite soon after the Ministerial movements a change in the enemy's method affected our general policy. Till now the hostile attack had been confined almost exclusively to high-explosive bombs; but with the full moon of October 15, when the heaviest attack of the month fell upon us, about 480 German aircraft dropped 386 tons of high explosive and in addition 70,000 incendiary bombs. Hitherto we had encouraged the Londoners to take cover, and every effort was being made to improve their protection. But now "To the basements" must be replaced by "To the roofs". It fell to the new Minister of Home Security to institute this policy. An organisation of fire-watchers and fire services on a gigantic scale and covering the whole of London (apart from measures taken in provincial cities) was rapidly brought into being. At first the fire-watchers were volunteers; but the numbers required were so great, and the feeling that every man should take his turn upon the roster so strong, that fire-watching soon became compulsory. This form of service had a bracing and buoyant effect upon all classes. Women pressed forward to take their share. Large-scale systems of training were developed to teach the fire-watchers how to deal with the various kinds of incendiaries which were used against us. Many became adept, and thousands of fires were extinguished before they took hold. The experience of remaining on the roof night after night under fire, with no protection but a tin hat, soon became habitual.

\* \* \* \* \*

Mr. Morrison presently decided to consolidate the fourteen hundred local fire brigades into a single National Fire Service, and to supplement this with a great Fire Guard of civilians trained and working in their spare time. The Fire Guard, like the roof-watchers, was at first recruited on a voluntary basis, but like them it became by general consent compulsory. The National Fire Service gave us the advantages of greater mobility, a universal standard of training and equipment, and formally recognised ranks. The other Civil Defence forces produced regional columns ready at a minute's notice to go anywhere. The name Civil Defence Service was substituted for the pre-war title of Air Raid Precautions (A.R.P.). Good uniforms were provided for large numbers, and they became conscious of being a Fourth Arm of the Crown. In all this work Herbert Morrison was ably assisted by a brave woman whose death we have lately mourned, Ellen Wilkinson. She was out and about in the shelters at all hours of the day and night and took a prominent part in the organisation of the Fire Guard. The Women's Voluntary Services, under the inspiring leadership of Lady Reading, also played an invaluable part.

\* \* \* \* \*

I was glad that, if any of our cities were to be attacked, the brunt should fall on London. London was like some huge prehistoric animal, capable of enduring terrible injuries, mangled and bleeding from many wounds, and yet preserving its life and movement. The Anderson shelters were widespread in the working-class districts of two-storey houses, and everything was done to make them habitable and to drain them in wet weather. Later the Morrison shelter was developed, which was no more than a heavy kitchen table made of steel with strong wire sides, capable of holding up the ruins of a small house and thus giving a measure of protection. Many owed their lives to it. For the rest, "London could take it". They took all they got, and could have taken more. Indeed, at this time we saw no end but the demolition of the whole Metropolis. Still, as I pointed out to the House of Commons at the time, the law of diminishing returns operates in the case of the demolition of large cities. Soon many of the bombs would only fall upon houses already ruined and only make the rubble jump. Over large areas there would be nothing more to burn or destroy, and yet human beings might make their homes here and there, and carry on their work with infinite resource and fortitude. At this time anyone would have been proud to be a Londoner. The admiration of the whole country was given to London, and all the other great cities in the land braced themselves to take their bit as and when it came and not to be outdone. Indeed, many persons seemed envious of London's distinction, and quite a number came up from the country in order to spend a night or two in town, share the risk, and "see the fun". We had to check this tendency for administrative reasons.

\*　　\*　　\*　　\*　　\*

As we could see no reason why the hostile bombing of London should not go on throughout the war, it was necessary to make long-term plans for safely housing the Central Government machine.

*Prime Minister to Sir Edward Bridges*　　　　　　　　　22.X.40

1. We now know the probable limits of the enemy air attack on London, and that it will be severe and protracted. It is probable indeed that the bombing of Whitehall and the centre of Government will be continuous until all old or insecure buildings have been demolished. It is therefore necessary to provide as soon as possible accommodation in the strongest houses and buildings that exist, or are capable of being fortified, for the large nucleus staffs and personnel connected with the governing machine and the essential Ministers and departments concerned in the conduct of the war. This becomes inevitable as a consequence of our decision not to be beaten out of London, and to release to the War Office or other departments the accommodation hitherto reserved in the West of England for the Black Move. We must do one thing or the other, and, having made our decision, carry it out thoroughly.

2. The accommodation at "Paddock" is quite unsuited to the conditions which have arisen. The War Cabinet cannot live and work there for weeks on end, while leaving the great part of their staffs less well provided for than they are now in Whitehall. Apart from the citadel of "Paddock," there is no adequate accommodation or shelter, and anyone living in Neville Court would have to be running to and fro on every Jim Crow warning. "Paddock" should be treated as a last resort, and in the meantime should be used by some department not needed in the very centre of London.

3. Nearly all the Government buildings and the shelters beneath them are either wholly unsafe or incapable of resisting a direct hit. The older buildings, like the Treasury, fall to pieces, as we have seen, and the shelters beneath them offer no trustworthy protection. The Foreign Office and Board of Trade blocks on either side of King Charles Street are strongly built and give a considerable measure of protection in their basements. I have approved the provision of a substantial

measure of overhead cover above the War Room and Central War Room offices, and Home Forces location in the Board of Trade building. This will take a month or six weeks, with perpetual hammering. We must press on with this. But even when finished it will not be proof. Richmond Terrace is quite inadequately protected, and essential work suffers from conditions prevailing there. The Board of Trade have been invited to move to new premises, and certainly the bulk of their staff should find accommodation out of London. However, this move of the Board of Trade must be considered as part of the general plan.

4. There are several strong modern buildings in London, of steel and cement construction, built with an eye to air-raid conditions. These should immediately be prepared to receive the War Cabinet and its Secretariat, and also to provide safe living accommodation for the essential Ministers. We need not be afraid of having too much accommodation, as increasing numbers will certainly have to be provided for. It is essential that the central work of the Government should proceed under conditions which ensure its efficiency.

5. I have already asked for alternative accommodation for Parliament. The danger to both Houses during their sessions is serious, and it is only a question of time before these buildings and chambers are struck. We must hope they will be struck when not occupied by their Members. The protection provided below the Houses of Parliament is totally inadequate against a direct hit. The Palace of Westminster and the Whitehall area is an obvious prime target of the enemy, and I dare say already more than fifty heavy bombs have fallen in the neighbourhood. The Cabinet has already favoured the idea of a trial trip being made by the Houses of Parliament in some alternative accommodation. I propose to ask for an adjournment from Thursday next for a fortnight, by which time it is hoped some plan can be made in London for their meeting.

6. I consider that a War Cabinet Minister, who should keep in close touch with the Chancellor of the Exchequer, should be entrusted with the general direction and supervision of the important and extensive works which are required, and that Lord Reith and his department should work for this purpose under Cabinet supervision. If my colleagues agree, I will ask Lord Beaverbrook, who has already concerned himself in the matter, to take general charge.

Lord Beaverbrook was thus entrusted with the task of making a large number of bomb-proof strongholds capable of housing the whole essential staffs of many departments of State, and a dozen of them, several connected by tunnels, survive in London to-day. Some of these were not finished till long after the aeroplane raids were over, and few were used during the pilotless-aircraft and rocket attacks which came in 1944 and 1945. However, although these buildings were never used for the purposes for which they were prepared, it was good to feel we had them under our lee. The Admiralty on their own constructed the vast monstrosity which weighs upon the Horse Guards Parade, and the demolition of whose twenty-foot-thick steel and concrete walls will be a problem for future generations when we reach a safer world.

*　　*　　*　　*　　*

Towards the middle of October Josiah Wedgwood began to make a fuss in Parliament about my not having an absolutely bomb-proof shelter for the night raids. He was an old friend of mine, and had been grievously wounded in the Dardanelles. He had always been a single-taxer. Later he broadened his views on taxation and joined the Labour Party. His brother was the Chairman of the Railway Executive Committee. Before the war they had had the foresight to construct a considerable underground office in Piccadilly. It was seventy feet below the surface and covered with strong, high buildings. Although one bomb had penetrated eighty feet in marshy subsoil, there was no doubt this depth with buildings overhead gave safety to anyone in it. I began to be pressed from all sides to resort to this shelter for sleeping purposes. Eventually I agreed, and from the middle of October till the end of the year I used to go there once the firing had started, to transact my evening business and sleep undisturbed. One felt a natural compunction at having much more safety than most other people; but so many pressed me that I let them have their way. After about forty nights in the railway shelter the Annexe became stronger, and I moved back to it. Here during the rest of the war my wife and I lived comfortably. We felt confidence in this solid stone building, and only on very rare occasions went down below the armour. My wife even hung up our few pictures in the sitting-room, which I had thought it better to keep bare. Her view prevailed and was justified by the event. From the roof near the cupola of the Annexe there was a splendid view of London on clear nights. They made a place for me with light overhead cover from splinters, and one could walk in the moonlight and watch the fireworks. In 1941 I used to take some of my American visitors up there from time to time after dinner. They were always most interested.

＊　　＊　　＊　　＊　　＊

On the night of November 3 for the first time in nearly two months no alarm sounded in London. The silence seemed quite odd to many. They wondered what was wrong. On the following night the enemy's attacks were widely dispersed throughout the Island; and this continued for a while. There had been another change in the policy of the German offensive. Although London was still regarded as the principal target, a major effort was now to be made to cripple the industrial centres of Britain. Special squadrons had been trained, with new navigational devices, to attack specific key centres. For instance, one formation was trained solely for the destruction of the Rolls-Royce aero-engine works at Hillington, Glasgow. All this was a makeshift and interim plan. The invasion of Britain had been temporarily abandoned, and the attack upon Russia had not yet been mounted, nor was expected outside Hitler's intimate circle. The remaining winter months were therefore to be for the German Air Force a period of experiment, both in technical devices in night-bombing and in attacks upon British sea-borne trade, together with an attempt to break down our production, military and civil. They would have done much better to have stuck to one thing at a time and pressed it to a conclusion. But they were already baffled and for the time being unsure of themselves.

These new bombing tactics began with the blitz on Coventry on the night of November 14. London seemed too large and vague a target for decisive results, but Goering hoped that provincial cities or munitions centres might be effectively obliterated. The raid started early in the dark hours of the 14th, and by dawn nearly five hundred German aircraft had dropped six hundred tons of high explosives and thousands of incendiaries. On the whole this was the most devastating raid which we sustained. The centre of Coventry was shattered, and its life for a spell completely disrupted. Four hundred people were killed and many more seriously injured. The German radio proclaimed that our other cities would be similarly "Coventrated". Nevertheless the all-important aero-engine and machine-tool factories were not brought to a standstill; nor was the population, hitherto untried in the ordeal of bombing, put out of action. In less than a week an emergency reconstruction committee did wonderful work in restoring the life of the city.

On November 15 the enemy switched back to London with a very heavy raid in full moonlight. Much damage was done, especially to churches and other monuments. The next target was Birmingham, and three successive raids from the 19th to the 22nd of November inflicted much destruction and loss of life. Nearly eight hundred people were killed and over two thousand injured; but the life and spirit of Birmingham survived this ordeal. When I visited the city a day or two later to inspect its factories, and see for myself what had happened, an incident, to me charming, occurred. It was the dinner-hour, and a very pretty young girl ran up to the car and threw a box of cigars into it. I stopped at once and she said: "I won the prize this week for the highest output. I only heard you were coming an hour ago." The gift must have cost her two or three pounds. I was very glad (in my official capacity) to give her a kiss. I then went on to see the long mass grave in which so many citizens and their children had been newly buried. The spirit of Birmingham shone brightly, and its million inhabitants, highly organised, conscious and comprehending, rode high above their physical suffering.

During the last week of November and the beginning of December the weight of the attack shifted to the ports. Bristol, Southampton, and above all Liverpool, were heavily bombed. Later on Plymouth, Sheffield, Manchester, Leeds, Glasgow, and other munitions centres passed through the fire undaunted. It did not matter where the blow struck, the nation was as sound as the sea is salt.

The climax raid of these weeks came once more to London, on Sunday, December 29. All the painfully-gathered German experience was expressed on this occasion. It was an incendiary classic. The weight of the attack was concentrated upon the City of London itself. It was timed to meet the dead-low-water hour. The water-mains were broken at the outset by very heavy high-explosive parachute-mines. Nearly fifteen hundred fires had to be fought. The damage to railway stations and docks was serious. Eight Wren churches were destroyed or damaged. The Guild-hall was smitten by fire and blast, and St. Paul's Cathedral was only saved by heroic exertions. A void of ruin at the very centre of the British world gapes upon us to this day. But when the King and Queen visited the scene they were received with enthusiasm far exceeding any Royal festival.

During this prolonged ordeal, of which several months were still to come, the King was constantly at Buckingham Palace. Proper shelters were being constructed in the basement, but all this took time. Also it happened several times that His Majesty arrived from Windsor in the middle of an air raid. Once he and the Queen had a very narrow escape. I have His Majesty's permission to record the incident in his own words:

*Friday, September* 13, 1940

We went to London [from Windsor] and found an air raid in progress. The day was very cloudy and it was raining hard. The Queen and I went upstairs to a small sitting-room overlooking the Quadrangle (I could not use my usual sitting-room owing to the broken windows by former bomb damage). All of a sudden we heard the zooming noise of a diving aircraft getting louder and louder, and then saw two bombs falling past the opposite side of Buckingham Palace into the Quadrangle. We saw the flashes and heard the detonations as they burst about eighty yards away. The blast blew in the windows opposite to us, and two great craters had appeared in the Quadrangle. From one of these craters water from a burst main was pouring out and flowing into the passage through the broken windows. The whole thing happened in a matter of seconds, and we were very quickly out into the passage. There were six bombs: two in the Forecourt, two in the Quadrangle, one wrecked the Chapel, and one in the garden.

The King, who as a sub-lieutenant had served in the Battle of Jutland, was exhilarated by all this, and pleased that he should be sharing the dangers of his subjects in the capital. I must confess that at the time neither I nor any of my colleagues were aware of the peril of this particular incident. Had the windows been closed instead of open the whole of the glass would have splintered into the faces of the King and Queen, causing terrible injuries. So little did they make of it all that even I, who saw them and their entourage so frequently, only realised long afterwards when making inquiries for writing this book what had actually happened.

In those days we viewed with stern and tranquil gaze the idea of going down fighting amid the ruins of Whitehall. His Majesty had a shooting-range made in the Buckingham Palace garden, at which he and other members of his family and his equerries practised assiduously with pistols and tommy-guns. Presently I brought the King an American short-range carbine, from a number which had been sent to me. This was a very good weapon.

About this time the King changed his practice of receiving me in a formal weekly audience at about five o'clock, which had prevailed during my first two months of office. It was now arranged that I should lunch with him every Tuesday. This was certainly a very agreeable method of transacting State business, and sometimes the Queen was present. On several occasions we all had to take our plates and glasses in our hands and go down to the shelter, which was making progress, to finish our meal. The weekly luncheons became a regular institution. After the first few months His Majesty decided that all servants should be excluded, and that we should help ourselves and help each other. During the four and a half years that this continued I became aware of the extraordinary diligence with which the King read all the telegrams and public documents submitted to him. Under the British constitutional system the Sovereign has a right to be made acquainted with everything for which his Ministers are responsible, and has an unlimited right of giving counsel to his Government. I was most careful that everything should be laid before the King, and at our weekly meetings he frequently showed that he had mastered papers which I had not yet dealt with. It was a great help to Britain to have so good a King and Queen in those fateful years, and as a convinced upholder of constitutional monarchy I valued as a signal honour the gracious intimacy with which I, as first Minister, was treated, for which I suppose there has been no precedent since the days of Queen Anne and Marlborough during his years of power.

\* \* \* \* \*

This brings us to the end of the year, and for the sake of continuity I have gone ahead of the general war. The reader will realise that all this clatter and storm was but an accompaniment to the cool processes by which our war effort was maintained and our policy and diplomacy conducted. Indeed, I must record that at the summit these injuries, failing to be mortal, were a positive stimulant to clarity of view, faithful comradeship, and judicious action. It would be unwise however to suppose that if the attack had been ten or twenty times as severe—or even perhaps two or three times as severe—the healthy reactions I have described would have followed.

# CHAPTER XIX

# THE WIZARD WAR

*A Hidden Conflict – Lindemann's Services – Progress of Radar – The German Beam – Mr. Jones's Tale – Principle of the Split Beam or Knickebein – Twisting the Beam – Goering's Purblind Obstinacy – The X Apparatus – Coventry, November 14–15 – The Decoy Fires – The Y Apparatus Forestalled – Frustration of the Luftwaffe – Triumph of British Science – Our Further Plans – The Rocket Batteries – General Pile's Command and the Air Defences of Great Britain – The Aerial Mine Curtains – The Proximity Fuze – The Prospect of Counter-attack – The Expansion of Our Defence Measure.*

DURING the human struggle between the British and German Air Forces, between pilot and pilot, between A.A. batteries and aircraft, between ruthless bombing and the fortitude of the British people, another conflict was going on step by step, month by month. This was a secret war, whose battles were lost or won unknown to the public; and only with difficulty is it comprehended, even now, by those outside the small high scientific circles concerned. No such warfare had ever been waged by mortal men. The terms in which it could be recorded or talked about were unintelligible to ordinary folk. Yet if we had not mastered its profound meaning and used its mysteries even while we saw them only in the glimpse, all the efforts, all the prowess of the fighting airmen, all the bravery and sacrifices of the people, would have been in vain. Unless British science had proved superior to German, and unless its strange, sinister resources had been effectively brought to bear on the struggle for survival, we might well have been defeated, and, being defeated, destroyed.

A wit wrote ten years ago: "The leaders of thought have reached the horizons of human reason, but all the wires are down,

and they can only communicate with us by unintelligible signals."
Yet upon the discerning of these signals, and upon the taking of
right and timely action on the impressions received, depended
our national fate and much else. I knew nothing about science,
but I knew something of scientists, and had had much practice as
a Minister in handling things I did not understand. I had, at any
rate, an acute military perception of what would help and what
would hurt, of what would cure and of what would kill. My
four years' work upon the Air Defence Research Committee had
made me familiar with the outlines of Radar problems. I therefore immersed myself so far as my faculties allowed in this
Wizard War, and strove to make sure that all that counted
came without obstruction or neglect at least to the threshold
of action. There were no doubt greater scientists than Frederick
Lindemann, though his credentials and genius command respect.
But he had two qualifications of vital consequence to me.
First, as these pages have shown, he was my trusted friend and
confidant of twenty years. Together we had watched the advance and onset of world disaster. Together we had done our
best to sound the alarm. And now we were in it, and I had the
power to guide and arm our effort. How could I have the
knowledge?

Here came the second of his qualities. Lindemann could
decipher the signals from the experts on the far horizons and
explain to me in lucid, homely terms what the issues were. There
are only twenty-four hours in the day, of which at least seven
must be spent in sleep and three in eating and relaxation. Anyone
in my position would have been ruined if he had attempted to
dive into depths which not even a lifetime of study could plumb.
What I had to grasp were the practical results, and just as Lindemann gave me his view for all it was worth in this field, so I made
sure by turning on my power-relay that some at least of these
terrible and incomprehensible truths emerged in executive
decisions.

*     *     *     *     *

Progress in every branch of Radar was constant and unceasing
during 1939, but even so the Battle of Britain, from July to
September 1940, was, as I have described, fought mainly by eye
and ear. I comforted myself at first in these months with the hope
that the fogs and mist and cloud which accompany the British

winter and shroud the Island with a mantle would at least give a great measure of protection against accurate bombing by day and still more in darkness.

For some time the German bombers had navigated largely by radio beacons. Scores of these were planted like lighthouses in various parts of the Continent, each with its own call-sign, and the Germans, using ordinary directional wireless, could fix their position by the angles from which any two of these transmissions came. To counter this we soon installed a number of stations which we called "Meacons". These picked up the German signals, amplified them, and sent them out again from somewhere in England. The result was that the Germans, trying to home on their beams, were often led astray, and a number of hostile aircraft were lost in this manner. Certainly one German bomber landed voluntarily in Devonshire thinking it was France.

However, in June I received a painful shock. Professor Lindemann reported to me that he believed the Germans were preparing a device by means of which they would be able to bomb by day or night whatever the weather. It now appeared that the Germans had developed a radio beam which, like an invisible searchlight, would guide the bombers with considerable precision to their target. The beacon beckoned to the pilot, the beam pointed to the target. They might not hit a particular factory, but they could certainly hit a city or town. No longer therefore had we only to fear the moonlight nights, in which our fighters could see at any rate as well as the enemy, but we must even expect the heaviest attacks to be delivered in cloud and fog.

Lindemann told me also that there was a way of bending the beam if we acted at once, but that I must see some of the scientists, particularly the Deputy Director of Intelligence Research at the Air Ministry, Dr. R. V. Jones, a former pupil of his at Oxford. Accordingly, with anxious mind I convened on June 21 a special meeting in the Cabinet Room, at which about fifteen persons were present, including Sir Henry Tizard and various Air Force commanders. A few minutes late, a youngish man—who, as I afterwards learned, had thought his sudden summons to the Cabinet Room must be a practical joke—hurried in and took his seat at the bottom of the table. According to plan, I invited him to open the discussion.

For some months, he told us, hints had been coming from all

sorts of sources on the Continent that the Germans had some novel mode of night-bombing on which they placed great hopes. In some way it seemed to be linked with the code-word Knicke-bein, which our Intelligence had several times mentioned without being able to explain. At first it had been thought that the enemy had got agents to plant beacons in our cities on which their bombers could home; but this idea had proved untenable. Some weeks before two or three curious squat towers had been photo-graphed in odd positions near the hostile coast. They did not seem the right shape for any known form of radio or Radar. Nor were they in places which could be explained on any such hypo-thesis. Recently a German bomber had been shot down with apparatus which seemed more elaborate than was required for night-landing by the ordinary Lorenz beam, which appeared to be the only known use for which it might be intended. For this and various other reasons, which he wove together into a cumula-tive argument, it looked as if the Germans might be planning to navigate and bomb on some sort of system of beams. A few days before under cross-examination on these lines a German pilot had broken down and admitted that he had heard that something of the sort was in the wind. Such was the gist of Dr. Jones's tale.

For twenty minutes or more he spoke in quiet tones, unrolling his chain of circumstantial evidence, the like of which for its con-vincing fascination was never surpassed by tales of Sherlock Holmes or Monsieur Lecoq. As I listened the *Ingoldsby Legends* jingled in my mind:

> But now one Mr. Jones
> Comes forth and depones
> That, fifteen years since, he had heard certain groans
> On his way to Stone Henge (to examine the stones
> Described in a work of the late Sir John Soane's),
> That he'd followed the moans,
> And led by their tones,
> Found a Raven a-picking a Drummer-boy's bones!

When Dr. Jones had finished there was a general air of incre-dulity. One high authority asked why the Germans should use a beam, assuming that such a thing was possible, when they had at their disposal all the ordinary facilities of navigation. Above twenty thousand feet the stars were nearly always visible. All our own pilots were laboriously trained in navigation, and it was thought they found their way about and to their targets very well. Others round the table appeared concerned.

\* \* \* \* \*

I will now explain in the kind of terms which I personally can understand how the German beam worked and how we twisted it. Like the searchlight beam, the radio beam cannot be made very sharp; it tends to spread; but if what is called the "split beam" method is used considerable accuracy can be obtained. Let us imagine two searchlight beams parallel to one another, both flickering in such a way that the left-hand beam comes on exactly when the right-hand beam goes out, and *vice versa*. If an attacking aircraft was exactly in the centre between the two beams, the pilot's course would be continuously illuminated, but if it got, say, a little bit to the right, nearer the centre of the right-hand beam, this would become the stronger and the pilot would observe the flickering light, which was no guide. By keeping in the position where he avoided the flickerings he would be flying exactly down the middle, where the light from both beams is equal. And this middle path would guide him to the target. Two split beams from two stations could be arranged to cross over any town in the Midlands or Southern England. The German airman had only to fly along one beam until he detected the second, and then to drop his bombs. Q.E.D.!

This was the principle of the split beam and the celebrated "Knickebein" apparatus, upon which Goering founded his hopes, and the Luftwaffe were taught to believe that the bombing of English cities could be maintained in spite of cloud, fog, and darkness, and with all the immunity, alike from guns and intercepting fighters, which these gave to the attacker. With their logical minds and deliberate large-scale planning, the German High Air Command staked their fortunes in this sphere on a device which, like the magnetic mine, they thought would do us in. Therefore they did not trouble to train the ordinary bomber pilots, as ours had been trained, in the difficult art of navigation. A far simpler and surer method, lending itself to drill and large numbers, producing results wholesale by irresistible science, attracted alike their minds and their nature. The German pilots followed the beam as the German people followed the Fuehrer. They had nothing else to follow.

But, duly forewarned, and acting on the instant, the simple British had the answer. By erecting the proper stations in good time in our own country we could jam the beam. This would of course have been almost immediately realised by the enemy. There was another and superior alternative. We could put a repeating device in such a position that it strengthened the signal from one half of the split beam and not from the other. Thus the hostile pilot, trying to fly so that the signals from both halves of the split beam were equal, would be deflected from the true course. The cataract of bombs which would have shattered, or at least tormented, a city would fall fifteen or twenty miles away in an open field. Being master, and not having to argue too much, once I was convinced about the principles of this queer and deadly game I gave all the necessary orders that very day in June for the existence of the beam to be assumed, and for all counter-measures to receive absolute priority. The slightest reluctance or deviation in carrying out this policy was to be reported to me. With so much going on I did not trouble the Cabinet, or even the Chiefs of Staff. If I had encountered any serious obstruction I should of course have appealed and told a long story to these friendly tribunals. This however was not necessary, as in this limited and at that time almost occult circle obedience was forth-coming with alacrity, and on the fringes all obstructions could be swept away.

About August 23 the first new Knickebein stations, near Dieppe and Cherbourg, were trained on Birmingham, and a large-scale night offensive began. We had of course our "teething troubles" to get through; but within a few days the Knickebein beams were deflected or jammed, and for the next two months, the critical months of September and October, the German bombers wandered around England bombing by guesswork, or else being actually led astray.

One instance happened to come to my notice. An officer in my Defence Office sent his wife and two young children to the country during the London raids. Ten miles away from any town they were much astonished to see a series of enormous explosions occurring three fields away. They counted over a hundred heavy bombs. They wondered what the Germans could be aiming at, and thanked God they were spared. The officer mentioned the incident the next day, but so closely was the secret kept, so narrow was the circle, so highly specialised the information, that no satisfactory explanation could be given to him, even in his intimate position. The very few who knew exchanged celestial grins.

The German air crews soon suspected that their beams were being mauled. There is a story that during these two months nobody had the courage to tell Goering that his beams were twisted or jammed. In his ignorance he pledged himself that this was impossible. Special lectures and warnings were delivered to the German Air Force, assuring them that the beam was infallible, and that anyone who cast doubt on it would be at once thrown out. We suffered, as has been described, heavily under the Blitz, and almost anyone could hit London anyhow. Of course there would in any case have been much inaccuracy, but the whole German system of bombing was so much disturbed by our counter-measures, added to the normal percentage of error, that not more than one-fifth of their bombs fell within the target areas. We must regard this as the equivalent of a considerable victory, because even the fifth part of the German bombing, which we got, was quite enough for our comfort and occupation.

\* \* \* \* \*

The Germans, after internal conflicts, at last revised their methods. It happened, fortunately for them, that one of their formations, Kampf Gruppe 100, was using a special beam of its own. It called its equipment the "X apparatus", a name of mystery which, when we came across it, threw up an intriguing challenge to our Intelligence. By the middle of September we had found out enough about it to design counter-measures, but this particular jamming equipment could not be produced for a further two months. In consequence Kampf Gruppe 100 could still bomb with accuracy. The enemy hastily formed a pathfinder group from it, which they used to raise fires in the target area by incendiary bombs, and these became the guide for the rest of the de-Knickebeined Luftwaffe.

Coventry, on November 14-15, was the first target attacked by the new method. Although our new jamming had now started, a technical error prevented it from becoming effective for another few months. Even so our knowledge of the beams was helpful. From the settings of the hostile beams and the times at which they played we could forecast the target and the time, route, and height of attack. Our night fighters had, alas! at this date neither the numbers nor the equipment to make much use of the information. It was nevertheless invaluable to our fire-fighting and other Civil Defence services. These could often be concentrated in the threatened area and special warnings given to the population before the attack started. Presently our counter-measures improved and caught up with the attack. Meanwhile decoy fires, code-named "Starfish", on a very large scale were lighted by us with the right timing in suitable open places to lead the main attack astray, and these sometimes achieved remarkable results.

By the beginning of 1941 we had mastered the "X apparatus"; but the Germans were also thinking hard, and about this time they brought in a new aid called the "Y apparatus". Whereas the two earlier systems had both used cross beams over the target, the new system used only one beam, together with a special method of range-finding by radio, by which the aircraft could be told how far it was along the beam. When it reached the correct distance it dropped its bombs. By good fortune and the genius and devotion of all concerned, we had divined the exact method of working the "Y apparatus" some months before the Germans were able to use it in operations, and by the time they were ready to make it their pathfinder we had the power to render it useless. On the very first night when the Germans committed themselves to the "Y apparatus" our new counter-measures came into action against them. The success of our efforts was manifest from the acrimonious remarks heard passing between the pathfinding aircraft and their controlling ground stations by our listening instruments. The faith of the enemy air crews in their new device was thus shattered at the outset, and after many failures the method was abandoned. The bombing of Dublin on the night of May 30, 1941, may well have been an unforeseen and unintended result of our interference with "Y".

General Martini, the German chief in this sphere, has since the war admitted that he had not realised soon enough that the "high-frequency war" had begun, and that he underrated the British Intelligence and counter-measures organisation. Our exploitation of the strategic errors which he made in the Battle of the Beams diverted enormous numbers of bombs from our cities during a period when all other means of defence had either failed or were still in their childhood. These were however rapidly improving under the pressure of potentially mortal attack. Since the beginning of the war we had brought into active production a form of air-borne Radar called A.I., on which the Air Defence Research Committee had fruitfully laboured from 1938 onwards, and with which it was hoped to detect and close on enemy bombers. This apparatus was too large and too complicated for a pilot to operate himself. It was therefore installed in two-seater Blenheims, and later in Beaufighters, in which the observer operated the Radar, and directed his pilot until the enemy aircraft became visible and could be fired on—usually at night about a hundred yards away. I had called this device in its early days "the Smeller", and longed for its arrival in action. This was inevitably a slow process. However, it began. A widespread method of ground-control interception grew up and came into use. The British pilots, with their terrible eight-gun batteries, in which cannon-guns were soon to play their part, began to close—no longer by chance but by system—upon the almost defenceless German bombers.

The enemy's use of the beams now became a positive advantage to us. They gave clear warning of the time and direction of the attacks, and enabled the night-fighter squadrons in the areas affected and all their apparatus to come into action at full force and in good time, and all the A.A. batteries concerned to be fully manned and directed by their own intricate science, of which more later. During March and April the steadily-rising rate of loss of German bombers had become a cause of serious concern to the German war chiefs. The "erasing" of British cities had not been found so easy as Hitler had imagined. It was with relief that the German Air Force received their orders in May to break off the night attacks on Great Britain and to prepare for action in another theatre.

Thus the three main attempts to conquer Britain after the fall of France were successively defeated or prevented. The first was the decisive defeat of the German Air Force in the Battle of Britain during July, August, and September. Instead of destroying the British Air Force and the stations and air factories on which it relied for its life and future, the enemy themselves, in spite of their preponderance in numbers, sustained losses which they could not bear. Our second victory followed from our first.

The German failure to gain command of the air prevented the cross-Channel invasion. The prowess of our fighter pilots, and the excellence of the organisation which sustained them, had in fact rendered the same service—under conditions indescribably different—as Drake and his brave little ships and hardy mariners had done three hundred and fifty years before, when, after the Spanish Armada was broken and dispersed, the Duke of Parma's powerful army waited helplessly in the Low Countries for the means of crossing the Narrow Seas.

The third ordeal was the indiscriminate night bombing of our cities in mass attacks. This was overcome and broken by the continued devotion and skill of our fighter pilots, and by the fortitude and endurance of the mass of the people, and notably the Londoners, who, together with the civil organisations which upheld them, bore the brunt. But these noble efforts in the high air and in the flaming streets would have been in vain if British science and British brains had not played the ever-memorable and decisive part which this chapter records.

*　　*　　*　　*　　*

There is a useful German saying, "The trees do not grow up to the sky." Nevertheless we had every reason to expect that the air attack on Britain would continue in an indefinite crescendo. Until Hitler actually invaded Russia we had no right to suppose it would die away and stop. We therefore strove with might and main to improve the measures and devices by which we had hitherto survived and to find new ones. The highest priority was assigned to all forms of Radar study and application. Scientists and technicians were engaged and organised on a very large scale. Labour and material was made available to the fullest extent. Other methods of striking down the hostile bomber were sought tirelessly, and for many months to come these efforts were spurred by repeated, costly, and bloody raids upon our ports and cities. I will mention three developments, constantly referred to in the Appendices to this volume, in which, at Lindemann's prompting and in the light of what we had studied together on the Air Defence Research Committee of pre-war years, I took special interest and used my authority. These were, first, the massed discharge of rockets, as a reinforcement of our A.A. batteries; secondly, the laying of aerial mine curtains in the path of a raiding force by means of bombs with long wires descending by parachutes; thirdly, the search for fuzes so sensitive that they did not need to hit their target, but would be set off by merely passing near an aircraft. Of these three methods, on which we toiled with large expenditure of our resources, some brief account must now be given.

None of these methods could come to fruition in 1940. At least a year stood between us and practical relief. By the time we were ready to go into action with our new apparatus and methods the enemy attack they were designed to meet came suddenly to an end, and for nearly three years we enjoyed almost complete immunity from it. Critics have therefore been disposed to underrate the value of these efforts, which could only be proved by major trial, and in any case in no way obstructed other developments in the same sphere.

* * * * *

By itself beam-distortion was not enough. Once having hit the correct target, it was easy for the German bombers, unless they were confused by our "Starfish" decoy fires, to return again to the glow of the fires they had lit the night before. Somehow they must be clawed down. For this we developed two new devices, rockets and aerial mines. By fitting our A.A. batteries with Radar it was possible to predict the position of an enemy aircraft accurately enough, provided it continued to fly in a straight line at the same speed, but this is hardly what experienced pilots do. Of course they zigzagged or "weaved", and this meant that in the twenty or thirty seconds between firing the gun and the explosion of the shell they might well be half a mile or so from the predicted point.

A wide yet intense burst of fire round the predicted point was an answer. Combinations of a hundred guns would have been excellent, if the guns could have been produced and the batteries manned and all put in the right place at the right time. This was beyond human power to achieve. But a very simple, cheap alternative was available in the rocket, or, as it had been called for secrecy, the Unrotated Projectile (U.P.). Even before the war Dr. Crow, in the days of the Air Defence Research Committee, had developed 2-inch and 3-inch rockets which could reach almost as high as our A.A. guns. The 3-inch rocket carried a much more powerful warhead than a 3-inch shell. It was not so accurate. On the other hand, rocket projectors had the inestimable advantage that they could be made very quickly and easily in enormous numbers without burdening our hard-driven gun factories. Thousands of these U.P. projectors were made, and some millions of rounds of ammunition. General Sir Frederick Pile, an officer of great distinction, who was in command of our anti-aircraft ground defences throughout the war, and who was singularly free from the distaste for novel devices so often found in professional soldiers, welcomed this accession to his strength. He formed these weapons into huge batteries of ninety-six projectors each, manned largely by the Home Guard, which could produce a concentrated volume of fire far beyond the power of A.A. artillery.

I worked in increasing intimacy throughout the war with General Pile, and always found him ingenious and serviceable in the highest degree. He was at his best not only in these days of expansion, when his command rose to a peak of over three hundred thousand men and women and two thousand four hundred guns, apart from the rockets, but also in the period which followed after the air attack on Britain had been beaten off. Here was a time when his task was to liberate the largest possible numbers of men from static defence by batteries, and, without diminishing the potential fire-power, to substitute the largest proportion of women and Home Guard for Regulars and technicians. But this is a story which must be told in its proper place.

The task of General Pile's command was not merely helped by the work of our scientists; as the battle developed their aid was the foundation on which all stood. In the daylight attacks of the Battle of Britain the guns had accounted for 296 enemy aircraft, and probably destroyed or damaged 74 more. But the night raids gave them new problems which with their existing equipment of only searchlights and sound locators could not be surmounted. In four months from October 1 only about seventy aircraft were destroyed. Radar came to the rescue. The first of these sets for directing gunfire was used in October, and Mr. Bevin and I spent most of the night watching them. The searchlight beams were not fitted till December. However, much training and experience were needed in their use, and many modifications and refinements in the sets themselves were found necessary. Great efforts were

made in all this wide field, and the spring of 1941 brought a full
reward.

During the attacks on London in the first two weeks of May—
the last of the German offensive—over seventy aircraft were
destroyed, or more than the four winter months had yielded. Of
course in the meanwhile the number of guns had grown. In
December there had been 1,400 heavy guns and 650 light; in May
there were 1,687 heavy guns and 790 light, with about 40 rocket
batteries.* But the great increase in the effectiveness of our gun
defences was due in its origin to the new inventions and technical
improvements which the scientists put into the soldiers' hands,
and of which the soldiers made such good use.

* * * * *

By the middle of 1941, when at last the rocket batteries began
to come into service in substantial numbers, air attack had much
diminished, so that they had few chances of proving themselves.
But when they did come into action the number of rounds needed
to bring down an aircraft was little more than that required by the
enormously more costly and scanty A.A. guns, of which we were
so short. The rockets were good in themselves, and also an
addition to our other means of defence.

Shells or rockets alike are of course only effective if they reach
the right spot and explode at the right moment. Efforts were
therefore made to produce aerial mines suspended on long wires
floating down on parachutes which could be laid in the path of
the enemy air squadrons. It was impossible to pack these into
shells. But a rocket, with much thinner walls, has more room. A
certain amount of 3-inch rocket ammunition which could lay
an aerial minefield on wires seven hundred feet long at heights
up to twenty thousand feet was made and held ready for use
against mass attacks on London. The advantage of such minefields over shell fire is of course that they remain lethal for anything up to a minute. For wherever the wing hits the wire it
pulls up the mine until it reaches the aircraft and explodes. There
is thus no need for exact fuze-setting, as with ordinary shells.

Aerial mines could of course be placed in position by rockets
laid by aircraft, or simply raised on small balloons. The last
method was ardently supported by the Admiralty. In fact how-

* See the table at the end of this chapter.

ever the rockets were never brought into action on any considerable scale. By the time they were manufactured in large numbers mass attacks by bombers had ceased. Nevertheless it was surprising and fortunate that the Germans did not develop this counter to our mass-bombing raids in the last three years of the war. Even a few mine-laying aircraft would have been able to lay and maintain a minefield over any German city, which would have taken a toll of our bombers the more deadly as numbers grew.

*   *   *   *   *

There was another important aspect. In 1940 the dive-bomber seemed to be a deadly threat to our ships and key factories. One might think that aircraft diving on a ship would be easy to shoot down, as the gunner can aim straight at them without making allowance for their motion. But an aeroplane end on is a very small target, and a contact fuze will work only in the rare event of a direct hit. To set a time fuze so that the shell explodes at the exact moment when it is passing the aircraft is almost impossible. An error in timing of one-tenth of a second causes a miss of many hundreds of feet. It therefore seemed worth while to try to make a fuze which would detonate automatically when the projectile passed near to the target, whether it actually hit it or not.

As there is little space in the head of a shell the roomier head of the 3-inch rocket was attractive. While I was still at the Admiralty in 1940 we pressed this idea. Photo-electric (P.E.) cells were used which produced an electrical impulse whenever there was a change of light, such as the shade of the enemy plane. By February 1940 we had a model which I took to the Cabinet and showed my colleagues after one of our meetings. When a match-box was thrown past the fuze it winked perceptibly with its demonstration lamp. The cluster of Ministers who gathered round, including the Prime Minister, were powerfully impressed. But there is a long road between a grimacing model and an armed mass-produced robot. We worked hard at the production of the so-called P.E. fuzes, but here again by the time they were ready in any quantity our danger and their hour had for the moment passed.

Attempts were made in 1941 to design a similar proximity fuze, using a tiny Radar set arranged to explode the warhead when the projectile passed near the aircraft. Successful preliminary experiments were made, but before this fuze was developed in England

350

the Americans, to whom we imparted our knowledge, actually succeeded not only in perfecting the instrument but in reducing its size so much that the whole thing could be put into the head not merely of a rocket but of a shell. These so-called "Proximity Fuzes", made in the United States, were used in great numbers in the last year of the war, and proved potent against the small unmanned aircraft (V1) with which we were assailed in 1944, and also in the Pacific against Japanese aircraft.

*　*　*　*　*

The final phase of the "Wizard War" was of course the Radar developments and inventions required for our counter-attack upon Germany. These suggested themselves to some extent from our own experiences and defensive efforts. The part they played will be described in future volumes. In September 1940 we had nearly nine long months ahead of us of heavy battering and suffering before the tide was to turn. It may be claimed that while struggling, not without success, against the perils of the hour we bent our thoughts steadily upon the future, when better times might come.

## AIR DEFENCE. GREAT BRITAIN
### EXPANSION 1940–1941

| JULY 1940 | DECEMBER 1940 | MAY 1941 |
|---|---|---|
| | HEAVY GUNS | |
| *Total:* 1,200 | *Total:* 1,450 | *Total:* 1,687 |
| Made up of: | Made up of: | Made up of: |
| 4.5-inch, 355 | Static, 1,040 | Static, 1,247 |
| 3.7-inch static, 313 | Mobile, 410 | Mobile, 440 |
| 3.7-inch mobile, 306 | | |
| 3-inch, 226 | | |
| | LIGHT GUNS | |
| *Total:* 587 | *Total:* 650 | *Total:* 790 |
| Made up of: | | |
| Bofors, 273 | | |
| 3-inch, 136 | | |
| (adapted for low | | |
| shooting) | | |
| 20-mm. Hispano, 38 | | |
| 2-pdrs., 140 | | |

| JULY 1940 | DECEMBER 1940 | MAY 1941 |
|---|---|---|
| | ROCKET BATTERIES | |
| *Nil* | *Nil* | *Total:* About 40 |
| | SEARCHLIGHTS | |
| *Total:* 3,932 | | *Total:* Over 4,500 (not fully manned) |
| | PERSONNEL STRENGTH | |
| *Total:* 157,319 | *Total:* 269,000, including 6,000 women (3,700 on batteries, 2,300 on H.Q. and administrative staffs) | *Total:* 312,500, including 6,500 women (3,500 on battery establishments, 3,000 on H.Q. and administrative staffs) |

# CHAPTER XX

# UNITED STATES DESTROYERS AND WEST INDIAN BASES

*My Appeal for Fifty American Destroyers – Lord Lothian's Helpfulness – My Telegram to the President of July 31 – Our Willingness to Lease Bases in the West Indies – My Objections to Bargaining about the Fleet – Further Telegram to the President of August 15 – The President's Statement – My Speech in Parliament of August 20 – Telegram to President of August 22 – And of August 25 – And of August 27 – Our Final Offer – My Assurance about the Fleet – Statement to Parliament of September 5.*

O N May 15, as already narrated, I had in my first telegram to the President after becoming Prime Minister asked for "the loan of forty or fifty of your older destroyers to bridge the gap between what we have now and the large new construction we put in hand at the beginning of the war. This time next year we shall have plenty. But if in the interval Italy comes in against us with another hundred submarines we may be strained to breaking-point". I recurred to this in my cable of June 11, after Italy had already declared war upon us. "Nothing is so important as for us to have the thirty or forty old destroyers you have already had reconditioned. We can fit them very rapidly with our Asdics. . . . The next six months are vital." At the end of July, when we were alone and already engaged in the fateful air-battle, with the prospect of imminent invasion behind it, I renewed my request. I was well aware of the President's goodwill and of his difficulties. For that reason I had endeavoured to put before him, in the blunt terms of various messages, the perilous position which the United States would occupy if British resistance collapsed and Hitler became master of Europe, with all its dockyards and navies.

*     *     *     *     *

It was evident as this discussion proceeded that the telegrams I had sent in June, dwelling on the grave consequences to the United States which might follow from the successful invasion and subjugation of the British Island, played a considerable part in high American circles. Assurances were requested from Washington that the British Fleet would in no circumstances be handed over to the Germans. We were very ready to give these assurances in the most solemn form. As we were ready to die, they cost nothing. I did not however wish, at this time, on what might be the eve of invasion and at the height of the air battle, to encourage the Germans with the idea that such contingencies had ever entered our minds. Moreover, by the end of August our position was vastly improved. The whole Regular Army was re-formed, and to a considerable extent rearmed. The Home Guard had come into active life. We were inflicting heavy losses on the German Air Force, and were far more than holding our own. Every argument about invasion that had given me confidence in June and July was doubled before September.

*     *     *     *     *

We had at this time in Washington a singularly gifted and influential Ambassador. I had known Philip Kerr, who had now succeeded as Marquess of Lothian, from the old days of Lloyd George in 1919 and before, and we had differed much and often from Versailles to Munich and later. As the tension of events mounted not only did Lothian develop a broad comprehension of the scene, but his eye penetrated deeply. He had pondered on the grave implications of the messages I had sent to the President during the collapse of France about the possible fate of the British Fleet if England were invaded and conquered. In this he moved with the ruling minds in Washington, who were deeply perturbed, not only by sympathy for Britain and her cause, but naturally even more by anxiety for the life and safety of the United States.

Lothian was worried by the last words of my speech in the House of Commons on June 4, when I had said: "We shall never surrender; and even if, which I do not for a moment believe, this Island or a large part of it were subjugated and starving, then our Empire beyond the seas, armed and guarded by the British Fleet, would carry on the struggle, until, in God's good time, the New

World, with all its power and might, steps forth to the rescue and the liberation of the Old." He thought these words had given encouragement "to those who believed that, even though Great Britain went under, the Fleet would somehow cross the Atlantic to them". The reader is aware of the different language I had been using behind the scenes. I had explained my position at the time to the Foreign Secretary and to the Ambassador.

*Prime Minister to Lord Lothian*                                              9.VI.40
My last words in my speech were of course addressed primarily to Germany and Italy, to whom the idea of a war of continents and a long war are at present obnoxious; also to [the] Dominions, for whom we are trustees. I have nevertheless always had in mind your point, and have raised it in various telegrams to President as well as to Mackenzie King. If Great Britain broke under invasion, a pro-German Government might obtain far easier terms from Germany by sur-rendering the Fleet, thus making Germany and Japan masters of the New World. This dastard deed would not be done by His Majesty's present advisers, but if some Quisling Government were set up it is exactly what they would do, and perhaps the only thing they could do, and the President should bear this very clearly in mind. You should talk to him in this sense and thus discourage any complacent assumption on United States' part that they will pick up the *débris* of the British Empire by their present policy. On the contrary, they run the terrible risk that their sea-power will be completely over-matched. Moreover, islands and naval bases to hold the United States in awe would certainly be claimed by the Nazis. If we go down Hitler has a very good chance of conquering the world.

I hope the foregoing will be a help to you in your conversations.

Nearly a month passed before any result emerged. Then came an encouraging telegram from the Ambassador. He said (July 5-6) that informed American opinion was at last beginning to realise that they were in danger of losing the British Fleet altogether if the war went against us and if they remained neutral. It would however be extremely difficult to get American public opinion to consider letting us have American destroyers unless it could be assured that in the event of the United States entering the war the British Fleet or such of it as was afloat would cross the Atlantic if Great Britain were overrun.

At the end of July, under the increasing pressure from so many angles at once, I took the matter up again.

*Former Naval Person to President Roosevelt* 31.VII.40

It is some time since I ventured to cable personally to you, and many things, both good and bad, have happened in between. It has now become most urgent for you to let us have the destroyers, motor-boats and flying-boats for which we have asked. The Germans have the whole French coastline from which to launch U-boats and dive-bomber attacks upon our trade and food, and in addition we must be constantly prepared to repel by sea action threatened invasion in the Narrow Waters, and also to deal with break-outs from Norway towards Ireland, Iceland, Shetlands, and Faroes. Besides this we have to keep control of the exits from the Mediterranean, and if possible the command of that inland sea itself, and thus to prevent the war spreading seriously into Africa.

We have a large construction of destroyers and anti-U-boat craft coming forward, but the next three or four months open the gap of which I have previously told you. Latterly the air attack on our shipping has become injurious. In the last ten days we have had the following destroyers sunk: *Brazen, Codrington, Delight, Wren,* and the following damaged: *Beagle, Boreas, Brilliant, Griffin, Montrose, Walpole, Whitshed*; total, eleven. All this in advance of any attempt which may be made at invasion! Destroyers are frightfully vulnerable to air bombing, and yet they must be held in the air-bombing area to prevent seaborne invasion. We could not sustain the present rate of casualties for long, and if we cannot get a substantial reinforcement the whole fate of the war may be decided by this minor and easily-remediable factor.

This is a frank account of our present situation, and I am confident, now that you know exactly how we stand, that you will leave nothing undone to ensure that fifty or sixty of your oldest destroyers are sent to me at once. I can fit them very quickly with Asdics and use them against U-boats on the Western Approaches, and so keep the more modern and better-gunned craft for the Narrow Seas against invasion. Mr. President, with great respect I must tell you that in the long history of the world this is a thing to do *now.* Large construction is coming to me in 1941, but the crisis will be reached long before 1941. I know you will do all in your power, but I feel entitled and bound to put the gravity and urgency of the position before you.

If the destroyers were given, the motor-boats and flying-boats, which would be invaluable, could surely come in behind them.

I am beginning to feel very hopeful about this war if we can get round the next three or four months. The air is holding well. We are hitting that man hard, both in repelling attacks and in bombing Germany. But the loss of destroyers by air attack may well be so serious as to break down our defence of the food and trade routes across the Atlantic.

To-night the latest convoys of rifles, cannon, and ammunition are coming in. Special trains are waiting to take them to the troops and Home Guard, who will take a lot of killing before they give them up. I am sure that, with your comprehension of the sea affair, you will not let this crux of the battle go wrong for want of these destroyers.

Three days later I telegraphed to our Ambassador:

3.VIII.40

[The] second alternative, *i.e.*, [granting of] bases [in British possessions], is agreeable, but we prefer that it should be on lease indefinitely and not sale. It is understood that this will enable us to secure destroyers and flying-boats at once. You should let Colonel Knox and others know that a request on these lines will be agreeable to us. . . . It is, as you say, vital to settle quickly. Now is the time when we want the destroyers. We can fit them with Asdics in about ten days from the time they are in our hands, all preparations having been made. We should also be prepared to give a number of Asdic sets to the United States Navy and assist in their installation and explain their working. Go ahead on these lines full steam.

Profound and anxious consultations had taken place at Washington, and in the first week of August the suggestion was made to us through Lord Lothian that the fifty old but reconditioned American destroyers which lay in the east coast Navy yards might be traded off to us in exchange for a series of bases in the West Indian islands, and also in Bermuda. There was of course no comparison between the intrinsic value of these antiquated and inefficient craft and the immense permanent strategic security afforded to the United States by the enjoyment of the island bases. But the threatened invasion, the importance of numbers in the Narrow Seas, made our need clamant. Moreover, the strategic value of these islands counted only against the United States. They were, in the old days, the stepping-stone by which America could be attacked from Europe or from England. Now, with air-power, it was all the more important for American safety that they should be in friendly hands, or in their own. But the friendly hands might fail in the convulsive battle now beginning for the life of Britain. Believing, as I have always done, that the survival of Britain is bound up with the survival of the United States, it seemed to me and to my colleagues that it was an actual

357

advantage to have these bases in American hands. I therefore did not look upon the question from any narrow British point of view.

There was another reason, wider and more powerful than either our need for the destroyers or the American need for the bases. The transfer to Great Britain of fifty American warships was a decidedly unneutral act by the United States. It would, according to all the standards of history, have justified the German Government in declaring war upon them. The President judged that there was no danger, and I felt there was no hope, of this simple solution of many difficulties. It was Hitler's interest and method to strike his opponents down one by one. The last thing he wished was to be drawn into war with the United States before he had finished with Britain. Nevertheless the transfer of the destroyers to Britain in September 1940 was an event which brought the United States definitely nearer to us and to the war, and it was the first of a long succession of increasingly unneutral acts in the Atlantic which were of the utmost service to us. It marked the passage of the United States from being neutral to being non-belligerent. Although Hitler could not afford to resent it, all the world, as will be seen, understood the significance of the gesture.

For all these reasons the War Cabinet and Parliament approved the policy of leasing the bases to obtain the destroyers, provided we could persuade the West Indian island Governments concerned to make what was to them a serious sacrifice and disturbance of their life for the sake of the Empire. On August 6 Lothian cabled that the President was anxious for an immediate reply about the future of the Fleet. He wished to be assured that if Britain were overrun the Fleet would continue to fight for the Empire overseas and would not either be surrendered or sunk. This was, it was said, the argument which would have the most effect on Congress in the question of destroyers. The prospects of legislative action, he thought, were steadily improving.

I expressed my own feelings to the Foreign Secretary:

7.VIII.40

The position is, I think, quite clear. We have no intention of surrendering the British Fleet, or of sinking it voluntarily. Indeed, such a fate is more likely to overtake the German Fleet—or what is left of it. The nation would not tolerate any discussion of what we should do if our Island were overrun. Such a discussion, perhaps on the eve of an invasion, would be injurious to public morale, now so high. Moreover, we must never get into a position where the United States Government might say: "We think the time has come for you to send your Fleet across the Atlantic in accordance with our understanding or agreement when we gave you the destroyers."

We must refuse any declaration such as is suggested, and confine the deal solely to the Colonial leases.

I now cabled to Lothian:

7.VIII.40

We need the fifty or sixty destroyers very much, and hope we shall obtain them. In no other way could the United States assist us so effectively in the next three or four months. We were, as you know, very ready to offer the United States indefinite lease facilities for naval and air bases in West Indian islands, and to do this freely on grounds of inevitable common association of naval and military interests of Great Britain and the United States. It was therefore most agreeable to us that Colonel Knox should be inclined to suggest action on these or similar lines as an accompaniment to the immediate sending of the said destroyers. But all this has nothing to do with any bargaining or declaration about the future disposition of the British Fleet. It would obviously be impossible for us to make or agree to any declaration being made on such a subject. I have repeatedly warned you in my secret telegrams and those to the President of the dangers United States would run if Great Britain were successfully invaded and a British Quisling Government came into office to make the best terms possible for the surviving inhabitants. I am very glad to find that these dangers are regarded as serious, and you should in no wise minimise them. We have no intention of relieving United States from any well-grounded anxieties on this point. Moreover, our position is not such as to bring the collapse of Britain into the arena of practical discussion. I have already several weeks ago told you that there is no warrant for discussing any question of the transference of the Fleet to American or Canadian shores. I should refuse to allow the subject even to be mentioned in any Staff conversations, still less that any technical preparations should be made or even planned. Above all, it is essential you should realise that no such declaration could ever be assented to by us for the purpose of obtaining destroyers or anything like that. Pray make it clear at once that we could never agree to the slightest compromising of our full liberty of action, nor tolerate any such defeatist announcement, the effect of which would be disastrous.

Although in my speech of June 4 I thought it well to open up to

German eyes the prospects of indefinite oceanic war, this was a suggestion in the making of which we could admit no neutral partner. Of course if the United States entered the war and became an ally we should conduct the war with them in common, and make of our own initiative and in agreement with them whatever were the best dispositions at any period in the struggle for the final effectual defeat of the enemy. You foresaw this yourself in your first conversation with the President, when you said you were quite sure that we should never send any part of our Fleet across the Atlantic except in the case of an actual war alliance.

To the President I telegraphed:

15.VIII.40

I need not tell you how cheered I am by your message, nor how grateful I feel for your untiring efforts to give us all possible help. You will, I am sure, send us everything you can, for you know well that the worth of every destroyer that you can spare to us is measured in rubies. But we also need the motor torpedo-boats which you mentioned, and as many flying-boats and rifles as you can let us have. We have a million men waiting for rifles.

The moral value of this fresh aid from your Government and people at this critical time will be very great and widely felt.

We can meet both the points you consider necessary to help you with Congress and with others concerned, but I am sure that you will not misunderstand me if I say that our willingness to do so must be conditional on our being assured that there will be no delay in letting us have the ships and flying-boats. As regards an assurance about the British Fleet, I am of course ready to reiterate to you what I told Parliament on June 4. We intend to fight this out here to the end, and none of us would ever buy peace by surrendering or scuttling the Fleet. But in any use you may make of this repeated assurance you will please bear in mind the disastrous effect from our point of view, and perhaps also from yours, of allowing any impression to grow that we regard the conquest of the British Islands and its naval bases as any other than an impossible contingency. The spirit of our people is splendid. Never have they been so determined. Their confidence in the issue has been enormously and legitimately strengthened by the severe air fighting in the past week. As regards naval and air bases, I readily agree to your proposals for ninety-nine-year leases, which are far easier for us than the method of purchase. I have no doubt that once the principle is agreed between us the details can be adjusted and we can discuss them at leisure. It will be necessary for us to consult the Governments of Newfoundland and Canada about the Newfoundland base, in which Canada has an interest. We are at once proceeding to seek their consent.

Once again, Mr. President, let me thank you for your help and encouragement, which mean so much to us.

Lothian thought this reply admirable, and said there was a real chance now that the President would be able to get the fifty destroyers without legislation. This was still uncertain, but he thought we should send some British destroyer crews to Halifax and Bermuda without any delay. It would create the worst impression in America if destroyers were made available and no British crews were ready to transport them across the Atlantic. Moreover, the fact that our crews were already waiting on the spot would help to impress the urgency of the case on Congress.

At his Press conference on August 16 the President made the following statement: "The United States Government is holding conversations with the Government of the British Empire with regard to acquisition of naval and air bases for the defence of the Western Hemisphere, and especially the Panama Canal. Furthermore, the United States Government is carrying on conversations with the Canadian Government towards the defence of the American hemisphere."

The President went on to say that the United States would give Great Britain something in return, but that he did not know what this would be. He emphasised more than once that the negotiations for the air bases were in no way connected with the question of destroyers. Destroyers were, he said, not involved in the prospective arrangements.

\*   \*   \*   \*   \*

The President, having always to consider Congress and also the Navy authorities in the United States, was of course increasingly drawn to present the transaction to his fellow-countrymen as a highly advantageous bargain whereby immense securities were gained in these dangerous times by the United States in return for a few flotillas of obsolete destroyers. This was indeed true; but not exactly a convenient statement for me. Deep feelings were aroused in Parliament and the Government at the idea of leasing any part of these historic territories, and if the issue were presented to the British as a naked trading away of British possessions for sake of the fifty destroyers it would certainly encounter vehement opposition. I sought therefore to place the transaction on the highest level, where indeed it had a right to stand, because it expressed and conserved the enduring common interests of the English-speaking world.

With the consent of the President I presented the question to Parliament on August 20, in words which have not perhaps lost their meaning with time:

Presently we learned that anxiety was also felt in the United States about the air and naval defence of their Atlantic seaboard, and President Roosevelt has recently made it clear that he would like to discuss with us, and with the Dominion of Canada and with Newfoundland, the development of American naval and air facilities in Newfoundland and in the West Indies. There is of course no question of any transference of sovereignty—that has never been suggested—or of any action being taken without the consent or against the wishes of the various Colonies concerned, but for our part His Majesty's Government are entirely willing to accord defence facilities to the United States on a ninety-nine years' leasehold basis, and we feel sure that our interests no less than theirs, and the interests of the Colonies themselves and of Canada and Newfoundland, will be served thereby. These are important steps. Undoubtedly this process means that these two great organisations of the English-speaking democracies, the British Empire and the United States, will have to be somewhat mixed up together in some of their affairs for mutual and general advantage. For my own part, looking out upon the future, I do not view the process with any misgivings. I could not stop it if I wished; no one can stop it. Like the Mississippi, it just keeps rolling along. Let it roll. Let it roll on—full flood, inexorable, irresistible, benignant, to broader lands and better days.

*Former Naval Person to President*                                22.VIII.40

1. I am most grateful for all you are doing on our behalf. I had not contemplated anything in the nature of a contract, bargain, or sale between us. It is the fact that we had decided in Cabinet to offer you naval and air facilities off the Atlantic coast quite independently of destroyers or any other aid. Our view is that we are two friends in danger helping each other as far as we can. We should therefore like to give you the facilities mentioned without stipulating for any return, and even if to-morrow you found it too difficult to transfer the destroyers, etc., our offer still remains open because we think it is in the general good.

2. I see difficulties, and even risks, in the exchange of letters now suggested or in admitting in any way that the munitions which you send us are a payment for the facilities. Once this idea is accepted people will contrast on each side what is given and received. The

money value of the armaments would be computed and set against the facilities, and some would think one thing about it and some another.

3. Moreover, Mr. President, as you well know, each island or location is a case by itself. If, for instance, there were only one harbour or site, how is it to be divided and its advantages shared? In such a case we should like to make you an offer of what we think is best for both, rather than to embark upon a close-cut argument as to what ought to be delivered in return for value received.

4. What we want is that you shall feel safe on your Atlantic seaboard so far as any facilities in possessions of ours can make you safe, and naturally, if you put in money and make large developments, you must have the effective security of a long lease. Therefore I would rather rest at this moment upon the general declaration made by me in the House of Commons yesterday, both on this point and as regards the future of the Fleet. Then, if you will set out in greater detail what you want, we will at once tell you what we can do, and thereafter the necessary arrangements, technical and legal, can be worked out by our experts. Meanwhile we are quite content to trust entirely to your judgment and the sentiments of the people of the United States about any aid in munitions, etc., you feel able to give us. But this would be entirely a separate, spontaneous act on the part of the United States, arising out of their view of the world struggle and how their own interests stand in relation to it and the causes it involves.

5. Although the air attack has slackened in the last few days and our strength is growing in many ways, I do not think that bad man has yet struck his full blow. We are having considerable losses in merchant ships on the North-Western Approaches, now our only channel of regular communication with the oceans, and your fifty destroyers, if they came along at once, would be a precious help.

Lothian now cabled that Mr. Sumner Welles had told him that the constitutional position made it "utterly impossible" for the President to send the destroyers as a spontaneous gift; they could come only as a *quid pro quo*. Under the existing legislation neither the Chief of the Staff nor the General Board of the Navy were able to give the certificate that the ships were not essential to national defence, without which the transfer could not be legally made, except in return for a definite consideration which they would certify added to the security of the United States. The President had tried to find another way out, but there was none.

*Former Naval Person to President*                    25.VIII.40
   1. I fully understand the legal and constitutional difficulties which

make you wish for a formal contract embodied in letters, but I venture to put before you the difficulties, and even dangers, which I foresee in this procedure. For the sake of the precise list of instrumentalities mentioned, which in our sore need we greatly desire, we are asked to pay undefined concessions in all the islands and places mentioned from Newfoundland to British Guiana, "as may be required in the judgment of the United States". Suppose we could not agree to all your experts asked for, should we not be exposed to a charge of breaking our contract, for which we had already received value? Your commitment is definite, ours unlimited. Much though we need the destroyers, we should not wish to have them at the risk of a misunderstanding with the United States, or, indeed, any serious argument. If the matter is to be represented as a contract, both sides must be defined, with far more precision on our side than has hitherto been possible. But this might easily take some time.

As I have several times pointed out, we need the destroyers chiefly to bridge the gap between now and the arrival of our new construction, which I set on foot on the outbreak of war. This construction is very considerable. For instance, we shall receive by the end of February new destroyers and new medium destroyers, 20; corvettes, which are a handy type of submarine-hunter adapted to ocean work, 60; motor torpedo-boats, 37; motor anti-submarine boats, 25; Fairmiles, a wooden anti-submarine patrol boat, 104; 72-foot launches, 29. An even greater inflow will arrive in the following six months. It is just in the gap from September to February inclusive, while this new crop is coming in and working up, that your fifty destroyers would be invaluable. With them we could minimise shipping losses in the North-Western Approaches and also take a stronger line against Mussolini in the Mediterranean. Therefore time is all-important. We should not however be justified, in the circumstances, if we gave a blank cheque on the whole of our transatlantic possessions merely to bridge this gap, through which, anyhow, we hope we make our way, though with added risk and suffering. This, I am sure you will see, sets forth our difficulties plainly.

2. Would not the following procedure be acceptable? I would offer at once certain fairly well defined facilities which will show you the kind of gift we have in mind, and your experts could then discuss these, or any variants of them, with ours—we remaining the final judge of what we can give. All this we will do freely, trusting entirely to the generosity and goodwill of the American people as to whether they on their part would like to do something for us. But anyhow, it is the settled policy of His Majesty's Government to offer you, and make available to you when desired, solid and effective means of

protecting your Atlantic seaboard. I have already asked the Admiralty and the Air Ministry to draw up in outline what we are prepared to offer, leaving your experts to suggest alternatives. I propose to send you this outline in two or three days and to publish it in due course. In this way there can be no possible dispute, and the American people will feel more warmly towards us, because they will see we are playing the game by the world's cause and that their safety and interests are dear to us.

3. If your law or your Admiral requires that any help you may choose to give us must be presented as a *quid pro quo*, I do not see why the British Government have to come into that at all. Could you not say that you did not feel able to accept this fine offer which we make unless the United States matched it in some way, and that therefore the Admiral would be able to link the one with the other?

4. I am so grateful to you for all the trouble you have been taking, and I am sorry to add to your burdens, knowing what a good friend you have been to us.

*Former Naval Person to President*                          27.VIII.40

1. Lord Lothian has cabled me the outline of the facilities you have in mind. Our naval and air experts studying the question from your point of view had reached practically the same conclusions, except that in addition they thought Antigua might be useful as a base for flying-boats. To this also you would be very welcome. Our settled policy is to make the United States safe on their Atlantic seaboard "beyond a peradventure", to quote a phrase you may remember.*

2. We are quite ready to make you a positive offer on these lines forthwith. There would of course have to be an immediate conference on details, but, for the reasons which I set out in my last telegram, we do not like the idea of an arbiter should any difference arise, because we feel that as donors we must remain the final judges of what the gift is to consist of within the general framework of the facilities which will have been promised, and always on the understanding that we shall do our best to meet United States wishes.

3. The two letters drafted by Lord Lothian to the Secretary of State are quite agreeable to us. The only reason why I do not wish the second letter to be published is that I think it is much more likely that the German Government will be the one to surrender or scuttle its Fleet, or what is left of it. In this, as you are aware, they have already had some practice. You will remember that I said some months ago in one of my private cables to you that any such action on our part would be a dastard act, and that is the opinion of every one of us.

4. If you felt able after our offer had been made to let us have the

* Used by President Wilson in 1917.

"instrumentalities"* which have been mentioned or anything else you think proper, this could be expressed as an act not in payment or consideration for, but in recognition of, what we had done for the security of the United States.

5. Mr. President, this business has become especially urgent in view of the recent menace which Mussolini is showing to Greece. If our business is put through on big lines and in the highest spirit of goodwill, it might even now save that small historic country from invasion and conquest. Even the next forty-eight hours are important.

*Prime Minister to General Ismay*                                27.VIII.40

Lord Lothian's account of President Roosevelt's request should now be put into the first person in case a public declaration is required in our name. For instance, "His Majesty's Government make the following offer to the President of the United States: 'We are prepared in friendship and goodwill to meet your representatives immediately in order to consider the provision of effective naval and air bases in the following islands,'" etc.

Let me have a draft on these lines, so that I can dictate a cable. The draft should be in my hands this morning.

Accordingly:

27.VIII.40

His Majesty's Government make the following offer to the President of the United States:

We are prepared in friendship and goodwill to meet your representatives forthwith, in order to consider the lease for ninety-nine years of areas for the establishment of naval and air bases in the following places:

|  |  |
|---|---|
| NEWFOUNDLAND | ANTIGUA |
| BERMUDA | ST. LUCIA |
| BAHAMAS | TRINIDAD |
| JAMAICA | BRITISH GUIANA |

Subject to later settlements on points of detail. . . .

At the same time I suggested the following text of the telegram for publication which the President might send me to elicit the assurance he desired.

The Prime Minister of Great Britain is reported to have stated on June 4, 1940, to Parliament, in effect, that if during the course of the present war in which Great Britain and British Colonies are engaged the waters surrounding the British Isles should become untenable for British ships of war, a British Fleet would in no event be surrendered

* Also a Wilsonian word.

or sunk, but would be sent overseas for the defence of other parts of the Empire.

The Government of the United States would respectfully inquire whether the foregoing statement represents the settled policy of the British Government.

The President adopted this version, and I sent him the following agreed reply:

31.VIII.40

You ask, Mr. President, whether my statement in Parliament on June 4, 1940, about Great Britain never surrendering or scuttling her Fleet "represents the settled policy of His Majesty's Government". It certainly does. I must however observe that these hypothetical contingencies seem more likely to concern the German Fleet, or what is left of it, than our own.

Thus all was happily settled, and on September 5, using the language of under-statement, I duly informed the House of Commons, and obtained their acquiescence and indeed general consent:

The memorable transactions between Great Britain and the United States which were foreshadowed when I last addressed the House have now been completed. As far as I can make out, they have been completed to the general satisfaction of the British and American peoples and to the encouragement of our friends all over the world. It would be a mistake to try to read into the official notes which have passed more than the documents bear on their face. The exchanges which have taken place are simply measures of mutual assistance rendered to one another by two friendly nations, in a spirit of confidence, sympathy, and goodwill. These measures are linked together in a formal agreement. They must be accepted exactly as they stand. Only very ignorant persons would suggest that the transfer of American destroyers to the British flag constitutes the slightest violation of international law, or affects in the smallest degree the non-belligerency of the United States.

I have no doubt that Herr Hitler will not like this transference of destroyers, and I have no doubt that he will pay the United States out, if he ever gets the chance. That is why I am very glad that the army, air, and naval frontiers of the United States have been advanced along a wide arc into the Atlantic Ocean, and that this will enable them to take danger by the throat while it is still hundreds of miles away from their homeland. The Admiralty tell us also that they are very glad to have these fifty destroyers, and that they will come in most conveniently to bridge the gap which, as I have previously explained to the House, inevitably intervenes before our considerable war-time programme of new construction comes into service.

I suppose the House realises that we shall be a good deal stronger next year on the sea than we are now, although that is quite strong enough for the immediate work in hand. There will be no delay in bringing the American destroyers into active service; in fact, British crews are already meeting them at the various ports where they are being delivered. You might call it the long arm of coincidence. I really do not think that there is any more to be said about the whole business at the present time. This is not the appropriate occasion for rhetoric. Perhaps I may however, very respectfully, offer this counsel to the House: When you have got a thing where you want it it is a good thing to leave it where it is.

Thus we obtained the fifty American destroyers. We granted ninety-nine-year leases of the air and naval bases specified in the West Indies and Newfoundland to the United States. And, thirdly, I repeated my declaration about not scuttling or surrendering the British Fleet in the form of an assurance to the President. I regarded all these as parallel transactions, and as acts of goodwill performed on their merits and not as bargains. The President found it more acceptable to present them to Congress as a connected whole. We neither of us contradicted each other, and both countries were satisfied. The effects in Europe were profound.

# CHAPTER XXI

## EGYPT
## AND THE MIDDLE EAST

### 1940

### June–July–August

*Mussolini Prepares to Invade Egypt – Our Competing Anxieties –*
*The Italian Strength in North Africa – Concentration towards the*
*Egyptian Frontier – Beads on the String – Initiative of Our Covering*
*Troops – My Complaints of Dispersion – The Kenya Front – Pales-*
*tine – The Mediterranean Short Cut – The Tanks Have to Go Round*
*the Cape – Plans for Cutting the Italian Coastal Road from the Sea –*
*Ministerial Committee on the Middle East – General Wavell Comes*
*Home for Conference – Hard and Tense Discussions With Him –*
*Directive of August 16 – Assembly of the Army of the Nile – Its*
*Tactical Employment – The Somaliland Episode – A Vexatious Re-*
*buff – Increase in Italian Forces in Albania – My Report on the General*
*Situation to the Prime Ministers of Australia and New Zealand.*

W ITH the disappearance of France as a combatant and
with Britain set on her struggle for life at home, Mussolini
might well feel that his dream of dominating the Mediterranean and rebuilding the former Roman Empire would come
true. Relieved from any need to guard against the French in
Tunis, he could still further reinforce the numerous army he had
gathered for the invasion of Egypt. The eyes of the world were
fixed upon the fate of the British Island, upon the gathering of the
invading German armies, and upon the drama of the struggle for
air mastery. These were of course our main preoccupations. In

369

many countries we were presumed to be at the last gasp. Our confident and resolute bearing was admired by our friends, but its foundations were deemed unsure. Nevertheless the War Cabinet were determined to defend Egypt against all comers with whatever resources could be spared from the decisive struggle at home. All the more was this difficult when the Admiralty declared themselves unable to pass even military convoys through the Mediterranean on account of the air dangers. All must go round the Cape. Thus we might easily rob the Battle of Britain without helping the Battle of Egypt. It is odd that while at the time everyone concerned was quite calm and cheerful, writing about it afterwards makes one shiver.

$$\star \quad \star \quad \star \quad \star \quad \star$$

When Italy declared war on June 10, 1940, the British Intelligence estimated—we now know correctly—that, apart from her garrisons in Abyssinia, Eritrea, and Somaliland, there were about 215,000 Italian troops in the North African coastal provinces. These were disposed as follows: in Tripolitania, six metropolitan and two militia divisions; in Cyrenaica, two metropolitan and two militia divisions, besides frontier forces equal to three divisions; a total of fifteen divisions. The British forces in Egypt consisted of the 7th Armoured Division, two-thirds of the 4th Indian Division, one-third of the New Zealand Division, and fourteen British battalions and two regiments of the Royal Artillery, ungrouped in higher formations; the whole amounting to perhaps fifty thousand men. From these both the defence of the western frontier and the internal security of Egypt had to be provided. We therefore had heavy odds against us in the field, and the Italians had also many more aircraft.

During July and August the Italians became active at many points. There was a threat from Kassala westwards towards Khartoum. Alarm was spread in Kenya by the fear of an Italian expedition marching four hundred miles south from Abyssinia towards the Tana River and Nairobi. Considerable Italian forces advanced into British Somaliland. But all these anxieties were petty compared with the Italian invasion of Egypt, which was obviously being prepared on the greatest scale. For some time past Mussolini had been steadily moving his forces eastwards towards Egypt. Even before the war a magnificent road had been made along the coast from the main base at Tripoli, through Tripolitania and Cyrenaica, to the Egyptian frontier. Along this road there had been for many months a swelling stream of military traffic. Large magazines were slowly established and filled at Benghazi, Derna, Tobruk, Bardia, and Sollum. The length of this road was over a thousand miles, and all these swarming Italian garrisons and supply depots were strung along it like beads on a string.

At the head of the road and near the Egyptian frontier an Italian army of seventy or eighty thousand men, with a good deal of modern equipment, had been patiently gathered and organised. Before this army glittered the prize of Egypt. Behind it stretched the long road back to Tripoli; and after that the sea! If this force, built up in driblets week by week for years, could advance continually eastward, conquering all who sought to bar the path, its fortunes would be bright. If it could gain the fertile regions of the Delta all worry about the long road back would vanish. On the other hand, if ill-fortune befell it only a few would ever get home. In the field army and in the series of great supply depots all along the coast there were by the autumn at least three hundred thousand Italians, who could, even if unmolested, retreat westward along the road only gradually or piecemeal. For this they required many months. And if the battle were lost on the Egyptian border, if the army's front were broken, and if time were not given to them, all were doomed to capture or death. However, in July 1940 it was not known who was going to win the battle.

Our foremost defended position at that time was the railhead at Mersa Matruh. There was a good road westward to Sidi Barrani, but thence to the frontier at Sollum there was no road capable of maintaining any considerable strength for long near the frontier. A small covering mechanised force had been formed of some of our finest Regular troops, consisting of the 7th Hussars (light tanks), the 11th Hussars (armoured cars), and two motor battalions of the 60th Rifles and Rifle Brigade, with two regiments of motorised Royal Horse Artillery. Orders had been given to attack the Italian frontier posts immediately on the outbreak of war. Accordingly, within twenty-four hours the 11th Hussars crossed the frontier, took the Italians, who had not heard that war had been declared, by surprise, and captured prisoners. The next night, June 12, they had a similar success, and on June 14, with the 7th Hussars and one company of the 60th Rifles, captured the frontier forts at Capuzzo and Maddalena, taking 220 prisoners. On the 16th they raided deeper, destroyed twelve tanks, intercepted a convoy on the Tobruk-Bardia road, and captured a general.

In this small but lively warfare our troops felt they had the advantage, and soon conceived themselves to be masters of the desert. Until they came up against large formed bodies or fortified posts they could go where they liked, collecting trophies from sharp encounters. When armies approach each other it makes all the difference which owns only the ground on which it stands or sleeps and which one owns all the rest. I saw this in the Boer War, where we owned nothing beyond the fires of our camps and bivouacs, whereas the Boers rode where they pleased all over the country.

Ever-growing enemy forces were now arriving from the west, and by the middle of July the enemy had re-established his frontier line with two divisions and elements of two more. Early in August our covering force was relieved by the Support Group of the 7th Armoured Division, comprising the 3rd Coldstream Guards, the 1st/60th Rifles, the 2nd Rifle Brigade, the 11th Hussars, one squadron of the 6th Royal Tank Battalion, and two mechanised batteries R.H.A., one of which was anti-tank. This small force, distributed over a front of sixty miles, continued to harass the enemy with increasing effect. The published Italian casualties for the first three months of war were nearly three thousand five hundred men, of whom seven hundred were prisoners. Our own losses barely exceeded one hundred and fifty. Thus the first phase in the war which Italy had declared upon the British Empire opened favourably for us.

*    *    *    *    *

It was proposed by the Middle East Command, under General Wavell, to await the shock of the Italian onslaught near the fortified position of Mersa Matruh. Until we could gather an army this seemed the only course open. I therefore proposed the following tasks. First, to assemble the largest fighting force possible to face the Italian invaders. For this it was necessary to run risks in many other quarters. I was pained to see the dispersions which were tolerated by the military authorities. Khartoum and the Blue Nile certainly required strengthening against the Italian-Abyssinian border, but what was the sense of keeping twenty-five thousand men, including the Union Brigade of South Africa and two brigades of excellent West African troops, idle in Kenya? I had ridden over some of this country, north of the Tana River, at the end of 1907. It is a very fine-looking country, but without much to eat. The idea of an Italian expedition of fifteen or twenty thousand men, with artillery and modern gear, traversing the four or five hundred miles before they could reach Nairobi seemed ridiculous. Behind the Kenya front would lie the metre-gauge Uganda railway. We had the command of the sea, and could move troops to and fro by sea and rail with a facility incomparable to anything that could be achieved by enemy land movements. On account of our superior communications it was our interest to fight an Italian expedition as near to Nairobi and the railway as possible. For this large numbers of troops were not required. They were more needed in the Egyptian Delta. I got something, but only after a prolonged hard fight against the woolly theme of being safe everywhere.

I did my utmost to draw upon Singapore and bring the Australian division which had arrived there, first to India for training and thence to the Western Desert. Palestine presented a different aspect. We had a mass of fine troops sprawled over Palestine: an Australian division, a New Zealand brigade, our own choice Yeomanry division, all in armoured cars or about to be; the Household Cavalry, still with horses, but longing for modern weapons; with lavish administrative services. I wished to arm the Jews at Tel Aviv, who with proper weapons would have made a good fight against all comers. Here I encountered every kind of resistance. My second preoccupation was to ensure that freedom of movement through the Mediterranean was fought for against the weak Italians and the grave air danger, in order that Malta might be made impregnable. It seemed to me most important to pass military convoys, especially of tanks and guns, through the Mediterranean instead of all round the Cape. This seemed a prize worth many hazards. To send a division from Britain round the Cape to Egypt was to make sure it could not fight anywhere for three months; but these were precious months, and we had very few divisions. Finally, there was our Island, now under pretty direct menace of invasion. How far could we denude our home and citadel for the sake of the Middle East?

* * * * *

In July 1940 I began, as the telegrams and minutes show, to concern myself increasingly about the Middle East. Always this long coastal road bulked in my mind. Again and again I recurred to the idea of cutting it by the landing of strong but light forces from the sea. We had not of course at that time proper tank landing-craft. Yet it should have been possible to improvise the necessary tackle for such an operation. If used in conjunction with a heavy battle it might have effected a valuable diversion of enemy troops from the front.

*Prime Minister to General Ismay*            10.VII.40
Bring the following before the C.O.S. Committee:
Have any plans been made in the event of large forces approaching the Egyptian border from Libya to cut the coastal motor road upon which they would be largely dependent for supplies of all kinds? It is not sufficient merely to bombard by air or from the sea. But if a couple of brigades of good troops could take some town or other suitable point on the communications, they might, with sea-power behind them, cause a prolonged interruption, require heavy forces to be moved against them, and then withdraw to strike again at some other point. Of course such an operation would not be effective until considerable forces of the enemy had already passed the point of interception. It may be however that the desert itself affords free movement to the enemy's supplies. I wonder whether this is so, and if so why the Italians were at pains to construct this lengthy road.

I still do not see why it should not have been possible to make a good plan. It is however a fact that none of our commanders, either in the Middle East or in Tunis, were ever persuaded to make the attempt. But General Patton in 1943 made several most successful turning movements of this character during the conquest of Sicily, and gained definite advantages thereby. It was not until Anzio in 1944 that I succeeded in having this experiment tried. This of course was on a far larger scale; nor did it, in spite of the success of the landing, achieve the decisive results for which we all hoped. But that is another story.

* * * * *

I was anxious that the case of the Middle East should be strongly presented by a group of Ministers, all experienced in war and deeply concerned in that theatre.

*Prime Minister to Sir Edward Bridges*                    10.VII.40

I think it would be well to set up a small standing Ministerial Committee, consisting of the Secretaries of State for War [Mr. Eden], India [Mr. Amery], and the Colonies [Lord Lloyd], to consult together upon the conduct of the war in the Middle East (in which they are all three concerned), and to advise me, as Minister of Defence, upon the recommendations I should make to the Cabinet. Will you kindly put this into the proper form. The Secretary of State for War has agreed to take the chair.

Mr. Eden reported to his Committee the shortage of troops, equipment, and resources in the Middle East, and that the C.I.G.S. was equally perturbed. The Committee urged the full equipment of the armoured division already in Egypt but far below strength, and also recommended the provision of a second armoured division at the earliest moment when it could be spared from home. The Chiefs of Staff endorsed these conclusions, the C.I.G.S. observing that the moment must be chosen in relation to declining risks at home and increasing risks abroad. On July 31 Mr. Eden considered that we might be able to spare some tanks in a few weeks' time, and that if they were to reach the Middle East by the end of September we might have to send them and the other equipment through the Mediterranean. In spite of the rising tension about invasion at home I was in full agreement with all this trend of thought, and brought the extremely harassing choice before the Cabinet several times.

The other aspects of the Middle East pressed upon me.

*Prime Minister to General Ismay*                    23.VII.40

Where is the South African Union Brigade of 10,000 men? Why is it playing no part in the Middle East? We have agreed to-day to send further reinforcements of Hurricanes and other modern aircraft to the South African Air Force. What is happening to the concert of the campaign in the Middle East? What has been done by the Committee of Ministers I recently set up? Now that large naval operations are contemplated in the Mediterranean, it is all the more essential that the attack on the Italian position in Abyssinia should be pressed and concerted by all means. Make sure I have a report about the position, which I can consider on Thursday morning.

I felt an acute need of talking over the serious events impending in the Libyan desert with General Wavell himself. I had not met this distinguished officer, on whom so much was resting, and I asked the Secretary of State for War to invite him over for a week for consultation when an opportunity could be found. He arrived on August 8. He toiled with the Staffs and had several long conversations with me and Mr. Eden. The command in the Middle East at that time comprised an extraordinary amalgam of military, political, diplomatic, and administrative problems of extreme complexity. It took nearly a year of ups and downs for me and my colleagues to learn the need of dividing the responsibilities of the Middle East between a Commander-in-Chief, a Minister of State, and an Intendant-General to cope with the supply problem. While not in full agreement with General Wavell's use of the resources at his disposal, I thought it best to leave him in command. I admired his fine qualities, and was impressed with the confidence so many people had in him.

The discussions, both oral and written, were severe. As usual I put my case in black and white.

*Prime Minister to General Ismay, for General Wavell*          10.VIII.40

I am very much obliged to you for explaining to me so fully the situation in Egypt and Somaliland. We have yet to discuss the position in Kenya and Abyssinia. I mentioned the very large forces which you have in Kenya, namely, the Union Brigade of 6,000 white South Africans, probably as fine material as exists for warfare in spacious countries; the East African settlers, who should certainly amount to 2,000 men, thoroughly used to the country; the two West African brigades, brought at much inconvenience from the West Coast, numbering 6,000; at least two brigades of King's African Rifles (K.A.R.); the whole at least 20,000 men—there may be more. Why should these all stand idle in Kenya waiting for an Italian invasion to make its way across the very difficult distances from Abyssinia to the south, or preparing themselves for a similar difficult inroad into Abyssinia, which must again entail long delays, while all the time the fate of the Middle East, and much else, may be decided at Alexandria or on the Canal?

Without of course knowing the exact conditions locally, I should suppose that a reasonable disposition would be to hold Kenya with the settlers and the K.A.R. and delay any Italian advance southwards, it being so much easier to bring troops round by sea than for the Italians to make their way overland. Thus we can always reinforce them unexpectedly and swiftly. This would allow the Union Brigade and the two

West African brigades to come round at once into the Delta, giving you a most valuable reinforcement in the decisive theatre at the decisive moment. What is the use of having the command of the sea if it is not to pass troops to and fro with great rapidity from one theatre to another? I am sure I could persuade General Smuts to allow this movement of the Union Brigade. Perhaps you will let me have your views on this by to-morrow night, as time is so short.

*Prime Minister to General Ismay, for General Wavell*　　　12.VIII.40

1. I am not at all satisfied about the Union Brigade and the West African brigade in Kenya. These forces as now disposed would play no part in the critical attacks now being developed against Egypt, Khartoum, and Somaliland. It is always considered a capital blemish on military operations that large bodies of troops should be standing idle while decisions are reached elsewhere. Without further information, I cannot accept the statement that the South African Brigade is so far untrained that it cannot go into action. The Natal Carbineers were much further advanced in training before the war than our British Territorials, and they have presumably been embodied since the declaration. I cannot see why the Union Brigade as a whole should be considered in any way inferior to British Territorial units. Anyhow, they are certainly good enough to fight Italians. I have asked for full particulars of their embodiment and training in each case.

2. I do not consider that proper use is being made of the large forces in Palestine. The essence of the situation depends on arming the Jewish colonists sufficiently to enable them to undertake their own defence, so that if necessary for a short time the whole of Palestine can be left to very small British forces. A proposal should be made to liberate immediately a large portion of the garrison, including the Yeomanry Cavalry Division. I do not understand why the Australians and New Zealanders, who have been training in Palestine for at least six months, should be able to provide only one brigade for service in Egypt. How many of them are there, and what are the facts of their training? These men were brought at great expense from Australia, having been selected as the first volunteers for service in Europe. Many of them had previous military training, and have done nearly a year's training since the war broke out. How disgraceful it would be if owing to our mishandling of this important force only one brigade took part in the decisive operations for the defence of Egypt!

3. The two West African brigades could certainly be brought to Khartoum via Port Soudan. It is a very good policy to mix native units from various sources, so that one lot can be used to keep the other in discipline. These two brigades ought to be moved immediately to the Soudan, so that the Indian division can be used in Egypt or Somaliland

as soon as it arrives. I do not know why these brigades were taken away from West Africa, if the only use to be made of them was to garrison Kenya.

4. Let me have a return of the white settlers of military age in Kenya. Are we to believe they have not formed any local units for the defence of their own province? If not, the sooner they are made to realise their position the better. No troops ought to be in Kenya at the present time other than the settlers and the K.A.R. Considering the risks and trouble we are taking to reinforce Egypt from home, it cannot be accepted that forces on the spot should not be used to the highest capacity at the critical moment.

5. Let me have a full account of the two British divisions in the Delta. It is misleading to think in divisions in this area; nor can any plea that they are not properly equipped in every detail be allowed to prejudice the employment of these fine Regular troops.

6. Surely the statement that the enemy's armoured forces and vehicles can move just as easily along the desert as along the coastal road requires further examination. This might apply to caterpillar vehicles, but these would suffer severely if forced to make long journeys over the rocky and soft deserts. Anyhow, wheeled transport would be hampered in the desert unless provided with desert-expanded india-rubber tyres of a special type. Are the Italian vehicles so fitted, and to what extent?

7. What arrangements have been made to "depotabilise" for long periods any wells or water supplies we do not require for ourselves? Has a store of delayed-action fuzes been provided for mines in roadways which are to be abandoned? Make sure that a supply of the longest delayed-action fuzes, i.e., up to at least a fortnight (but I hope they run longer now), are sent to Egypt by the first ship to go through. Examine whether it is not possible to destroy the asphalt of the tarmac road as it is abandoned by chemical action of heavy petroleum oil, or some other treatment.

8. Let me have a statement in full and exact detail of all units in the Middle East, including Polish and French volunteers and arrivals.

I should be glad to discuss all these points to-night.

*     *     *     *     *

As a result of the Staff discussions on August 10 Dill, with Eden's ardent approval, wrote me that the War Office were arranging to send immediately to Egypt one cruiser tank battalion of fifty-two tanks, one light tank regiment (fifty-two tanks), and one Infantry tank battalion of fifty tanks, together with forty-eight anti-tank guns, twenty Bofors light A.A., forty-eight 25-

pounder field guns, five hundred Bren guns, and two hundred and fifty anti-tank rifles, with the necessary ammunition. These would start as soon as they could be loaded. The only question open was whether they should go round the Cape or take a chance through the Mediterranean. I pressed the Admiralty hard, as will be seen in a later chapter, for direct convoy through the Mediterranean. Much discussion proceeded on this latter point. Meanwhile the Cabinet approved the embarkation and dispatch of the armoured force, leaving the final decision about which way they should go till the convoy approached Gibraltar. This option remained open to us till August 26, by which time we should know a good deal more about the imminence of any Italian attack. No time was lost. The decision to give this blood-transfusion while we braced ourselves to meet a mortal danger was at once awful and right. No one faltered.

\*     \*     \*     \*     \*

The following directive, which we had thrashed out together, was finally drafted by me, and the Cabinet approved it without amendment in accord with the Chiefs of Staff.

*Prime Minister to Secretary of State for War and C.I.G.S.*     16.VIII.40

### GENERAL DIRECTIVE FOR COMMANDER-IN-CHIEF, MIDDLE EAST

1. A major invasion of Egypt from Libya must be expected at any time now. It is necessary therefore to assemble and deploy the largest possible army upon and towards the western frontier. All political and administrative considerations must be set in proper subordination to this.

2. The evacuation of Somaliland is enforced upon us by the enemy, but is none the less strategically convenient. All forces in or assigned to Somaliland should be sent to Aden, to the Soudan via Port Soudan, or to Egypt, as may be thought best.

3. The defence of Kenya must rank *after* the defence of the Soudan. There should be time after the crisis in Egypt and the Soudan is passed to reinforce Kenya by sea and rail before any large Italian expedition can reach the Tana River. We can always reinforce Kenya faster than Italy can pass troops thither from Abyssinia or Italian Somaliland.

4. Accordingly either the two West African brigades or two brigades of the K.A.R. should be moved forthwith to Khartoum. General Smuts is being asked to allow the Union Brigade, or a large part of it, to move to the Canal Zone and the Delta for internal

security purposes. Arrangements should be made to continue their training. The Admiralty are being asked to report on shipping possibilities in the Indian Ocean and Red Sea.

5. In view of the increased air attack which may be expected in the Red Sea following upon the Italian conquest of British Somaliland, the air reinforcement of Aden becomes important.

6. The two brigades, one of Regulars and the other Australian, which are held ready in Palestine should now move into the Delta in order to clear the Palestine communications for the movement of further reserves, as soon as they can be equipped for field service or organised for internal security duties.

7. However, immediately three or four regiments of British cavalry, without their horses, should take over the necessary duties in the Canal Zone, liberating the three Regular battalions there for general reserve of the Field Army of the Delta.

8. The rest of the Australians in Palestine, numbering six battalions, will thus [also] be available at five days' notice to move into the Delta for internal security or other emergency employment. The Polish Brigade and the French Volunteer Unit should move to the Delta from Palestine as may be convenient and join the general reserve

9. The movement of the Indian division now embarking or in transit should be accelerated to the utmost. Unless some of the troops evacuated from Somaliland and not needed for Aden are found sufficient to reinforce the Soudan, in addition to reinforcements from Kenya, this whole division, as is most desirable, should proceed to Suez to join the Army of the Delta [later called the Army of the Nile]. In addition to the above at least three batteries of British artillery, although horse-drawn, must be embarked immediately from India for Suez. Admiralty to arrange transport.

10. Most of the above movements should be completed between September 15 and October 1, and on this basis the Army of the Delta should comprise:

(a) The British Armoured Force in Egypt.
(b) The four British battalions at Mersa Matruh, the two at Alexandria, and the two in Cairo—total, eight.
(c) The three battalions from the Canal Zone.
(d) The reserve British brigade from Palestine—total, fourteen British Regular infantry battalions.
(e) The New Zealand Brigade.
(f) The Australian Brigade from Palestine.
(g) The Polish Brigade.
(h) Part of the Union Brigade from East Africa.
(i) The 4th Indian Division, now in rear of Mersa Matruh.

(j)  The new Indian division in transit.
(k)  The 11,000 men in drafts arriving almost at once at Suez.
(l)  All the artillery (150 guns) now in the Middle East or *en route* from India.
(m)  The Egyptian Army so far as it can be used for field operations.

11. The above should constitute by October 1, at the latest, 39 battalions, together with the armoured forces; a total of 56,000 men and 212 guns. This is exclusive of internal security troops.

Part II

12. It is hoped that the armoured brigade from England of three regiments of tanks will be passed through the Mediterranean by the Admiralty. If this is impossible their arrival round the Cape may be counted upon during the first fortnight in October. The arrival of this force in September must be deemed so important as to justify a considerable degree of risk in its transportation.

Part III

Tactical employment of the above force:

13. The Mersa Matruh position must be fortified completely and with the utmost speed. The sector held by the three Egyptian battalions must be taken over by three British battalions, making the force homogeneous. This must be done even if the Egyptian Government wish to withdraw the artillery now in the hands of these three battalions. The possibility of reinforcing by sea the Mersa Matruh position and cutting enemy communications, once they have passed by on their march to the Delta, must be studied with the Naval Commander-in-Chief, Mediterranean Fleet. Alternatively a descent upon the communications at Sollum or farther west may be preferred.

14. All water supplies between Mersa Matruh and the Alexandria defences must be rendered "depotable".* A special note on this is attached. No attempt should be made to leave small parties to defend the wells near the coast in this region. The 4th Indian Division should withdraw upon Alexandria when necessary or be taken off by sea. The road from Sollum to Mersa Matruh, and still more the tarmac road from Mersa Matruh to Alexandria, must be rendered impassable, as it is abandoned, by delayed-action mines or by chemical treatment of the asphalt surface.

15. A main line of defence to be held by the whole Army of the Delta, with its reserves suitably disposed, must be prepared (as should long ago have been done) from Alexandria along the edge of the

---

\* This was the wretched word used at this time for "undrinkable". I am sorry.

cultivated zone and irrigation canals of the Delta. For this purpose the strongest concrete and sandbag works and pill-boxes should be built or completed from the sea to the cultivated zone and the main irrigation canal. The pipe-line forward of this line should be extended as fast as possible. The Delta zone is the most effective obstacle to tanks of all kinds, and can be lightly held by sandbag works to give protection to Egypt and form a very strong extended flank for the Alexandria front. A broad strip, four or five miles wide, should be inundated from the flood waters of the Nile, controlled at Assouan. Amid or behind this belt a series of strong posts armed with artillery should be constructed.

16. In this posture, then, the Army of the Delta will await the Italian invasion. It must be expected that the enemy will advance in great force, limited only, but severely, by the supply of water and petrol. He will certainly have strong armoured forces in his right hand to contain and drive back our weaker forces, unless these can be reinforced in time by the armoured regiment from Great Britain. He will mask, if he cannot storm, Mersa Matruh. But if the main line of the Delta is diligently fortified and resolutely held he will be forced to deploy an army whose supply of water, petrol, food, and ammunition will be difficult. Once the army is deployed and seriously engaged, the action against his communications, from Mersa Matruh, by bombardment from the sea, by descent at Sollum, or even much farther west, would be a deadly blow to him.

17. The campaign for the defence of the Delta therefore resolves itself into *strong defence with the left arm from Alexandria inland, and a reaching out with the right hand, using sea-power upon his communications.* At the same time it is hoped that the [our] reinforcements [acting] from Malta will hamper the sending of further reinforcements—Italian or German—from Europe into Africa.

18. All this might be put effectively in train by October 1, *provided we are allowed the time.* If not, we must do what we can. All trained or Regular units, whether fully equipped or not, must be used in defence of the Delta. All armed white men and also Indian or foreign units must be used for internal security. The Egyptian Army must be made to play its part in support of the Delta front, thus leaving only riotous crowds to be dealt with in Egypt proper.

Pray let the above be implemented, and be ready to discuss it in detail with me at 4.30 p.m., August 16.

With this General Wavell returned to Cairo in the third week of August.

\*　　\*　　\*　　\*　　\*

I now have to record a small but at the time vexatious military episode. The Italians, using vastly superior forces, drove us out of Somaliland. This story requires to be told.

Until December 1939 our policy in a war with Italy was to evacuate Somaliland; but in that month General Ironside, C.I.G.S., declared for defence of the territory, and in the last resort to hold Berbera. Defences were to be prepared to defend the Tug Argan Gap through the hills. One British battalion (the Black Watch), two Indian, and two East African battalions, with the Somaliland Camel Corps and one African light battery, with small detachments of anti-tank and anti-aircraft units, were gathered by the beginning of August. General Wavell on July 21 telegraphed to the War Office that withdrawal without fighting would be disastrous for our influence, and that Somaliland might be a valuable base for further offensive action. Fighting began during his visit to London, and he told the Middle East Ministerial Committee that although the strategic disadvantages of the loss of Somaliland would be slight it would be a blow to our prestige.

The Italians entered British Somaliland on August 3 with three battalions of Italian infantry, fourteen of colonial infantry, two groups of pack artillery, and detachments of medium tanks, light tanks, and armoured cars. These large forces advanced upon us on August 10, and a new British commander, General Godwin-Austen, arrived on the night of the 11th. In his instructions he had been told: "Your task is to prevent any Italian advance beyond the main position. . . . You will take the necessary steps for withdrawal if necessary." Fighting took place on the 12th and 13th, and one of our four key positions was captured from us after heavy artillery bombardment. On the night of the 15th General Godwin-Austen determined to withdraw. This, he said, "was the only course to save us from disastrous defeat and annihilation". The Middle East Headquarters authorised evacuation, and this was successfully achieved under a strong rearguard of the Black Watch.

I was very much disappointed with this affair, which remains on record as our only defeat at Italian hands. This in no way reflects upon the officers or men of the British and Somali troops in the Protectorate, who had to do their best with what equipment they were allotted and obey the orders they received.

There was much jubilation in Italy, and Mussolini exulted in the prospects of his attack on the Nile Valley. General Wavell however defended the local commander, affirming that the fighting had been severe.

In view of the great business we had together, I did not press my view further either with the War Office or with General Wavell.

\* \* \* \* \*

Our information at this time showed a rapid increase in the Italian forces in Albania and a consequent menace to Greece. As the German preparations for the invasion of Britain grew in scale and became more evident it would have been particularly inconvenient to lessen our bombing attack on the German and Dutch river-mouths and French ports, where barges were being collected. I had formed no decision in my own mind about moving bomber squadrons away from home. It is often wise however to have plans worked out in detail. Strange as it may seem, the Air Force, except in the air, is the least mobile of all the services. A squadron can reach its destination in a few hours, but its establishments, depots, fuel, spare parts, and workshops take many weeks, and even months, to develop.

*(Action this Day)*
*Prime Minister to C.A.S. and General Ismay* 28.VIII.40
Pray let me have proposals for moving at least four heavy bombing squadrons to Egypt in addition to anything now in progress. These squadrons will operate from advanced bases in Greece as far as may be convenient should Greece be forced into the war by Italy. They would refuel there before attacking Italy. Many of the finest targets, including the Italian Fleet, will be open to such attacks. It is better to operate from Greece, should she come in, than from Malta in its present undefended state. The report should be brief, and should simply show the method, the difficulties, and the objectives, together with a timetable. It is not necessary to argue the question of policy, which will be decided by the Defence Committee of the Cabinet. Making the best plan possible will not commit the Air Ministry or anyone else to the adoption of the plan, but every effort is to be made to solve its difficulties.

\* \* \* \* \*

I cannot better end this chapter than with the report I gave of the situation in August to the Prime Ministers of Australia and New Zealand. This followed up my message of June 16.

*Prime Minister to the Prime Ministers of Australia*
*and New Zealand*                                                  11.VIII.40

The combined Staffs are preparing a paper on the Pacific situation, but I venture to send you in advance a brief foreword. We are trying our best to avoid war with Japan, both by conceding on points where the Japanese military clique can perhaps force a rupture, and by standing up where the ground is less dangerous, as in arrests [by the Japanese] of individuals. I do not think myself that Japan will declare war unless Germany can make a successful invasion of Britain. Once Japan sees that Germany has either failed or dares not try I look for easier times in the Pacific. In adopting against the grain a yielding policy towards Japanese threats we have always in mind your interests and safety.

Should Japan nevertheless declare war on us her first objective outside the Yellow Sea would probably be the Dutch East Indies. Evidently the United States would not like this. What they would do we cannot tell. They give no undertaking of support, but their main fleet in the Pacific must be a grave preoccupation to the Japanese Admiralty. In this first phase of an Anglo-Japanese war we should of course defend Singapore, which if attacked—which is unlikely—ought to stand a long siege. We should also be able to base on Ceylon a battle-cruiser and a fast aircraft-carrier, which, with all the Australian and New Zealand cruisers and destroyers, which would return to you, would act as a very powerful deterrent upon the hostile raiding cruisers.

We are about to reinforce with more first-class units the Eastern Mediterranean Fleet. This fleet could of course at any time be sent through the Canal into the Indian Ocean, or to relieve Singapore. We do not want to do this, even if Japan declares war, until it is found to be vital to your safety. Such a transference would entail the complete loss of the Middle East, and all prospect of beating Italy in the Mediterranean would be gone. We must expect heavy attacks on Egypt in the near future, and the Eastern Mediterranean Fleet is needed to help in repelling them. If these attacks succeed the Eastern Fleet would have to leave the Mediterranean either through the Canal or by Gibraltar. In either case a large part of it would be available for your protection. We hope however to maintain ourselves in Egypt and to keep the Eastern Fleet at Alexandria during the first phase of an Anglo-Japanese war, should that occur. No one can lay down beforehand what is going to happen. We must just weigh events from day to day, and use our available resources to the utmost.

A final question arises: whether Japan, having declared war, would attempt to invade Australia or New Zealand with a considerable army. We think this very unlikely, first because Japan is absorbed in China, secondly, would be gathering rich prizes in the Dutch East Indies, and, thirdly, would fear very much to send an important part of her Fleet far to the southward, leaving the American Fleet between it and home. If however, contrary to prudence and self-interest, Japan set about invading Australia or New Zealand on a large scale, I have the explicit authority of the Cabinet to assure you that we should then cut our losses in the Mediterranean and sacrifice every interest, except only the defence and feeding of this Island, on which all depends, and would proceed in good time to your aid with a fleet able to give battle to any Japanese force which could be placed in Australian waters, and able to parry any invading force, or certainly cut its communications with Japan.

We hope however that events will take a different turn. By gaining time with Japan the present dangerous situation may be got over. We are vastly stronger here at home than when I cabled to you in May. We have a large army, now beginning to be well equipped. We have fortified our beaches. We have a strong reserve of mobile troops, including our Regular Army and Australian, New Zealand, and Canadian contingents, with several armoured divisions or brigades ready to strike in counter-attack at the head of any successful lodgment. We have ferried over from the United States their grand aid of nearly a thousand guns and six hundred thousand rifles, with ammunition complete. Relieved of the burden of defending France, our Army is becoming daily more powerful and munitions are gathering. Besides this, we have the Home Guard of 1,500,000 men, many of them war veterans, and most with rifles or other arms.

The Royal Air Force continues to show that same individual superiority over the enemy on which I counted so much in my cable to you of June 16. Yesterday's important action in the Channel showed that we could attack against odds of three to one, and inflict losses of three and a half to one. Astounding progress has been made by Lord Beaverbrook in output of the best machines. Our fighter and bomber strength is nearly double what it was when I cabled you, and we have a very large reserve of machines in hand. I do not think the German Air Force has the numbers or quality to overpower our air defences.

The Navy increases in strength each month, and we are now beginning to receive the immense programme started at the declaration of war. Between June and December 1940 over five hundred vessels, large and small, but many most important, will join the Fleet. The German Navy is weaker than it has ever been. *Scharnhorst* and *Gneisenau* are both in dock damaged, *Bismarck* has not yet done her trials, *Tirpitz* is three months behind *Bismarck*. There are available now in this critical fortnight, after which the time for invasion is getting very late, only one pocket-battleship, a couple of 8-inch-gun *Hippers*, two light cruisers, and perhaps a score of destroyers. To try to transport a large army, as would now be needed for success, across the seas virtually without escort in the face of our Navy and Air Force, only to meet our powerful military force on shore, still more to maintain such an army and nourish its lodgments with munitions and supplies, would be a very unreasonable act. On the other hand, if Hitler fails to invade and conquer Britain before the weather breaks he has received his first and probably fatal check.

We therefore feel a sober and growing conviction of our power to defend ourselves successfully, and to persevere through the year or two that may be necessary to gain victory.

# CHAPTER XXII

# THE MEDITERRANEAN PASSAGE

*The New Situation – France Out, Italy In – Admiral Cunningham at Alexandria – Successful Action off Calabria – Increasing Naval Burdens – Mediterranean Inhibitions – My Minute of July 12 and the First Sea Lord's Reply – My Minute of July 15 – Admiral Cunningham's View on Sending Reinforcements through the Mediterranean – The First Sea Lord's Minute of July 23 – The Plan for Operation "Hats" – My Efforts to Send the Tanks through the Mediterranean – My Minute of August 13 – Failure to Persuade the Admiralty – Conduct of Operation "Hats" – Successful Daring – My Telegram to Admiral Cunningham of September 8 – Hard Efforts to Succour Malta Air Defence – Admiral Somerville's Excursions – The Takoradi Route to Egypt Opened – A Vital Trickle – Malta Still in the Foreground.*

NTIL the French collapse the control of the Mediterranean had been shared between the British and French Fleets. At Gibraltar we had maintained a small force of cruisers and destroyers watching the Straits. In the Eastern Basin lay our Mediterranean Fleet, based on Alexandria. This had been reinforced earlier in the year, when the Italian attitude became menacing, to a force of four battleships, seven cruisers, twenty-two destroyers, one aircraft-carrier, and twelve submarines. The French Mediterranean Fleet comprised five capital ships, one aircraft-carrier, fourteen cruisers, and many smaller ships. Now France was out and Italy was in. The numerically powerful Italian Fleet included six battleships, including two of the latest type (*Littorios*), mounting 15-inch guns, but two of the older ships were being reconstructed and were not immediately ready for service. Besides this their Fleet comprised nineteen modern cruisers, seven of which were of the 8-inch-gun type, one hundred

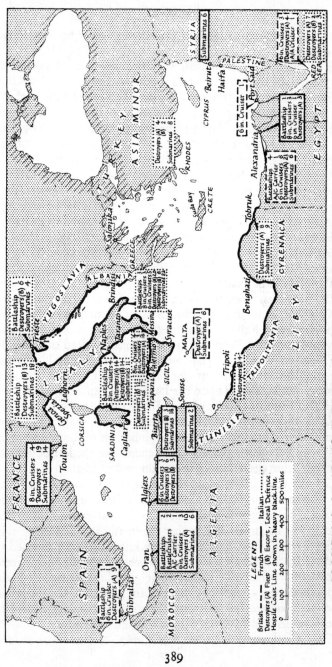

DISPOSITION OF MAIN FLEETS IN MEDITERRANEAN  June 14 1940

and twenty destroyers and torpedo-boats, and over a hundred submarines.

In addition, a strong Italian Air Force was ranged against us. So formidable did the situation appear at the end of June that Admiralty first thoughts contemplated the abandonment of the Eastern Mediterranean and concentration at Gibraltar. I resisted this policy, which, though justified on paper by the strength of the Italian Fleet, did not correspond to my impressions of the fighting values, and also seemed to spell the doom of Malta. It was resolved to fight it out at both ends. On July 3 the Chiefs of Staff prepared a paper about the Mediterranean in which they stressed the importance of the Middle East as a war theatre, but recognised that for the time being our policy must be generally defensive. The possibility of a German attack on Egypt must be taken seriously, but so long as the Fleet could be retained in the Eastern Mediterranean our existing forces were enough to deal with purely local attack.

We have seen how, at the end of June, Force H was constituted at Gibraltar under Admiral Somerville. It comprised the *Hood*, *Resolution*, and *Valiant*, the carrier *Ark Royal*, two cruisers, and eleven destroyers. With this we had done the deed at Oran. In the Eastern Mediterranean we found in Admiral Andrew Cunningham an officer of the highest qualities and dauntless courage. Immediately upon the Italian declaration of war he put to sea in search of the enemy. The Royal Air Force attacked Tobruk and sank the old Italian cruiser *San Giorgio*. The Fleet bombarded Bardia from the sea. Submarines on both sides were active, and we destroyed ten of the enemy for the loss of three of our own from deep mining before the end of June.

On July 8, whilst covering the passage of a convoy from Malta to Alexandria, Admiral Cunningham became aware of the presence of powerful Italian forces. It was evident from the intensity of Italian air attacks that the enemy also was engaged in an important operation, and we now know that they planned to lead the British Admiral into an area where he could be subjected to concentrated attack by the full weight of the Italian Air Force and submarines. Admiral Cunningham at once seized the initiative, and, despite his inferiority in numbers, boldly turned to interpose his fleet between the enemy and their base.

On the following day contact was made, and an action at long range ensued, in which one enemy battleship and two cruisers were hit without any damage being suffered by the British fleet. The enemy refused to stand and fight, and, thanks to superior speed, was able to escape, pursued by Admiral Cunningham to a point within twenty-five miles of the Italian mainland. Throughout this and the next two days the intense air attacks continued without success, and the convoy, though frequently bombed, reached Alexandria safely. This spirited action established the ascendancy of the British Fleet in the Mediterranean, and Italian prestige suffered a blow from which it never recovered. Ten days later the *Sydney*, an Australian cruiser, with a British destroyer flotilla, sank an Italian cruiser. Our first contacts with the new enemy were therefore by no means discouraging.

The burdens which lay upon the Admiralty at this time were however heavy in the extreme. The invasion danger required a high concentration of flotillas and small craft in the Channel and North Sea. The U-boats, which had by August begun to work from Biscayan ports, took severe toll of our Atlantic convoys without suffering many losses themselves. Until now the Italian Fleet had never been tested. The possibility of a Japanese declaration of war, with all that it would bring upon our Eastern Empire, could never be excluded from our thoughts. It is therefore not strange that the Admiralty viewed with the deepest anxiety all risking of warships in the Mediterranean, and were sorely tempted to adopt the strictest defensive at Gibraltar and Alexandria. I, on the other hand, did not see why the large numbers of ships assigned to the Mediterranean should not play an active part from the outset. Malta had to be reinforced both with air squadrons and troops. Although all commercial traffic was rightly suspended, and all large troop convoys to Egypt must go round the Cape, I could not bring myself to accept the absolute closure of the inland sea. Indeed I hoped that by running a few special convoys we might arrange and provoke a trial of strength with the Italian Fleet. I hoped that this might happen and Malta be properly garrisoned and equipped with aeroplanes and A.A. guns before the appearance, which I already dreaded, of the Germans in this theatre. All through the summer and autumn months I engaged in friendly though tense discussion with the Admiralty upon this part of our war effort.

*Prime Minister to First Lord and First Sea Lord*      12.VII.40

I thought that *Illustrious* might well go to the Mediterranean and exchange with *Ark Royal*. In this case *Illustrious* could take perhaps a good lot of Hurricanes to Malta. As we have a number of Hurricanes surplus at the moment, could not the Malta Gladiator pilots fly the Hurricanes themselves? This would not diminish our flying strength in this country.

The operation against Luleå [in the Baltic] has become less important now that the Germans have control of all the French and Belgian orefields. We must look to the Mediterranean for action.

You were going to let me have your plan for exchanging destroyers of more endurance with the Mediterranean flotilla. Could I have this, with dates?

To this Admiral Pound replied through the First Lord the same day:

We have now gained experience of the air conditions in the Western Mediterranean, and as soon as the present operation on which the Eastern Fleet is employed is completed we shall know pretty well what we are faced with in the Eastern Mediterranean.

There is no doubt that both Force H and the Eastern Mediterranean Fleet work under a grave disadvantage, inasmuch as it is not possible to give them fighter protection, as we do in the North Sea when ships are in the bombing area.

At the moment we are faced with the immediate problem of getting aircraft and A.A. guns to Malta, and aircraft to Alexandria. I am not at all certain that the risk of passing a ship with all these available stores through the Mediterranean is not too great, and that it might not be better to accept the delay of sending her round the Cape.

There is also the question of *Illustrious* to be considered, but this need not be settled immediately, as she must first come home to embark a full complement of Fulmar fighters.

Arrangements are being made to replace some of the destroyers at Gibraltar by others with longer endurance, but the date on which they leave will probably be dependent on the escorting of the ship I have referred to above to Gibraltar.

*Prime Minister to the First Sea Lord*      15.VII.40

1. It is now three weeks since I vetoed the proposal to evacuate the Eastern Mediterranean and bring Admiral Cunningham's fleet to Gibraltar. I hope there will be no return to that project. Anyone can see the risk from air attack which we run in the Central Mediterranean. From time to time and for sufficient objects this risk will have to be

faced. Warships are meant to go under fire. Our position would be very different if I had been assisted in my wish in October of last year to reconstruct the *Royal Sovereign* class with heavy anti-aircraft armour on their decks at a cost to their speed through increased bulging. The difficulties which were presented at every stage were such as to destroy this proposal, and we are no further on than we were a year ago. If we had the *Royal Sovereigns* armoured, and their guns cocked up, or some of them, we could assault the Italian coasts by bombardment with comparative impunity.* The various Boards of Admiralty which preceded this war altogether underrated the danger of air attack, and authorised sweeping statements to Parliament on the ability of ships of war to cope with it. Now there is a tendency to proceed to the other extreme, and consider it wrong to endanger His Majesty's ships by bringing them under air bombardment, as must from time to time be necessary in pursuance of operations. . . .

It may be taken for certain that the scale of the enemy's air attack will increase in the Mediterranean as the Germans come there.

2. It becomes of high and immediate importance to build up a very strong anti-aircraft defence at Malta, and to base several squadrons of our best fighter aircraft there. This will have to be done under the fire of the enemy. I should be glad to know the full scale of defence which was proposed in various papers I have seen. The emplacements should be made forthwith. I understand that a small consignment of A.A. guns and Hurricanes is now being procured, and that the main equipment is to follow later. It may well be possible at the end of this month to detach the larger consignment from our home defence. The urgent first consignment should reach Malta at the earliest moment. The stores may be divided between several ships, so as to avoid losing all if one is hit. The immense delay involved in passing these ships round the Cape cannot be accepted. So far as Malta is concerned, it is not seen how the dangers will be avoided by this *détour*, the voyage from Alexandria to Malta being, if anything, more dangerous than the voyage from Gibraltar to Malta.

3. *Illustrious.* Considering that in the North Sea and Atlantic we are on the defensive and that no one would propose to bring *Illustrious* into the narrow waters north and south of Dover, where we have already good shore-based aircraft, our aircraft-carriers in home waters will be able to operate some distance from the enemy's coast. In the Mediterranean, on the other hand, we must take the offensive against Italy, and endeavour especially to make Malta once again a Fleet base for special occasions. *Illustrious*, with her armoured deck, would seem to be better placed in the Mediterranean, and the *Ark Royal* in the

*The subject is discussed in Vol. I, Chapter XXV.

home theatre. The delays in bringing *Illustrious* into service have been very great, and I should be glad to know when the Fulmars [fast fighter aircraft] will be embarked and she be ready to exchange with *Ark Royal*.

4. I am very glad that arrangements will be made to send out destroyers of longer radius to Gibraltar, and to bring home the short-radius vessels to the Narrow Seas.

\* \* \* \* \*

Meanwhile Admiralty policy had again been most carefully scrutinised, and on July 15 the intention to maintain a strong force in the Eastern Mediterranean was reiterated in a signal to the Commander-in-Chief. This message stated that in the east the chief British task would be to destroy the enemy naval forces, although they had a numerical preponderance. In the west Force H would control the western exit from the Mediterranean and undertake offensive operations against the coast of Italy. I was in general agreement with this strong policy. The Commander-in-Chief was invited to say what heavy ships he considered necessary for the two forces, and if redistribution was thought desirable to advise whether the exchange should take place through the Mediterranean or round the Cape.

In reply he asked that both the *Valiant* and the *Barham* should join him. This would give him four battleships with the best available gun-range and speed. He could then dispense with the *Royal Sovereign*, as with her poor deck protection and inferior speed she was a constant source of anxiety. Furthermore he required two carriers, including the *Illustrious*, and two 8-inch-gun cruisers. He agreed with the First Sea Lord that in the Western Mediterranean a force comprising the *Hood* and the *Ark Royal* with either one or two "R." class battleships would meet requirements. With these forces he considered that the Mediterranean could be dominated and the Eastern Basin held indefinitely provided that Malta was adequately protected by fighters and that his resources at Alexandria were built up. In conclusion he said: "By carrying out a concerted movement it should be possible to pass reinforcements through the Mediterranean, but it would probably be desirable to do it all in one operation."

We thus reached a considerable measure of agreement in our talks at the Admiralty. It was common ground between us that Admiral Cunningham's fleet should be reinforced by a battleship, an aircraft-carrier, and two cruisers, and at the same time the opportunity should be taken to run a convoy of store ships to Malta *from Alexandria*. Thereafter on July 23 the First Sea Lord in the course of a minute to the First Lord and me said:

Full consideration has been given as to whether it is possible to pass through the Mediterranean not only the additional fighting ships which are being added to the Eastern Mediterranean force, but also merchant vessels containing spare ammunition for the Fleet, high-angle guns for Malta, and aircraft for Malta and the Middle East. The Commander-in-Chief is definitely of the opinion that under existing conditions it will be unsound to attempt to pass through the Central Mediterranean merchant vessels containing valuable cargoes, as if one or more ships were damaged in such a way as to reduce their speed it would be necessary to scuttle them. I am entirely in agreement with the Commander-in-Chief.

Thus it came about that the plan for the important operation which followed under the code name "Hats" did not include provision for the passage of merchant ships. None the less, with the full support of Admiral Cunningham it challenged the Italian Fleet and Air Force in the Central Mediterranean. I was now well content with the main decisions which the Admiralty were taking, and hoped that they might lead to a trial of strength. All preparations therefore went forward.

A few weeks later the bold and far-reaching step of the War Cabinet, with the full agreement of the Chiefs of Staff, to send nearly half our best available tanks to Egypt in spite of the invasion menace raised the question of the Mediterranean passage again and in a sharper way. I was of course in full accord with sending the tanks, but I feared that if they went round the Cape they might arrive too late for the battle on which the fate of Egypt depended. The First Sea Lord was at first inclined to run the risk, but on further study he thought it would complicate Operation "Hats", which now held the centre of the Admiralty stage. It involved sending at least two fast mechanical transport ships (16 knots) from Gibraltar to Malta, and this was regarded as more dangerous than sending them by the route from Alexandria. This led to further discussion.

*Prime Minister to General Ismay, for C.O.S. Committee*        11.VIII.40
I cannot accept this proposal [*i.e.*, to use the Cape route for sending

the tanks to Egypt], which deprives us of invaluable resources [fifty Infantry tanks or "I" tanks] during a most critical period, without making them available for the Middle East at the moment when they are most needed there. I must ask the Admiralty to make further proposals and overcome the difficulties. If necessary, could not the personnel be distributed among the destroyers, a larger force of destroyers being sent through from Force H to the Eastern Mediterranean, and returned thereafter in the same way as the six destroyers are now being sent westward by Admiral Cunningham?

There is no objection to the 3rd Hussars (the personnel of a tank regiment) going by the Cape, as General Wavell can make temporary arrangements for manning [the tanks] in the meanwhile, so long as he gets their light tanks. I am prepared to risk the fifty Infantry tanks in the Mediterranean, provided their personnel is distributed among H.M. ships; but there can be no question of them or their personnel going by the Cape, thus making sure they are out of everything for two months. The personnel sent through the Mediterranean must be cut down to essentials, the balance going round.

Pray let me have further proposals by to-morrow (Monday).

*Prime Minister to First Lord and First Sea Lord*          13.VIII.40

1. Just before the French went out of the war Admiral Darlan bombarded Genoa in full daylight without any Asdic destroyer protection or any aircraft protection and returned to Toulon unscathed. The Eastern Mediterranean Fleet has three times advanced to the centre of the Mediterranean and returned to Alexandria with only one ship— *Gloucester*—hit by one bomb. A few weeks ago a fast and a slow convoy were conducted uninjured from Malta to Alexandria—two days of their voyage being beset by Italian aircraft.

2. The Admiralty now propose to send six destroyers from Alexandria to meet Force H. These destroyers, which will certainly be detected from the air, will be within air-attacking distance of the very numerous, fast Italian cruiser forces in their home bases. This movement should be rightly condemned as hazardous in the extreme but for the just estimation in which Italian naval enterprise is held by C.-in-C. Mediterranean and the Admiralty.

3. We are now told that it is too dangerous for the powerful forces we shall have in motion in the near future to carry through to the Eastern Mediterranean two M.T. [mechanical transport] ships steaming in company at only 15 knots. Yet at the same time we are asked to spend vast sums fortifying a large part of the western coasts of Britain against what the Admiralty declare is a possible invasion by twelve thousand men embarked and shipped [from the river] Gironde [or from] St. Nazaire, who are to be sent to their destination

without any warship protection of any kind. If it is held to be a feasible operation to move twelve thousand men unescorted on to the Irish or British western coasts in the face of the full British sea-power, can this be reconciled with the standard of danger-values now adopted in the Mediterranean?

4. No one can see where or when the main attack on Egypt will develop. It seems however extremely likely that if the Germans are frustrated in an invasion of Great Britain or do not choose to attempt it they will have great need to press and aid the Italians to the attack of Egypt. The month of September must be regarded as critical in the extreme.

5. In these circumstances it is very wrong that we should attempt to send our armoured brigade round the Cape, thus making sure that during September it can play no part either in the defence of England or Egypt.

6. I request that the operation of passing at least two M.T. ships through with the Eastern reinforcements may be re-examined. The personnel can be distributed in the warships, and it is a lesser risk, from the point of view of the general war, to pass the M.T. ships through the Mediterranean than to have the whole armoured brigade certainly out of action going round the Cape. So long as the personnel are properly distributed among the warships, I am prepared to take the full responsibility for the possible loss of the armoured vehicles.

I was not able to induce the Admiralty to send the armoured brigade, or at the least their vehicles, through the Mediterranean. I was both grieved and vexed at this. Though my friendship for Admiral Pound and confidence in his judgment were never affected, sharp argument was maintained. The professional responsibility was his, and no naval officer with whom I ever worked would run more risks than he. We had gone through a lot together. If he would not do it, no one else would. If I could not make him, no one else could. I knew the Admiralty too well to press them or my great friend and comrade, Pound, or the First Lord, for whom I had high esteem, beyond a certain point. My relations with the Admiralty were too good to be imperilled by a formal appeal to the Cabinet against them.

When on August 15 I brought the question before the Cabinet finally I said that I had hoped to persuade the Admiralty to fit the two armoured regiments into Operation "Hats". If the tank units proceeded through the Mediterranean they would arrive in Alexandria about September 5; if by the Cape about three weeks later. However, the Chief of the Imperial General Staff did not consider that an attack in force by the Italians was imminent, and this was also General Wavell's view. Having done my utmost in favour of the short cut, I thought that the War Cabinet ought not to take the responsibility of overruling the judgment of the commanders, and I acquiesced in the longer voyage round the Cape with regret. The Chiefs of Staff however prepared an alternative plan for the short cut should the position in the Middle East suddenly worsen before Operation "Hats" was actually launched. Two fast M.T. ships carrying cruiser and "I" tanks would accompany the naval forces through the Mediterranean. The decision was to be taken before the reinforcements passed Gibraltar. In the event reports received from the Middle East were not considered to justify putting the alternative plan into force, and the whole convoy continued on its way round the Cape.

Operation "Hats" was carried out successfully and without loss between August 30 and September 5. Admiral Cunningham left Alexandria on August 30, and on the evening of the 31st his aircraft reported the approach of an enemy force of two battleships and seven cruisers. Hopes of an engagement were raised, but evidently the Italians were not seeking trouble and nothing happened. The following evening our aircraft again made contact with the enemy, who were now retiring to Taranto. Thereafter Admiral Cunningham's ships moved about with complete freedom to the east and south of Malta and were not seriously molested from the air. The convoy reached Malta safely, only one ship being damaged by air attack. Meanwhile the reinforcements, consisting of the *Valiant* but not her unreconstructed sister-ship the *Barham*, the aircraft-carrier *Illustrious*, and two antiaircraft cruisers, accompanied by Admiral Somerville with Force H, were approaching from Gibraltar. The *Valiant* and the cruisers had no difficulty in landing much-needed guns and ammunition in Malta, and then joined Admiral Cunningham to the eastward on September 3. During the return passage to Alexandria the fleet attacked Rhodes and Scarpanto and easily repulsed an E-boat attack. Admiral Somerville's force returned to Gibraltar without being molested in any way.

All this convinced me that it would have been a fair risk, especially compared with those we were resolutely running in seriously depleting our armour at home in the teeth of the enemy's invasion preparations, to transport the armoured brigade through the Malta channel, and that it would now be in Egypt, instead of more than three weeks away. No serious disaster did in fact occur in Egypt during those three weeks. Nevertheless an exaggerated fear of Italian aircraft had been allowed to hamper naval operations. I thought, and think, the event proved my case. Towards the end of November Admiral Somerville with Force H did in fact successfully escort a convoy to Malta from the westward, and on the way fought a partial action near Sardinia with that part of the Italian Fleet which had escaped damage at Taranto. One ship of this convoy passed on to Alexandria, together with three more store ships from Malta, escorted by further fleet reinforcements for the Eastern Mediterranean. This was the first time that a merchant ship made the complete passage of the Mediterranean after the Italian entry into the war. The reader will see in the next volume how a still more hazardous exploit was performed by the Navy in sending tanks to Egypt in 1941, *when the German Air Force was fully established in Sicily.*

*Prime Minister to First Lord*                           7.IX.40

1. The course of Operation "Hats" makes me quite sure that it was wrong to recede from the idea of passing the armoured vehicles through the Mediterranean. If you will read my minute reciting all the reasons why this course should be adopted you will see that they are reinforced by new facts now. . . .

*Prime Minister to First Lord*                           7.IX.40

I should be glad if you would let me have a short *résumé* of the different occasions when I pressed, as First Lord, for the preparation of the *Ramillies* class ships to withstand air bombardment by thick deck armour and larger bulges. If those ships had been put in hand when I repeatedly pressed for them to be, we should now have the means of attacking the Italian shores, which might be productive of the highest political and military results. Even now there is a disposition to delay taking this most necessary step, and no substitute is offered.

I have not yet heard from you in reply to the minute I sent you renewing this project of reconstruction in the hope that we may not be equally destitute of bombarding vessels next year. I shall be glad to have a talk with you on this subject when I have refreshed my mind with the papers.

This issue could never at any moment be decided without

balancing other bitter needs in new construction. It was on this rock, and not on differences of principle, that my wishes finally foundered.

*Prime Minister to General Ismay*                                    8.IX.40

Following for Sir Andrew Cunningham, C.-in-C. Eastern Mediterranean, from Prime Minister and Minister of Defence:

I congratulate you on the success of the recent operation in the Eastern and Central Mediterranean, and upon the accession to your fleet of two of our finest units, with other valuable vessels. I am sorry however that the armoured brigade which is so necessary to the defence of Egypt and Alexandria is still separated by more than three weeks from its scene of action. I hope you will find it possible to review the naval situation in the light of the experience gained during "Hats" and the arrival of *Illustrious* and *Valiant*. Not only the paper strength of the Italian Navy, but also the degree of resistance which they may be inclined to offer, should be measured. It is of high importance to strike at the Italians this autumn, because as time passes the Germans will be more likely to lay strong hands upon the Italian war machine, and then the picture will be very different. We intend to strengthen the anti-aircraft defences of Malta by every possible means, and some novel weapons of which I have high hopes will shortly be sent there for experiment. I trust that Malta may become safe for temporary visits of the Fleet at an earlier date than April 1941. If in the meanwhile you have any proposals for offensive action to make, they should be transmitted to the Admiralty. I shall be glad if you will also concert with the Army and Air Force plans for an operation against the Italian communications in Libya, which at the right time could be used to hamper any large-scale offensive against Egypt. The advantages of gaining the initiative are obviously very great. I hope the Fulmars [the fast fighter planes, which had at last reached our aircraft-carriers] have made a good impression. The battle here for air mastery continues to be severe, but firm confidence is felt in its eventual outcome.

It is surprising that the violent impact of the air upon our control of the Mediterranean had not been more plainly foreseen by the British Government before the war and by their expert advisers. In any case however we had fallen so far behind in the air race with Germany that the defence of Britain made an overwhelming demand on the already outnumbered forces we possessed. Until the Battle of Britain had been decisively won every reinforcement of aircraft to the Mediterranean and Egypt had been an act of acute responsibility. Even in the winter months, when we felt we were masters of our own daylight air at home, it was very hard under the full fury of the Blitz to send away fighter aircraft either to Malta or to Egypt. It was also most painful to take from bombarded British cities and vital seaports and munitions factories the anti-aircraft guns and shells sorely needed for their protection, and to send these either all round the Cape to Egypt or at much peril direct to Malta.

The reinforcement of Malta's hitherto neglected air defences was pressed forward in spite of losses and disappointments. Among the tasks of Admiral Somerville's force at Gibraltar was the convoying of fighter aircraft in a carrier to within flying distance of Malta. The first of these efforts was made in the beginning of August, when twelve Hurricanes were flown into the island from the aircraft-carrier *Argus*. Until their arrival the air defence of Malta consisted of three Gladiators, known locally by the affectionate names of "Faith", "Hope", and "Charity". We made a second attempt in November; but there was a tragedy. Nine aircraft out of fourteen, which had been launched from the *Argus*, four hundred miles to the westward of the island, ran out of fuel on the way through a change of wind, and perished at sea with their devoted pilots. Never again were the margins cut so fine, and though many similar operations took place in the future never did such a catastrophe recur.

★   ★   ★   ★   ★

It had also become necessary to find a way of sending aircraft to the Middle East which would avoid both the dangers of the Mediterranean and the fearful delay around the Cape. An overland route from West Africa would save many vital days and some shipping. The machines had either to be flown ashore from an aircraft-carrier, or dismantled and crated for the voyage and then reassembled at some port for their flight. The choice lay between Lagos and Takoradi.

After careful examination Takoradi was chosen, and as early as August 21, 1940, an operating party arrived. The course lay by Kano to Khartoum and eventually to Cairo, a total distance of three thousand seven hundred miles. Considerable workshops and accommodation had to be built at Takoradi, and various refuelling and rest stations provided along the route. A dozen crated Hurricanes and Blenheims arrived by sea on September 5.

followed next day by thirty Hurricanes landed from the carrier *Argus*. The first delivery flight left Takoradi on September 20, and arrived at Khartoum four days later. By the end of the year a trickle of one hundred and seven planes had reached Egypt in this way.

Although a quick start had been made, many months' work was needed before the route was organised. The climate at Takoradi and the local malaria harassed the men erecting the crated aircraft. The use of the carriers was limited by other clamant needs. Weather hampered the air convoys. The number of aircraft unserviceable awaiting spares along the route piled up. The heavy wear on engines in their flight over vast barren sandy spaces reduced their fighting life. Terrible teething troubles had to be overcome. None of this aircraft supply was effective in 1940. But if we had not begun in good time the Army of the Nile and all its ventures could not have lived through the tragic events of 1941.

*     *     *     *     *

By the close of 1940 the British Navy had once more firmly established itself in the Mediterranean. The defences of Malta had been considerably strengthened by Admiral Somerville's excursion to carry in A.A. and other equipment. Admiral Cunningham's offensive policy in the Eastern Basin had also yielded excellent results. Everywhere, despite the Italian air strength, we held the initiative, and Malta remained in the foreground of events as an advanced base for offensive operations against the Italian communications with their forces in Africa.

# CHAPTER XXIII

# SEPTEMBER TENSIONS

*Climax of the Air Battle - Intense Strain upon the Fighter Pilots - Evidences of the Impending Invasion - Disappointing Bombing on Concentrations of Barges - Britain Braced - Munitions Policy - My General Directive - A Survey of 1941 Requirements in Material - An Eight Months' Programme - My October Note on Priorities - Laggards - Climax at Home and in Egypt - The Perils of Fog - Need for De Wilde Ammunition - Achievements of the Ministry of Aircraft Production - Policy of Creating Commandos Enforced - Advance of Marshal Graziani's Army, September 13 - Their Halt at Sidi Barrani - Parlous Conditions at Malta - Troubles that Never Happened.*

$S$EPTEMBER, like June, was a month of extreme opposing stresses for those who bore the responsibility for British war direction. The air battle, already described, on which all depended, raged with its greatest fury and rose steadily to its climax. The victory of the Royal Air Force on September 15 is seen now in retrospect to have marked its decisive turning-point. But this was not apparent at the time, nor could we tell whether even heavier attacks were not to be expected or how long they would go on. The fine weather facilitated daylight fighting on the largest scale. Hitherto we had welcomed this, but when I visited Air Vice-Marshal Park at No. 11 Group in the third week of September I noticed a slight but definite change in outlook. I asked about the weather, and was told it was set fair for some days to come. This however did not seem to be as popular a prospect as it had been at the beginning of the month. I had the distinct feeling that a break in the weather would no longer be regarded as a misfortune.

It happened while I was there in Park's room with several officers that an officer brought in a notification from the Air Ministry that all supplies of De Wilde ammunition were exhausted. This was the favourite of the fighter pilots. The factory on which it depended had been bombed. I saw that this hit Park hard; but after a gulp and a pause he replied magnificently: "We fought them without it before, and we can fight them without it again."

In my talks with Air Chief Marshal Dowding, who usually motored over from Uxbridge to Chequers during the week-ends, the sense of Fighter Command being at its utmost strain was evident. The weekly figures over which I pored showed we had adequate numbers, provided the weight of the hostile attack did not increase. But the physical and mental stresses upon the pilots were not reflected on the paper charts. For all their sublime devotion, often facing odds of five and six to one, for all the sense of superiority which their continued success and the enemy's heavy losses created, there are limits to human endurance. There is such a thing as sheer exhaustion, both of the spirit and the animal. I thought of Wellington's mood in the afternoon of the Battle of Waterloo: "Would God that night or Blücher would come". This time we did not want Blücher.

Meanwhile all the evidences of impending German invasion multiplied. Upwards of three thousand self-propelled barges were counted on our air photographs in the Dutch, Belgian, and French ports and river-mouths. We could not tell exactly what reserves of larger vessels might not be gathered in the Rhine estuary, or in the Baltic, from which the Kiel Canal was still open. In my examination of the invasion problem I have set forth the reasoning on which I based my confidence that we should beat them if they came, and consequently that they would not come, and continued to contemplate the issue with a steady gaze. All the same it was impossible to watch these growing preparations, week after week, in the photographs and reports of agents, without a sense of awe. A thing like this gets hold of you bit by bit. The terrible enemy would not come unless he had solid assurance of victory and plans made with German thoroughness. Might there not also be surprises? Might there be tank landing-craft or some clever improvisation of them? What else might there not be? All our night-bombing was concentrated on the invasion ports, where every night German rehearsal exercises of marching on and off the barges and other vessels seemed to be taking place. The results of our bombing of the masses of barges which crowded the basins or lay along the quays, judged by the photographs, had several times disappointed me.

*Prime Minister to Secretary of State for Air*          **23.IX.40**

What struck me about these photographs was the apparent inability of the bombers to hit these very large masses of barges. I should have thought that sticks of explosive bombs thrown along these oblongs would have wrought havoc, and it is very disappointing to see that they all remained intact and in order, with just a few apparently damaged at the entrance.

Can nothing be done to improve matters?

As already mentioned, the Chiefs of Staff were on the whole of the opinion that invasion was imminent, while I was sceptical and expressed a contrary view. Nevertheless, it was impossible to quell that inward excitement which comes from the prolonged balancing of terrible things. Certainly we strained every nerve to be ready. Nothing was neglected that could be achieved by the care and ingenuity of our commanders, the vigilance of our now large and formidable armies, and the unquenchable and fearless spirit of our whole people.

\*　\*　\*　\*　\*

The whole of our war production and its priorities now required to be reviewed in the light of our exclusion from the Continent. In this I worked in consultation with the Minister of Supply and others concerned. At the beginning of this month, after much labour in my small circle, and careful checking, I prepared for the Cabinet a general directive upon munitions, which was intended to govern our affairs in 1941.

## THE MUNITIONS SITUATION

MEMORANDUM BY THE PRIME MINISTER

*September* 3, 1940

1. The Navy can lose us the war, but only the Air Force can win it. Therefore our supreme effort must be to gain overwhelming mastery in the air. The Fighters are our salvation, but the Bombers alone provide the means of victory. We must therefore develop

405

the power to carry an ever-increasing volume of explosives to Germany, so as to pulverise the entire industry and scientific structure on which the war effort and economic life of the enemy depend, while holding him at arm's length from our Island. In no other way at present visible can we hope to overcome the immense military power of Germany, and to nullify the further German victories which may be apprehended as the weight of their force is brought to bear upon African or Oriental theatres. The Air Force and its action on the largest scale must therefore, subject to what is said later, claim the first place over the Navy or the Army.

2. The weapon of blockade has become blunted, and rendered, as far as Germany is concerned, less effectual, on account of their land conquests and power to rob captive or intimidated peoples for their own benefit. There remain no very important special commodities the denial of which will hamper their war effort. The Navy is at present somewhat pressed in its task of keeping open the communications, but as this condition is removed by new Admiralty measures, by the arrival of the American destroyers, and by the increasing output of anti-U-boat craft from our own yards, we may expect a marked improvement. It is of the utmost importance that the Admiralty should direct their attention to aggressive schemes of war, and to the bombardment of enemy or enemy-held coasts, particularly in the Mediterranean. The production of anti-U-boat craft must proceed at the maximum until further orders, each slip being filled as it is vacated. The Naval Programme does not impinge markedly upon the Air, and should cede some of its armour-plate to tank production.

3. The decision to raise the Army to a strength of fifty-five divisions as rapidly as possible does not seem to require any reconsideration. Within this, we should aim at ten armoured divisions, five by the spring, seven by the summer, and ten by the end of 1941. The execution of these programmes of armament supply will tax our munitions factories to the full. I agree in principle with the proposals of the Minister of Supply [Mr. Herbert Morrison] for handling the ammunition supply problem, and also that firings on the 1917-18 scale are not to be expected in the present war.

4. Intense efforts must be made to complete the equipment of our Army at home and of our Army in the Middle East. The most serious weak points are tanks and small-arms ammunition, particularly the special types; anti-tank guns and rifles, and even more their ammunition; trench mortars, and still more their ammunition; and rifles. We hope to obtain an additional 250,000 rifles from the United States, but it is lamentable that we should be told that no more than half a million additional rifles can be manufactured here before the

end of 1941. Surely, as large numbers of our Regular Army proceed abroad the need of the Home Guard and of garrison troops for home defence on a far larger scale than at present will be felt. A substantial increase in rifle-making capacity is necessary.

5. The danger of invasion will not disappear with the coming of winter, and may confront us with novel possibilities in the coming year. The enemy's need to strike down this country will naturally increase as the war progresses, and all kinds of appliances for crossing the seas that do not now exist may be devised. Actual invasion must be regarded as perpetually threatened, but unlikely to materialise as long as strong forces stand in this Island. Apart from this, the only major theatre of war which can be foreseen in 1940-41 is the Middle East. Here we must endeavour to bring into action British, Austral-asian, and Indian forces, on a scale which should only be limited by sea transport and local maintenance. We must expect to fight in Egypt and the Soudan, in Turkey, Syria, or Palestine, and possibly in Iraq and Persia. Fifteen British divisions, six Australasian, and at least six Indian divisions should be prepared for these theatres, these forces not being, however, additional to the fifty-five divisions which have been mentioned. One would not imagine that the ammunition expenditure would approach the last-war scale. Air-power and mechanised troops will be the dominant factors.

6. There remain the possibilities of amphibious aggressive warfare against the enemy or enemy-held territory in Europe or North Africa. But the needs of such operations will be provided by the arms and supplies already mentioned in general terms.

7. Our task, as the Minister of Supply rightly reminds us, is indeed formidable when the gigantic scale of German military and aviation equipment is considered. This war is not however a war of masses of men hurling masses of shells at each other. It is by devising new weapons, and above all by scientific leadership, that we shall best cope with the enemy's superior strength. If, for instance, the series of inventions now being developed to find and hit enemy aircraft, both from the air and from the ground, irrespective of visibility, realise what is hoped from them, not only the strategic but the munitions situation would be profoundly altered. And if the U.P. [Unrotated Projectile] weapon can be provided with ammunition, predictors, and other aids which realise an accuracy of hitting three or four times as great as that which now exists, the ground will have taken a long step towards the re-conquest of the air. The Navy will regain much of its old freedom of movement and power to take offensive action. And the Army will be able to land at many points without the risk of being "Namsossed".*

---

* Defenceless from air attack, as at Namsos.

We must therefore regard the whole sphere of R.D.F. [Radar], with its many refinements and measureless possibilities, as ranking in priority with the Air Force, of which it is in fact an essential part. The multiplication of the high-class scientific personnel, as well as the training of those who will handle the new weapons and research work connected with them, should be the very spear-point of our thought and effort. Very great reliefs may be expected in anti-aircraft guns and ammunition, although it is at present too soon to alter present plans.

8. Apart from a large-scale invasion, which is unlikely, there is no prospect of any large expenditure or wastage of military munitions before the spring of 1941. Although heavy and decisive fighting may develop at any time in the Middle East, the difficulties of transport, both of reinforcements and of supplies, will restrict numbers and expenditure. We have therefore before us, if not interrupted, a period of eight months in which to make an enormous improvement in our output of warlike equipment, and in which steady and rapid accumulations may be hoped for. It is upon this purpose that all our resources of credit, materials, and above all of skilled labour, must be bent.

This policy was generally accepted by my colleagues, and the action of all departments conformed to it.

\* \* \* \* \*

I found it necessary in October to add a further note about priorities, which were a source of fierce contention between the different departments, each striving to do their utmost.

## PRIORITIES

### NOTE BY THE PRIME MINISTER

*October 15, 1940*

1. The very highest priority in personnel and material should be assigned to what may be called the Radio sphere. This demands scientists, wireless experts, and many classes of highly-skilled labour and high-grade material. On the progress made much of the winning of the war and our future strategy, especially naval, depends. We must impart a far greater accuracy to the A.A. guns, and a far better protection to our warships and harbours. Not only research and experiments, but production, must be pushed hopefully forward from many directions, and after repeated disappointments we shall achieve success.

2. The 1A priority must remain with aircraft production, for the purpose of executing approved target programmes. It must be an obligation upon them to contrive by every conceivable means not to let this priority be abused and needlessly hamper other vital depart-

ments. For this purpose they should specify their requirements in labour and material beforehand quarter by quarter, or, if practicable, month by month, and make all surplus available for others immediately. The priority is not to be exercised in the sense that aircraft production is completely to monopolise the supplies of any limited commodity. Where the condition prevails that the approved M.A.P. demands absorb the total supply, a special allocation must be made, even at prejudice to aircraft production, to provide the minimum essential needs of other departments or branches. This allocation, if not agreed, will be decided on the Cabinet level.

3. At present we are aiming at five armoured divisions, and armoured brigades equivalent to three more. This is not enough. We cannot hope to compete with the enemy in numbers of men, and must therefore rely upon an exceptional proportion of armoured fighting vehicles. Ten armoured divisions is the target to aim for to the end of 1941. For this purpose the Army must searchingly review their demands for mechanised transport, and large purchases of M.T. must be made in the United States. The home Army, working in this small Island with highly-developed communications of all kinds, cannot enjoy the same scale of transport which divisions on foreign service require. Improvisation and makeshift must be their guides. A Staff officer renders no service to the country who aims at ideal standards, and thereafter simply adds and multiplies until impossible totals are reached. A report should be furnished of Mechanical Transport, 1st, 2nd, and 3rd line, of British divisions—

    (a)  for foreign service,
    (b)  for home service,
    (c)  for troops on the beaches.

Any attempt to make heavy weather out of this problem is a failure to aid us in our need.

Wherever possible in England, horse transport should be used to supplement M.T. We improvidently sold a great many of our horses to the Germans, but there are still a good many in Ireland.

4. Special aid and occasional temporary priorities must be given to the Laggard elements. Among these stand out the following:

    (a)  Rifles.
    (b)  Small arms ammunition—above all, the special types.

Intense efforts must be made to bring the new factories into production. The fact that scarcely any improvement is now expected until the end of the year—i.e., sixteen months after the outbreak of war—is grave. Twelve months should suffice for a cartridge factory. We have been mercifully spared from the worst consequences of this failure through the armies not being in action as was anticipated.

Trench mortar ammunition and A.T. gun ammunition are also in a shocking plight, and must be helped.

All these Laggards must be the subject of weekly reports to the Production Council and to me.

5. The Navy must exercise its existing priorities in respect of small craft and anti-U-boat building. This applies also to merchant ship-building, and to craft for landing operations. Delay must be accepted upon all larger vessels that cannot finish in 1941. Plans must be made to go forward with all processes and parts which do not clash with prior needs. The utmost possible steel and armour-plate must be ordered in America.

By the middle of September the invasion menace seemed sufficiently glaring to arrest further movement of vital units to the East, especially as they had to go round the Cape. After a visit to the Dover sector, where the electric atmosphere was compulsive, I suspended for a few weeks the dispatch of the New Zealanders and the remaining two tank battalions to the Middle East. At the same time I kept our three fast transports, "the Glen [Line] ships" as they were called, in hand for an emergency dash through the Mediterranean.

*Prime Minister to General Ismay, for Chiefs of Staff Committee*                                      17.IX.40

In all the circumstances it would be impossible to withdraw the New Zealand Brigade from their forward position on the Dover promontory. The two cruiser-tank battalions cannot go. Would it not be better to keep the Australians back and delay the whole convoy until the third week in October? After all, none of these forces going round the Cape can possibly arrive in time to influence the impending battle in Egypt. But they may play a big part here. Perhaps by the third week in October the Admiralty will be prepared to run greater risks. Anyhow, we cannot afford to make sure that the New Zealanders and the tank battalions are out of action throughout October in either theatre.

*Prime Minister to General Ismay*                                      19.IX.40

Be careful that the Glen ships are not got out of the way so that it will be impossible to take the armoured reinforcements through the Mediterranean if the need is sufficient to justify the risk. I don't want to be told there are no suitable vessels available.

Let me know what other ships would be available if we should decide to run a convoy from west to east through the Mediterranean about the third week in October.

Although it was a fine September, I was frightened of fog.

*Prime Minister to Colonel Jacob*                                    16.IX.40

Pray send a copy of this report by First Sea Lord [about invasion in fog] to the Chiefs of the Staff for C.-in-C. Home Forces, adding: "I consider that fog is the gravest danger, as it throws both Air Forces out of action, baffles our artillery, prevents organised naval attack, and specially favours the infiltration tactics by which the enemy will most probably seek to secure his lodgments. Should conditions of fog prevail, the strongest possible air barrage must be put down upon the invasion ports during the night and early morning. I should be glad to be advised of the proposed naval action by our flotillas, both in darkness and at dawn: (*a*) if the fog lies more on the English than the French side of the Channel; (*b*) if it is uniform on both sides.

"Are we proposing to use radio aids to navigation?

"Prolonged conditions of stand-by under frequent air bombardment will be exhausting to the enemy. None the less, fog is our foe."

In spite of all the danger it was important not to wear the men out.

*Prime Minister to General Ismay*                                    18.IX.40

Inquire from the C.O.S. Committee whether in view of the rough weather Alert No. 1 might not be discreetly relaxed to the next grade.
Report to me.

*Prime Minister to General Ismay*                                    18.IX.40

Make inquiries whether there is no way in which a sheet of flaming oil can be spread over one or more of the invasion harbours. This is no more than the old fire-ship story, with modern improvements, that was tried at Dunkirk in the days of the Armada. The Admiralty can surely think of something.

*Prime Minister to Minister of Supply*                               18.IX.40

The De Wilde ammunition is of extreme importance. At No. 11 Group the bombing of its factory was evidently considered a great blow. I can quite understand the output dropping to thirty-eight thousand rounds in the week while you are moving from Woolwich and getting reinstated, but I trust it will revive again. Pray let me know your forecast for the next four weeks. If there is revival in prospect we might perhaps draw a little upon our reserve.

*Prime Minister to Minister of Supply*                               25.IX.40

I must show you the comments made upon the latest returns of small arms ammunition by my Statistical Department. They cause me the greatest anxiety. In particular the De Wilde ammunition, which is the most valuable, is the most smitten. It seems to me that a most tremendous effort must be made, not only on the whole field of Marks 7 and 8, but on De Wilde and armour-piercing. I am well aware of your difficulties. Will you let me know if there is any way in which I can help you to overcome them?

The reader must pardon this next minute.

*Prime Minister to First Lord*                                                18.IX.40
Surely you can run to a new Admiralty flag. It grieves me to see the present dingy object every morning.

\*     \*     \*     \*     \*

I was relieved by the results produced by the new Ministry of Aircraft Production.

*Prime Minister to Lord Beaverbrook*                                          21.IX.40
The figures you gave me of the improvement in operational types between May 10 and August 30 are magnificent. If similar figures could be prepared down to September 30, which is not far off, I should prefer to read them to the Cabinet rather than circulate them. If however the September figures cannot be got until late in October, I will read [what I now have] to the Cabinet.

The country is your debtor, and of your Ministry.

*Prime Minister to Lord Beaverbrook*                                          25.IX.40
These wonderful results, achieved under circumstances of increasing difficulty, make it necessary for me to ask you to convey to your department the warmest thanks and congratulations from His Majesty's Government.

\*     \*     \*     \*     \*

Throughout the summer and autumn I wished to help the Secretary of State for War in his conflict with War Office and Army prejudice about the Commandos, or storm troops.

*Prime Minister to Secretary of State for War*                                25.VIII.40
I have been thinking over our very informal talk the other night, and am moved to write to you because I hear that the whole position of the Commandos is being questioned. They have been told "no more recruiting" and that their future is in the melting-pot. I thought therefore I might write to let you know how strongly I feel that the Germans have been right, both in the last war and in this, in the use they have made of storm troops. In 1918 the infiltrations which were so deadly to us were by storm troops, and the final defence of Germany in the last four months of 1918 rested mainly upon brilliantly-posted and valiantly-fought machine-gun nests. In this war all these factors

are multiplied. The defeat of France was accomplished by an incredibly small number of highly-equipped *élite*, while the dull mass of the German Army came on behind, made good the conquest and occupied it. If we are to have any campaign in 1941 it must be amphibious in its character, and there will certainly be many opportunities for minor operations, all of which will depend on surprise landings of lightly-equipped, nimble forces accustomed to work like packs of hounds instead of being moved about in the ponderous manner which is appropriate to the regular formations. These have become so elaborate, so complicated in their equipment, so vast in their transport, that it is very difficult to use them in any operations in which time is vital.

For every reason therefore we must develop the storm troop or Commando idea. I have asked for five thousand parachutists, and we must also have at least ten thousand of these small "bands of brothers" who will be capable of lightning action. In this way alone will those positions be secured which afterwards will give the opportunity for highly-trained regular troops to operate on a larger scale.

I hope therefore that you will let me have an opportunity of discussing this with you before any action is taken to reverse the policy hitherto adopted or to throw into uncertainty all the volunteers who have been gathered together.

The resistances of the War Office were obstinate, and increased as the professional ladder was descended. The idea that large bands of favoured "irregulars" with their unconventional attire and free-and-easy bearing should throw an implied slur on the efficiency and courage of the Regular battalions was odious to men who had given all their lives to the organised discipline of permanent units. The colonels of many of our finest regiments were aggrieved. "What is there they can do that my battalion cannot? This plan robs the whole Army of its prestige and of its finest men. We never had it in 1918. Why now?" It was easy to understand these feelings without sharing them. The War Office responded to their complaints. But I pressed hard.

*Prime Minister to Secretary of State for War*                                8.IX.40

You told me that you were in entire agreement with the views I put forward about the Special Companies and ending the uncertainty in which they were placed. Unhappily, nothing has happened so far of which the troops are aware. They do not know they are not under sentence of disbandment. All recruiting has been stopped, although there is a waiting list, and they are not even allowed to call up the men who want to join and have been vetted and approved. Although these companies comprise many of the best and most highly trained of our personnel, they are at present only armed with rifles, which seems a shocking waste should they be thrown into the invasion *mêlée*. I hope you will make sure that when you give an order it is obeyed with promptness. Perhaps you could explain to me what has happened to prevent your decision from being made effective. In my experience of Service departments, which is a long one, there is always a danger that anything contrary to Service prejudices will be obstructed and delayed by officers of the second grade in the machine. The way to deal with this is to make signal examples of one or two. When this becomes known you get a better service afterwards.

Perhaps you will tell me about this if you can dine with me to-night.

*Prime Minister to Secretary of State for War*                     21.IX.40

I am not happy about the equipment position of the Commandos. It is a waste of this fine material to leave them without sufficient equipment for training purposes, much less for operations.

Pray let me have a statement showing:

1. What equipment has already been issued to the various Commandos.
2. What is the scale of equipment which these units are to have.
3. What can be issued to them immediately for training purposes.

I should like to have a return each week showing the precise position as regards the equipment of the various Commandos.

\*     \*     \*     \*     \*

*Prime Minister to C.-in-C. Home Forces [Sir Alan Brooke]*        21.IX.40

We often hear tales of how the Germans will invade on an enormous front, trying to throw, say, a quarter of a million men ashore anyhow, and trusting afterwards to exploit lodgments which are promising. For an attack of this kind our beach defence system seems admirably devised. The difficulty of defending an island against overseas attack has always consisted in the power of the invader to concentrate a very superior force at one point or another. But if he is going to spread himself out very widely, the bulk of his forces, if they reach shore, will come up against equal or superior forces spread along the coast. It will be a case of one thin line against another. Whereas I can readily imagine a concentrated attack pressed forward with tremendous numbers succeeding against our thin line, I find it difficult to see what would be the good of his landing large numbers of small parties, none of which would be strong enough to break our well-organised shore defence. If he is going to lose, say, one hundred thousand in the passage, and another hundred and fifty thousand are to be brought up short at the beaches, the actual invasion would be rather an expensive process, and the enemy would have sustained enormous losses before we had even set our reserves in motion. If therefore there is anything in this alleged German plan, it seems to me it should give us considerable satisfaction. Far more dangerous would be the massed attack on a few particular selected points.

Perhaps you will talk to me about this when we next meet.

\* \* \* \* \*

Our anxieties about the Italian invasion of Egypt were, it now appears, far surpassed by those of Marshal Graziani, who commanded it. Ciano notes in his diary:

*August* 8, 1940. Graziani has come to see me. He talks about the attack on Egypt as a very serious undertaking, and says that our present preparations are far from perfect. He attacks Badoglio, who does not check the Duce's aggressive spirit—a fact which, "for a man who knows Africa, means that he must suffer from softening of the brain, or, what is worse, from bad faith. The water supply is entirely insufficient. We move towards a defeat which, in the desert, must inevitably develop into a rapid and total disaster."

I reported this to the Duce, who was very much upset about it, because in his last conversation with Graziani he had received the impression that the offensive would start in a few days. Graziani did not set any date with me. He would rather not attack at all, or, at any rate, not for two or three months. Mussolini concluded that "one should only give jobs to people who are looking for at least one promotion. Graziani's only anxiety is to remain a Marshal."\*

A month later the Commander-in-Chief asked for a further month's postponement. Mussolini however replied that if he did not attack on Monday he would be replaced. The Marshal answered that he would obey. "Never," says Ciano, "has a military operation been undertaken so much against the will of the commanders."

On September 13 the main Italian army began its long-expected advance across the Egyptian frontier.† Their forces amounted to six infantry divisions and eight battalions of tanks. Our covering troops consisted of three battalions of infantry, one battalion of tanks, three batteries, and two squadrons of armoured cars. They were ordered to make a fighting withdrawal, an

\**Ciano's Diaries*, p. 281.
†See map on page 545.

operation for which their quality and desert-worthiness fitted them. The Italian attack opened with a heavy barrage on our positions near the frontier town of Sollum. When the dust and smoke cleared the Italian forces were seen ranged in a remarkable order. In front were motor-cyclists in precise formation from flank to flank and front to rear; behind them were light tanks and many rows of mechanical vehicles. In the words of a British colonel, the spectacle resembled "a birthday party in the Long Valley at Aldershot". The 3rd Coldstream Guards, who confronted this imposing array, withdrew slowly, and our artillery took its toll of the generous targets presented to them.

Farther south two large enemy columns moved across the open desert south of the long ridge that runs parallel to the sea and could be crossed only at Halfaya—the "Hellfire Pass" which played its part in all our later battles. Each Italian column consisted of many hundreds of vehicles, with tanks, anti-tank guns, and artillery in front, and with lorried infantry in the centre. This formation, which was several times adopted, we called the "Hedgehog". Our forces fell back before these great numbers, taking every opportunity to harass the enemy, whose movements seemed erratic and indecisive. Graziani afterwards explained that at the last moment he decided to change his plan of an enveloping desert movement and "concentrate all my forces on the left to make a lightning movement along the coast to Sidi Barrani". Accordingly the great Italian mass moved slowly forward along the coast road by two parallel tracks. They attacked in waves of infantry carried in lorries, sent forward in fifties. The Coldstream Guards fell back skilfully at their convenience from Sollum to successive positions for four days, inflicting severe punishment as they went.

On the 17th the Italian army reached Sidi Barrani. Our casualties were forty killed and wounded, and the enemy's about ten times as many, including one hundred and fifty vehicles destroyed. Here, with their communications lengthened by sixty miles, the Italians settled down to spend the next three months. They were continually harassed by our small mobile columns, and suffered serious maintenance difficulties. Mussolini at first was "radiant with joy. He has taken the entire responsibility of the offensive on his shoulders," says Ciano, "and is proud that he was right". As the weeks lengthened into months his satisfaction diminished. It seemed however certain to us in London that in two or three months an Italian army far larger than any we could gather would renew the advance to capture the Delta. And then there were always the Germans who might appear! We could not of course expect the long halt which followed Graziani's advance. It was reasonable to suppose that a major battle would be fought at Mersa Matruh. The weeks that had already passed had enabled our precious armour to come round the Cape without the time-lag so far causing disadvantage.

★    ★    ★    ★    ★

*Prime Minister to Secretary of State for War*      14.IX.40

I hope the Armoured Brigade will be in time. I have no doubt it could have been conducted safely through the Mediterranean and the present danger that it will be too late averted. It must however be remembered that General Wavell himself joined in the declaration of the Commanders-in-Chief of the Navy, Army, and Air that the situation in Egypt did not warrant the risk. It was this declaration that made it impossible for me to override the Admiralty objections, as I would otherwise have done.

*(Action this Day)*
*Prime Minister to Secretary of State for War*      19.IX.40
(General Ismay to see.)

The armoured reinforcements are now in the Gulf of Aden. We have been assured that of course General Wavell has made all arrangements to get them into action as quickly as possible. I hope this is so. I am sorry that someone like Lord Beaverbrook is not waiting on the quay to do the job of passing them to the fighting line. We must do the best we can. Has it been considered whether it would be better to carry these vehicles through the Canal to Alexandria and debark them there close to the front, or have special trains and railway cars, cranes and other facilities been accumulated at Suez? Let the alternatives be examined *here*. Without waiting for this, let a telegram be drafted inquiring about the alternatives and the arrangements now made by General Wavell. Every day and even every hour counts in this matter.

All the time I had a fear for Malta, which seemed almost defenceless.

*Prime Minister to General Ismay, for C.I.G.S.*      21.IX.40

This telegram [from Governor and C.-in-C. Malta] confirms my apprehensions about Malta. Beaches defended on an average battalion front of fifteen miles, and no reserves for counter-attack worth speaking of, leave the island at the mercy of a landing force. You must remember that we do not possess the command of the sea around Malta. The danger therefore appears to be extreme. I should have thought four battalions were needed, but owing to the difficulty of moving transports from the West we must be content with two for the moment. We must find two good ones. Apparently there is no insuperable difficulty in accommodation.

★    ★    ★    ★    ★

When I look back on all these worries I remember the story of the old man who said on his deathbed that he had had a lot of trouble in his life, most of which had never happened. Certainly this is true of my life in September 1940. The Germans were beaten in the Air Battle of Britain. The overseas invasion of Britain was not attempted. In fact, by this date Hitler had already turned his glare upon the East. The Italians did not press their attack upon Egypt. The Tank Brigade sent all round the Cape arrived in good time, not indeed for a defensive battle of Mersa Matruh in September, but for a later operation incomparably more advantageous. We found means to reinforce Malta before any serious attack from the air was made upon it, and no one dared to try a landing upon the island fortress at any time. Thus September passed.

# CHAPTER XXIV

# DAKAR

*Importance of Aiding General de Gaulle – Plan for Liberating Dakar – Need to Support the Free French Forces – My Minute of August 8, 1940 – The War Cabinet Approves Operation "Menace" – Dangers of Delay and Leakage – Message from "Jacques" – Our Second String – The French Cruisers Sighted – A Failure at Whitehall – Too Late – I Advise the War Cabinet to Abandon the Project – Strong Desire of the Commanders to Attack – General de Gaulle's Persistence – The War Cabinet Gives Full Discretion to the Commanders – My Telegrams to General Smuts and President Roosevelt – The Attack on Dakar – Ships Versus Forts – Stubborn Resistance of the Vichy French – We Suffer Appreciable Naval Losses – Cabinet and Commanders Agree to Break Off – Changes of Rôle at Home and on the Spot – Justification of the Commanders – Parliament Requires No Explanations.*

AT this time H.M. Government attached great importance to aiding General de Gaulle and the Free French to rally the African possessions and colonies of France, especially those upon the Atlantic coast. Our information was that a large portion of the French officers, officials, and traders in all these territories had not despaired. They were stunned by the sudden collapse of their motherland, but being still free from Hitler's force and Pétain's fraud were in no mood to surrender. To them General de Gaulle shone as a star in the pitch-black night. Distance gave them time, and time gave them opportunity.

Once it was clear that Casablanca was beyond our strength my mind naturally turned to Dakar. In all this the small handling committee I formed to advise me personally on French affairs was convinced and active. On the evening of August 3, 1940, I sent my general approval from Chequers to a proposal for landing Free French forces in West Africa. General de Gaulle, Major-General Spears, and Major Morton had evolved a plan in outline, of which the object was to raise the Free French flag in West Africa, to occupy Dakar, and thus consolidate the French colonies in West and Equatorial Africa for General de Gaulle, and later to rally the French colonies in North Africa. General Catroux was to come from Indo-China to England and eventually take command of the French North African colonies, should these be liberated later on.

On August 4 the Chiefs of Staff Committee considered the details of this plan, as worked out further by the Joint Planning Sub-Committee, and drew up their report for the War Cabinet. The proposals of the Chiefs of Staff were based on the three following assumptions: first, that the force must be equipped and loaded so that it could land in any French West African port; secondly, that the expedition should consist entirely of Free French troops and have no British elements, except the ships in which it moved and their naval escort; thirdly, that the matter should be settled as between Frenchmen, so that the expedition would land without effective opposition.

The strength of the Free French force would be about two thousand five hundred men, comprising two battalions, a company of tanks, sections of artillery and engineers, and a bomber and a fighter flight, for which we should supply the Hurricanes. This force could be ready at Aldershot on August 10, and it was estimated that transports and store-ships could sail from Liverpool on August 13 and troopships between the 19th and 23rd, arriving at Dakar on the 28th, or at the other ports, Konakri and Duala, a few days later. The War Cabinet approved these proposals at their meeting on August 5.

It soon became clear that General de Gaulle required more British support than the Chiefs of Staff had contemplated. They represented to me that this would involve commitments larger and more enduring than those which had been foreseen, and also that the expedition was beginning to lose its Free French character. Our resources were at this time so severely strained that this extension could not be lightly accepted. However, on August 6 I conferred with General de Gaulle, and at 11 p.m. on August 7 I presided over a meeting of the Chiefs of Staff Committee on the project. It was agreed that the best place to land the Free French force was Dakar. I stated that the expedition must be sufficiently backed by British troops to ensure its success, and asked for a larger plan on these lines. The Chiefs of Staff dwelt upon the conflict between a policy of improving our relations with Vichy and our interests in marshalling the French colonies against Germany. They set forth the danger that General de Gaulle's movement might lead to war with Metropolitan France and also with the French colonies. If nevertheless reports from the Free French agents on the spot and from our own representatives in the area were favourable, they recommended that the expedition should go forward. Accordingly, in the early hours of August 8 I issued the following directive:

*Prime Minister to General Ismay, for C.O.S. Committee*    8.VIII.40

1. The telegram from the Governor of Nigeria shows the danger of German influence spreading quickly through the West African colonies of France with the connivance or aid of the Vichy Government. Unless we act with celerity and vigour, we may find effective U-boat bases, supported by German aviation, all down this coast, and it will become barred to us but available for the Germans in the same way as the western coast of Europe.

2. It is now six weeks since the Cabinet was strongly disposed to action at Casablanca, and Mr. Duff Cooper and Lord Gort were dispatched. Nothing however came of this. The local French were hostile. The Chiefs of Staff were not able to make any positive proposals, and the situation has markedly deteriorated.

3. It would seem extremely important to British interests that General de Gaulle should take Dakar at the earliest moment. If his emissaries report that it can be taken peaceably so much the better. If their report is adverse an adequate Polish and British force should be provided and full naval protection given. The operation, once begun, must be carried through. De Gaulle should impart a French character to it, and of course, once successful, his administration will rule. But we must provide the needful balance of force.

4. The Chiefs of Staff should make a plan for achieving the capture of Dakar. For this purpose they should consider available: (a) de Gaulle's force and any French warships which can be collected; (b) ample British naval force, both to dominate French warships in the neighbourhood and to cover the landing; (c) a brigade of Poles properly equipped; (d) the Royal Marine Brigade which was being held available for the Atlantic islands, but might well help to put de Gaulle ashore first, or alternatively commandos from Sir Roger Keyes' force;

(e) proper air support, either by carrier or by machines working from a British West African colony.

5. Let a plan be prepared forthwith, and let the dates be arranged in relation to the Mediterranean operation.

6. It is not intended, after Dakar is taken, that we shall hold it with British forces. General de Gaulle's administration would be set up, and would have to maintain itself, British assistance being limited to supplies on a moderate scale, and of course preventing any sea-borne expedition from Germanised France. Should de Gaulle be unable to maintain himself permanently against air attack or air-borne troops, we will take him off again after destroying all harbour facilities. We should of course in any case take over *Richelieu* under the French flag and have her repaired. The Poles and the Belgians would also have their gold, which was moved before the armistice to Africa by the French Government for safety, recovered for them.

7. In working out the above plan time is vital. We have lost too much already. British ships are to be used as transports whenever convenient, and merely hoist French colours. No question of Orders in Council or legislation to transfer British transports to the French flag need be considered.

8. The risk of a French declaration of war and whether it should be courted is reserved for the Cabinet.

★   ★   ★   ★   ★

On August 13 I brought the matter before the War Cabinet, explaining that it went further than the original plan of a purely French expedition. The details of a landing of six different parties at dawn on the beaches near Dakar and thus dispersing the efforts of the defenders, assuming there was opposition, were examined by my colleagues. The War Cabinet approved the plan, subject to consideration by the Foreign Secretary upon the chances of Vichy France declaring war. Measuring the situation as far as I could, I did not believe this would happen. I had now become set upon this venture. I approved the appointment of Vice-Admiral John Cunningham and Major-General Irwin as the commanders of the expedition. They visited me at Chequers on the night of August 12, and we went through all the aspects of this doubtful and complex affair. I drafted their instructions myself.

I thus undertook in an exceptional degree the initiation and advocacy of the Dakar expedition, to which the code name "Menace" was assigned. Of this, although I cannot feel we were well served on all occasions and certainly had bad luck, I never at

422

any time repented. Dakar was a prize; rallying the French colonial empire a greater. There was a fair chance of gaining these results without bloodshed, and I felt in my finger-tips that Vichy France would not declare war. The stubborn resistance of Britain, the stern mood of the United States, had lit new hope in French hearts. If we won, Vichy could shrug its shoulders. If we lost, they could trade off their resistance with their German masters as a virtue. The most serious danger was prolonged fighting. But these were days in which far more serious risks were the commonplaces of our daily life. I conceived that our resources, albeit strained to the last inch and ounce, could just manage it. With invasion looming up ever nearer and more imminent, we had not shrunk from lending half our tanks to Wavell for the defence of Egypt. Compared to that, this was a pup. Our national War Cabinet, Tory, Labour, and Liberal, were hard, resolute men, imbued with an increasing sense of playing a winning hand. So all the orders were given, and everything went forward under unchallengeable authority.

Our two dangers were now delay and leakage, and the first aggravated the second. At this time the Free French forces in England were a band of exiled heroes in arms against the reigning Government of their country. They were ready to fire on their own fellow-countrymen, and accept the sinking of French warships by British guns. Their leaders lay under sentence of death. Who can wonder at, still less blame them for, a tenseness of emotion or even for indiscretion? The War Cabinet could give orders to our own troops without anyone but the commanders and the Chiefs of Staff circle having to be informed of our intentions. But General de Gaulle had to carry his gallant band of Frenchmen with him. Many got to know. Dakar became common talk among the French troops. At a dinner in a Liverpool restaurant French officers toasted "Dakar!" Our assault landing-craft had to travel on trolleys across England from near Portsmouth to Liverpool, and their escort wore tropical kit. We were all in our war-time infancy. The sealing of the Island was not to be compared with what we achieved later in the supreme operations of "Torch" and "Overlord".

Then there were delays. We had hoped to strike on September 8, but now it appeared that the main force must first go to Freetown to refuel and make their final poise. The plan was based upon the French troopships reaching Dakar in sixteen days at twelve knots. It was found however that the ships carrying the mechanical transport could only make eight to nine knots, and this discovery was reported only at a stage of loading when the time lost in re-loading into faster ships offered no gain. In all ten days' delay from the original date became inevitable: five days for the miscalculation of the speed of the ships, three days for unforeseen loading troubles, two days for the refuelling at Freetown. We must now be content with September 18.

I presided over a meeting of the Chiefs of Staff and General de Gaulle on August 20 at 10.30 p.m., and am on record as summing up the plan as follows:

The Anglo-French armada would arrive at Dakar at dawn, aircraft would drop streamers and leaflets over the town, the British squadron would remain on the horizon, and French ships would come towards the port. An emissary, in a picket-boat flying the Tricolour and a white flag, would go into the harbour with a letter to the Governor saying that General de Gaulle and his Free French troops had arrived. General de Gaulle would stress in the letter that he had come to free Dakar from the danger of imminent German aggression and was bringing food and succour to the garrison and inhabitants. If the Governor was amenable all would be well; if not, and the coast defences opened fire, the British squadrons would close in. If the opposition continued the British warships would open fire on the French gun positions, but with the utmost restraint. If determined opposition was met with the British forces would use all means to break down resistance. It was essential that the operation should be completed, and General de Gaulle master of Dakar, by nightfall.

General de Gaulle expressed his agreement.

On the 22nd we met again, and a letter was read from the Foreign Secretary to me disclosing a leakage of information. Exactly what this leakage amounted to no one could tell. The advantage of sea-power used offensively is that when a fleet sails no one can be sure where it is going to strike. The seas are broad and the oceans broader. Tropical kit was a clue no more definite than the continent of Africa. The wife of a Frenchman in Liverpool who was suspected of Vichy contacts was known to be convinced that the Mediterranean was the destination of the troopships which were gathering in the Mersey. Even the word "Dakar", if bruited carelessly, might be a blind. Such forms of "cover" were carried to remarkable refinements as we became more experienced and wily. I was worried by the delays and beat against them. As to the leakage, none could tell. At any rate, on August 27 the Cabinet gave their final general approval for going ahead. Our target date was then September 19.

\*    \*    \*    \*    \*

At 6.24 p.m. on September 9 the British Consul-General at Tangier cabled to Admiral North, commanding the North Atlantic Station, a shore appointment at Gibraltar, and repeated to the Foreign Office:

Following received from "Jacques". French squadron may try to pass the Straits, proceeding westward for unknown destination. This attempt may be timed to take place within the next seventy-two hours.

The Admiral was *not* in the Dakar circle, and took no special action. The telegram was repeated from Tangier simultaneously to the Foreign Office and received at 7.50 a.m. on the 10th. At this time we were under almost continuous bombardment in London. Owing to the recurrent stoppages of work through the air raids, arrears had accumulated in the cipher branch. The message was not marked "Important", and was deciphered only in its turn. It was not ready for distribution until September 14, when at last it reached the Admiralty.

But we had a second string. At 6 p.m. on September 10 the British Naval Attaché in Madrid was officially informed by the French Admiralty that three French cruisers, type *Georges Leygues*, and three destroyers, had left Toulon and intended to pass the Straits of Gibraltar on the morning of the 11th. This was the normal procedure accepted at this time by the Vichy Government, and was a measure of prudence taken by them only at the latest moment. The British Naval Attaché reported at once to the Admiralty, and also to Admiral North at Gibraltar. The signal was received in the Admiralty at 11.50 p.m. on September 10. It was deciphered and sent to the Duty Captain, who passed it on to the Director of Operations Division (Foreign). It should have been obvious to this officer, who was himself fully informed of the Dakar expedition, that the message was of decisive importance. He took no instant action on it, but let it go forward in the ordinary way with the First Sea Lord's telegrams. For this mistake he received in due course the expression of their lordships' displeasure.

However, the destroyer *Hotspur*, on patrol in the Mediterranean, sighted the French ships at 5.15 a.m. on September 11, fifty miles to the east of Gibraltar, and reported to Admiral North. Admiral Somerville, who commanded Force H, which was based on Gibraltar, had also received a copy of the Naval Attaché's signal at eight minutes past midnight that same morning. He brought the *Renown* to one hour's notice for steam at 7 a.m. and awaited instructions from the Admiralty. In consequence of the error in the Director of Operations Division, and of the delay at the Foreign Office upon the other message from the Consul-General, the First Sea Lord knew nothing about the passage of the French warships till *Hotspur's* signal was brought to him during the Chiefs of Staff meeting before the Cabinet. He at once telephoned the Admiralty to order *Renown* and her destroyers to raise steam. This had already been done. He then came to the War Cabinet. But through the coincidence of this failure of two separate communications—one from the Consul-General in Tangier and the other from the Naval Attaché in Madrid—and through lack of appreciation in various quarters, all was too late. If the Consul-General had marked the first message "Important", or if either of the Admirals at Gibraltar, even though not in the secret, had so considered it themselves, or if the Foreign Office had been working normally, or if the Director of Operations had given the second message the priority which would have ensured the First Sea Lord's being woken up to read it immediately, the *Renown* could have stopped and parleyed with the French squadron pending decisive orders, which would certainly have been given by the War Cabinet, or, till they could be summoned, by me.

In the event all our network of arrangements broke down, and three French cruisers and three destroyers passed the Straits at full speed (25 knots) at 8.35 a.m. on the 11th and turned southwards down the African coast. The War Cabinet, on being apprised, instantly instructed the First Lord to order the *Renown* to get in touch with the French ships, ask for their destination, and make it clear that they would not be allowed to proceed to any German-occupied ports. If they replied that they were going south they were to be told they could proceed to Casablanca, and in this case they were to be shadowed. If they tried to go beyond Casablanca to Dakar they were to be stopped. But the cruisers were never caught. A haze lay over Casablanca on the 12th and 13th. One of the reconnoitring British aircraft was shot down; reports about the presence of additional warships in Casablanca harbour were conflicting; and the *Renown* and her destroyers waited all day and night south of Casablanca to intercept the French squadron. At 4.20 on the afternoon of the 13th the *Renown* received an air report that there were no cruisers in Casablanca. In fact they were already far to the southward, steaming for Dakar at full speed.

There seemed however to be still another chance. Our expedition and its powerful escort was by now itself south of Dakar, approaching Freetown. At 12.16 a.m. on September 14 the Admiralty signalled to Admiral John Cunningham telling him that the French cruisers had left Casablanca at a time unknown and ordering him to prevent them entering Dakar. He was to use every ship available, including the *Cumberland*; and the *Ark Royal* should operate her aircraft without a destroyer screen if this were unavoidable. The cruisers *Devonshire*, *Australia*, and *Cumberland*, and the *Ark Royal* thereupon turned back at maximum speed to establish a patrol line to the north of Dakar. They did not reach their stations until evening on September 14. The French squadron was already anchored in the port with awnings spread.

This chapter of accidents sealed the fate of the Franco–British expedition to Dakar. I had no doubt whatever that the enterprise should be abandoned. The whole scheme of a bloodless landing and occupation by General de Gaulle seemed to me ruined by the arrival of the French squadron, probably carrying reinforcements, good gunners, and bitter-minded Vichy officers, to decide the Governor, to pervert the garrison, and man the batteries. It was possible however to cancel the plan without any loss of prestige, so important to us at this time, and indeed without anyone knowing anything about it. The expedition could be diverted to Duala and cover General de Gaulle's operations against the French Cameroons, and thereafter the ships and transports could be dispersed or return home.

Accordingly, at the meeting of the War Cabinet at noon on

September 16, after outlining the history of the Dakar operation from its inception, the serious results of the postponement of the date, originally fixed for September 13, the leakage of information from various sources, and the misfortune of the French warships having slipped through the Straits, I declared that the whole situation was altered and that the operation was now out of the question. The Cabinet adopted my advice, and the following orders were dispatched to the Dakar force at 2 p.m. that day:

His Majesty's Government have decided that presence of French cruisers at Dakar renders the execution of Dakar operation impracticable. Alternative plans have been examined here. Landing at Konakri does not appear to offer any chance of success in view of difficulty of communications to Bomako, the lack of transport with the force, and the probability that forces from Dakar would forestall. Moreover, close blockade of Dakar from seaward is not possible with the naval forces available, and therefore presence of de Gaulle's force at Bomako would not appreciably influence situation at Dakar. Best plan appears to be for General de Gaulle's force to land at Duala with the object of consolidating the Cameroons, Equatorial Africa, and Chad, and extending influence of de Gaulle to Libreville. The British portion of the force would remain for the present at Freetown.

Unless General de Gaulle has any strong objections to the latter course it should be put into operation forthwith.

\*    \*    \*    \*    \*

The expedition arrived at Freetown on September 17. All the leaders reacted vehemently against the idea of abandoning the enterprise. The Admiral and the General argued that until it was known to what extent the arrival of the Vichy cruisers had raised local morale their presence did not materially alter the previous naval situation. At present, they said, the cruisers had awnings spread, and two were so berthed as to be virtually impotent, while presenting excellent bombing targets.

Here was another twist in the situation. It was very rare at this stage in the war for commanders on the spot to press for audacious courses. Usually the pressure to run risks came from home. In this case the General, General Irwin, had carefully put all his misgivings on paper before he started. I was therefore agreeably surprised at the evident zeal to put this complicated and semi-political operation to the test. If the men on the spot thought it was a time to do and dare, we should certainly give them a free

hand. I therefore sent at 11.52 p.m. on September 16 the follow--
ing:

You are fully at liberty to consider the whole situation yourselves
and consult de Gaulle, and we shall carefully consider then any advice
you may give.

There soon arrived a vehement protest from General de Gaulle,
who wished to carry out the plan. "At the very least," he said,
"should the British Government uphold its new and negative
decision concerning direct action upon Dakar by sea, I request
immediate co-operation of British naval and air forces here
present to support and cover an operation which I personally shall
conduct with my own troops against Dakar from the interior."*

Our commanders now reported:†

At meeting to-day de Gaulle insisted upon necessity for early action
at Dakar. . . . He is advised that substantial support for him is likely to
be found in Dakar if agents are sent to foster it, action is not unduly
deferred, and a too-British complexion of the operation avoided. His
agents are ready at Bathurst and have their instructions. De Gaulle
now proposes original plan to enter harbour unopposed should go
forward, but that if this fails Free French troops should attempt landing
at Rufisque, supported by naval and air action if necessary, and thence
advance on Dakar. British troops only to be landed in support if
called upon after bridgehead has been established. . . .

After careful consideration of all factors, we are of the opinion that
the presence of these three cruisers has not sufficiently increased the
risks, which were always accepted, to justify the abandonment of the
enterprise. We accordingly recommend acceptance of de Gaulle's new
proposal, and that, should he fail, landing of British troops should be
undertaken to install him as previously contemplated. Increased
strength in [our] naval forces is however considered essential.

The operation should be carried out four days after decision of
His Majesty's Government is received.

And, finally, from Major-General Irwin to the C.I.G.S.:

As you know, I have already accepted risks in this operation not
fully justified on purely military grounds. New information possibly
increases those risks, but I consider them worth accepting in view of
obvious results of success. De Gaulle has also committed himself to

* September 17, 1940; received at 11.55 a.m.
† Received by the Admiralty at 7.56 a.m. on September 18, 1940.

complete co-operation with British troops in case of need, and he has not shirked responsibility for fighting between Frenchmen.

The War Cabinet met for the second time on the 17th at 9 p.m. Everyone was agreed to let the commanders go ahead as they wished. Final decision was postponed till noon the next day, it being plain that no time was being lost, as there was still nearly a week before the blow could be struck. At the request of the Cabinet I drafted the following message to the commanders of the Dakar force:

We cannot judge relative advantages of alternative schemes from here. We give you full authority to go ahead and do what you think is best in order to give effect to the original purpose of the expedition. Keep us informed.

This was dispatched at 1.20 p.m. September 18.

*    *    *    *    *

There was nothing to do now but await results. On the 19th the First Sea Lord reported that the French squadron, or parts of it, were leaving Dakar for the south. This made it pretty clear that it had carried Vichy-minded troops, technicians, and authorities to Dakar. The probabilities of a vigorous resistance were increased out of all proportion to the new forces involved. There would certainly be sharp fighting. My colleagues, who were tough, and also nimble to change with circumstances, as is right in war, shared my instinct to let things rip, and the various reports were heard in silence.

On the 20th Admiral Pound told us that the French cruiser *Primauguet*, intercepted by the *Cornwall* and *Delhi*, had agreed to go to Casablanca and was now being escorted thither. The three French warships sighted by the *Australia* turned out to be the cruisers *Georges Leygues*, *Montcalm*, and *Gloire*. At noon on the 19th the *Australia* had been joined by the *Cumberland*, and they continued to shadow the Vichy ships till evening. These now turned to the northward, and increased their speed from 15 to 31 knots. A chase ensued. We were not able to overtake them. At 9.0 p.m. however the *Gloire* had an engine breakdown and could steam no more than 15 knots. Her captain agreed to return to Casablanca, escorted by the *Australia*. This pair were due to pass Dakar about midnight, and the captain of the *Australia* told the *Gloire* that if he were attacked by submarines he would at

430

once sink her. She no doubt spoke to Dakar, and all passed off pleasantly. The *Cumberland*, shadowing the other two Vichy warships, lost touch in a heavy rainstorm, and both, though sighted, got back into Dakar without fire being made upon them. The *Poitiers* when challenged at sea on the 17th had already scuttled herself.

\*     \*     \*     \*     \*

I kept General Smuts fully informed.

*Prime Minister to General Smuts*                                            22.IX.40

You will have seen my message about Dakar. I have been thinking a great deal about what you said in your various messages about not neglecting the African sphere. The de Gaulle movement to rescue the French colonies has prospered in Equatoria and the Cameroons. We could not allow these solid gains to be destroyed by French warships and personnel from Vichy, sent probably at German dictation. If Dakar fell under German control and became a U-boat base the consequences to the Cape route would be deadly. We have therefore set out upon the business of putting de Gaulle into Dakar, peaceably if we can, forcibly if we must, and the expedition now about to strike seems to have the necessary force.

Naturally the risk of a bloody collision with the French sailors and part of the garrison is not a light one. On the whole I think the odds are heavily against any serious resistance, having regard to the low morale and unhappy plight of this French colony, and the ruin and starvation which faces them through our sea control. Still, no one can be sure till we try. The argument that such a risk ought not to be run at a time when French opinion, encouraged by British resistance, is veering towards us even at Vichy, and that anything like a second Oran would be a great set-back, has weighed heavily with us. Nevertheless we came to the united conclusion that this objection might not turn out to be valid, and must in any case be surpassed by the dangers of doing nothing and of allowing Vichy to prevail against de Gaulle. If Vichy did not declare war after Oran, or under the pressure of our blockade, there is no reason why they should do so if there is a fight at Dakar. Besides the strategical importance of Dakar and political effects of its capture by de Gaulle, there are sixty or seventy millions of Belgian and Polish gold wrongfully held in the interior, and the great battleship *Richelieu*, by no means permanently disabled, would indirectly come into our hands. Anyhow, the die is cast.

We do not intend to disturb Morocco at present on account of the German pressure on Spain and Spanish interests there. We are very hopeful about Syria, whither General Catroux will go next week.

An important battle is now impending at Mersa Matruh, and I hope our armoured reinforcements will arrive in time.

I am not particularly impressed with the dangers in Kenya, especially if we lie back and fight from the railway, leaving the enemy the difficult communications. I am trying to send a few suitable tanks to this theatre, which otherwise I feel is overstocked with troops needed in the Soudan and in the Delta.

It gives me so much pleasure and confidence to be trekking with you along the path we have followed together for so many years.

To Roosevelt I telegraphed:

*Former Naval Person to President*                                        23.IX.40
I was encouraged by your reception of information conveyed by Lord Lothian about Dakar. It would be against our joint interests if strong German submarine and aircraft bases were established there. It looks as if there might be a stiff fight. Perhaps not, but anyhow orders have been given to ram it through. We should be delighted if you would send some American warships to Monrovia and Freetown, and I hope by that time to have Dakar ready for your call. But what really matters now is that you should put it across the French Government that a war declaration would be very bad indeed for them in all that concerns United States. If Vichy declares war, that is the same thing as Germany, and Vichy possessions in the Western Hemisphere must be considered potentially German possessions.

Many thanks also for your hint about invasion. We are all ready for them. I am very glad to hear about the rifles.

*       *       *       *       *

It is not necessary here to narrate in detail all that happened during the three days in which Dakar was attacked. These deserve their place in military chronicles, and are a further good example of bad luck. The meteorologists at the Air Ministry had of course carefully studied climatic conditions on the West African coast. A long survey of records reveals uniform, regular bright sunlight and clear weather at this season of the year. On September 23, when the Anglo-French armada approached the fortress, with de Gaulle and his French ships well in the van, fog reigned supreme. We had hoped, since the great majority of the population, French and native, was on our side, that the appearance of all these ships with the British lying far back on the horizon would have decided the action of the Governor. It soon proved however that the Vichy partisans were masters, and there can be no doubt that the arrival of the Vichy cruisers had blotted out any hope of Dakar joining the Free French movement. De Gaulle's two aeroplanes landed on the local airfield, and their pilots were immediately arrested. One of them had on his person a list of the leading Free French adherents. De Gaulle's emissaries, sent under the Tricolour and the white flag, were rebuffed, and others who entered later in launches were fired upon and two of them wounded. All hearts were hardened, and the British Fleet approached through the mist to within five thousand yards. At 10 a.m. a harbour battery opened fire on one of our wing destroyers. The fire was returned, and the engagement soon became general. The destroyers *Inglefield* and *Foresight* were slightly damaged, and the *Cumberland* was struck in the engine-room and had to quit. One French submarine was bombed by an aircraft at periscope depth, and one French destroyer set on fire.

There is an age-long argument about ships *versus* forts. Nelson said that a six-gun battery could fight a 100-gun ship-of-the-line. Mr. Balfour, in the Dardanelles inquiry, said in 1916: "If the ship has guns which can hit the fort at ranges where the fort cannot reply, the duel is not necessarily so unequal." On this occasion the British Fleet, with proper spotting, could in theory engage and after a certain number of rounds destroy the Dakar batteries of 9.4-inch guns at twenty-seven thousand yards. But the Vichy forces had at this time also the battleship *Richelieu*, which proved capable of firing two-gun salvos from 15-inch artillery. This had to be taken into account by the British Admiral. Above all there was the fog. The firing therefore died away at about 11.30 a.m., and all British and Free French ships retired.

In the afternoon General de Gaulle tried to land his troops at Rufisque, but the fog and the confusion had now become so dense that the attempt was abandoned. By 4.30 p.m. the commanders decided to withdraw the troopships and resume the operation next day. The signal with this information reached London at 7.19 p.m., and I thereupon sent the following personal message to the commander timed at fourteen minutes past ten o'clock on September 23:

Having begun we must go on to the end. Stop at nothing.

An ultimatum was sent that night to the Governor of Dakar, to which reply was made that he would defend the fortress to the last. The commanders answered that they intended continuing the operation. Visibility on the 24th was better than on the previous day, but still poor. The shore batteries opened on our ships as they closed, and *Barham* and *Resolution* engaged *Richelieu* at thirteen thousand six hundred yards. Shortly afterwards *Devonshire* and *Australia* engaged a cruiser and a destroyer, damaging the latter. The bombardment ended at about ten o'clock, by which time *Richelieu* had been hit by a 15-inch shell, as also had Fort Manuel, and a light cruiser was on fire. Moreover, one enemy submarine which had tried to interfere with our approach had been forced to the surface by a depth charge, the crew surrendering. None of our ships was hit. In the afternoon the bombardment was renewed for a short time. On this occasion *Barham* was hit four times without serious damage. The bombardment was inconclusive except to indicate that the defences were strong and the garrison determined to resist.

On September 25 the action was resumed. The weather was clear, and our fleet bombarded at twenty-one thousand yards' range, when they were replied to, not only by the very accurate coastal batteries, but by double salvos from the 15-inch guns of the *Richelieu*. A smoke-screen used by the Dakar commander baffled our aim. Soon after 9 a.m. the battleship *Resolution* was hit by a torpedo from a Vichy submarine. After this the Admiral decided to withdraw to seaward, "in view of the condition of the *Resolution*, the continued danger from submarines, and the great accuracy and determination of the shore defences".

Meanwhile the Defence Committee, which met at 10 a.m. without me, had formed the opinion that no pressure should be brought to bear on the commanders to take any action against their better judgment. The Cabinet met at 11.30 a.m., and news of the results of the morning's operations reached us during the meeting. On these tidings it seemed clear that the matter had been pressed as far as prudence and our resources would allow. Several good ships had been severely damaged. It was obvious that Dakar would be defended to the death. No one could be sure that the fierce passions of protracted fighting would not provoke a French declaration of war from Vichy. We therefore, after a painful discussion, were all agreed to push no more.

Accordingly I sent the following telegram (1.27 p.m., September 25) to the commanders:

On all the information now before us, including damage to *Resolution*, we have decided that the enterprise against Dakar should be abandoned, the obvious evil consequences being faced. Unless something has happened which we do not know, which makes you wish to attempt landing in force, you should forthwith break off. You should inform us "Most Immediate" whether you concur, but unless the position has entirely changed in our favour you should not actually begin landing till you receive our reply.

Assuming enterprise abandoned, we shall endeavour to cover Duala by naval force, but we cannot safeguard de Gaulle's forces [if they remain] at Bathurst. Question of reinforcing Freetown with troops is being considered. Instructions regarding disposal of remainder of forces will be given on receipt of your reply.

The commanders made the following reply:

Concur in breaking off.

\* \* \* \* \*

*Former Naval Person to President Roosevelt* 25.IX.40

I much regret we had to abandon Dakar enterprise. Vichy got in before us and animated defence with partisans and gunnery experts. All friendly elements were gripped and held down. Several of our ships were hit, and to persist with landing in force would have tied us to an undue commitment, when you think of what we have on our hands already.

\* \* \* \* \*

In the three days' bombardment no British ships were sunk, but the battleship *Resolution* was disabled for several months, and two destroyers sustained damage which required considerable repairs in home dockyards. Two Vichy submarines were sunk, the crew of one being saved, two destroyers were burnt out and beached, and the battleship *Richelieu* was hit by a 15-inch shell and damaged by two near misses of 250-lb. bombs. There was of course no means at Dakar of repairing this formidable vessel, which had already been rendered temporarily immobile in July, and it could now be definitely dismissed as a hostile factor from our calculations.

It is interesting to note the changes of *rôle* of the War Cabinet and of its commanders in the enterprise. The commanders were at first by no means enthusiastic, and General Irwin protected himself by a lengthy reasoned memorandum to the V.C.I.G.S., in which all the difficulties were stressed. After the expedition had got south of the Canary Islands, the French cruiser squadron, with its reinforcements of Vichy partisans, carrying with it in physical as well as moral form the authority of the French Republic, slipped through the Straits of Gibraltar. I had no doubt from that moment that the situation had been transformed; and the War Cabinet, on my advice, supported by the Chiefs of Staff, agreed that we should stop the enterprise while time remained and no loss had been incurred and no failure would be exposed.

Then the commanders on the spot came forward with their strong desire to take action, and the War Cabinet, quite rightly in my view, felt that the commanders should be the judges and be given a free hand. Accordingly the attempt was made, and it was immediately apparent, by the efficient and vehement resistance of Dakar, that the War Cabinet had been right and rightly advised.

Although the fighting at Dakar had been far more serious than had been expected, we were not wrong in our judgment that the Vichy Government would not declare war upon Great Britain. They contented themselves with air retaliation upon Gibraltar from North Africa. On September 24 and 25 successive raids were made upon the harbour and dockyard; in the first 150 bombs were dropped, and in the second, in which about one hundred aircraft took part, twice as many. The French aviators did not seem to have their hearts in the business, and most of the bombs fell in the sea. Some damage was done, but there were very few casualties. Our A.A. batteries shot down three aircraft. Fighting at Dakar having ended in a Vichy success, the incident was tacitly treated as "quits".

No blame attached to the British naval and military commanders, and both were constantly employed until the end of the war, the Admiral attaining the highest distinction. It was one of my rules that *errors towards the enemy* must be lightly judged. They were quite right to try, if with their knowledge on the spot they thought they could carry the matter through; and the fact that they under-estimated the effect produced on the Vichy garrison by the arrival of the cruisers and their reinforcements was in no way counted against them. Of General de Gaulle I said in the

House of Commons that his conduct and bearing on this occasion had made my confidence in him greater than ever.

The story of the Dakar episode deserves close study, because it illustrates in a high degree not only the unforeseeable accidents of war, but the interplay of military and political forces, and the difficulties of combined operations, especially where allies are involved. To the world at large it seemed a glaring example of miscalculation, confusion, timidity, and muddle. In the United States, where special interest was taken on account of the proximity of Dakar to the American continent, there was a storm of unfavourable criticism. The Australian Government was distressed. At home there were many complaints of faulty war direction. I decided however that no explanations should be offered, and Parliament respected my wish.*

\* \* \* \* \*

In retrospect a brighter view may perhaps be taken of these events. Students of naval history may be struck by the resemblance of this affair to one which occurred nearly three centuries ago. In 1655 Cromwell dispatched a joint naval and military expedition to seize San Domingo, in the West Indies. The attack did not succeed, but the commanders, instead of returning empty-handed, turned failure into success by going on to capture Jamaica.

Although we failed at Dakar, we succeeded in arresting the onward progress of the French cruisers and frustrating their determined efforts to suborn the garrisons in French Equatorial Africa. Within a fortnight General de Gaulle was enabled to establish himself at Duala, in the Cameroons, which became a rallying-point for the Free French cause. Free French activities in these regions played their part not only in halting the penetration of the Vichy virus, but in making possible, through their control of Central Africa, the later development of our trans-continental air transport route from Takoradi to the Middle East.

* See Appendix D for my correspondence with Mr. Menzies.

# CHAPTER XXV

## MR. EDEN'S MISSION

### October 1940

*Retirement of Mr. Chamberlain – Cabinet Changes – The Leadership of the Conservative Party – Reasons for My Decision to Accept the Vacant Post – We Reopen the Burma Road – My Telegram to the President – Growth of Our Strength on the Desert Front – My Complaints about the Middle East Administration – Malta Anxieties – Mr. Eden Flies to the Middle East – My Appreciation of October 13, 1940 – Mr. Eden's Conferences with the Generals at Cairo – His Report and Requests – Our Growing Strength at Mersa Matruh – Proposed Meeting of Mr. Eden and General Smuts at Khartoum – My Desire for a Forestalling Offensive against the Italians – Need for Better Use of Our Resources in the Middle East.*

AT the end of September Mr. Chamberlain's health got far worse. The exploratory operation to which he had subjected himself in July and from which he had returned so courageously to duty had revealed to the doctors that he was suffering from cancer and that there was no surgical remedy. He now became aware of the truth and that he would never be able to return to his work. He therefore placed his resignation in my hands. In view of the pressure of events I felt it necessary to make the changes in the Government which have been mentioned in an earlier chapter. Sir John Anderson became Lord President of the Council and presided over the Home Affairs Committee of the Cabinet. Mr. Herbert Morrison succeeded him as Home Secretary and Minister of Home Security, and Sir Andrew Duncan became Minister of Supply. These changes were effective on October 3.

Mr. Chamberlain also thought it right to resign the Leadership of the Conservative Party, and I was invited to take his place. I had to ask myself the question—about which there may still be various opinions—whether the Leadership of one great party was compatible with the position I held from King and Parliament as Prime Minister of an Administration composed of, and officially supported by, all parties. I had no doubt about the answer. The Conservative Party possessed a very large majority in the House of Commons over all other parties combined. Owing to the war conditions no election appeal to the nation was available in case of disagreement or deadlock. I should have found it impossible to conduct the war if I had had to procure the agreement in the compulsive days of crisis and during long years of adverse and baffling struggle not only of the Leaders of the two minority parties but of the Leader of the Conservative majority. Whoever had been chosen and whatever his self-denying virtues, he would have had the real political power. For me there would have been only the executive responsibility.

These arguments do not apply in the same degree in time of peace; but I do not feel I could have borne such a trial successfully in war. Moreover, in dealing with the Labour and Liberal Parties in the Coalition it was always an important basic fact that as Prime Minister and at this time Leader of the largest party I did not depend upon their votes and I could in the ultimate issue carry on in Parliament without them. I therefore accepted the position of Leader of the Conservative Party which was pressed upon me, and I am sure that without it, and all the steady loyalties which attached to it, I should not have been able to discharge my task until victory was won. Lord Halifax, who might have been an alternative choice of the party if I had declined, himself proposed the motion, which was unanimously adopted.

\* \* \* \* \*

The summer had crashed its way along with massive, rending shocks, but with growing assurance of survival. Autumn and winter plunged us into a maze of complications, less mortal but more puzzling. The invasion challenge had definitely weakened. The Battle of Britain in the air was won. We had bent the German beam. Our Home Army and Home Guard had grown vastly more powerful. The equinoctial gales of October stretched rough, capricious hands across the Channel and the

Narrow Seas. All the arguments from which I had formerly drawn comfort were justified and strengthened. In the Far East the danger of a Japanese declaration of war seemed to have receded. They had waited to see what would happen about the invasion; and nothing had happened. The Japanese war lords had looked for a certainty. But certainties are rare in war. If they had not thought it worth while to strike in July, why should they do so now when the light of the British Empire burned brighter and fiercer and world conditions were less favourable to them? We felt ourselves strong enough to reopen the Burma Road when its three months' closure had elapsed. The Japanese were experienced in sea war, and probably thought about it along the same lines as the British Admiralty. None the less, it was not without anxiety that the decision to open the Burma Road and allow supplies to flow along it into China was taken. In this broad measurement of the unknowable our judgment was not proved wrong.

I was glad to telegraph to the President news which I was sure would be agreeable to him and to the United States.

*Former Naval Person to President*                                      4.X.40
After prolonged consideration of all the issues involved we to-day decided to let the Burma Road be reopened when the three months period expires on October 17. The Foreign Secretary and I will announce this to Parliament on Tuesday, 8th. I shall say that our hopes of a just settlement being reached between Japan and China have not borne fruit, and that the Three-Power Pact revives the Anti-Comintern Pact of 1939 and has a clear pointer against the United States. I know how difficult it is for you to say anything which would commit the United States to any hypothetical course of action in the Pacific. But I venture to ask whether at this time a simple action might not speak louder than words. Would it not be possible for you to send an American squadron, the bigger the better, to pay a friendly visit to Singapore? There they would be welcomed in a perfectly normal and rightful way. If desired, occasion might be taken of such a visit for a technical discussion of naval and military problems in those and Philippine waters, and the Dutch might be invited to join. Anything in this direction would have a marked deterrent effect upon a Japanese declaration of war upon us over the Burma Road opening. I should be very grateful if you would consider action along these lines, as it might play an important part in preventing the spreading of the war.

In spite of the Dakar fiasco the Vichy Government is endeavouring to enter into relations with us, which shows how the tides are flowing in France now that they feel the German weight and see we are able to hold our own.

Although our position in the air is growing steadily stronger both actually and relatively, our need for aircraft is urgent. Several important factories have been seriously injured, and the rate of production is hampered by air alarms. On the other hand, our losses in pilots have been less than we expected, because in fighting over our own soil a very large proportion get down safely or only wounded. When your officers were over here we were talking in terms of pilots. We are now beginning to think that aeroplanes will be the limiting factor so far as the immediate future is concerned.

I cannot feel that the invasion danger is past. The gent has taken off his clothes and put on his bathing-suit, but the water is getting colder and there is an autumn nip in the air. We are maintaining the utmost vigilance.

\* \* \* \* \*

These welcome events at opposite ends of the world cleared the way for stronger action in the Middle East. Every nerve had to be strained to make headway against Italy, whose movements were slower than I had expected. Strong reinforcements had reached General Wavell. The two tank regiments had arrived in the desert. General Maitland Wilson, who commanded the "Army of the Nile", as it was now called, formed a high opinion of the possibilities of the "Matildas"—as the Infantry or "I" tanks were nicknamed by the troops. Our defence position at Mersa Matruh was now far more solid, and—though this I did not yet know—new thoughts began to stir in staff and planning circles at the Middle East Headquarters. Obviously our next main task was to strengthen our forces in the Middle East, and especially in the Western Desert, both from Britain and from India.

I was still in argument with the Admiralty about military convoys attempting the passage of the Mediterranean, I saying: "You can now see that we ought to have tried it," and they: "There was not so much hurry after all." I still remained extremely dissatisfied with the distribution of our forces already in the Middle East, and with the disparity, as I judged it, between ration and fighting strength. I feared greatly for Malta. I pressed General Wavell and the Secretary of State, both directly and through the Chiefs of Staff, on all these points. To Mr. Eden I wrote:

*Prime Minister to Secretary of State for War*        24.IX.40

There is no difference between us in principle; but the application of the principle raises issues of detail, and this is especially true of the denudation of this Island in the face of the imminent threat of invasion. Meanwhile the General Staff continue to press for diversions from the Middle East, such as the 7th Australian Division to be used for garrisoning the Malay peninsula. Now the two Indian brigades are to be employed in these jungles against a possible war with Japan, and a still more unlikely Japanese siege of Singapore. The paper on Indian reinforcements was considered last night by me and the Chiefs of Staff. You will see in it that a division is to be provided for Malaya, another for Basra, and a corps for Iraq, thus absorbing all the Indian reinforcements available in 1941. This geographical distribution or dispersion of our forces shows the ideas prevailing, which are altogether erroneous in a strategic sense. However, it was explained to me that, although these forces were earmarked for particular theatres, they could all go to the Middle East if required. I therefore agreed to words being inserted making this clear. None the less, the paragraph dispersing these divisions without regard to war needs made an unfavourable impression upon me.

We have next to consider the increasing waste of troops in Kenya, and the continued waste in Palestine. Some improvement has been made in Palestine, but Kenya, on the contrary, is at this moment to have a mountain battery sent there instead of to the Soudan. I fear that when General Smuts goes there he will naturally be influenced by the local situation. However, I hope to keep in touch with him by cable.

Lastly, there is the shocking waste of British Regular troops on mere police duty in the Canal Zone, in Cairo, and at Alexandria, and the general slackness of the Middle East Command in concentrating the maximum for battle and in narrowing the gap between ration strength and fighting strength. I have not had any answer to my request for figures on this point.

My idea, like yours, is to gather the strongest army in the Middle East possible in the next few months, and I have indicated on other papers the number of divisions I hope can be assembled there. But I think the first thing would be for the War Office and the Egyptian Command to make the best use possible of the very large number of troops they have already, and for which we are paying heavily.

Further, I am much disquieted about the position at Malta. It is now agreed that two battalions shall be sent as reinforcements; but after how much haggling and boggling, and excuses that they could not be accommodated in the island! Have you read General Dobbie's appreciation and his statement that he has his battalions all spread on fifteen-mile fronts each, with no reserves not already allocated to the defence of aerodromes? Do you realise there is no command of the sea at Malta, and that it might be attacked at any time by an expeditionary force of twenty or thirty thousand men from Italy, supported by the Italian Fleet? Yet it was proposed that these two battalions should go to Freetown to complete the brigade there, although no enemy can possibly attack Freetown while we have the command of the Atlantic Ocean. You will, I am sure, excuse my putting some of these points to you, because they illustrate tendencies which appear ill-related to the very scheme of war which you have in mind.

*Prime Minister to General Ismay*                                        6.x.40

Whenever the Fleet is moving from Alexandria to the Central Mediterranean reinforcements should be carried in to Malta, which I consider to be in grievous danger at the present time. These reinforcements should be found by taking battalions from the Canal Zone and replacing them by dismounted Yeomanry or Australian details now in Palestine, or by South African units presently to be moved from Kenya. Pray let me have proposals on these lines, and make sure that at least one battalion goes to Malta on the next occasion. We cannot waste Regular battalions on internal security duties in Egypt. If they were needed for the field army they would of course be irremovable, but that is not what they are being used for.

\*     \*     \*     \*     \*

I was in such close agreement with the Secretary of State for War, and felt so much the need of having our views put forward on the spot, instead of through endless telegrams, that I now asked him whether he would not make a personal inspection of the Middle East. He was delighted, and started immediately. He made a thorough tour of the whole theatre. In his absence I took over the War Office.

I also at this time laid the whole military situation as I saw it before the Chiefs of Staff.

*Prime Minister to General Ismay, for C.O.S. Committee*              13.x.40

1. First in urgency is the reinforcement of Malta:

   (a) by further Hurricane aircraft, flown there as can best be managed;

   (b) by the convoy now being prepared, which should carry the largest anti-aircraft outfit possible, as well as the battalions and the battery—I understand another M.T. ship can be made available;

(c) by one or, better still, two more battalions released from police duty on the Canal or in Palestine, and carried to Malta when next the Fleet moves thither from Alexandria. General Dobbie's latest appreciation bears out the grievous need of strengthening the garrison. Every effort should be made to meet his needs, observing that once Malta becomes a thorn in the Italian side the enemy's force may be turned upon it. The movement of these reinforcements should therefore precede any marked activity from Malta.

(d) Even three Infantry tanks at Malta would be important, not only in actual defence, but as a deterrent if it were known that they were there. Some mock-up tanks also might be exhibited where they would be detected from the air.

2. The movement of the Fleet to Malta must await this strengthening of the air defences. It is however a most needful and profoundly advantageous step. I welcome the possibility of basing even light forces upon Malta, as they immediately increase its security. I understand it is intended they shall sally forth by day and only lie in harbour as a rule at night. It must be observed that a strong ship like the *Valiant* can far better withstand a hit from a bomb than light craft, and in addition she carries a battery of twenty very high-class A.A. guns. Apart from the stake being higher, it is not seen why, if light forces can be exposed in Malta harbour, well-armoured and well-armed ships cannot use it too. The multiple aerial mine U.P. weapon gives considerable security against dive-bombing.

I should be glad to be more fully informed by the Admiralty about this.

Occasional visits by the whole Battle Fleet would be an immense deterrent on hostile attack, and also a threat to the [enemy] Libyan communications while they last.

Let me have the number of A.A. guns now in position, and the whole maximum content [of them in] the new convoy, together with estimated dates for their being mounted.

3. Relations with Vichy. We cannot accept the position that we must yield to the wishes of Vichy out of fear lest they make air raids upon Gibraltar, for there would be no end to that. We must reassert our blockade of the Straits, dealing with vessels whether escorted or unescorted, though without violating Spanish territorial waters. We should assemble a sufficient force at Gibraltar for this purpose at the earliest date possible. Meanwhile we must maintain as good a blockade of Dakar as possible, and protect Duala, etc., from a counter-stroke by the French cruisers in Dakar. The conversations with Vichy, if they take place, may reach a *modus vivendi* falling some-

what short of these desiderata. Of course, if we could be assured that Vichy, or part of Vichy, was genuinely moving in our direction we could ease up on them to a very large extent. It seems probable that they will be increasingly inclined to move as we desire, and I personally do not believe that hard pressure from us will prevent this favourable movement. It is becoming more difficult every day for Vichy to lead France into war with us. We must not be too much afraid of checking this process, because the tide in our favour will master and overwhelm the disturbing eddies of the blockade and possible sea incidents. I do not believe that any trouble will arise with the French which will prevent the impending movement of our convoy to Malta. The chance is there, but it is remote and must be faced.

4. The greatest prize open to Bomber Command is the disabling of *Bismarck* and *Tirpitz*. If *Bismarck* could be set back for three or four months the *King George V* could go to the Eastern Mediterranean to work up, and could therefore play a decisive part in the occupation of Malta by the Fleet. This would speedily transform the strategic situation in the Mediterranean.

5. Should October pass without invasion we should begin the reinforcement of the Middle East by the Cape route to the utmost extent our shipping permits, sending, as arranged, the armoured units, the Australians, and New Zealanders in November, another British division before Christmas, and at least four more during January, February, and March. All this would be in addition to the necessary drafts. Let me know how far your present programme of sailings conforms to this.

6. The time has also come for a further strong reinforcement of the Middle East by bombers and by fighters. I should be glad to know how far the Chiefs of Staff would be prepared to go, observing that though the risk is very great so also is the need.

7. Let me see the programme for reinforcing the Mediterranean Fleet during the next six months. It should be possible by the end of the year to send three flotillas of destroyers to the Eastern Mediterranean, and one additional to Gibraltar. If *King George V* must be kept to watch *Bismarck*, *Nelson* or *Rodney* should go to Alexandria, and either *Barham* or *Queen Elizabeth*. What cruiser reinforcements are contemplated? Will it be possible to send *Formidable* [an aircraft-carrier] thither also, and when?

8. Agreeably to the dispatch of divisions to the Middle East, the Home Army and the Home Guard will be developed to fill the gap. A minimum of twelve mobile divisions must lie in reserve [at home], apart from the troops on the beaches, at any time.

9. It should be possible also to provide by the end of July a striking

force for amphibious warfare of six divisions, of which two should be armoured. The various alternative plans for the employment of such a force are being studied.

\*　\*　\*　\*　\*

Meanwhile Mr. Eden was on his journey. He "was deeply impressed with the rapid progress in recent work on the defences of Gibraltar", which he said had "been driven forward with energy, determination, and ingenuity". The morale of the troops was high and the garrison confident. He was more anxious about the position at Malta, and pressed for at least another battalion and a battery of 25-pounders, together of course with continued air reinforcements. The Governor, General Dobbie, thought it important that an offensive policy which would provoke retaliation should be avoided at Malta until April 1941, by which time the various programmes of reinforcement in aircraft and A.A. guns would be fulfilled.

On the 15th Mr. Eden reached Cairo. He held searching discussions with Generals Wavell and Maitland Wilson, who commanded the Desert Army. There was good confidence about repelling an Italian offensive. General Wilson estimated that the maximum strength the Italians could deploy against Matruh was three divisions, the limiting factors being maintenance, particularly water, and communications. Against this he had the 7th Armoured Division, with its newly arrived tank regiments, the 4th Indian Division, the Matruh garrison of five rifle battalions, a machine-gun battalion, and eight or nine batteries. The 16th British Brigade Group and the New Zealand Brigade Group had arrived from Palestine. An Australian Brigade Group lay west of Alexandria; a second Australian brigade was moving thither. There was also a Polish brigade. The concentration of these forces, wrote Eden, was considered by General Wilson to be sufficient to meet the threat of the enemy and to enable him to defeat it, provided he was assured of adequate air support. Eden added that inundations for which I had asked had been carried out and anti-tank obstacles created. He sent a lengthy list of requirements, particularly aircraft. This last was easier asked for than given at the time when the bombing of London was rising to its peak. He urged that a company of Infantry tanks should be included in the November convoy, destination Port Soudan, in order to take the offensive against the Italian threat from Kassala.

Eden also raised at Cairo a pertinent question: What action would be taken by our forces, supposing the Italian attack did not take place? Upon this the Generals first spoke of their own offensive hopes. "It has emerged from our discussion this morning", Eden cabled, "that Infantry tanks [Matildas] can play a much more important *rôle* in the fighting in this theatre than we had thought. General Wavell would much like a second battalion of I tanks, and a Brigade Recovery Section, especially important to maintain full serviceability."

Although no reference had been made in the Secretary of State for War's telegram to our taking the offensive, I was very glad to learn all the good news, and urged him to continue his inspection.

*Prime Minister to Secretary of State for War*  16.X.40

I have read all your telegrams with deepest interest and realisation of the value of your visit. We are considering how to meet your needs. Meanwhile, continue to master the local situation. Do not hurry your return.

Eden further arranged for a Turkish Mission to join our Army, and proposed to General Smuts a meeting at Khartoum to discuss the whole situation, and particularly our Soudan offensive project, and my complaints about the overcrowding in Kenya. This meeting was fixed for October 28, a date which later acquired significance. I need scarcely add that requests for all kinds of equipment, including ten thousand rifles to aid the rebellion in Abyssinia, and above all for anti-tank guns, anti-tank rifles, A.A. batteries, and air reinforcements, flowed to us in a broadening stream. We did our utmost to meet these needs at the expense of home defence at this time. There was not half enough for everybody, and whatever was given to one man had to be denied or taken from another also in danger.

Mr. Eden proposed to fly back by Lagos immediately after his conference at Khartoum, preferring to make a full verbal report of all he had seen and done. I was so much encouraged by the picture as to become hungry for a turn to the offensive in the Western Desert. I therefore telegraphed to him:

26.X.40

Before leaving you should consider searchingly with your Generals possibilities of a forestalling offensive. I cannot form any opinion about it from here, but if any other course was open it would not be sound strategy to await the concentration and deployment of overwhelming forces. I thought the existing plans for repelling an attack by a defensive battle and counter-stroke very good, but what happens if the enemy do not venture until the Germans arrive in strength? Do not send any answer to this, but examine it thoroughly and discuss it on return.

Please examine in detail the field state of the Middle Eastern Army in order to secure the largest proportion of fighting men and units for the great numbers on our ration strength. Study improvisation from White details for the Canal Zone and internal security. All British battalions should be mobile and capable of taking part in battle. I fear that the proportion of fighting compared with ration strength is worse in the Middle East than anywhere else. Please do not be content with the stock answers. Even Army Ordnance and Service Corps depots and other technical details can all help in keeping order where they are, and should be organised for use in an emergency. Not only the best, but the second and third best, must be made to play their part.

Thus on the main issue our minds at home and on the spot were moving forward in harmony.

# CHAPTER XXVI

# RELATIONS WITH VICHY AND SPAIN

I N spite of the Armistice and Oran and the ending of our diplomatic relations with Vichy, I never ceased to feel a unity with France. People who have not been subjected to the personal stresses which fell upon prominent Frenchmen in the awful ruin of their country should be careful in their judgments of individuals. It is beyond the scope of this story to enter the maze of French politics. But I felt sure that the French nation would do its best for the common cause according to the facts presented to it. When they were told that their only salvation lay in following the advice of the illustrious Marshal Pétain, and that England, which had given them so little help, would soon be conquered or give in, very little choice was offered to the masses. But I was sure they wanted us to win, and that nothing would give them more joy than to see us continue the struggle with vigour. It was our first duty to give loyal support to General de Gaulle in his valiant constancy. On August 7 I signed a military agreement with him which dealt with practical needs. His stirring addresses were made known to France and the world by the British broadcast. The sentence of death which the Pétain Government passed upon him glorified his name. We did everything in our power to aid him and magnify his movement.

At the same time it was necessary to keep in touch not only with France, but even with Vichy. I therefore always tried to make the best of them. I was very glad when at the end of the year the United States sent an Ambassador to Vichy of so much influence and character as Admiral Leahy, who was himself so close to the President. I repeatedly encouraged Mr. Mackenzie King to keep his representative, the skilful and accomplished M. Dupuy, at Vichy. Here at least was a window upon a courtyard to which we had no other access. On July 25 I sent a minute to the Foreign Secretary in which I said: "I want to promote a kind of collusive conspiracy in the Vichy Government whereby certain members of that Government, perhaps with the consent of those who remain, will levant to North Africa in order to make a better bargain for France from the North African shore and from a position of independence. For this purpose I would use both food and other inducements, as well as the obvious arguments." It was in this spirit that I was to receive in October a certain M. Rougier, who represented himself as acting on the personal instructions of Marshal Pétain. This was not because I or my colleagues had any respect for Marshal Pétain, but only because no road that led to France should be incontinently barred. Our consistent policy was to make the Vichy Government and its members feel that, so far as we were concerned, it was never too late to mend. Whatever had happened in the past, France was our comrade in tribulation, and nothing but actual war between us should prevent her being our partner in victory.

This mood was hard upon de Gaulle, who had risked all and kept the flag flying, but whose handful of followers outside France could never claim to be an effective alternative French Government. Nevertheless we did our utmost to increase his influence, authority, and power. He for his part naturally resented any kind of truck on our part with Vichy, and thought we ought to be exclusively loyal to him. He also felt it to be essential to his position before the French people that he should maintain a proud and haughty demeanour towards "perfidious Albion", although an exile, dependent upon our protection and dwelling in our midst. He had to be rude to the British to prove to French eyes that he was not a British puppet. He certainly carried out this policy with perseverance. He even one day explained this technique to me, and I fully comprehended the extraordinary difficulties of his problem. I always admired his massive strength.

\* \* \* \* \*

On October 21 I made an appeal by radio to the French people. I took great pains to prepare this short address, as it had to be given in French. I was not satisfied with the literal translation at first provided, which did not give the spirit of what I could say in English and could feel in French, but M. Duchesne, one of the Free French staff in London, made a far better rendering, which I rehearsed several times and delivered from the basement of the Annexe, amid the crashes of an air raid.

Frenchmen!

For more than thirty years in peace and war I have marched with you, and I am marching still along the same road. To-night I speak to you at your firesides wherever you may be, or whatever your fortunes are. I repeat the prayer around the *louis d'or:* "*Dieu protège la France.*" Here at home in England, under the fire of the Boche, we do not forget the ties and links that unite us to France, and we are persevering steadfastly and in good heart in the cause of European freedom and fair dealing for the common people of all countries, for which, with you, we drew the sword. When good people get into trouble because they are attacked and heavily smitten by the vile and wicked, they must be very careful not to get at loggerheads with one another. The common enemy is always trying to bring this about, and, of course, in bad luck a lot of things happen which play into the enemy's hands. We must just make the best of things as they come along.

Here in London, which Herr Hitler says he will reduce to ashes, and which his aeroplanes are now bombarding our people are bearing up unflinchingly. Our Air Force has more than held its own. We are waiting for the long-promised invasion. So are the fishes. But, of course, this for us is only the beginning. Now in 1940, in spite of occasional losses, we have, as ever, command of the seas. In 1941 we shall have the command of the air. Remember what that means. Herr Hitler with his tanks and other mechanical weapons, and also by Fifth Column intrigue with traitors, has managed to subjugate for the time being most of the finest races in Europe, and his little Italian accomplice is trotting along hopefully and hungrily, but rather wearily and very timidly, at his side. They both wish to carve up France and her Empire as if it were a fowl: to one a leg, to another a wing or perhaps part of the breast. Not only the French Empire will be devoured by these two ugly customers, but Alsace-Lorraine will go once again under the German yoke, and Nice, Savoy, and Corsica—Napoleon's Corsica—will be torn from the fair realm of France. But Herr Hitler is not thinking only of stealing other people's territories, or flinging gobbets of them to his little confederate. I tell you truly what you must believe when I say that this evil man, this monstrous abortion of hatred and defeat, is resolved on nothing less than the complete wiping out of the French nation, and the disintegration of its whole life and future. By all kinds of sly and savage means he is plotting and working to quench for ever the fountain of characteristic French culture and of French inspiration to the world. All Europe, if he has his way, will be reduced to one uniform Boche-land, to be exploited, pillaged, and bullied by his Nazi gangsters. You will excuse my speaking frankly, because this is not a time to mince words. It is not defeat that France will now be made to suffer at German hands, but the doom of complete obliteration. Army, Navy, Air Force, religion, law, language, culture, institutions, literature, history, tradition, all are to be effaced by the brute strength of a triumphant army and the scientific low-cunning of a ruthless Police Force.

Frenchmen—rearm your spirits before it is too late. Remember how Napoleon said before one of his battles: "These same Prussians who are so boastful to-day were three to one at Jena, and six to one at Montmirail." Never will I believe that the soul of France is dead. Never will I believe that her place amongst the greatest nations of the world has been lost for ever! All these schemes and crimes of Herr Hitler's are bringing upon him and upon all who belong to his system a retribution which many of us will live to see. The story is not yet finished, but it will not be so long. We are on his track, and so are our friends across the Atlantic Ocean, and your friends across the Atlantic Ocean. If he cannot destroy us, we will surely destroy him and all his gang, and all their works. Therefore have hope and faith, for all will come right.

Now what is it we British ask of you in this present hard and bitter time? What we ask at this moment in our struggle to win the victory which we will share with you, is that if you cannot help us, at least you will not hinder us. Presently you will be able to weight the arm that strikes for you, and you ought to do so. But even now we believe that Frenchmen, wherever they may be, feel their hearts warm and a proud blood tingle in their veins when we have some success in the air or on the sea, or presently—for that will come—upon the land.

Remember we shall never stop, never weary, and never give in, and that our whole people and Empire have vowed themselves to the task of cleansing Europe from the Nazi pestilence and saving the world from the new Dark Ages. Do not imagine, as the German-controlled wireless tells you, that we English seek to take your ships and colonies. We seek to beat the life and soul out of Hitler and Hitlerism. That alone, that all the time, that to the end. We do not covet anything from any nation except their respect. Those Frenchmen who are in the French Empire, and those who are in so-called Unoccupied France, may see their way from time to time to useful action. I will not go into details. Hostile ears are listening. As for those, to whom English hearts go out in full, because they see them under the sharp discipline, oppression, and spying of the Hun—as to those Frenchmen in the occupied regions, to them I say, when they think of the future let them remember the words which Gambetta, that great Frenchman, uttered after 1870 about the future of France and what was to come: "Think of it always: speak of it never."

Good night then: sleep to gather strength for the morning. For the morning will come. Brightly will it shine on the brave and true, kindly upon all who suffer for the cause, glorious upon the tombs of heroes. Thus will shine the dawn. *Vive la France!* Long live also the forward march of the common people in all the lands towards their just and true inheritance, and towards the broader and fuller age.

There is no doubt that this appeal went home to the hearts of millions of Frenchmen, and to this day I am reminded of it by men and women of all classes in France, who always treat me with the utmost kindness in spite of the hard things I had to do—sometimes to them—for our common salvation.

\*     \*     \*     \*     \*

At this time it was necessary to insist upon essentials. We could not relax the Blockade of Europe, and particularly of France, while they remained under Hitler's domination. Although from time to time to meet American wishes we allowed a few specified ships with medical stores to pass into Unoccupied France, we did not hesitate to stop and search all other ships seeking or coming out of French ports. Whatever Vichy might do for good or ill, we would not abandon de Gaulle or discourage accessions to his growing colonial domain. Above all we would not allow any portion of the French Fleet, now immobilised in French colonial harbours, to return to France. There were times when the Admiralty were deeply concerned lest France should declare war upon us and thus add to our many cares. I always believed that once we had proved our resolve and ability to fight on indefinitely the spirit of the French people would never allow the Vichy Government to take so unnatural a step. Indeed, there was by now a strong enthusiasm and comradeship for Britain, and French hopes grew as the months passed. This was recognised even by M. Laval when he presently became Foreign Minister to Marshal Pétain.

As the autumn drew into winter I was concerned with the danger of the two great French battleships attempting to make their way back to Toulon, where they could be completed. President Roosevelt's envoy, Admiral Leahy, had established intimate relations with Marshal Pétain. It was to Roosevelt therefore that I turned, and not in vain.

*Former Naval Person to President Roosevelt*                    20.X.40

We hear rumours from various sources that the Vichy Government are preparing their ships and colonial troops to aid the Germans against us. I do not myself believe these reports, but if the French fleet at Toulon were turned over to Germany it would be a very heavy blow. It would certainly be a wise precaution, Mr. President, if you would speak in the strongest terms to the French Ambassador, emphasising the disapprobation with which the United States would view such a betrayal of the cause of democracy and freedom. They will pay great heed in Vichy to such a warning.

You will have seen what very heavy losses we have suffered in the North-Western Approaches to our last two convoys.* This is due to our shortage of destroyers in the gap period I mentioned to you. Thank God your fifty are now coming along, and some will soon be in action. We ought to be much better off by the end of the year, as we have a lot of our own anti-U-boat vessels completing, but naturally we are passing through an anxious and critical period, with so many small

* From October 17 to 19 (inclusive) thirty-three ships, twenty-two of them British, were sunk by U-boats in the North-Western Approaches. These figures include twenty ships out of one convoy.

craft having to guard against invasion in the Narrow Waters, and with the very great naval effort we are making in the Mediterranean, and the immense amount of convoy work.

The President in consequence sent a very severe personal message to the Pétain Government about the Toulon fleet. "The fact," he said, "that a Government is a prisoner of war of another Power does not justify such a prisoner in serving its conqueror in operations against its former ally." He reminded the Marshal of the solemn assurances he had received that the French Fleet would not be surrendered. If the French Government attempted to permit the Germans to use the French Fleet in hostile operations against the British Fleet, such action would constitute a flagrant and deliberate breach of faith with the United States Government. Any agreement of that character would most definitely wreck the traditional friendship between the French and American peoples. It would create a wave of bitter indignation against France in American public opinion and would permanently end all American aid to the French people. If France pursued such a policy the United States could make no effort when the proper time came to secure for France the retention of her oversea possessions.

*Former Naval Person to President Roosevelt*                              26.X.40
Your cable with terms of splendid warning you gave the French crossed mine to you about a suggested message to Pétain. Most grateful for what you have already done, but everything still in balance. Foreign Office tell me they have cabled you our latest information of German terms, which Pétain is said to be resisting. In this connection the surrender of bases on the African shores for air or U-boats would be just as bad as surrender of ships. In particular Atlantic bases in bad hands would be a menace to you and a grievous embarrassment to us. I hope therefore you will make it clear to the French that your argument about ships applies also to the betrayal of bases.

In spite of the invasion threats and air attacks of the last five months, we have maintained a continuous flow of reinforcements round the Cape to the Middle East, as well as sending modern aircraft and major units of the Fleet. I do not think the invasion danger is yet at an end, but we are now augmenting our eastern transferences. The strain is very great in both theatres, and all contributions will be thankfully received.

At this time the Admiralty were so deeply concerned about the dangers of a rupture with Vichy that they were inclined to underrate the disadvantages of letting the two French battleships return to Toulon. On this I gave directions.

*Prime Minister to First Lord and First Sea Lord*
*(From the train)*                                                2.XI.40

After the defection of France it was considered vital not to allow the *Jean Bart* and the *Richelieu* to fall into enemy hands, or to reach harbours where they could be completed. For this purpose you attacked the *Richelieu*, and claimed to have disabled her to a very large extent. The *Jean Bart* is in an unfinished state, and neither ship can be fitted for action in the African harbours on the Atlantic, where they now lie. It is our decided policy not to allow these ships to pass into bad hands. I was therefore surprised to hear the First Sea Lord demur to the idea that the *Jean Bart* should be prevented from returning to Toulon, and argue in the sense that she might safely be allowed to do so. Toulon has always been judged by us to be an enemy-controlled harbour. It was for this reason that the most extreme efforts were made, unhappily without success, to prevent the *Strasbourg* reaching Toulon. I cannot reconcile this action with the apparent readiness to allow the *Jean Bart* to proceed there.

The Admiralty is held responsible for preventing the return of either of these two ships to French ports on the Atlantic, or to the Mediterranean, where they could be repaired and completed at Toulon, and then at any time betrayed to the Germans or captured by them.

*Prime Minister to Foreign Secretary (From the train)*            2.XI.40

I do not know how imminent the movement of the *Jean Bart* may be. I have informed the Admiralty that they are responsible for stopping her from entering the Mediterranean. It would seem therefore very important that you should give a clear warning to Vichy that the ship in question will be stopped, and if necessary sunk, if she attempts to go either to a German-controlled port in the Atlantic, or to a Mediterranean port which may at any time fall into German hands. My Private Office in London is sending you a copy of the minute I have sent to the First Lord and the First Sea Lord.

*Former Naval Person to President Roosevelt*                     10.XI.40

1. We have been much disturbed by reports of intention of French Government to bring *Jean Bart* and *Richelieu* to Mediterranean for completion. It is difficult to exaggerate [the] potential danger if this were to happen, and so open the way for these ships to fall under German control. We should feel bound to do our best to prevent it.

2. We conveyed a warning to French Government through Ambassador at Madrid a few days ago, on the following lines:

Such a step would greatly increase the temptation to the Germans and Italians to seize the French Fleet. We doubt, not the good faith of the French Government, but their physical ability to implement their assurances that they will not let the Fleet fall into enemy hands. We particularly wish to avoid any clash between British and French naval forces, and therefore hope that if they had thought of moving the ships they will now refrain from doing so.

3. As we said to French Government, we should not question good faith of assurances, but even if we accept assurances we can feel no security that they will in fact be able to maintain them once the ships are in French ports in the power or reach of the enemy, and I must confess that the desire of French Government to bring these ships back, if this turns out to be well founded, seems to me to give cause for some suspicion.

4. It would be most helpful if you felt able to give a further warning at Vichy on this matter, for if things went wrong it might well prove of extreme danger for us both.

\*   \*   \*   \*   \*

I kept in close touch with General de Gaulle.

*Prime Minister to General de Gaulle (Libreville)*                    10.XI.40

I feel most anxious for consultation with you. Situation between France and Britain has changed remarkably since you left. A very strong feeling has grown throughout France in our favour, as it is seen that we cannot be conquered and that war will go on. We know Vichy Government is deeply alarmed by the very stern pressure administered to them by United States. On the other hand, Laval and revengeful Darlan are trying to force French declaration of war against us and rejoice in provoking minor naval incidents. We have hopes of Weygand in Africa, and no one must underrate advantage that would follow if he were rallied. We are trying to arrive at some *modus vivendi* with Vichy which will minimise the risk of incidents and will enable favourable forces in France to develop. We have told them plainly that if they bomb Gibraltar or take other aggressive action we shall bomb Vichy, and pursue the Vichy Government wherever it chooses to go. So far we have had no response. You will see how important it is that you should be here. I therefore hope you will be able to tidy up at Libreville and come home as soon as possible. Let me know your plans.

On November 13 the President replied to my message of the 10th about the possible transfer of the *Jean Bart* and *Richelieu* to the Mediterranean for completion. He had immediately instructed the American Chargé d'Affaires at Vichy to obtain a confirmation or denial of this report and to point out that it was of vital interest to the Government of the United States that these vessels should remain in stations where they would not be exposed to control or seizure by a Power which might employ them to ends in conflict with the interests of the United States in the future of the French Fleet. Any such step on the part of France would inevitably seriously prejudice Franco-American relations. He also offered to buy the ships from the French Government if they would sell them.

The President also informed me that Pétain had stated to the American Chargé d'Affaires that the most solemn assurances had been given by him that the French Fleet, including the two battleships, would never fall into the hands of Germany. The Marshal said he had given those assurances to the United States Government, to the British Government, and even to me personally. "Again I reiterate them," he said. "These ships will be used to defend the possessions and territories of France. Unless we are attacked by the British, they will never be used against England. Even if I wanted to, I cannot sell those ships. It is impossible under the terms of the armistice, and even if it were possible it would never be permitted by the Germans. France is under Germany's heel and impotent. I would gladly sell them, if I were free, on condition that they be returned to us after the war, and save them for France in this way. I must repeat I have neither the right nor the possibility of selling them under present circumstances." Marshal Pétain had made this statement with great seriousness, but with no sign of either surprise or resentment at the suggestion. President Roosevelt had further instructed the Chargé d'Affaires to inform Marshal Pétain that the American offer remained open both about these vessels as well as about any others in the French Navy.

On November 23 the President sent me further reassurances. Marshal Pétain had stated categorically that he would keep the vessels now at Dakar and Casablanca where they were, and that if there was any change in this plan he would give the President previous notice.

\* \* \* \* \*

The attitude of Spain was of even more consequence to us than that of Vichy, with which it was so closely linked. Spain had much to give and even more to take away. We had been neutral in the sanguinary Spanish Civil War. General Franco owed little or nothing to us, but much—perhaps life itself—to the Axis Powers. Hitler and Mussolini had come to his aid. He disliked and feared Hitler. He liked and did not fear Mussolini. At the beginning of the World War Spain had declared, and since then strictly observed, neutrality. A fertile and needful trade flowed between our two countries, and the iron ore from Biscayan ports was important for our munitions. But now in May the "Twilight War" was over. The might of Nazi Germany was proved. The French front was broken. The Allied armies of the North were in peril. It was at this moment that I had gladly offered to a former colleague, displaced by the Ministerial changes, a new sphere of responsibility, for which his gifts and temperament were suited. On May 17 Sir Samuel Hoare had been appointed Ambassador to Spain, and certainly I believe that no one could have carried out better this wearing, delicate, and cardinal five years' mission. Thus we were very well represented at Madrid, not only by the Ambassador and by the Counsellor of the Embassy, Mr. Arthur Yencken,* but also by the Naval Attaché, Captain Hillgarth, who had retired from the Navy and lived in Majorca, but now returned to duty equipped with profound knowledge of Spanish affairs.

General Franco's policy throughout the war was entirely selfish and cold-blooded. He thought only of Spain and Spanish interests. Gratitude to Hitler and Mussolini for their help never entered his head. Nor, on the other hand, did he bear any grudge against England for the hostility of our Left Wing parties. This narrow-minded tyrant only thought about keeping his blood-drained people out of another war. They had had enough of war. A million men had been slaughtered by their brothers' hands. Poverty, high prices, and hard times froze the stony peninsula. No more war for Spain and no more war for Franco! Such were the commonplace sentiments with which he viewed and met the awful convulsion which now shook the world.

His Majesty's Government was quite content with this unheroic outlook. All we wanted was the neutrality of Spain. We wanted to trade with Spain. We wanted her ports to be denied to German and Italian submarines. We wanted not only an un-

* Mr. Yencken was killed in an air accident in 1944.

molested Gibraltar, but the use of the anchorage of Algeciras for our ships and the use of the ground which joins the Rock to the mainland for our ever-expanding air base. On these facilities depended in large measure our access to the Mediterranean. Nothing was easier than for the Spaniards to mount or allow to be mounted a dozen heavy guns in the hills behind Algeciras. They had a right to do so at any time, and, once mounted, they could at any moment be fired, and our naval and air bases would become unusable. The Rock might once again stand a long siege, but it would be only a rock. Spain held the key to all British enterprises in the Mediterranean, and never in the darkest hours did she turn the lock against us. So great was the danger that for nearly two years we kept constantly at a few days' notice an expedition of over five thousand men and their ships, ready to seize the Canary Islands, by which we could maintain air and sea control over the U-boats, and contact with Australasia round the Cape, if ever the harbour of Gibraltar were denied to us by the Spaniards.

There was another very simple manner in which the Franco Government could have struck us this destructive blow. They could have allowed Hitler's troops to traverse the Peninsula, besiege and take Gibraltar for them, and meanwhile themselves occupy Morocco and French North Africa. This became a deep anxiety after the French Armistice, when on June 27, 1940, the Germans reached the Spanish frontier in force, and proposed fraternal ceremonial parades in San Sebastian and in towns beyond the Pyrenees. Some German troops actually entered Spain. However, as the Duke of Wellington wrote in April 1820: "There is no country in Europe in the affairs of which foreigners can interfere with so little advantage as in those of Spain. There is no country in which foreigners are so much disliked, and even despised, and whose manners and habits are so little congenial with those of other countries in Europe." Now, a hundred and twenty years later, the Spaniards, reeling and quivering under the self-inflicted mutilations of the Civil War, were even less sociable. They did not wish to have foreign armies marching about their country. Even if they were Nazi and Fascist in their ideology, these morose people would rather have the foreigners' room than their company. Franco shared these feelings to the full, and in a most crafty manner he managed to give effect to them. We could admire his astuteness, especially as it was helpful to us.

\* \* \* \* \*

Like everyone else, the Spanish Government was staggered by the sudden downfall of France and the expected collapse or destruction of Britain. Lots of people all over the world had reconciled themselves to the idea of the "New Order in Europe", the "Herrenvolk", and all that. Franco therefore indicated in June that he was prepared to join the victors and share in the distribution of the spoils. Partly from appetite, and partly also from prudence, he made it clear that Spain had large claims. But at this moment Hitler did not feel the need of allies. He, like Franco, expected that in a few weeks or even days general hostilities would cease and England would be suing for terms. He therefore showed little interest in the gestures of active solidarity from Madrid.

By August the scene had changed. It was certain that Britain would fight on and probable that the war would be lengthy. With the contemptuous British rejection of his "Peace Offer" of July 19 Hitler sought allies, and to whom should he turn but to the dictator he had helped and who had so lately offered to join him? But Franco also had a different outlook, arising from the same causes. On August 8 the German Ambassador in Madrid informed Berlin that the Caudillo still held the same view, but that he had certain requests to make. First, the assurance that Gibraltar, French Morocco, and part of Algeria, including Oran, should be given to Spain, together with various expansions of territory in the Spanish African colonies. Adequate military and economic assistance would also be necessary, because Spain had only enough grain for eight months. Finally, Franco felt that the intervention of Spain should not take place until after the German landing in England, "in order to avoid too premature an entry into the war, and thus a duration which would be unbearable to Spain and in certain conditions a fountain of danger for the *régime*". At the same time Franco wrote to Mussolini recapitulating Spanish claims and asking for his support. Mussolini replied on August 25 by urging the Caudillo "not to cut himself off from the history of Europe". Hitler was embarrassed by the size of the Spanish claims, some of which would embroil him anew with Vichy. The taking of Oran from France would almost certainly lead to the setting up of a hostile French Government in North Africa. He balanced the issue.

Meanwhile the days were passing. During September Great Britain seemed to be holding her own against the German air offensive. The transfer of the fifty American destroyers made a profound impression throughout Europe, and to Spain it seemed that the United States was moving nearer to the war. Franco and his Spaniards therefore pursued the policy of raising and defining their claims and making it clear that these must be agreed in advance. Supplies also must be provided, particularly a number of 15-inch howitzers for the Spanish batteries facing Gibraltar. Meanwhile they paid the Germans in small coin. All the Spanish newspapers were Anglophobe. German agents were allowed to flaunt themselves all over Madrid. As the Spanish Foreign Minister, Beigbeder, was suspected of lack of enthusiasm for Germany, a special envoy, Serrano Suñer, head of the Falange, was sent on a formal visit to Berlin to smooth things over and preserve a sense of comradeship. Hitler harangued him at length, dwelling on the Spanish prejudices against the United States. The war, he suggested, might well turn into a war of continents —America against Europe. The islands off West Africa must be made secure. Later in the day Ribbentrop asked for a military base for Germany in the Canaries. Suñer, the pro-German and Falangist, refused even to discuss this, but dwelt incessantly upon Spanish needs for modern weapons and food and petrol, and for the satisfaction of her territorial demands at the expense of France. All this was necessary before Spain could realise her hopes of entering the war.

Ribbentrop went to Rome on September 19 to report and confer. He said that the Fuehrer thought the British attitude was "dictated by desperation, and also a complete failure to understand realities, as well as the hope of intervention by the Russians and the Americans". Mussolini observed that "the United States are for all practical purposes at the side of England". The sale of the fifty destroyers proved this. He advised an alliance with Japan to paralyse American action. "Although the American Navy can be considered large in the quantitative sense, it must be regarded as a dilettante organisation, like the British Army. . . ." The Duce continued: "There remains the problem of Yugoslavia

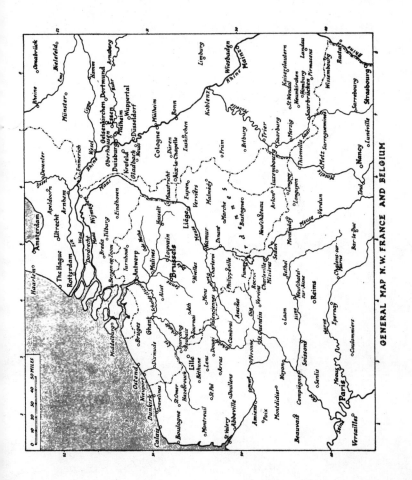

GENERAL MAP N.W. FRANCE AND BELGIUM

and Greece. Italy has half a million men on the Yugoslav frontier, and two hundred thousand on the Greek frontier. The Greeks represent for Italy what the Norwegians represented for Germany before the action of April. It is necessary for us to proceed with the liquidation of Greece, all the more so as when our land forces will have advanced into Egypt the English Fleet will not be able to remain at Alexandria, and will seek refuge in Greek ports."

At this point they both agreed that the principal object was to defeat England. The only question was, how? "Either the war," said Mussolini, "will finish before the spring or be protracted into next year." The second alternative now seemed to him the more probable, and the Spanish card must be played in the most effective way. Ribbentrop affirmed that a declaration of war by Spain following upon the alliance with Japan would be a new and formidable blow for England. But Suñer had not fixed any date.

\*　　\*　　\*　　\*　　\*

While the Spaniards became less ardent and more acquisitive Hitler felt an increased desire for their help. As early as August 15 General Jodl had pointed out that there were other means besides direct invasion by which England could be defeated, namely, prolonged air warfare, the stepping up of U-boat warfare, the capture of Egypt and the capture of Gibraltar. Hitler was strongly in favour of the assault on Gibraltar. But the Spanish terms were too high, and also by the end of September other ideas stirred his mind. On September 27 the Tripartite Pact between Germany, Italy, and Japan was signed in Berlin. This opened wider fields.

\*　　\*　　\*　　\*　　\*

The Fuehrer now decided to throw his personal influence into the scale. On October 4 he met Mussolini at the Brenner Pass. He spoke of the high demands and dilatory procedure of the Spanish Government. He feared that to give Spain what she asked would have two immediate consequences: an English occupation of the Spanish bases in the Canaries and the adhesion of the French Empire in North Africa to de Gaulle's movement. This, he said, would force the Axis seriously to extend their own sphere of operations. On the other hand, he did not exclude the possibility of having the French armed forces on his side in a

European campaign against Great Britain. Mussolini dilated on his plans for the conquest of Egypt. Hitler offered him special units for this attack. Mussolini did not think he needed them, at least before the final phase. On the Russian question Hitler remarked: "It is necessary to realise that my distrust of Stalin is equalled by his distrust of me." In any case, Molotov was coming in a short time to Berlin, and it would be the Fuehrer's task to direct Russian dynamism towards India.

On October 23 Hitler went all the way to the Franco-Spanish frontier at Hendaye to meet the Spanish dictator. Here the Spaniards, instead of being flattered by his condescension, demanded, according to Hitler's account to Mussolini, "objectives absolutely out of proportion to their strength". Spain demanded rectifications of the Pyrenees frontier, the cession of French Catalonia (French territory, once historically linked with Spain, but actually *north* of the Pyrenees), of Algeria from Oran to Cape Blanco, and virtually the whole of Morocco. The conversations, conducted through interpreters, lasted nine hours. They produced only a vague protocol and an arrangement for military conversations. "Rather than go through it again," Hitler told Mussolini later at Florence, "I would prefer to have three or four of my teeth out."*

On the way back from Hendaye the Fuehrer summoned Marshal Pétain to meet him at Montoire, near Tours. This interview had been prepared by Laval, who two days earlier had met Ribbentrop, and to his surprise Hitler, at this very place. Hitler and Laval both hoped to rally France to the defeat of Britain. The Marshal and most of his circle were at first shocked at this. But Laval portrayed the proposed meeting in glowing terms. When asked whether Hitler had initiated the idea, or whether it had been suggested to him, Laval replied: "What do you take him for? Do you think that Hitler needs a nurse? He has his own ideas, that man. He wants to see the Marshal. Besides, he has a great respect for him. This interview between the heads of the two States will be an historic event. In any case, something very different from a luncheon at Chequers."† Pétain was converted to the plan. He thought that his personal prestige might weigh with Hitler, and that it was worth while giving him the

* Ciano, *Diplomatic Papers*, p. 402.
† Du Moulin de Labarthète, *Le Temps des Illusions*, pp. 43–4.

impression that France would not be unwilling to "collaborate". At ease in the West, Hitler might turn his thoughts and armies eastwards.

The meeting took place in Hitler's armoured train, near a tunnel, on the afternoon of October 24. "I am happy," said the Fuehrer, "to shake hands with a Frenchman who is not responsible for this war."

Little more than shameful civilities resulted. The Marshal regretted that close relations had not been developed between France and Germany before the war. Perhaps it was not yet too late. Hitler pointed out that France had provoked the war and was defeated. But his aim now was to crush England. Before the United States could help her effectively, Britain would be occupied or else reduced to a heap of ruins. His object was to end the war as quickly as possible, for there was no business less profitable than war. All Europe would have to pay the cost, and so all Europe had the same interest. To what extent would France help? Pétain conceded the principle of collaboration, but pleaded that he could not define its limits. A *procès-verbal* was drawn up by which, "in accord with the Duce, the Fuehrer manifested his determination to see France occupy in the New Europe the place to which she is entitled". The Axis Powers and France had an identical interest in seeing the defeat of England accomplished as soon as possible. Consequently the French Government would support, within the limits of its ability, the measures which the Axis Powers might take for defence. Questions of detail would be settled by the armistice commission in concert with the French delegation. The Axis Powers would undertake that at the conclusion of peace with England France would retain in Africa a colonial domain "essentially equivalent to what she possessed at the moment".

According to the German record, Hitler was disappointed. Even Laval had begged him not to press France to make war against Britain before French opinion was duly prepared. Hitler afterwards spoke of Laval as "a dirty little democratic politico"; but he carried away a more favourable impression of Marshal Pétain. The Marshal however is reported to have said, when he got back to Vichy: "It will take six months to discuss this programme, and another six months to forget it." But the infamous transaction is not forgotten yet in France.

In October I had telegraphed to our Ambassador in Madrid:

*Prime Minister to Sir Samuel Hoare*                                19.X.40

We admire the way in which you are dealing with your baffling task. I hope you will manage to convey to Vichy, through the French Ambassador, two root ideas. First, that we will let bygones go and work with anyone who convinces us of his resolution to defeat the common foes. Secondly, that as we are fighting for our lives as well as for a victory which will relieve simultaneously all the captive States, we shall stop at nothing. Try to make Vichy feel what we here all take for certain, namely, that we have got Hitler beat, and though he may ravage the Continent and the war may last a long time his doom is certain. It passes my comprehension why no French leaders secede to Africa, where they would have an empire, the command of the seas, and all the frozen French gold in the United States. If this had been done at the beginning we might well have knocked out Italy by now. But surely the opportunity is the most splendid ever offered to daring men. Naturally one would not expect precise responses to such suggestions, but try to put it into their heads if you see any opening.

The various reports which we received of Montoire did not alter my general view of what our attitude towards Vichy should be. Now in November I expressed my views to my colleagues in a memorandum.

14.XI.40

Although revenge has no part in politics, and we should always be looking forward rather than looking back, it would be a mistake to suppose that a solution of our difficulties with Vichy will be reached by a policy of mere conciliation and forgiveness. The Vichy Government is under heavy pressure from Germany, and there is nothing that they would like better than to feel a nice, soft, cosy, forgiving England on their other side. This would enable them to win minor favours from Germany at our expense, and hang on as long as possible to see how the war goes. We, on the contrary, should not hesitate, when our interests require it, to confront them with difficult and rough situations, and make them feel that *we* have teeth as well as Hitler.

It must be remembered that these men have committed acts of baseness on a scale which has earned them the lasting contempt of the world, *and that they have done this without the slightest authority from the French people.* Laval is certainly filled by the bitterest hatred of England, and is reported to have said that he would like to see us *"écrabouillés"*, which means squashed so as to leave only a grease-spot. Undoubtedly, if he had had the power, he would have marketed the unexpected British resistance with his German masters to secure a better price for French help in finishing us off. Darlan is mortally envenomed by the injury we have done to his fleet. Pétain has always been an anti-British defeatist, and is now a dotard. The idea that we can build on such men is vain. They may however be forced by rising opinion in France and by German severities to change their line in our favour. Certainly we should have contacts with them. But in order to promote such favourable tendencies we must make sure the Vichy folk are kept well ground between the upper and nether millstones of Germany and Britain. In this way they are most likely to be brought into a more serviceable mood during the short run which remains to them.

\* \* \* \* \*

Marshal Pétain became increasingly resentful of Laval's prodding him along the road which would lead to war with Britain and German occupation of the North African colonies. On December 13 Laval arrived at Vichy with the proposal that Pétain should come to Paris to be present at the ceremonial transfer of the ashes of Napoleon's son, the Duke of Reichstadt ("l'Aiglon"), to the Invalides. This was Hitler's flowery idea of a solemn consecration of the entente reached at Montoire.

Pétain was not however attracted by a parade where the victor of Verdun would be exhibited on French soil with German guards of honour before the tomb of the Emperor Napoleon. He was moreover both wearied and fearful of Laval's methods and aims. Members of Pétain's staff therefore arranged the arrest of Laval. Energetic German intervention procured his release, but Pétain refused to accept him back as Minister. Laval retired in wrath to German-occupied Paris. I was glad that M. Flandin took his place as Foreign Minister. These events marked a change at Vichy. It seemed that the limits of collaboration had at last been reached. There were at this moment hopes of better French relations with Britain and of more sympathetic understanding for Vichy from the United States.

\* \* \* \* \*

It is convenient to carry the Spanish story forward at this point. Franco, now convinced of a long war and of the Spanish abhorrence of any more war, and by no means sure of a German victory, used every device of exasperating delay and exorbitant demands. He was by this time so sure of Suñer that on October 18 he made him Foreign Minister, representing the removal of Beigbeder as a proof of his devotion to the Axis. In November Suñer was summoned to Berchtesgaden, and Hitler expressed his impatience with Spain's delay in coming into the war. By now the Battle of Britain had been lost by the German Air Force. Italy was already involved in Greece and in North Africa. Serrano Suñer did not respond as was wished. He dwelt lengthily instead upon the economic difficulties of the Peninsula. Three weeks later Admiral Canaris, Chief of the German Secret Service, was sent to Madrid to arrange the details of Spain's entry into the war. He suggested that the German troops should pass the Spanish frontier on January 10, in preparation for an attack on Gibraltar on January 30. The Admiral was surprised when Franco told him that it was impossible for Spain to enter the war on the date mentioned. It seemed that the Caudillo feared the loss of Atlantic islands and Spanish colonies to the British Navy. He also emphasised the lack of food and the inability of Spain to stand a protracted war. As the German landing in England seemed indefinitely postponed Franco introduced a new condition. He would not move at any rate *until Suez was in Axis hands*, since not till then would he feel sure that Spain would not be involved in long-drawn-out hostilities.

On February 6, 1941, Hitler wrote a letter to Franco, appealing in strong and urgent terms that he should play the man without further delay. Franco replied, expressing his undying loyalty. He urged that preparations for the attack on Gibraltar should be continued with renewed vigour. As another new point he declared that only Spanish troops with German equipment must be used for this enterprise. Even if all this was arranged, Spain could not enter the war for economic reasons. Ribbentrop thereupon reported to the Fuehrer that Franco had no intention of making war. Hitler was scandalised, but, being now set upon the invasion of Russia, he did not perhaps like the idea of trying Napoleon's other unsuccessful enterprise, the invasion of Spain, at the same time. Considerable Spanish forces were now gathered along the Pyrenees, and he felt it was wiser to stick to his method with nations, "One by One". Thus by subtlety and trickery and blandishments of all kinds Franco succeeded in tiding things over and keeping Spain out of the war, to the inestimable advantage of Britain when she was all alone.

We could not count upon this at the time, and I urged the President to do all in his power to help forward the policy of conciliation.

*Former Naval Person to President Roosevelt*           **23.XI.40**

Our accounts show that situation in Spain is deteriorating and that the Peninsula is not far from starvation point. An offer by you of food month by month so long as they keep out of the war might be decisive. Small things do not count now, and this is a time for very plain talk to them. The occupation by Germany of both sides of the Straits would be a grievous addition to our naval strain, already severe. The Germans would soon have batteries working by Radar [*i.e.*, they could aim in the darkness], which would close the Straits both by night and day. With a major campaign developing in the Eastern Mediterranean and the need to reinforce and supply our armies there all round the Cape, we could not contemplate any military action on the mainland at or near the Straits. The Rock of Gibraltar will stand a long siege, but what is the good of that if we cannot use the harbour or pass the Straits? Once in Morocco the Germans will work southwards, and U-boats and aircraft will soon be operating freely from Casablanca and Dakar. I need not, Mr. President, enlarge upon the trouble this will cause to us, or the approach of trouble to the Western Hemisphere. We must gain as much time as possible.

This great danger had in fact passed away, and, though we did not know it, it passed for ever. It is fashionable at the present time to dwell on the vices of General Franco, and I am therefore glad to place on record this testimony to the duplicity and ingratitude of his dealings with Hitler and Mussolini. I shall presently record even greater services which these evil qualities in General Franco rendered to the Allied cause.

# CHAPTER XXVII

# MUSSOLINI ATTACKS GREECE

## October–November 1940

*Mussolini's Decision to Attack Greece – His Letter to Hitler of October 19 – The Florence Conference – The Italian Invasion of Greece, October 28, 1940 – Reinforcement of Admiral Cunningham's Fleet – The Arrival of the "Illustrious" – Our Obligations – Importance of Crete – Air Support for Greece – Minute to Chief of the Air Staff, November 2, 1940 – Wavell-Wilson Plans for an Offensive in Libya – Secrecy Causes Misunderstanding – Further Telegrams to Mr. Eden – Greek Need for the Cretan Division – Mr. Eden's Latest Telegrams – His Return – He Unfolds Operation "Compass" – General Agreement – War Cabinet Approves – The Fleet Air Arm Attack the Italian Fleet – Gallant Exploit at Taranto – Half the Italian Fleet Disabled for Six Months – Naval Dispositions – My Desire for an Amphibious Feature in "Compass" – My Telegram to Wavell of November 26 – Policy towards Turkey – An Improved Situation – Shortcomings at Suda Bay – The Abortive Italian Invasion of Greece from Albania – Death of Mr. Chamberlain – A Tribute to His Memory.*

A FRESH though not entirely unexpected outrage by Mussolini, with baffling problems and far-reaching consequences to all our harassed affairs, now broke upon the Mediterranean scene.

The Duce took the final decision to attack Greece on October 15, 1940. That morning a meeting of the Italian war leaders was held in the Palazzo Venezia. He opened the proceedings in the following words:

The object of this meeting is to define the course of action—in general terms—which I have decided to initiate against Greece. In the first instance, this action will have aims of both a maritime and territorial character. The territorial aims will be based on the possession

of the whole coast of Southern Albania . . . and the Ionian islands—
Zante, Cephalonia, and Corfu—and the occupation of Salonika. When
we have attained these objectives we shall have improved our position
*vis-à-vis* England in the Mediterranean. In the second instance . . . the
complete occupation of Greece, in order to put her out of action and
to assure that in all circumstances she will remain in our politic-
economic sphere.

Having thus defined the question, I have laid down the date—which
in my opinion must not be postponed even for an hour—and that is for
the 26th of this month. This is an action which I have matured at
length for months, before our entry into the war and before the
beginning of the conflict. . . . I would add that I foresee no complications in the north. Yugoslavia has every interest to keep quiet. . . . I also
exclude complications from the side of Turkey, particularly since
Germany has established herself in Roumania and since Bulgaria has
increased her strength. The latter can play a part in our game, and I
shall take the necessary steps so as not to miss the present unique opportunity for achieving her aspirations in Macedonia and for an outlet to
the sea. . . .*

On October 19 Mussolini wrote to Hitler telling him of the
decision to which he had come. Hitler was then on his journey
to Hendaye and Montoire. The letter (the text of which has not
come to light) seems to have followed him round. When it
finally reached him he at once proposed to Mussolini a meeting
to discuss the general political situation in Europe. This meeting
took place in Florence on October 28. That morning the Italian
attack on Greece had begun.

It seems however that Hitler did not choose to make an issue
of the Greek adventure. He said politely that Germany was in
accord with the Italian action in Greece, and then proceeded to
tell the tale of his meetings with Franco and Pétain. There can
be no doubt that he did not like what had been done by his
associate. A few weeks later, after the Italian attack was checked,
he wrote to Mussolini in his letter of November 20: "When I
asked you to receive me at Florence I began the journey with the
hope of being able to expound my views *before* the threatened
action against Greece had been taken, about which I had heard
only in general terms." In the main however he accepted the
decision of his ally.

* * * * *

* *Hitler and Mussolini: Letters and Documents*, p. 61.

Before dawn on October 28 the Italian Minister in Athens presented an ultimatum to General Metaxas, the Premier of Greece. Mussolini demanded that the whole of Greece should be opened to Italian troops. At the same time the Italian army in Albania invaded Greece at various points. The Greek Government, whose forces were by no means unready on the frontier, rejected the ultimatum. They also invoked the guarantee given by Mr. Chamberlain on April 13, 1939. This we were bound to honour. By the advice of the War Cabinet, and from his own heart, His Majesty replied to the King of the Hellenes: "Your cause is our cause; we shall be fighting against a common foe." I responded to the appeal of General Metaxas: "We will give you all the help in our power. We will fight a common foe and we will share a united victory." This undertaking was during a long story made good.

\* \* \* \* \*

Although we were still heavily outnumbered on paper by the Italian Fleet, marked improvements had been made in our Mediterranean strength. During September the *Valiant*, the armoured-deck aircraft-carrier *Illustrious*, and two A.A. cruisers had come safely through the Mediterranean to join Admiral Cunningham at Alexandria. Hitherto his ships had always been observed and usually bombed by the greatly superior Italian Air Force. The *Illustrious*, with her modern fighters and latest Radar equipment, by striking down patrols and assailants gave a new secrecy to the movements of our Fleet. This advantage was timely. Apart from a few air squadrons, a British mission, and perhaps some token troops, we had nothing to give; and even these trifles were a painful subtraction from ardent projects already lighting in the Libyan theatre. One salient strategic fact leaped out upon us—CRETE! The Italians must not have it. We must get it first—and at once. It was fortunate that at this moment Mr. Eden was in the Middle East, and that I thus had a ministerial colleague on the spot with whom to deal. He was about to return home after his conference with General Smuts at Khartoum. I telegraphed to him:

29.X.40

I recognise importance of your conference with Smuts, but hope first Wavell, and thereafter you, will return at earliest to Cairo.

We here are all convinced an effort should be made to establish ourselves in Crete, and that risks should be run for this valuable prize. You will have seen the Service telegrams on this subject.

*Prime Minister to Mr. Eden (at Khartoum)*                          29.X.40

It seems of prime importance to hold the best airfield possible and a naval fuelling base at Suda Bay. Successful defence of Crete is invaluable aid to defence of Egypt. Loss of Crete to the Italians would be a grievous aggravation of all Mediterranean difficulties. So great a prize is worth the risk, and almost equal to a successful offensive in Libya. Pray after an examination of whole problem with Wavell and Smuts, do not hesitate to make proposals for action on large scale at expense of other sectors, and ask for any further aid you require from here, including aircraft and anti-aircraft batteries. We are studying how to meet your need. Consider your return to Cairo indispensable.

At the invitation of the Greek Government, Suda Bay, the best harbour in Crete, was occupied by our forces two days later.

*Prime Minister to C.I.G.S.*                                        30.X.40

What steps are we taking to get news from the Greek front? Have we observers there? What is our attaché there doing?

Why do you not send one of your generals from Egypt at the head of a military mission to be at the headquarters of the Greek Field Army? Let them go and see the fighting and give us some close-up information about the relative merits of the two armies. I expect to have a good wire every day or so, telling us exactly what is happening, as far as the Greeks will allow it.

*Prime Minister to General Ismay, for C.O.S. Committee*            30.X.40

There is no objection to two battalions going to Freetown, pending their relief by the West African Brigade, after which they can go on to Egypt. They are not to leave England until it is agreed that the West African Brigade is to go to West Africa.

Both Crete and Malta come before Freetown in A.A. guns, and I cannot approve of this diversion at the present time. Neither can I agree to the diversion of a fighter squadron [for Freetown] at this stage. The Navy is responsible for preventing any sea-borne expedition attacking our West African colonies. As to the air attack, if the French bomb Freetown or Bathurst we will bomb Vichy. I do not think this will happen.

*Prime Minister to Air Vice-Marshal Longmore**                      1.XI.40

[In dispatching a Blenheim squadron to Greece] you have taken a very bold and wise decision. I hope to reinforce you as soon as possible.

* Commander-in-Chief Air Forces, Middle East.

*Prime Minister to General Ismay, for C.A.S. and for*
*C.O.S. Committee*                                           1.XI.40

I should propose to make immediate arrangements to send four additional heavy bomber squadrons (including the one already sent to Malta) to the Middle East at once, and also four Hurricane fighter squadrons. Let me see plans for this movement. I should like to have a report on this to-day.

*Prime Minister to General Ismay, for C.O.S. Committee*      1.XI.40

Mr. Eden has asked for ten thousand rifles for the Middle East. Can we not supply these out of the American packet, or is there any small parcel of rifles anywhere in the world to be picked up?

*Prime Minister to C.A.S.*                                   2.XI.40

1. I had in mind that the four bomber squadrons would fly to Crete or Greece via Malta. The personnel and ground stores would have to be carried through by cruiser. It is essential to have these squadrons operating at the earliest from bases in Greek territory upon the Italian fleet at Taranto, and generally against Southern Italy. For so vital an operation of war the Navy would have to make special exertions, and you should not assume that a ship will not be forthcoming, at any rate for such ground personnel, stores, etc., as are necessary to come into action at this very critical time. I see more difficulty in the vehicles, but perhaps some could come from Egypt, and the rest be improvised.

2. The fighters are, of course, more difficult, but I should hope that they could fly from a carrier to Malta, as was done last time. If necessary, the *Furious* would have to help the *Ark Royal*. Could they fly from Malta to an aerodrome in Greece? If not, could they fly on to a carrier to refuel, and thence to Greece? In the case of the fighters the same arrangements would have to be made about stores, ground personnel, etc., as with the bombers.

*Prime Minister to Mr. Eden (at G.H.Q., Middle East)*       2.XI.40

Greek situation must be held to dominate others now. We are well aware of our slender resources. Aid to Greece must be attentively studied lest whole Turkish position is lost through proof that England never tries to keep her guarantees. I invite you to stay in Cairo for at least another week while these questions are being studied and we make sure we have done our best from both ends. Meanwhile another thirty thousand men are reaching you by November 15, which must affect local situation in Egypt.

During Mr. Eden's earlier conferences and talks with General Wavell and also with General Wilson he posed the question, what action was intended if the Italian offensive did not develop. He was told in extreme secrecy that a plan was being made to attack the Italians in the Western Desert instead of waiting for them to open their offensive against Mersa Matruh. Neither he nor Wavell imparted these ideas to me or to the Chiefs of Staff. General Wavell begged the Secretary of State for War not to send any telegram on this subject, but to tell us verbally about it when he got home. Thus for some weeks we remained without knowledge of the way their minds were moving. It is clear from my message of October 26 that any forestalling operation on a large scale in the Western Desert would command my keen support. We were all however until Mr. Eden's return left under the impression that Wavell and Wilson were still wedded to the defensive battle at Mersa Matruh, and would wait there until they were attacked. The only action they seemed to contemplate in this extremely serious crisis was to send a battalion or so to Crete, a few air squadrons to Greece, and make some minor diversions against the Dodecanese and a small though desirable offensive in the Soudan. This seemed by no means good enough employment for the very large forces with which at great risk, exertion, and cost we had furnished them.

Our correspondence during this period was thus on both sides based upon misunderstanding. Wavell and the Secretary of State thought that for the sake of giving ineffectual aid to Greece we were pressing them to dissipate the forces they were gathering for an offensive in the Western Desert. We, on the other hand, not crediting them with offensive intentions, objected to their standing idle or trifling at such a crucial moment. In fact, as will presently be seen, we were all agreed. On November 1, indeed, Mr. Eden telegraphed cryptically:

We cannot from Middle East forces send sufficient air or land reinforcements to have any decisive influence upon course of fighting in Greece. To send such forces from here, or to divert reinforcements now on their way or approved, would imperil our whole position in the Middle East *and jeopardise plans for an offensive operation now being laid in more than one theatre.*\* After much painful effort and at the cost of grave risks we have, so far as our land forces are concerned, now built up a reasonably adequate *defensive*\* force here. We should presently be in a position to undertake certain offensive operations which if successful may have far-reaching effects on the course of the war as a

\* Author's italics.

whole. It would surely be bad strategy to allow ourselves to be diverted from this task, and unwise to employ our forces in fragments in a theatre of war where they cannot be decisive. . . . The best way in which we can help Greece is by striking at Italy, and we can do that most effectively from areas where our strength has been developed and where our plans are laid. I am anxious to put before you in detail at the earliest date the dispositions and plans which have been worked out here, and propose . . . to return home by the shortest route, leaving on the 3rd.

This telegram crossed one from me to him at Khartoum which afterwards had to be repeated back to Cairo, whither he had repaired.

*Prime Minister to Mr. Eden (at G.H.Q., Middle East)*      3.XI.40
Gravity and consequence of Greek situation compels your presence in Cairo. However unjust it may be, collapse of Greece without any effort by us will have deadly effect on Turkey and on future of war. . . . The Germans are not yet on the spot. Establishment of fuelling base and airfield in Crete to be steadily developed into permanent war fortresses [is] indispensable. This is being done. But surely effort must be made to aid Greece directly even if only with token forces. Quite understand how everyone with you is fixed on idea of set-piece battle at Mersa Matruh. For that very reason it is unlikely to occur. Enemy will await completion pipe-line and development of larger forces than are now concentrated. Your difficulties in attacking across the desert obvious, but if you have no major offensive of your own in Libya possible during next two months then you should run risks to stimulate Greek resistance. Over seventy thousand men sent to Middle East Command since June and thirty thousand reaching you before November 15, fifty-three thousand by end of the year. Armoured regiments have started in big convoy yesterday. Cannot therefore believe that various minor offensives of which you speak, plus major defence at Mersa Matruh, will outvalue need of effective action in Greece.

No one will thank us for sitting tight in Egypt with ever-growing forces while Greek situation and all that hangs on it is cast away. Loss of Athens far greater injury than Kenya and Khartoum, but no necessity for such a price to be paid. Read carefully Palairet's [our Minister in Athens] telegrams. New emergencies must be met in war as they come, and local views must not subjugate main issue. No one expected Italy so late in the year would attack Greece. Greece, resisting vigorously, with reasonable aid from Egypt and England, might check invaders. I am trying to send substantial bomber and fighter reinforcements to

Crete and Greece, flying from England, with stores by cruiser. If this proves feasible details will be cabled to-morrow or Monday. Trust you will grasp situation firmly, abandoning negative and passive policies and seizing opportunity which has come into our hands. "Safety first" is the road to ruin in war, even if you had the safety, which you have not. Send me your proposals earliest, or say you have none to make.

And again:

4.XI.40

We are sending you air reinforcements, arriving as fully explained in accompanying message from Chiefs of Staff. Send at once to Greece one Gladiator squadron and two more Blenheim squadrons, three in all. If necessary send a second battalion to Crete. Agreeably with arrival of our air reinforcements aforesaid and at earliest send one more Gladiator squadron. A.A. guns for airfields in Greece should precede arrival of squadrons.

It was proposed at this time to ask the Greeks to keep their Cretan division in the island. I therefore minuted:

*Prime Minister to C.I.G.S.*                6.XI.40

It will be difficult to deny the Greeks the use of this Cretan division. If that be so, we shall certainly have to put more troops on the island. It is important that there should be a certain number of troops, and that it should be thought by the enemy that we are landing considerable numbers. The area to be watched is very extensive, and the consequences of a counter-attack would be most disastrous.

Pray let me know your views.

*Prime Minister to C.I.G.S.*                7.XI.40

We shall render poor service to Greece if in consequence of our using Crete for our own purposes we deny them the use of two-thirds of their 5th Division. The defence of Crete depends on the Navy, but nevertheless there must be a certain deterrent force of troops on shore. I doubt if the two battalions of British and the three remaining Greek battalions will be sufficient. I am much obliged to you for telegraphing as I asked to General Wavell. He must provide in meal or in malt:

(a) Three or four thousand additional British troops and a dozen guns. These need not be fully equipped or mobile.

(b) He must do this from forces which he will not be using in the possibly impending battle.

(c) We must tell the Greeks we release [for service with their main army] the six battalions and the artillery of the 5th Greek Division.

Every effort should be made to rush arms or equipment to enable a reserve division of Greeks to be formed in Crete. Rifles and machine-guns are quite sufficient in this case. To keep a Greek division out of the battle on the Epirus front would be very bad, and to lose Crete because we had not sufficient bulk of forces there would be a crime.

It was time Mr. Eden should come home to report to us as he earnestly desired. The following telegrams are self-explanatory.

*Mr. Eden to Prime Minister* 3.XI.40

All strongly of the opinion I should return home as rapidly as possible in order to put whole position as seen from here before you. Earnestly hope you will agree to this. Propose to leave to-morrow morning. Perfectly prepared to fly back here if required after I have seen you, but am convinced that this meeting between us is most urgent. It is impossible to explain position and plans fully by telegram.

Please reply urgently.

Assent was given, and the Secretary of State began his journey. The following points were made in his simultaneous telegrams to me:

Conference in Cairo discussed situation in Crete. Admiral Cunningham emphasised the value of possession of Crete to us as a means of securing Eastern Mediterranean and of interfering with Italian transit traffic to North Africa. It would not however be possible to base fleet on Suda Bay for more than a few hours at a time at present owing to lack of anti-submarine protection.

He does not consider Italian attempt to take Crete is to be anticipated in the near future, nor unless and until Greece is overrun. He and Wavell have concerted arrangements for sending at once to Crete a part of the reinforcements referred to in my telegram of November 1. Admiral Cunningham does not consider it is necessary to keep any large British military garrison in Crete, and is convinced that once the Cretans are organised, one battalion, together with A.A. defences, would suffice. We then discussed the general question of help to Greece. As we said on September 22, "any assistance we may be able to give to Greece cannot be given until German-Italian threat to Egypt is finally liquidated, the security of Egypt being vital to our strategy and incidentally to the future of Greece. . . ."

Chief cry for help is for air reinforcements. No. 30 Blenheim Squadron left to-day for Athens. Longmore again emphasised his extreme reluctance to add any more squadrons to the Greek commitment in present conditions. He feels that to do so would lead to a large wastage of his aircraft from Italian attack whilst the aircraft are on

Greek or Cretan aerodromes unprepared with protecting pens, adequate ground A.A. defence, and other precautions of such nature, which are difficult to improvise at short notice. . . . In general all Commanders-in-Chief were strongly of the opinion that the defence of Egypt is of paramount importance to our whole position in the Middle East. They consider that from the strategical point of view the security of Egypt is the most urgent commitment, and must take precedence of attempts to prevent Greece being overrun. It is also essential if we are to retain the support of Turkey. . . .

Mr. Eden added in my private cipher the following:

5.XI.40

Although reinforcements ordered in Chiefs of Staff's telegrams involve additional risks in Western Desert and probably increased casualties, these risks must be faced in view of political commitments to aid Greece. Withdrawal, though it will hamper arrangements made in Western Desert, will not entirely dislocate them. But any increase in commitment or attempt to hasten rate of dispatch to Greece beyond that now laid down will mean serious risk to our position in Egypt. Uncertain factor still remains date by which air reinforcements, particularly fighters, arrive in Egypt to replace those sent to Greece. Experience hitherto shows that previous forecasts have not been fulfilled and time-table is sadly behind. Now feel that there is nothing further I can do here, and propose leave to-morrow morning by air.

\*    \*    \*    \*    \*

The Secretary of State for War got back home on November 8, and came that evening after the usual raid had begun to see me in my temporary underground abode in Piccadilly. He brought with him the carefully-guarded secret which I wished I had known earlier. Nevertheless no harm had been done. Mr. Eden unfolded in considerable detail to a select circle, including the C.I.G.S. and General Ismay, the offensive plan which General Wavell and General Wilson had conceived and prepared. No longer were we to await in our fortified lines at Mersa Matruh an Italian assault, for which defensive battle such long and artful preparations had been made. On the contrary, within a month or so we were ourselves to attack. The operation was to be called "Compass".

As will be seen from the map,\* Marshal Graziani's Italian army, now above eighty thousand strong, which had crossed the Egyptian frontier, was spread over a fifty-mile front in a series of fortified

\* On page 545.

camps, which were separated by wide distances and not mutually supporting, and with no depth in the system. Between the enemy's right flank at Sofafi and his next camp at Nibeiwa there was a gap of over twenty miles. The plan was to make an offensive spring through this gap, and, turning towards the sea, attack Nibeiwa camp and the Tummar group of camps in succession from the west—that is to say, from the rear. Meanwhile both the Sofafi camps and the camp at Meiktila, on the coast, were to be contained by light forces. For this purpose there were to be employed the 7th Armoured Division, the 4th Indian Division, now complete, and the 16th British Infantry Brigade, together with a composite force from the garrison of Mersa Matruh. This plan involved a serious risk, but also offered a glittering prize. The risk lay in the launching of all our best troops into the heart of the enemy's position by a move of seventy miles on two successive nights over the open desert, and with the peril of being observed and attacked from the air during the intervening day. Besides this, the food and petrol had to be nicely calculated, and if the time-scale went wrong the consequences must be grave.

The prize was worthy of the hazard. The arrival of our vanguard on the sea at Buq Buq or thereabouts would cut the communications of three-quarters of Marshal Graziani's army. Attacked by surprise from the rear, they might well be forced as a result of vigorous fighting into mass surrenders. In this case the Italian front would be irretrievably broken. With all their best troops captured or destroyed, no force would be left capable of withstanding a further onslaught, nor could any organised retreat be made to Tripoli along the hundreds of miles of coastal road.

Here, then, was the deadly secret which the Generals had talked over with their Secretary of State. This was what they had not wished to telegraph. We were all delighted. I purred like six cats. Here was something worth doing. It was decided there and then, subject to the agreement of the Chiefs of Staff and the War Cabinet, to give immediate sanction and all possible support to this splendid enterprise, and that it should take first place in all our thoughts and have, amid so many other competing needs, first claim upon our strained resources.

In due course these proposals were brought before the War

Cabinet. I was ready to state the case or have it stated. But when my colleagues learned that the Generals on the spot and the Chiefs of Staff were in full agreement with me and Mr. Eden, they declared that they did not wish to know the details of the plan, that the fewer who knew them the better, and that they whole-heartedly approved the general policy of the offensive. This was the attitude which the War Cabinet adopted on several important occasions, and I record it here that it may be a model, should similar dangers and difficulties arise in future times.

*　　*　　*　　*　　*

The Italian Fleet had not reacted in any way against our occupation of Crete, but Admiral Cunningham had for some time been anxious to strike a blow at them with his now augmented naval air forces as they lay in their main base at Taranto. The attack was delivered on November 11 as the climax of a well-concerted series of operations, during which Malta received troops, and further naval reinforcements, including the battleship *Barham*, two cruisers, and three destroyers, reached Alexandria. Taranto lies in the heel of Italy three hundred and twenty miles from Malta. Its magnificent harbour was heavily defended against all modern forms of attack. The arrival at Malta of some fast reconnaissance machines enabled us to discern our prey. The British plan was to fly two waves of aircraft from the *Illustrious*, the first of twelve and the second of nine, of which eleven were to carry torpedoes, and the rest either bombs or flares. The *Illustrious* released her aircraft shortly after dark from a point about a hundred and seventy miles from Taranto. For an hour the battle raged amid fire and destruction among the Italian ships. Despite the heavy flak only two of our aircraft were shot down. The rest flew safely back to the *Illustrious*.

By this single stroke the balance of naval power in the Mediterranean was decisively altered. The air photographs showed that three battleships, one of them the new *Littorio*, had been torpedoed, and in addition one cruiser was reported hit and much damage inflicted on the dockyard. Half the Italian battle fleet was disabled for at least six months, and the Fleet Air Arm could rejoice at having seized by their gallant exploit one of the rare opportunities presented to them.

An ironic touch is imparted to this event by the fact that on this very day the Italian Air Force at the express wish of Mussolini had taken part in the air attack on Great Britain. An Italian bomber force, escorted by about sixty fighters, attempted to bomb Allied convoys in the Medway. They were intercepted by our fighters, eight bombers and five fighters being shot down. This was their first and last intervention in our domestic affairs. They might have found better employment defending their fleet at Taranto.

I kept the President well informed.

*Former Naval Person to President*                                       16.XI.40
I am sure you will have been pleased about Taranto. The three uninjured Italian battleships have quitted Taranto to-day, which perhaps means they are withdrawing to Trieste.

And again:

*Former Naval Person to President*                                       21.XI.40
You may be interested to receive the following naval notes on the action at Taranto which I have asked the Admiralty to prepare:
1. This attack had been in Commander-in-Chief Mediterranean's mind for some time; he had intended to carry it out on October 21 (Trafalgar Day), when the moon was suitable, but a slight mishap to *Illustrious* led to a postponement. During his cruise in the Central Mediterranean on October 31 and November 1 it was again considered, but the moon did not serve and it was thought an attack with parachute flares would be less effective. Success in such an attack was believed to depend on state of moon, weather, an undetected approach by the Fleet, and good reconnaissance. The latter was provided by flying boats and a Glen Martin squadron working from Malta. On the night of November 11-12 all the above conditions were met. Unfavourable weather in the Gulf of Taranto prevented a repetition on 12th-13th.
2. Duplex pistols were used, and probably contributed to the success of the torpedo attack.
3. The Greek Ambassador at Angora reported on November 11 that Italian Fleet was concentrating at Taranto in preparation for an attack on Corfu. Reconnaissance on November 13 shows that undamaged battleships and 8-inch-gun cruisers have left Taranto—presumably owing to the attack on 11th-12th.

★　　★　　★　　★　　★

I now addressed General Wavell.

*Prime Minister to General Wavell*                                           14.XI.40

Chiefs of Staff, Service Ministers, and I have examined general situation in the light of recent events. Italian check on Greek front; British naval success against battle fleet at Taranto; poor showing Italian airmen have made over here; encouraging reports received of low morale in Italy; Gallabat; your own experiences by contacts in Western Desert; above all, the general political situation, make it very desirable to undertake operation of which you spoke to Secretary of State for War.

It is unlikely that Germany will leave her flagging ally unsupported indefinitely. Consequently it seems that now is the time to take risks and strike the Italians by land, sea, and air. You should act accordingly in concert with other Commanders-in-Chief.

*Prime Minister to General Wavell*                                           26.XI.40

News from every quarter must have impressed on you the importance of "Compass" in relation to whole Middle East position, including Balkans and Turkey, to French attitude in North Africa, to Spanish attitude, now trembling on the brink, to Italy, in grievous straits, and generally to the whole war. Without being over-sanguine, I cannot repress strong feeling of confidence and hope, and feel convinced risks inseparable from great deeds are fully justified.

Have asked Admiralty to inquire about part assigned to Fleet. If success is achieved, presume you have plans for exploiting it to the full. I am having a Staff study made of possibilities open to us, if all goes well, for moving fighting troops and also reserve forward by sea in long hops along the coast, and setting up new supply bases to which pursuing armoured vehicles and units might resort. Without wishing to be informed on details, I should like to be assured that all this has been weighed, explored, and as far as possible prepared.

It seems difficult to believe that Hitler will not be forced to come to the rescue of his partner, and obviously German plans may be far advanced for a drive through Bulgaria at Salonika. From several quarters we have reports in that Germans do not approve of Mussolini's adventure, and are inclined to let him pay the price himself. This makes me all the more suspicious that something bad is banking up ready to be let off soon. Every day's delay is in our favour. It might be that "Compass" would in itself determine action of Yugoslavia and Turkey, and anyhow, in event of success, we should be able to give Turkey far greater assurances of early support than it has been in our power to do so far. One may indeed see possibility of centre of gravity in Middle East shifting suddenly from Egypt to the Balkans, and from Cairo to Constantinople. You are no doubt preparing your mind for this, and a Staff study is being made here.

As we told you the other day, we shall stand by you and Wilson in any well-conceived action irrespective of result, because no one can guarantee success in war, but only deserve it.

Tell Longmore that I much admire his calling in of the southern squadrons and accepting the risk of punishment there. If all is well *Furious* and her outfit should reach Takoradi to-morrow. This should make amends for all the feathers we have had to pull out of him for Greece, where the part played by R.A.F. in Greek victories has been of immense military and political consequence. All good wishes to you both, and to the Admiral, who is doing so splendidly. I rejoice to hear that he finds Suda Bay "an inestimable benefit".

*Prime Minister to Foreign Secretary*         26.XI.40

I suggest the following to our Ambassador in Turkey:

(*Begins.*) We have placed before you the various arguments for and against Turkish intervention which have occurred to the Staff officers who have reported upon the matter, but we do not wish to leave you in any doubt of what our own opinion and your instructions are. We want Turkey to come into the war as soon as possible. We are not pressing her to take any special steps to help the Greeks, except to make it clear to Bulgaria that any move by Germany through Bulgaria to attack Greece, or any hostile movement by Bulgaria against Greece, will be followed by immediate Turkish declaration of war. We should like Turkey and Yugoslavia now to consult together so as, if possible, to have a joint warning ready to offer Bulgaria and Germany at the first sign of a German movement towards Bulgaria. In the event of German troops traversing Bulgaria with or without Bulgarian assistance, it is vital that Turkey should fight there and then. If she does not, she will find herself left absolutely alone, the Balkans will have been eaten up one by one, and it will be beyond our power to help her. You may mention that by the summer of 1941 we hope to have at least fifteen divisions operating in the Middle East, and by the end of the year nearly twenty-five. We do not doubt our ability to defeat Italy in Africa.

6 p.m.—The Chiefs of Staff are in general agreement with the above.

*Prime Minister to First Lord, First Sea Lord, and*
*General Ismay, for C.O.S. Committee*

(C.A.S. to see.)         30.XI.40

*Furious* should return home at once, and carry another load of aircraft and pilots as reinforcement for the Middle East. Every effort should be made to put off her refit till after she has carried this force. C.A.S. should say what composition of force is best.

*Prime Minister to General Ismay*                                    1.XII.40

Exactly what have we got and done at Suda Bay [Crete]—*i.e.*, troops, A.A. guns, coast defence guns, lights, wireless, R.D.F., nets, mines, preparation of aerodromes, etc.?

I hope to be assured that many hundreds of Cretans are working at strengthening the defences and lengthening and improving the aerodromes.

*General Ismay, for C.O.S. Committee*                                 1.XII.40

The continued retreat of the Italians in Albania, and the reports which we have received to-day of difficulties of feeding and watering their forces in the Libyan Desert, together with other reports of aircraft being moved back to Tripoli to be safer from our attacks, combined with safe arrival at Takoradi of thirty-three Hurricanes with first-class pilots, all constitute new facts entitling us to take a more confident view of the situation, which should be communicated to General Wavell.

The enormous advantage of being able, once an enemy is on the run, to pull supplies and fighting troops forward eighty miles in a night by sea, and bring fresh troops up to the advance-guard, is very rarely offered in war. General Wavell's reply to my telegram does not seem to take any account of this, and, considering how much we have ourselves at stake, I do not think we should be doing our duty if we did not furnish him with the results of our Staff study. It is a crime to have amphibious power and leave it unused. Therefore I wish the study, if favourable, to be telegraphed. It must however be ready by the 3rd at latest.

I add the following general observation: The fact that we now have established ourselves at Suda Bay entitles us to feel much easier about Malta. While the Fleet is or may be at Suda it will be most unlikely that any large landing will be attempted at Malta, which we have already reinforced by tanks and guns from Middle East. . . . The possession of Suda Bay has made an enormous change in the Eastern Mediterranean.

The story of Suda Bay is sad. The tragedy was not reached until 1941. I believe I had as much direct control over the conduct of the war as any public man had in any country at this time. The knowledge I possessed, the fidelity and active aid of the War Cabinet, the loyalty of all my colleagues, the ever-growing efficiency of our war machine, all enabled an intense focusing of constitutional authority to be achieved. Yet how far short was the action taken by the Middle East Command of what was ordered and what we all desired! In order to appreciate the limitations of human action, it must be remembered how much was going on in every direction at the same time. Nevertheless it remains astonishing to me that we should have failed to make Suda Bay the amphibious citadel of which all Crete was the fortress. Everything was understood and agreed, and much was done; but all was half-scale effort. We were presently to pay heavily for our shortcomings.

\* \* \* \* \*

The Italian invasion of Greece from Albania was another heavy rebuff to Mussolini. The first assault was repulsed with heavy loss, and the Greeks immediately counter-attacked. In the northern (Macedonian) sector the Greeks advanced into Albania, capturing Koritza on November 22. In the central sector of the northern Pindus an Italian Alpini division was annihilated. In the coastal zone, where the Italians had at first succeeded in making deep penetrations, they hastily retreated from the Kalamas river. The Greek army, under General Papagos, showed superior skill in mountain warfare, out-manoeuvring and outflanking their enemy. By the end of the year their prowess had forced the Italians thirty miles behind the Albanian frontier along the whole front. For several months twenty-seven Italian divisions were pinned in Albania by sixteen Greek divisions. The remarkable Greek resistance did much to hearten the other Balkan countries and Mussolini's prestige sank low.

\* \* \* \* \*

On November 9 Mr. Neville Chamberlain died at his country home in Hampshire. I had obtained the King's permission to have him supplied with the Cabinet papers, and until a few days before the end he followed our affairs with keenness, interest, and tenacity. He met the approach of death with a steady eye. I think he died with the comfort of knowing that his country had at least turned the corner.

As soon as the House met on November 12 I paid a tribute to his character and career:

At the lychgate we may all pass our own conduct and our own judgments under a searching review. It is not given to human beings, happily for them, for otherwise life would be intolerable, to foresee or to predict to any large extent the unfolding course of events. In one phase men seem to have been right, in another they seem to have been wrong. Then again, a few years later, when the perspective of time has lengthened, all stands in a different setting. There is a new proportion. There is another scale of values. History with its flickering lamp stumbles along the trail of the past, trying to reconstruct its scenes, to revive its echoes, and kindle with pale gleams the passion of former days. What is the worth of all this? The only guide to a man is his conscience; the only shield to his memory is the rectitude and sincerity of his actions. It is very imprudent to walk through life without this shield, because we are so often mocked by the failure of our hopes and the upsetting of our calculations; but with this shield, however the fates may play, we march always in the ranks of honour.

Whatever else history may or may not say about these terrible, tremendous years, we can be sure that Neville Chamberlain acted with perfect sincerity according to his lights and strove to the utmost of his capacity and authority, which were powerful, to save the world from the awful, devastating struggle in which we are now engaged. . . . Herr Hitler protests with frantic words and gestures that he has only desired peace. What do these ravings and outpourings count for before the silence of Neville Chamberlain's tomb? Long, hard, and hazardous years lie before us, but at least we enter upon them united and with clean hearts. . . .

He was, like his father and his brother Austen before him, a famous Member of the House of Commons, and we here assembled this morning, members of all parties, without a single exception, feel that we do ourselves and our country honour in saluting the memory of one whom Disraeli would have called "an English worthy".

# CHAPTER XXVIII

# LEND-LEASE

ABOVE the roar and clash of arms there now loomed upon us a world-fateful event of a different order. The Presidential Election took place on November 5. In spite of the tenacity and vigour with which these four-yearly contests are conducted, and the bitter differences on domestic issues which at this time divided the two main parties, the Supreme Cause was respected by the responsible leaders, Republicans and Democrats alike. At Cleveland on November 2 Mr. Roosevelt said: "Our policy is to give all possible material aid to the nations which still resist aggression across the Atlantic and Pacific Oceans." His opponent, Mr. Wendell Willkie, declared the same day at Madison Square Garden: "All of us—Republicans, Democrats, and Independents—believe in giving aid to the heroic British people. We must make available to them the products of our industry."

This larger patriotism guarded both the safety of the American Union and our life. Still, it was with profound anxiety that I awaited the result. No new-comer into power could possess or soon acquire the knowledge and experience of Franklin Roosevelt. None could equal his commanding gifts. My own relations with him had been most carefully fostered by me, and seemed already to have reached a degree of confidence and friendship which was a vital factor in all my thought. To close the slowly-built-up comradeship, to break the continuity of all our discussions, to begin again with a new mind and personality, seemed to me a repellent prospect. Since Dunkirk I had not been conscious of the same sense of strain. It was with indescribable relief that I received the news that President Roosevelt had been re-elected.

*Former Naval Person to President Roosevelt*     **6.XI.40**
I did not think it right for me as a foreigner to express any opinion upon American politics while the election was on, but now I feel you will not mind my saying that I prayed for your success and that I am truly thankful for it. This does not mean that I seek or wish for anything more than the full, fair, and free play of your mind upon the world issues now at stake in which our two nations have to discharge their respective duties. We are entering upon a sombre phase of what must evidently be a protracted and broadening war, and I look forward to being able to interchange my thoughts with you in all that confidence and goodwill which has grown up between us since I went to the Admiralty at the outbreak. Things are afoot which will be remembered as long as the English language is spoken in any quarter of the globe, and in expressing the comfort I feel that the people of the United States have once again cast these great burdens upon you I must avow my sure faith that the lights by which we steer will bring us all safely to anchor.

Curiously enough, I never received any answer to this telegram. It may well have been engulfed in the vast mass of congratulatory messages which were swept aside by urgent work.

Up till this time we had placed our orders for munitions in the United States separately from, though in consultation with, the American Army, Navy, and Air Services. The ever-increasing volume of our several needs had led to overlapping at numerous points, with possibilities of friction arising at lower levels in spite of general goodwill. "Only a single, unified Government procurement policy for all defence purposes", writes Mr. Stettinius,* "could do the tremendous job that was now ahead." This meant that the United States Government should place all the orders for weapons in America. Three days after his re-election the President publicly announced a "rule of thumb" for the division of American arms output. As weapons came off the production line they were to be divided roughly fifty-fifty between the United States forces and the British and Canadian forces. That same day the Priorities Board approved a British request to order twelve thousand more aeroplanes in the United States in addition to the eleven thousand we had already booked. But how was all this to be paid for?

\*        \*        \*        \*        \*

In mid-November Lord Lothian, who had recently flown home from Washington, spent two days with me at Ditchley. I had been advised not to make a habit of staying at Chequers every week-end, especially when the moon was full, in case the enemy should pay me special attention. Mr. Ronald Tree and his wife made me and my staff very welcome many times at their large and charming house near Oxford. Ditchley is only four or five miles away from Blenheim. In these agreeable surroundings I received the Ambassador. Lothian seemed to me a changed man. In all the years I had known him he had given me the impression of high intellectual and aristocratic detachment from vulgar affairs. Airy, viewy, aloof, dignified, censorious, yet in a light and gay manner, he had always been good company. Now, under the same hammer that smote upon us all, I found an earnest, deeply-stirred man. He was primed with every aspect and detail of the American attitude. He had won nothing but goodwill and confidence in Washington by his handling of the Destroyer-cum-Bases negotiations. He was fresh from intimate contact with the President, with whom he had established a warm personal friendship. His mind was now set upon the Dollar Problem; this was grim indeed.

Before the war the United States was governed by the Neu-

*Stettinius, *Lend-Lease*, p. 62.

trality Act, which obliged the President on September 3, 1939, to place an embargo on all shipments of arms to any of the belligerent nations. Ten days later he had called Congress to a special session to consider the removal of this ban, which, under the appearance of impartiality, virtually deprived Great Britain and France of all the advantages of the command of the seas in the transport of munitions and supplies. It was not until the end of November 1939, after many weeks of discussion and agitation, that the Neutrality Act was repealed and the new principle of "Cash and Carry" substituted. This still preserved the appearance of strict neutrality on the part of the United States, for Americans were as free to sell weapons to Germany as to the Allies. In fact, however, our sea-power prevented any German traffic, while Britain and France could "Carry" freely as long as they had "Cash". Three days after the passage of the new law our Purchasing Commission, headed by Mr. Arthur Purvis, a man of outstanding ability, began its work.

\* \* \* \* \*

Britain entered the war with about 4,500 millions in dollars, or in gold and in United States investments that could be turned into dollars. The only way in which these resources could be increased was by new gold-production in the British Empire, mainly of course in South Africa, and by vigorous efforts to export goods, principally luxury goods, such as whisky, fine woollens, and pottery, to the United States. By these means an additional 2,000 million dollars were procured during the first sixteen months of the war. During the period of the "Twilight War" we were torn between a vehement desire to order munitions in America and gnawing fear as our dollar resources dwindled. Always in Mr. Chamberlain's day the Chancellor of the Exchequer, Sir John Simon, would tell us of the lamentable state of our dollar resources and emphasise the need for conserving them. It was more or less accepted that we should have to reckon with a rigorous limitation of purchases from the United States. We acted, as Mr. Purvis once said to Stettinius, "as if we were on a desert island on short rations which we must stretch as far as we could".\*

This had meant elaborate arrangements for eking out our

\* Stettinius, *Lend-Lease*, p. 60.

money. In peace we imported freely and made payments as we liked. When war came we had to create a machine which mobilised gold and dollars and other private assets, which stopped the ill-disposed from remitting their funds to countries where they felt things were safer, and which cut out wasteful imports and other expenditures. On top of making sure that we did not waste our money, we had to see that others went on taking it. The countries of the sterling area were with us: they adopted the same kind of exchange control policy as we did and were willing takers and holders of sterling. With others we made special arrangements by which we paid them in sterling, which could be used anywhere in the sterling area, and they undertook to hold any sterling for which they had no immediate use and to keep dealings at the official rates of exchange. Such arrangements were originally made with the Argentine and Sweden, but were extended to a number of other countries on the Continent and in South America. These arrangements were completed after the spring of 1940, and it was a matter of satisfaction—and a tribute to sterling—that we were able to achieve and maintain them in circumstances of such difficulty. In this way we were able to go on dealing with most parts of the world in sterling, and to conserve most of our precious gold and dollars for our vital purchases in the United States.

When the war exploded into hideous reality in May 1940, we were conscious that a new era had dawned in Anglo-American relations. From the time I formed the new Government, and Sir Kingsley Wood became Chancellor of the Exchequer, we followed a simpler plan, namely, to order everything we possibly could and leave future financial problems on the lap of the Eternal Gods. Fighting for life and presently alone, under ceaseless bombardment, with invasion glaring upon us, it would have been false economy and misdirected prudence to worry too much about what would happen when our dollars ran out. We were conscious of the tremendous changes taking place in American opinion, and of the growing belief, not only in Washington but throughout the Union, that their fate was bound up with ours. Moreover, at this time an intense wave of sympathy and admiration for Britain surged across the American nation. Very friendly signals were made to us from Washington direct, and also through Canada, encouraging our boldness and indicating that somehow or other a way would be found. In Mr. Morgenthau, Secretary of the Treasury, the cause of the Allies had a tireless champion. The taking over of the French contracts in June had almost doubled our rate of spending across the Exchange. Besides this, we placed new orders for aeroplanes, tanks, and merchant ships in every direction, and promoted the building of great new factories both in the United States and Canada.

★　★　★　★　★

Up till November 1940 we had paid for everything we had received. We had already sold 335 million dollars' worth of American shares requisitioned for sterling from private owners in Britain. We had paid out over 4,500 million dollars in cash. We had only 2,000 millions left, the greater part in investments, many of which were not readily marketable. It was plain that we could not go on any longer in this way. Even if we divested ourselves of all our gold and foreign assets, we could not pay for half we had ordered, and the extension of the war made it necessary for us to have ten times as much. We must keep something in hand to carry on our daily affairs.

Lothian was confident that the President and his advisers were earnestly seeking the best way to help us. Now that the election was over the moment to act had come. Ceaseless discussions on behalf of the Treasury were proceeding in Washington between their representative, Sir Frederick Phillips, and Mr. Morgenthau. The Ambassador urged me to write a full statement of our position to the President. Accordingly that Sunday at Ditchley I drew up, in consultation with him, a personal letter. On November 16 I telegraphed to Roosevelt: "I am writing you a very long letter on the outlook for 1941 which Lord Lothian will give you in a few days." As the document had to be checked and rechecked by the Chiefs of Staff and the Treasury, and approved by the War Cabinet, it was not completed before Lothian's return to Washington. On November 26 I sent him a message, "I am still struggling with my letter to the President, but hope to cable it to you in a few days." In its final form the letter was dated December 8, and was immediately sent to the President. As it gives a view of the whole situation agreed to by all concerned in London, and as it played a recognisable part in our fortunes, it deserves study.

10 DOWNING STREET, WHITEHALL
December 8, 1940

My dear Mr. President,

1. As we reach the end of this year I feel you will expect me to lay before you the prospects for 1941. I do so with candour and confidence, because it seems to me that the vast majority of American citizens have recorded their conviction that the safety of the United States as well as the future of our two Democracies and the kind of civilisation for which they stand are bound up with the survival and independence of the British Commonwealth of Nations. Only thus can those bastions of sea-power upon which the control of the Atlantic and Indian Oceans depend be preserved in faithful and friendly hands. The control of the Pacific by the United States Navy and of the Atlantic by the British Navy is indispensable to the security and trade routes of both our countries, and the surest means of preventing war from reaching the shores of the United States.

2. There is another aspect. It takes between three and four years to convert the industries of a modern state to war purposes. Saturation-point is reached when the maximum industrial effort that can be spared from civil needs has been applied to war production. Germany certainly reached this point by the end of 1939. We in the British Empire are now only about half-way through the second year. The United States, I should suppose, is by no means so far advanced as we. Moreover, I understand that immense programmes of naval, military, and air defence are now on foot in the United States, to complete which certainly two years are needed. It is our British duty in the common interest, as also for our own survival, to hold the front and grapple with the Nazi power until the preparations of the United States are complete. Victory may come before two years are out; but we have no right to count upon it to the extent of relaxing any effort that is humanly possible. Therefore I submit with very great respect for your good and friendly consideration that there is a solid identity of interest between the British Empire and the United States while these conditions last. It is upon this footing that I venture to address you.

3. The form which this war has taken, and seems likely to hold, does not enable us to match the immense armies of Germany in any theatre where their main power can be brought to bear. We can however, by the use of sea-power and air-power, meet the German armies in regions where only comparatively small forces can be brought into action. We must do our best to prevent the German domination of Europe spreading into Africa and into Southern Asia. We have also to maintain in constant readiness in this Island armies strong enough to make

the problem of an oversea invasion insoluble. For these purposes we are forming as fast as possible, as you are already aware, between fifty and sixty divisions. Even if the United States were our ally, instead of our friend and indispensable partner, we should not ask for a large American expeditionary army. Shipping, not men, is the limiting factor, and the power to transport munitions and supplies claims priority over the movement by sea of large numbers of soldiers.

4. The first half of 1940 was a period of disaster for the Allies and for Europe. The last five months have witnessed a strong and perhaps unexpected recovery by Great Britain fighting alone, but with the invaluable aid in munitions and in destroyers placed at our disposal by the great Republic of which you are for the third time the chosen Chief.

5. The danger of Great Britain being destroyed by a swift, overwhelming blow has for the time being very greatly receded. In its place there is a long, gradually-maturing danger, less sudden and less spectacular, but equally deadly. This mortal danger is the steady and increasing diminution of sea tonnage. We can endure the shattering of our dwellings and the slaughter of our civil population by indiscriminate air attacks, and we hope to parry these increasingly as our science develops, and to repay them upon military objectives in Germany as our Air Force more nearly approaches the strength of the enemy. The decision for 1941 lies upon the seas. Unless we can establish our ability to feed this Island, to import the munitions of all kinds which we need, unless we can move our armies to the various theatres where Hitler and his confederate Mussolini must be met, and maintain them there, and do all this with the assurance of being able to carry it on till the spirit of the Continental Dictators is broken, we may fall by the way, and the time needed by the United States to complete her defensive preparations may not be forthcoming. It is therefore in shipping and in the power to transport across the oceans, particularly the Atlantic Ocean, that in 1941 the crunch of the whole war will be found. If on the other hand we are able to move the necessary tonnage to and fro across salt water indefinitely, it may well be that the application of superior air-power to the German homeland and the rising anger of the German and other Nazi-gripped populations will bring the agony of civilisation to a merciful and glorious end.

But do not let us underrate the task.

6. Our shipping losses, the figures for which in recent months are appended, have been on a scale almost comparable to those of the worst year of the last war. In the five weeks ending November 3 losses reached a total of 420,300 tons. Our estimate of annual tonnage which ought to be imported in order to maintain our effort at full strength is 43 million tons; the tonnage entering in September was only at the

rate of 37 million tons, and in October of 38 million tons. Were this diminution to continue at this rate it would be fatal, unless indeed immensely greater replenishment than anything at present in sight could be achieved in time. Although we are doing all we can to meet this situation by new methods, the difficulty of limiting losses is obviously much greater than in the last war. We lack the assistance of the French Navy, the Italian Navy, and the Japanese Navy, and above all of the United States Navy, which was of such vital help to us during the culminating years. The enemy commands the ports all around the northern and western coasts of France. He is increasingly basing his submarines, flying-boats, and combat planes on these ports and on the islands off the French coast. We are denied the use of the ports or territory of Eire in which to organise our coastal patrols by air and sea. In fact, we have now only one effective route of entry to the British Isles, namely, the Northern Approaches, against which the enemy is increasingly concentrating, reaching ever farther out by U-boat action and long-distance aircraft bombing. In addition, there have for some months been merchant-ship raiders both in the Atlantic and Indian Oceans. And now we have the powerful warship raider to contend with as well. We need ships both to hunt down and to escort. Large as are our resources and preparations, we do not possess enough.

7. The next six or seven months [will] bring relative battleship strength in home waters to a smaller margin than is satisfactory. *Bismarck* and *Tirpitz* will certainly be in service in January. We have already *King George V*, and hope to have *Prince of Wales* in the line at the same time. These modern ships are of course far better armoured, especially against air attack, than vessels like *Rodney* and *Nelson*, designed twenty years ago. We have recently had to use *Rodney* on transatlantic escort, and at any time when numbers are so small a mine or a torpedo may alter decisively the strength of the line of battle. We get relief in June, when *Duke of York* will be ready, and shall be still better off at the end of 1941, when *Anson* also will have joined. But these two first-class modern 35,000-ton* 15-inch-gun German battleships force us to maintain a concentration never previously necessary in this war.

8. We hope that the two Italian *Littorios* will be out of action for a while, and anyway they are not so dangerous as if they were manned by Germans. Perhaps they might be! We are indebted to you for your help about the *Richelieu* and *Jean Bart*, and I daresay that will be all right. But, Mr. President, as no one will see more clearly than you, we have during these months to consider for the first time in this war a fleet action in which the enemy will have two ships at least as good as

* Actually they were nearer 45,000 tons.

our two best and only two modern ones. It will be impossible to reduce our strength in the Mediterranean, because the attitude of Turkey, and indeed the whole position in the Eastern Basin, depends upon our having a strong fleet there. The older, unmodernised battleships will have to go for convoy. Thus even in the battleship class we are at full extension.

9. There is a second field of danger. The Vichy Government may, either by joining Hitler's New Order in Europe or through some manœuvre, such as forcing us to attack an expedition dispatched by sea against the Free French colonies, find an excuse for ranging with the Axis Powers the very considerable undamaged naval forces still under its control. If the French Navy were to join the Axis the control of West Africa would pass immediately into their hands, with the gravest consequences to our communications between the Northern and Southern Atlantic, and also affecting Dakar and of course thereafter South America.

10. A third sphere of danger is in the Far East. Here it seems clear that Japan is thrusting southward through Indo-China to Saigon and other naval and air bases, thus bringing them within a comparatively short distance of Singapore and the Dutch East Indies. It is reported that the Japanese are preparing five good divisions for possible use as an overseas expeditionary force. We have to-day no forces in the Far East capable of dealing with this situation should it develop.

11. In the face of these dangers we must try to use the year 1941 to build up such a supply of weapons, particularly of aircraft, both by increased output at home in spite of bombardment and through ocean-borne supplies, as will lay the foundations of victory. In view of the difficulty and magnitude of this task, as outlined by all the facts I have set forth, to which many others could be added, I feel entitled, nay bound, to lay before you the various ways in which the United States could give supreme and decisive help to what is, in certain aspects, the common cause.

12. The prime need is to check or limit the loss of tonnage on the Atlantic approaches to our island. This may be achieved both by increasing the naval forces which cope with the attacks, and by adding to the number of merchant ships on which we depend. For the first purpose there would seem to be the following alternatives:

(1) The reassertion by the United States of the doctrine of the freedom of the seas from illegal and barbarous methods of warfare, in accordance with the decisions reached after the late Great War, and as freely accepted and defined by Germany in 1935. From this, United States ships should be free to trade with countries against which there is not an effective legal blockade.

(2) It would, I suggest, follow that protection should be given to this lawful trading by United States forces, *i.e.*, escorting battleships, cruisers, destroyers, and air flotillas. The protection would be immensely more effective if you were able to obtain bases in Eire for the duration of the war. I think it is improbable that such protection would provoke a declaration of war by Germany upon the United States, though probably sea incidents of a dangerous character would from time to time occur. Herr Hitler has shown himself inclined to avoid the Kaiser's mistake. He does not wish to be drawn into war with the United States until he has gravely undermined the power of Great Britain. His maxim is "One at a time".

The policy I have ventured to outline, or something like it, would constitute a decisive act of constructive non-belligerency by the United States, and, more than any other measure, would make it certain that British resistance could be effectively prolonged for the desired period and victory gained.

(3) Failing the above, the gift, loan, or supply of a large number of American vessels of war, above all destroyers, already in the Atlantic is indispensable to the maintenance of the Atlantic route. Further, could not the United States Naval Forces extend their sea control of the American side of the Atlantic so as to prevent the molestation by enemy vessels of the approaches to the new line of naval and air bases which the United States is establishing in British islands in the Western Hemisphere? The strength of the United States Naval Forces is such that the assistance in the Atlantic that they could afford us, as described above, would not jeopardise the control of the Pacific.

(4) We should also then need the good offices of the United States and the whole influence of its Government, continually exerted, to procure for Great Britain the necessary facilities upon the southern and western shores of Eire for our flotillas, and, still more important, for our aircraft, working to the westward into the Atlantic. If it were proclaimed an American interest that the resistance of Great Britain should be prolonged and the Atlantic route kept open for the important armaments now being prepared for Great Britain in North America, the Irish in the United States might be willing to point out to the Government of Eire the dangers which its present policy is creating for the United States itself.

His Majesty's Government would of course take the most effective measures beforehand to protect Ireland if Irish action exposed it to German attack. It is not possible for us to compel the people of Northern Ireland against their will to leave the United Kingdom and join Southern Ireland. But I do not doubt that if the Government of Eire would show its solidarity with the democracies of the English-speaking world at this crisis a Council for Defence of all Ireland could be set up out of which the unity of the island would probably in some form or other emerge after the war.

13. The object of the foregoing measures is to reduce to manageable proportions the present destructive losses at sea. In addition, it is indispensable that the merchant tonnage available for supplying Great Britain, and for the waging of the war by Great Britain with all vigour, should be substantially increased beyond the one and a quarter million tons per annum which is the utmost we can now build. The convoy system, the *détours*, the zigzags, the great distances from which we now have to bring our imports, and the congestion of our western harbours, have reduced by about one-third the fruitfulness of our existing tonnage. To ensure final victory not less than three million tons of additional merchant shipbuilding capacity will be required. Only the United States can supply this need. Looking to the future, it would seem that production on a scale comparable to that of the Hog Island scheme of the last war ought to be faced for 1942. In the meanwhile we ask that in 1941 the United States should make available to us every ton of merchant shipping, surplus to its own requirements, which it possesses or controls, and to find some means of putting into our service a large proportion of merchant shipping now under construction for the National Maritime Board.

14. Moreover, we look to the industrial energy of the Republic for a reinforcement of our domestic capacity to manufacture combat aircraft. Without that reinforcement reaching us in substantial measure we shall not achieve the massive preponderance in the air on which we must rely to loosen and disintegrate the German grip on Europe. We are at present engaged on a programme designed to increase our strength to seven thousand first-line aircraft by the spring of 1942. But it is abundantly clear that this programme will not suffice to give us the weight of superiority which will force open the doors of victory. In order to achieve such superiority it is plain that we shall need the greatest production of aircraft which the United States of America is capable of sending us. It is our anxious hope that in the teeth of continuous bombardment we shall realise the greater part of the production which we have planned in this country. But not even with the addition to our squadrons of all the aircraft which, under present arrangements, we may derive from planned output in the United States can we hope to achieve the necessary ascendancy. May I invite you then, Mr. President, to give earnest consideration to an immediate order on joint account for a further two thousand combat aircraft a month? Of these aircraft, I would submit, the highest possible proportion should

be heavy bombers, the weapon on which, above all others, we depend to shatter the foundations of German military power. I am aware of the formidable task that this would impose upon the industrial organisation of the United States. Yet, in our heavy need, we call with confidence to the most resourceful and ingenious technicians in the world. We ask for an unexampled effort, believing that it can be made.

15. You have also received information about the needs of our armies. In the munitions sphere, in spite of enemy bombing, we are making steady progress here. Without your continued assistance in the supply of machine tools and in further releases from stock of certain articles, we could not hope to equip as many as fifty divisions in 1941. I am grateful for the arrangements, already practically completed, for your aid in the equipment of the Army which we have already planned, and for the provision of the American type of weapons for an additional ten divisions in time for the campaign of 1942. But when the tide of Dictatorship begins to recede many countries trying to regain their freedom may be asking for arms, and there is no source to which they can look except the factories of the United States. I must therefore also urge the importance of expanding to the utmost American productive capacity for small arms, artillery, and tanks.

16. I am arranging to present you with a complete programme of the munitions of all kinds which we seek to obtain from you, the greater part of which is of course already agreed. An important economy of time and effort will be produced if the types selected for the United States Services should, whenever possible, conform to those which have proved their merit under the actual conditions of war. In this way reserves of guns and ammunition and of aeroplanes become interchangeable, and are by that very fact augmented. This is however a sphere so highly technical that I do not enlarge upon it.

17. Last of all, I come to the question of Finance. The more rapid and abundant the flow of munitions and ships which you are able to send us, the sooner will our dollar credits be exhausted. They are already, as you know, very heavily drawn upon by the payments we have made to date. Indeed, as you know, the orders already placed or under negotiation, including the expenditure settled or pending for creating munitions factories in the United States, many times exceed the total exchange resources remaining at the disposal of Great Britain. The moment approaches when we shall no longer be able to pay cash for shipping and other supplies. While we will do our utmost, and shrink from no proper sacrifice to make payments across the Exchange, I believe you will agree that it would be wrong in principle and mutually disadvantageous in effect if at the height of this struggle Great

Britain were to be divested of all saleable assets, so that after the victory was won with our blood, civilisation saved, and the time gained for the United States to be fully armed against all eventualities, we should stand stripped to the bone. Such a course would not be in the moral or economic interests of either of our countries. We here should be unable, after the war, to purchase the large balance of imports from the United States over and above the volume of our exports which is agreeable to your tariffs and industrial economy. Not only should we in Great Britain suffer cruel privations, but widespread unemployment in the United States would follow the curtailment of American exporting power.

18. Moreover, I do not believe that the Government and people of the United States would find it in accordance with the principles which guide them to confine the help which they have so generously promised only to such munitions of war and commodities as could be immediately paid for. You may be certain that we shall prove ourselves ready to suffer and sacrifice to the utmost for the Cause, and that we glory in being its champions. The rest we leave with confidence to you and to your people, being sure that ways and means will be found which future generations on both sides of the Atlantic will approve and admire.

19. If, as I believe, you are convinced, Mr. President, that the defeat of the Nazi and Fascist tyranny is a matter of high consequence to the people of the United States and to the Western Hemisphere, you will regard this letter not as an appeal for aid, but as a statement of the minimum action necessary to achieve our common purpose.

A table was added showing the losses by enemy action of British, Allied, and neutral merchant tonnage for the periods given.*

The letter, which was one of the most important I ever wrote, reached our great friend when he was cruising, on board an American warship, the *Tuscaloosa*, in the sunlight of the Caribbean Sea. He had only his own intimates around him. Harry Hopkins, then unknown to me, told me later that Mr. Roosevelt read and re-read this letter as he sat alone in his deck-chair, and that for two days he did not seem to have reached any clear conclusion. He was plunged in intense thought, and brooded silently.

From all this there sprang a wonderful decision. It was never a question of the President not knowing what he wanted to do.

* See Appendix B.

His problem was how to carry his country with him and to persuade Congress to follow his guidance. According to Stettinius, the President, as early as the late summer, had suggested at a meeting of the Defence Advisory Commission on Shipping Resources that "It should not be necessary for the British to take their own funds and have ships built in the United States, or for us to loan them money for this purpose. There is no reason why we should not take a finished vessel and lease it to them for the duration of the emergency". It seems that this idea had originated in the Treasury Department, whose lawyers, especially Oscar S. Cox, of Maine, had been stirred by Secretary Morgenthau. It appeared that by a Statute of 1892 the Secretary for War, "when in his discretion it will be for the public good", could lease Army property if not required for public use for a period of not longer than five years. Precedents for the use of this Statute, by the *lease* of various Army items, from time to time were on record.

Thus the word "lease" and the idea of applying the lease principle to meeting British needs had been in President Roosevelt's mind for some time as an alternative to a policy of indefinite loans which would soon far outstrip all possibilities of repayment. Now suddenly all this sprang into decisive action, and the glorious conception of Lend-Lease was proclaimed.

The President returned from the Caribbean on December 16, and broached his plan at his Press Conference next day. He used a simple illustration. "Suppose my neighbour's house catches fire and I have a length of garden hose four or five hundred feet away. If he can take my garden hose and connect it up with his hydrant, I may help him to put out the fire. Now what do I do? I don't say to him before that operation, 'Neighbour, my garden hose cost me fifteen dollars; you have to pay me fifteen dollars for it.' No! What is the transaction that goes on? I don't want fifteen dollars—I want my garden hose back after the fire is over." And again: "There is absolutely no doubt in the mind of a very overwhelming number of Americans that the best immediate defence of the United States is the success of Great Britain defending itself; and that therefore, quite aside from our historic and current interest in the survival of Democracy in the world as a whole, it is equally important from a selfish point of view and of American defence that we should do everything possible to help the British Empire to defend itself." Finally:

"I am trying to eliminate the dollar mark."

On this foundation the ever-famous Lend-Lease Bill was at once prepared for submission to Congress. I described this to Parliament later as "the most unsordid act in the history of any nation". Once it was accepted by Congress it transformed immediately the whole position. It made us free to shape by agreement long-term plans of vast extent for all our needs. There was no provision for repayment. There was not even to be a formal account kept in dollars or sterling. What we had was lent or leased to us because our continued resistance to the Hitler tyranny was deemed to be of vital interest to the great Republic. According to President Roosevelt, the defence of the United States and not dollars was henceforth to determine where American weapons were to go.

\* \* \* \* \*

It was at this moment, the most important in his public career, that Philip Lothian was taken from us. Shortly after his return to Washington he fell suddenly and gravely ill. He worked unremittingly to the end. On December 12, in the full tide of success, he died. This was a loss to the nation and to the Cause. He was mourned by wide circles of friends on both sides of the ocean. To me, who had been in such intimate contact with him a fortnight before, it was a personal shock. I paid my tribute to him in a House of Commons united in deep respect for his work and memory.

\* \* \* \* \*

I had now to turn immediately to the choice of his successor. It seemed that our relations with the United States at this time required as Ambassador an outstanding national figure and a statesman versed in every aspect of world politics. Having ascertained from the President that my suggestion would be acceptable, I invited Mr. Lloyd George to take the post. He had not felt able to join the War Cabinet in July, and was not happily circumstanced in British politics. His outlook on the war and the events leading up to it was from a different angle from mine. There could be no doubt however that he was our foremost citizen, and that his incomparable gifts and experience would be devoted to the success of his mission. I had a long talk with him in the Cabinet Room, and also at luncheon on a second day. He showed genuine pleasure at having been invited. "I tell my friends," he said, "I have had honourable offers made to me by the Prime Minister." He was sure that at the age of seventy-seven he ought not to undertake so exacting a task. As a result of my long conversations with him I was conscious that he had aged even in the months which had passed since I had asked him to join the War Cabinet, and with regret but also with conviction I abandoned my plan.

I next turned to Lord Halifax, whose prestige in the Conservative Party stood high, and was enhanced by his being at the Foreign Office. For a Foreign Secretary to become an Ambassador marks in a unique manner the importance of the mission. His high character was everywhere respected, yet at the same time his record in the years before the war and the way in which events had moved left him exposed to much disapprobation and even hostility from the Labour side of our National Coalition. I knew that he was conscious of this himself.

When I made him this proposal, which was certainly not a personal advancement, he contented himself with saying in a simple and dignified manner that he would serve wherever he was thought to be most useful. In order to emphasise still further the importance of his duties, I arranged that he should resume his function as a member of the War Cabinet whenever he came home on leave. This arrangement worked without the slightest inconvenience, owing to the qualities and experience of the personalities involved, and for six years thereafter, both under the National Coalition and the Labour-Socialist Government, Halifax discharged the work of Ambassador to the United States with conspicuous and ever-growing influence and success.

President Roosevelt, Mr. Hull, and other high personalities in Washington were extremely pleased with the selection of Lord Halifax. Indeed it was at once apparent to me that the President greatly preferred it to my first proposal. The appointment of the new Ambassador was received with marked approval both in America and at home, and was judged in every way adequate and appropriate to the scale of events.

\*     \*     \*     \*     \*

I had no doubt who should fill the vacancy at the Foreign Office. On all the great issues of the past four years I had, as these pages have shown, dwelt in close agreement with Anthony Eden. I have described my anxieties and emotions when he parted company with Mr. Chamberlain in the spring of 1938. Together we had abstained from the vote on Munich. Together we had resisted the party pressures brought to bear upon us in our constituencies during the winter of that melancholy year. We had been united in thought and sentiment at the outbreak of the war and as colleagues during its progress. The greater part of Eden's public life had been devoted to the study of foreign affairs. He had held the splendid office of Foreign Secretary with distinction, and had resigned it when only forty-two years of age for reasons which are in retrospect, and at this time, viewed with the approval of all parties in the State. He had played a fine part as Secretary of State for War during this terrific year, and his conduct of Army affairs had brought us very close together. We thought alike, even without consultation, on a very great number of practical issues as they arose from day to day. I looked forward to an agreeable and harmonious comradeship between the Prime Minister and the Foreign Secretary, and this hope was certainly fulfilled during the four and a half years of war and policy which lay before us. Eden was sorry to leave the War Office, in all the stresses and excitements of which he was absorbed; but he returned to the Foreign Office like a man going home.

\* \* \* \* \*

I filled Mr. Eden's place as Secretary of State for War by submitting to the King the name of Captain Margesson, at that time the Chief Whip to the National Government. This choice excited some adverse comment. David Margesson had been for nearly ten years at the head of the Government Whip's Office in the House of Commons, and it had fallen to him to marshal and to stimulate the patient and solid Conservative majorities which had so long sustained the Baldwin and Chamberlain Administrations. I had, as a leading figure among the Conservative dissentients from the India Bill, had many sharp passages with him. In the course of those eleven years of my exclusion from office my contacts with him had been not infrequent and generally hostile. I had formed the opinion that he was a man of high ability, serving his chief, whoever he was, with unfaltering loyalty, and treating his opponents with strict good faith. This opinion was also held by the Whips of the Labour and Liberal Parties, and such a reputation is of course essential to the discharge of this particular office. When I became Prime Minister it was generally expected that I should find someone else for the task, but I was quite sure that I should receive from Margesson the same skilful and faithful service that he had given to my predecessors; and in this I had been in no way disappointed. He had served in the first World War, and through much of the worst of it as a regimental officer, gaining the Military Cross. He thus had a strong soldierly background as well as a complete knowledge of the House of Commons.

In Margesson's place I appointed Captain James Stuart, with whom also I had had many differences, but for whose character I had high respect.

\* \* \* \* \*

The interval between November 1940 and the passage of Lend-Lease in March 1941 was marked by an acute stringency in dollars. Every kind of expedient was devised by our friends. The American Government bought from us some of the war plants which they had built to our order in the United States. They assigned them to the American defence programme, but bade us go on using them to the full. The War Department placed orders for munitions that it did not need immediately, so that when finished they could be released to us. On the other hand, certain things were done which seemed harsh and painful to us. The President sent a warship to Capetown to carry away all the gold we had gathered there. The great British business of Courtaulds in America was sold by us at the request of the United States Government at a figure much below its intrinsic worth. I had a feeling that these steps were taken to emphasise the hardship of our position and raise feeling against the opponents of Lend-Lease. Anyhow, in one way or another we came through.

On December 30 the President gave a "fireside chat" on the radio, urging his policy upon his countrymen. "There is danger ahead—danger against which we must prepare. But we well know that we cannot escape danger by crawling into bed and pulling the covers over our heads. . . . If Britain should go down, all of us in all the Americas would be living at the point of a gun, a gun loaded with explosive bullets, economic as well as military. We must produce arms and ships with every energy and resource we can command. . . . *We must be the great arsenal of Democracy.*"

*Former Naval Person to President Roosevelt*                31.XII.40

We are deeply grateful for all you said yesterday. We welcome especially the outline of your plans for giving us the aid without which Hitlerism cannot be extirpated from Europe and Asia. We can readily guess why you have not been able to give a precise account of how your proposals will be worked out. Meanwhile some things make me anxious.

First, sending the warship to Capetown to take up the gold lying there may produce embarrassing effects. It is almost certain to become known. This will disturb public opinion here and throughout the Dominions and encourage the enemy, who will proclaim that you are sending for our last reserves. If you feel this is the only way, directions will be given for the available Capetown gold to be loaded on the ship. But we should avoid it if we can. Could we, for instance, by a technical operation, exchange gold in South Africa for gold held for others at Ottawa and make the latter available for movement to New York? We must know soon, because the ship is on its way.

My second anxiety is because we do not know how long Congress will debate your proposals and how we should be enabled to place orders for armaments and pay our way if this time became protracted. Remember, Mr. President, we do not know what you have in mind, or exactly what the United States is going to do, and we are fighting for our lives. What would be the effect upon the world situation if we had to default in payments to your contractors, who have their workmen to pay? Would not this be exploited by the enemy as a complete breakdown in Anglo-American co-operation? Yet a few weeks' delay might well bring this upon us.

Thirdly, apart from the interim period, there arises a group of problems about the scope of your plan after being approved by Congress. What is to be done about the immense heavy payments still due to be made under existing orders before delivery is completed? Substantial advance payments on these same orders have already denuded our resources. We have continued need for various American commodities not definitely weapons—for instance, raw materials and oil. Canada and other Dominions, Greece and refugee allies, have clamant dollar needs to keep their war effort alive. I do not seek to know immediately how you will solve these latter questions. We shall be entirely ready, for our part, to lay bare to you all our resources and our liabilities around the world, and we shall seek no more help than the common cause demands. We naturally wish to feel sure that the powers with which you propose to arm yourself will be sufficiently wide to deal with these larger matters, subject to all proper examination.

Sir Frederick Phillips is discussing these matters with Mr. Secretary Morgenthau, and he will explain the war commitments we have in many parts of the world for which we could not ask your direct help, but for which gold and dollars are necessary. This applies also to the Dutch and Belgian gold, which we may become under obligation to return in specie in due course.

They burned a large part of the City of London last night, and the scenes of widespread destruction here and in our provincial centres are shocking; but when I visited the still-burning ruins to-day the spirit of the Londoners was as high as in the first days of the indiscriminate bombing in September, four months ago.

I thank you for testifying before all the world that the future safety and greatness of the American Union are intimately concerned with the upholding and the effective arming of that indomitable spirit.

All my heartiest good wishes to you in the New Year of storm that is opening upon us.

# CHAPTER XXIX

# GERMANY AND RUSSIA

*Hitler Turns Eastward – Stalin's Attempts to Placate Germany – Communist Machinations in the British Factories – Soviet Miscalculations – Molotov's Visit to Berlin – His Meeting with Ribbentrop – And with the Fuehrer – Soviet-Nazi Negotiations – Projects of Dividing the British Empire – Further Argument with the Fuehrer – A British Air Raid Intervenes – Talks in a Dug-out – Stalin's Account Given to Me in August 1942 – Hitler's Final Resolve to Invade Russia – Military Preparations – The Draft Agreement – The Soviets Ask for More – Ambassador Schulenburg's Efforts to Reach an Agreement – Operation "Barbarossa", December 1940.*

HITLER had failed to quell or conquer Britain. It was plain that the Island would persevere to the end. Without the command of the sea or the air it had been deemed impossible to move German armies across the Channel. Winter with its storms had closed upon the scene. The German attempt to cow the British nation or shatter their war-making capacity and will-power by bombing had been foiled, and the Blitz was costly. There must be many months' delay before "Sea Lion" could be revived, and with every week that passed the growth, ripening, and equipment of the British home armies required a larger "Sea Lion", with aggravated difficulties of transportation. Even three-quarters of a million men with all their furnishings would not be enough in April or May 1941. What chance was there of finding by then the shipping, the barges, the special landing-craft necessary for so vast an oversea stroke? How could they be assembled under ever-increasing British air-power? Meanwhile this air-power, fed by busy factories in Britain and the United States, and by immense training schemes for pilots in the

Dominions centred in Canada, would perhaps in a year or so make the British Air Force superior in numbers, as it was already in quality, to that of Germany. Can we wonder then that Hitler, once convinced that Goering's hopes and boasts had been broken, should turn his eyes to the East? Like Napoleon in 1804, he recoiled from the assault of the Island until at least the Eastern danger was no more. He must, he now felt, at all costs settle with Russia before staking everything on the invasion of Britain. Obeying the same forces and following the same thoughts as Napoleon when he marched the Grand Army from Boulogne to Ulm, Austerlitz, and Friedland, Hitler abandoned for the moment his desire and need to destroy Great Britain. That must now become the final act of the drama.

There is no doubt that he had made up his own mind by the end of September 1940. From that time forth the air attacks on Britain, though often on a larger scale through the general multiplication of aircraft, took second place in the Fuehrer's thoughts and German plans. They might be maintained as effective cover for other designs, but Hitler no longer counted on them for decisive victory. Eastward ho! Personally, on purely military grounds, I should not have been averse from a German attempt at the invasion of Britain in the spring or summer of 1941. I believed that the enemy would suffer the most terrific defeat and slaughter that any country had ever sustained in a specific military enterprise. But for that very reason I was not so simple as to expect it to happen. In war what you don't dislike is not usually what the enemy does. Still, in the conduct of a long struggle, when time seemed for a year or two on our side, and mighty allies might be gained, I thanked God that the supreme ordeal was to be spared our people. As will be seen from my papers written at the time, I never seriously contemplated a German descent upon England in 1941. By the end of 1941 the boot was on the other leg; we were no longer alone; three-quarters of the world were with us. But tremendous events, measureless before they happened, were to mark that memorable year.

While to uninformed continentals and the outer world our fate seemed forlorn, or at best in the balance, the relations between Nazi Germany and Soviet Russia assumed the first position in world affairs. The fundamental antagonisms between the two despotic Powers resumed their sway once it was certain that Britain could not be stunned and overpowered like France and the Low Countries. To do him justice, Stalin tried his very best to work loyally and faithfully with Hitler, while at the same time gathering all the strength he could in the enormous mass of Soviet Russia. He and Molotov sent their dutiful congratulations on every German victory. They poured a heavy flow of food and essential raw materials into the Reich. Their Fifth Column Communists did what they could to disturb our factories. Their radio diffused its abuse and slanders against us. They were at any time ready to reach a permanent settlement with Nazi Germany upon the numerous important questions open between them, and to accept with complacency the final destruction of the British power. But all the while they recognised that this policy might fail. They were resolved to gain time by every means, and had no intention, as far as they could measure the problem, of basing Russian interests or ambitions solely upon a German victory. The two great totalitarian empires, equally devoid of moral restraints, confronted each other, polite but inexorable.

There had of course been disagreements about Finland and Roumania. The Soviet leaders had been shocked at the fall of France, and the end of the Second Front for which they were so soon to clamour. They had not expected so sudden a collapse, and had counted confidently on a phase of mutual exhaustion on the Western Front. Now there was no Western Front! Still, it would be foolish to make any serious change in their collaboration with Germany till it could be seen whether Britain would give in or be crushed in 1940. As it gradually became apparent to the Kremlin that Britain was capable of maintaining a prolonged and indefinite war, during which anything might happen about the United States and also in Japan, Stalin became more conscious of his danger and more earnest to gain time. Nevertheless it is remarkable, as we shall see, what advantages he sacrificed and what risks he ran to keep on friendly terms with Nazi Germany. Even more surprising were the miscalculations and the ignorance which he displayed about what was coming to him. He was indeed from September 1940 to the moment of Hitler's assault in June 1941 at once a callous, a crafty, and an ill-informed giant.

\*    \*    \*    \*    \*

With these preliminaries we may come to the episode of Molotov's visit to Berlin on November 12, 1940. Every compliment was paid and all ceremony shown to the Bolshevik envoy when he reached the heart of Nazi Germany. During the next two days long and tense discussions took place between Molotov and Ribbentrop, and also with Hitler. All the essential facts of these formidable interchanges and confrontations have been laid bare in the selection of captured documents published early in 1948 by the State Department in Washington under the title *Nazi-Soviet Relations, 1939-1941*. On this it is necessary to draw if the story is to be told or understood.

Molotov's first meeting was with Ribbentrop.*

November 12, 1940

The Reich Foreign Minister said that in the letter to Stalin he had already expressed the firm conviction of Germany that no power on earth could alter the fact that the beginning of the end had now arrived for the British Empire. England was beaten, and it was only a question of time when she would finally admit her defeat. It was possible that this would happen soon, because in England the situation was deteriorating daily. Germany would of course welcome an early conclusion of the conflict, since she did not wish under any circumstances to sacrifice human lives unnecessarily. If however the British did not make up their minds in the immediate future to admit their defeat, they would definitely ask for peace during the coming year. Germany was continuing her bombing attacks on England day and night. Her submarines would gradually be employed to the full extent and would inflict terrible losses on England. Germany was of the opinion that England could perhaps be forced by these attacks to give up the struggle. A certain uneasiness was already apparent in Great Britain, which seemed to indicate such a solution. If however England were not forced to her knees by the present mode of attack, Germany would, as soon as weather conditions permitted, resolutely proceed to a large-scale attack and thereby definitely crush England. This large-scale attack had thus far been prevented only by abnormal weather conditions. . . .

Any attempt at a landing or at military operations on the European continent by England or by England backed by America was doomed to complete failure at the start. This was no military problem at all. This the English had not yet understood, because apparently there was some degree of confusion in Great Britain and because the country was

* See *Nazi-Soviet Relations*, pp. 218 ff.

led by a political and military dilettante by the name of Churchill, who throughout his previous career had completely failed at all decisive moments and who would fail again this time.

Furthermore, the Axis completely dominated its part of Europe militarily and politically. Even France, which had lost the war and had to pay for it (of which the French, incidentally, were quite aware), had accepted the principle that France in the future would never again support England and de Gaulle, the quixotic conqueror of Africa. Because of the extraordinary strength of their position, the Axis Powers were not therefore considering how they might win the war, but rather how rapidly they could end the war which was already won.

     ★    ★    ★    ★    ★

After luncheon the Soviet Envoy was received by the Fuehrer, who dilated further upon the total defeat of Britain. The war, he said, had led to complications which were not intended by Germany, but which had compelled her from time to time to react militarily to certain events.

The Fuehrer then outlined to Molotov the course of military operations up to the present, which had led to the fact that England no longer had an ally on the Continent. . . . The English retaliatory measures were ridiculous, and the Russian gentlemen could convince themselves at first hand of the fiction of alleged destruction in Berlin. As soon as atmospheric conditions improved Germany would be poised for the great and final blow against England. At the moment, then, it was her aim to try not only to make military preparations for this final struggle, but also to clarify the political issues which would be of importance during and after this showdown. He had therefore re-examined the relations with Russia, and not in a negative spirit, but with the intention of organising them positively—if possible, for a long period of time. In so doing he had reached several conclusions:

1. Germany was not seeking to obtain military aid from Russia.

2. Because of the tremendous extension of the war, Germany had been forced, in order to oppose England, to penetrate into territories remote from her and in which she was not basically interested politically or economically.

3. There were nevertheless certain requirements, the full importance of which had become apparent only during the war, but which were absolutely vital to Germany. Among them were certain sources of raw materials, which were considered by Germany as most vital and absolutely indispensable.

     ★    ★    ★    ★    ★

To all this Molotov gave a non-committal assent.

Molotov asked about the Tripartite Pact.* What was the meaning of the New Order in Europe and in Asia, and what *rôle* would the U.S.S.R. be given in it? These issues must be discussed during the Berlin conversations and during the contemplated visit of the Reich Foreign Minister to Moscow, on which the Russians were definitely counting. Moreover, there were issues to be clarified regarding Russia's Balkan and Black Sea interests, about Bulgaria, Roumania, and Turkey. It would be easier for the Russian Government to give specific replies to the questions raised by the Fuehrer if it could obtain the explanations just requested. The Soviet would be interested in the New Order in Europe, and particularly in the tempo and the form of this New Order. It would also like to have an idea of the boundaries of the so-called Greater East Asian Sphere.

The Fuehrer replied that the Tripartite Pact was intended to regulate conditions in Europe as to the natural interests of the European countries, and consequently Germany was now approaching the Soviet Union in order that she might express herself regarding the areas of interest to her. In no case was a settlement to be made without Soviet-Russian co-operation. This applied not only to Europe, but also to Asia, where Russia herself was to co-operate in the definition of the Greater East Asian Sphere and where she was to designate her claims there. Germany's task in this case was that of a mediator. Russia by no means was to be confronted with a *fait accompli*.

When the Fuehrer undertook to try to establish the above-mentioned coalition of Powers it was not the German-Russian relationship which appeared to him to be the most difficult point, but the question of whether a collaboration between Germany, France, and Italy was possible. Only now . . . had he thought it possible to contact Soviet Russia for the purpose of settling the questions of the Black Sea, the Balkans, and Turkey.

In conclusion, the Fuehrer summed up by stating that the discussion, to a certain extent, represented the first concrete step towards a comprehensive collaboration, with due consideration for the problems of Western Europe, which were to be settled between Germany, Italy, and France, as well as for the issues of the East, which were essentially the concern of Russia and Japan, but in which Germany offered her good offices as mediator. It was a matter of opposing any attempt on the part of America to "make money on Europe". The United States had no business in Europe, in Africa, or in Asia.

Molotov expressed his agreement with the statements of the Fuehrer

---

* Signed between Germany, Italy, and Japan on September 27, 1940.

upon the *rôle* of America and England. The participation of Russia in the Tripartite Pact appeared to him entirely acceptable in principle, provided that Russia was to co-operate as a partner and not be merely an object. In that case he saw no difficulties in the matter of participation of the Soviet Union in the common effort. But the aim and the significance of the Pact must first be more closely defined, particularly with regard to the delimitation of the Greater East Asian Sphere.

     ★   ★   ★   ★   ★

When the conferences were resumed on November 13:

Molotov mentioned the question of the strip of Lithuanian territory and emphasised that the Soviet Government had not received any clear answer yet from Germany on this question. However, it awaited a decision. Regarding the Bukovina, he admitted that this involved an additional territory, one not mentioned in the Secret Protocol. Russia had at first confined her demands to Northern Bukovina. Under the present circumstances however Germany must understand the Russian interest in Southern Bukovina. But Russia had not received an answer to her question regarding this subject either. Instead, Germany had guaranteed the entire territory of Roumania and completely disregarded Russia's wishes with regard to Southern Bukovina.

The Fuehrer replied that it would mean a considerable concession on the part of Germany if even part of Bukovina were to be occupied by Russia. . . .

Molotov however persisted in the opinion previously stated: that the revisions desired by Russia were insignificant.

The Fuehrer replied that if German-Russian collaboration was to show positive results in the future the Soviet Government would have to understand that Germany was engaged in a life-and-death struggle, which at all events she wanted to conclude successfully. . . . Both sides agreed in principle that Finland belonged to the Russian sphere of influence. Instead therefore of continuing a purely theoretical discussion, they should rather turn to more important problems.

After the conquest of England the British Empire would be apportioned as a gigantic world-wide estate in bankruptcy of forty million square kilometres. In this bankrupt estate there would be for Russia access to the ice-free and really open ocean. Thus far a minority of forty-five million Englishmen had ruled six hundred million inhabitants of the British Empire. He was about to crush this minority. Even the United States was actually doing nothing but picking out of this bankrupt estate a few items particularly suitable to the United States. Germany of course would like to avoid any conflict which would divert her from her struggle against the heart of the Empire, the British

Isles. For that reason, he (the Fuehrer) did not like Italy's war against Greece, as it diverted forces to the periphery instead of concentrating them against England at one point. The same would occur during a Baltic war. The conflict with England would be fought to the last ditch, and he had no doubt that the defeat of the British Isles would lead to the dissolution of the Empire. It was a chimera to believe that the Empire could possibly be ruled and held together from Canada. Under those circumstances there arose world-wide perspectives. During the next few weeks they would have to be settled in joint diplomatic negotiations with Russia, and Russia's participation in the solution of these problems would have to be arranged. All the countries which could possibly be interested in the bankrupt estate would have to stop all controversies among themselves and concern themselves exclusively with the partition of the British Empire. This applied to Germany, France, Italy, Russia, and Japan.

Molotov replied that he had followed the arguments of the Fuehrer with interest, and that he was in agreement with everything that he had understood.

\* \* \* \* \*

Hitler then retired for the night. After supper at the Soviet Embassy there was a British air raid on Berlin. We had heard of the conference beforehand, and though not invited to join in the discussion did not wish to be entirely left out of the proceedings. On the "Alert" all moved to the shelter, and the conversation was continued till midnight by the two Foreign Secretaries in safer surroundings. The German official account says:

Because of the air raid the two Ministers went into the Reich Foreign Minister's air raid shelter at 9.40 p.m. in order to conduct the final conversation. . . .

The time was not yet ripe, said Ribbentrop, for discussing the new order of things in Poland. The Balkan issue had already been discussed extensively. In the Balkans Germany had solely an economic interest, and she did not want England to disturb her there. The granting of the German guarantee to Roumania had apparently been misconstrued by Moscow. . . . In all its decisions the German Government was guided solely by the endeavour to preserve peace in the Balkans and to prevent England from gaining a foothold there and from interfering with supplies to Germany. Thus German action in the Balkans was motivated exclusively by the circumstances of the war against England. As soon as England conceded her defeat and asked for peace German interests in the Balkans would be confined exclusively to the

economic field, and German troops would be withdrawn from Roumania. Germany had, as the Fuehrer had repeatedly declared, no
territorial interests in the Balkans. He could only repeat again and
again that the decisive question was whether the Soviet Union was prepared and in a position to co-operate with Germany in the great
liquidation of the British Empire. On all other questions Germany and
the Soviet Union would easily reach an understanding if they could
succeed in extending their relations and in defining the spheres of influence. Where the spheres of influence lay had been stated repeatedly.
It was therefore—as the Fuehrer had so clearly put it—a matter of the
interests of the Soviet Union and Germany requiring that the partners
stand not breast to breast but back to back, in order to support each
other in the achievement of their aspirations. . . .

In his reply Molotov stated that the Germans were assuming that
the war against England had already actually been won. If therefore, as
had been said in another connection, Germany was waging a life-and-
death struggle against England, he could only construe this as meaning
that Germany was fighting "for life" and England "for death". As to
the question of collaboration, he quite approved of it, but he added that
they had to come to a thorough understanding. This idea had also been
expressed in Stalin's letter. A delimitation of the spheres of influence
must also be sought. On this point however he (Molotov) could not
take a definitive stand at this time, since he did not know the opinion
of Stalin and of his other friends in Moscow in the matter. However, he had to state that all these great issues of to-morrow could not
be separated from the issues of to-day and the fulfilment of existing
agreements. . . .

Thereupon Herr Molotov cordially bade farewell to the Reich
Foreign Minister, stressing that he did not regret the air raid alarm,
because he owed to it such an exhaustive conversation with the Reich
Foreign Minister.

*     *     *     *     *

When in August 1942 I first visited Moscow I received from
Stalin's lips a shorter account of this conversation which in no
essential differs from the German record, but may be thought
more pithy.

"A little while ago," said Stalin, "the great complaint against
Molotov was that he was too pro-German. Now everyone says
he is too pro-British. But neither of us ever trusted the Germans.
For us it was always life and death." I interjected that we had
been through this ourselves, and so knew how they felt. "When
Molotov," said the Marshal, "went to see Ribbentrop in Berlin

in November of 1940 you got wind of it and sent an air raid."
I nodded. "When the alarm sounded Ribbentrop led the way
down many flights of stairs to a deep shelter sumptuously
furnished. When he got inside the raid had begun. He shut the
door and said to Molotov: 'Now here we are alone together.
Why should we not divide?' Molotov said: 'What will England
say?' 'England,' said Ribbentrop, 'is finished. She is no more use
as a Power.' 'If that is so,' said Molotov, 'why are we in this
shelter, and whose are these bombs which fall?' "

\*     \*     \*     \*     \*

The Berlin conversations made no difference to Hitler's deep
resolve. During October Keitel, Jodl, and the German General
Staff had under his orders been forming and shaping the plans
for the eastward movement of the German armies and for the
invasion of Russia in the early summer of 1941. It was not
necessary at this stage to decide on the exact date, which might
also be affected by the weather. Having regard to the distances
to be traversed after the frontiers were crossed, and the need of
taking Moscow before the winter began, it was obvious that the
beginning of May offered the best prospects. Moreover, the
assembly and deployment of the German Army along the two-
thousand-mile front from the Baltic to the Black Sea, and the
provision of all the magazines, camps, and railway sidings, was in
itself one of the largest military tasks ever undertaken, and no
delay either in planning or in action could be tolerated. Over all
hung the vital need for concealment and deception.

For this purpose two separate forms of cover were used by
Hitler, each of which had advantages of its own. The first was an
elaborate negotiation about a common policy based on the partition and distribution of the British Empire in the East. The
second was the domination of Roumania, Bulgaria, and Greece,
with Hungary on the way, by a steady influx of troops. This
offered important military gains, and at the same time masked or
presented an explanation for the building up of the German
armies on the southern flank of the front to be developed against
Russia.

The negotiations took the form of draft proposals by Germany
for the accession of Soviet Russia to the Three-Power Pact at the
expense of British interests in the Orient. If Stalin had accepted

this scheme events might for a time have taken a different course. It was possible at any moment for Hitler to suspend his plans for invading Russia. We cannot attempt to describe what might have happened as the result of an armed alliance between the two great empires of the Continent, with their millions of soldiers, to share the spoil in the Balkans, Turkey, Persia, and the Middle East, with India always in the background and with Japan as an eager partner in the "Greater East Asian Scheme". But Hitler's heart was set on destroying the Bolsheviks, for whom his hatred was mortal. He believed that he had the force to gain his main life-aim. Thereafter all the rest would be added unto him. He must have known from the conversations at Berlin and other contacts that the proposals which he made Ribbentrop send to Moscow fell far short of Russian ambitions.

A draft, bearing no date, of a Four-Power Pact was found in the captured correspondence of the German Foreign Office with the German Embassy in Moscow. This apparently formed the basis for Schulenburg's conversation with Molotov reported on November 26, 1940. By this Germany, Italy, and Japan were to agree to respect each other's natural spheres of influence. In so far as these spheres of interest came into contact with each other, they would constantly consult each other in an amicable way with regard to the problems arising therefrom.

Germany, Italy, and Japan declared on their part that they recognised the present extent of the possessions of the Soviet Union and would respect them.

The Four Powers undertook to join no combination of Powers and to support no combination of Powers which was directed against one of the Four Powers. They would assist each other in economic matters in every way and would supplement and extend the agreements existing among themselves. The agreement would continue for a period of ten years.

To this there was to be a Secret Protocol by which Germany declared that, apart from the territorial revisions in Europe to be carried out at the conclusion of peace, her territorial aspirations centred in the territories of Central Africa; Italy declared that, apart from territorial revisions in Europe, her territorial aspirations centred in the territories of Northern and North-Eastern Africa; Japan declared that her territorial aspirations centred in the area of Eastern Asia to the south of the Island Empire of Japan;

and the Soviet Union declared that its territorial aspirations centred south of the national territory of the Soviet Union in the direction of the Indian Ocean.

The Four Powers declared that, reserving the settlement of specific questions, they would mutually respect these territorial aspirations and would not oppose their achievement.*

* * * * *

As was expected, the Soviet Government did not accept the German project. They were alone with Germany in Europe, and at the other side of the world Japan lay heavy upon them. Nevertheless they had confidence in their growing strength and in their vast expanse of territory, amounting to one-sixth of the land-surface of the globe. They therefore bargained toughly. On November 26, 1940, Schulenburg sent to Berlin the draft of the Russian counter-proposals. These stipulated that the German troops should be immediately withdrawn from Finland, which, under the compact of 1939, belonged to the Soviet Union's sphere of influence; that within the next few months the security of the Soviet Union in the Straits should be assured by the conclusion of a mutual assistance pact between the Soviet Union and Bulgaria, which geographically is situated inside the security zone of the Black Sea boundaries of the Soviet Union, and by the establishment of a base for land and naval forces of the U.S.S.R. within range of the Bosphorus and the Dardanelles by means of a long-term lease; that the area south of Batum and Baku in the general direction of the Persian Gulf should be recognised as the centre of the aspirations of the Soviet Union; that Japan should renounce her rights to concessions for coal and oil in northern Sakhalin.

No effective answer was returned to this document. No attempt was made by Hitler to split the difference. Issues so grave as these might well justify a prolonged and careful study in a friendly spirit by both sides. The Soviets certainly expected and

---

* It is worth noting that though in Berlin the main emphasis of Hitler and Ribbentrop was on sharing British territory, in the draft agreement the British Empire is not mentioned by name, while the colonial possessions of France, Holland, and Belgium are obviously included in the areas to be shared under the Secret Protocol. Both at Berlin and in the negotiations in Moscow the British Empire, though offering the most conspicuous and valuable booty, was not the only intended victim of Hitler. He was seeking an even wider redistribution of the colonial possessions in Africa and Asia of all the countries with which he was or had been at war.

awaited an answer. Meanwhile on both sides of the frontier the forces, already heavy, began to grow, and Hitler's right hand reached out towards the Balkans.

★   ★   ★   ★   ★

The plans prepared on his instructions by Keitel and Jodl had by now reached sufficient maturity to enable the Fuehrer to issue from his headquarters on December 18, 1940, his historic Directive No. 21.

## OPERATION BARBAROSSA

The German Armed Forces must be prepared *to crush Soviet Russia in a quick campaign* even before the conclusion of the war against England.

For this purpose the *Army* will have to employ all available units, with the reservation that the occupied territories must be secured against surprise attacks.

For the *Air Force* it will be a matter of releasing such strong forces for the Eastern campaign in support of the Army that a quick completion of the ground operations may be expected and that damage to Eastern German territory by enemy air attacks will be as slight as possible. This concentration of the main effort in the East is limited by the requirement that the entire combat and armament area dominated by us must remain adequately protected against enemy air attacks, and that the offensive operations against England, particularly her supply lines, must not be permitted to break down.

The main effort of the *Navy* will remain unequivocally directed against *England* even during an Eastern campaign.

I shall order the *concentration* against Soviet Russia possibly eight weeks before the intended beginning of operations.

Preparations requiring more time to begin are to be started now—if this has not yet been done—and are to be completed by *May 15, 1941*.

It is to be considered of decisive importance however that the intention to attack is not discovered.

The preparations of the High Commands are to be made on the following basis:

### I. *General Purpose*

The mass of the Russian *Army* in Western Russia is to be destroyed in daring operations, by driving forward deep armoured wedges, and the retreat of units capable of combat into the vastness of Russian territory is to be prevented.

521

In quick pursuit a line is then to be reached from which the Russian Air Force will no longer be able to attack German Reich territory. The ultimate objective of the operation is to establish a defence line against Asiatic Russia from a line running approximately from the Volga river to Archangel. Then, in case of necessity, the last industrial area left to Russia in the Urals can be eliminated by the Luftwaffe.

In the course of these operations the Russian *Baltic Sea Fleet* will quickly lose its bases and thus will no longer be able to fight.

Effective intervention by the Russian *Air Force* is to be prevented by powerful blows at the very beginning of the operation.

## II. *Probable Allies and their Tasks*

1. On the flanks of our operation we can count on the active participation of *Roumania* and *Finland* in the war against Soviet Russia. The High Command will in due time concert and determine in what form the armed forces of the two countries will be placed under German command at the time of their intervention.

2. It will be the task of *Roumania*, together with the force concentrating there, to pin down the enemy facing her, and in addition to render auxiliary services in the rear area.

3. *Finland* will cover the concentration of the redeployed German *North Group* (parts of the XXI Group) coming from Norway, and will operate jointly with it. Besides, Finland will be assigned the task of eliminating Hango.

4. It may be expected that *Swedish* railroads and highways will be available for the concentration of the German North Group, from the start of operations at the latest.

## III. *Direction of Operations*

A. *Army* (hereby approving the plans presented to me):

In the zone of operations divided by the Pripet Marshes into a southern and northern sector the main effort will be made *north* of this area. Two Army Groups will be provided here.

The southern group of these two Army Groups—the centre of the entire front—will be given the task of annihilating the forces of the enemy in White Russia by advancing from the region around and north of Warsaw with especially strong armoured and motorised units. . . . Only a surprisingly fast collapse of Russian resistance could justify aiming at both objectives simultaneously. . . .

*The Army Group employed south of the Pripet Marshes* is to make its main effort in the area from Lublin in the general direction of Kiev, in order to penetrate quickly with strong armoured units into the deep flank and rear of the Russian forces and then to roll them up along the Dnieper river.

The German-Roumanian groups on the right flank are assigned the task of:

(a) protecting Roumanian territory and thereby the southern flank of the entire operation;

(b) pinning down the opposing enemy forces while Army Group South is attacking on its northern flank and, according to the progressive development of the situation and in conjunction with the Air Force, preventing their orderly retreat across the Dniester during the pursuit;

[and] *in the north*, of reaching Moscow quickly.

The capture of this city means a decisive success politically and economically, and, beyond that, the elimination of the most important railway centre.

B. *Air Force*

Its task will be to paralyse and to eliminate as far as possible the intervention of the Russian Air Force, as well as to support the Army at its main points of effect, particularly those of Army Group Centre and, on the flank, those of Army Group South. The Russian railroads, in the order of their importance for the operations, will be cut or the most important near-by objectives (river crossings) seized by the bold employment of parachute and airborne troops.

In order to concentrate all forces against the enemy Air Force and to give immediate support to the Army the armament industry will not be attacked during the main operations. Only after the completion of the mobile operations may such attacks be considered—primarily against the Ural region. . . .

IV. All orders to be issued by the Commanders-in-Chief on the basis of this directive must clearly indicate that they are *precautionary measures* for the possibility that Russia should change her present attitude towards us. The number of officers to be assigned to the preparatory work at an early date is to be kept as small as possible; additional personnel should be briefed as late as possible, and only to the extent required for the activity of each individual. Otherwise, through the discovery of our preparations—the date of their execution has not even been fixed—there is danger that most serious political and military disadvantages may arise.

V. I expect reports from the Commanders-in-Chief concerning their further plans based on this directive.

The contemplated preparations of all branches of the Armed Forces, including their progress, are to be reported to me through the High Command.

ADOLF HITLER*

\* \* \* \* \*

From this moment the moulds had been shaped for the supreme events of 1941. We of course had no knowledge of the bargainings between Germany and Russia for dividing the spoils of our Empire and for our destruction; nor could we measure the as yet unformed intentions of Japan. The main troop movements of the German armies eastwards had not yet become apparent to our active Intelligence Service. Only the infiltration and gradual massing in Bulgaria and Roumania could be discerned. Had we known what is set forth in this chapter we should have been greatly relieved. The combination against us of Germany, Russia, and Japan was the worst of our fears. But who could tell? Meanwhile: "Fight on!"

\* *Nazi–Soviet Relations*, pp. 260 ff.

# CHAPTER XXX

## OCEAN PERIL

*Disguised Surface Raiders – Excursion of the "Scheer" – The "Jervis Bay" Saves the Convoy – Further Depredations of the "Scheer" – A Surprise for the "Hipper" – Disproportionate Strains – The U-Boat Peril Dominates – Increasing Stranglehold upon the North-Western Approaches – The Diver's Anxieties – Grievous Losses – Need to Shift the Control of the North-Western Approaches from Plymouth to Liverpool – Sharp Contraction of Imports – Losses off the Bloody Fore-land – Withdrawal of the Irish Subsidies – My Telegram to the President of December 13 – A Sombre Admiralty Proposal – The Dynamite Carpet – Reinforcement and Stimulation of the Air Force Coastal Command – Eventual Success of their Counter-offensive.*

*T*HE destruction of the *Graf Spee* in the action off the Plate in December 1939 had brought to an abrupt end the first German campaign against our shipping in the wide oceans. The fighting in Norway had, as we have seen, paralysed for the time being the German Navy in home waters. What was left of it was necessarily reserved for the invasion project. Admiral Raeder, whose ideas on the conduct of the German war at sea were technically sound, had some difficulty in carrying his views in the Fuehrer's councils. He had even at one time to resist a proposal made by the Army to disarm all his heavy ships and use their guns for long-range batteries on shore. During the summer however he had fitted out a number of merchant ships as disguised raiders. They were more powerfully armed, were generally faster than our armed merchant-cruisers, and were provided with reconnaissance aircraft. Five ships of this type evaded our patrols and entered the Atlantic between April and June 1940, whilst a sixth undertook the hazardous north-east passage to the

Pacific along the north coasts of Russia and Siberia. Assisted by a Russian ice-breaker, she succeeded in making the passage in two months, and emerged into the Pacific through the Bering Sea in September. The object which Admiral Raeder laid down for the conduct of these ships was threefold: first, to destroy or capture enemy ships; secondly, to dislocate shipping movements; and, thirdly, to force the dispersion of British warships for escort and patrol to counter the menace. These well-conceived tactics caused us both injury and embarrassment. By the first weeks of September these five disguised raiders were loose upon our trade routes. Two of them were working in the Atlantic, two others in the Indian Ocean, and the fifth, after laying mines off Auckland, New Zealand, was in the Pacific. Only two contacts were made with them during the whole year. On July 29 "Raider E" was engaged in the South Atlantic by the armed merchant-cruiser *Alcantara*, but escaped after an inconclusive action. In December another armed merchant-cruiser, the *Carnarvon Castle*, attacked her again off the Plate River, but she escaped after some damage. Up till the end of September 1940 these five raiders sank or captured thirty-six ships, amounting to 235,000 tons.

At the end of October 1940 the pocket-battleship *Scheer* was at last ready for service. When the invasion of England had been shelved she left Germany on October 27, and broke out into the Atlantic through the Denmark Strait north of Iceland. She was followed a month later by the 8-inch-gun cruiser, *Hipper*. The *Scheer* had orders to attack the North Atlantic convoys, from which the battleship escorts had been withdrawn to reinforce the Mediterranean. Captain Krancke believed that a homeward-bound convoy had left Halifax on October 27, and he hoped to intercept it about November 3. On the 5th his aircraft reported eight ships in the south-east, and he set off in pursuit. At 2.27 p.m. he sighted a single ship, the *Mopan*, which he sank by gunfire, after taking on board the crew of sixty-eight. By threats he had been able to prevent any wireless reports being made by the *Mopan*. At 4.50 p.m., whilst thus occupied, the masts of the convoy H.X. 84, consisting of thirty-seven ships, appeared over the horizon. In the centre of the convoy was the ocean escort, the armed merchant-cruiser *Jervis Bay*. Her commanding officer, Captain Fegen, R.N., realised at once that he was faced with hopeless odds. His one thought, after reporting the presence of the enemy by wireless, was to engage the pocket-battleship for as long as possible, and thus gain time for the convoy to disperse. Darkness approached, and there would then be a chance of many escaping. While the convoy scattered the *Jervis Bay* closed his overwhelming antagonist at full speed. The *Scheer* opened fire at eighteen thousand yards. The shots of the old 6-inch guns of the *Jervis Bay* fell short. The one-sided fight lasted till 6 p.m., when the *Jervis Bay*, heavily on fire and completely out of control, was abandoned. She finally sank about eight o'clock with the loss of over two hundred officers and men. With them perished Captain Fegen, who went down with the ship. He was awarded the Victoria Cross posthumously for his heroic conduct, which takes an honoured place in the records of the Royal Navy.

Not until the end of the fight did the *Scheer* pursue the convoy; but the wintry night had now closed in. The ships had scattered and she was able to overtake and sink only five before darkness fell. She could not afford, now that her position was known, to remain in the area, on which she expected that powerful British forces would soon converge. The great majority of this valuable convoy was therefore saved by the devotion of the *Jervis Bay*. The spirit of the merchant seamen was not unequal to that of their escort. One ship, the tanker *San Demetrio*, carrying seven thousand tons of petrol, was set on fire and abandoned. But the next morning part of the crew reboarded the ship, put out the fire, and then, after gallant efforts, without compasses or navigational aids, brought the ship into a British port with her precious cargo. In all however 47,000 tons of shipping and 206 merchant seamen were lost.

The *Scheer*, determined to place as many miles as possible between herself and her pursuers, steamed south, where ten days later she met a German supply ship and replenished her fuel and stores. On November 24 she appeared in the West Indies, where she sank the *Port Hobart*, outward bound to Curaçao, and then doubled back to the Cape Verde Islands. Her later activities were spread over the South Atlantic and Indian Oceans, and not till April 1941 did she return to Kiel, after again successfully traversing the Denmark Strait. Her five months' cruise had yielded a harvest of sixteen ships, amounting to 99,000 tons, sunk or captured.

\*     \*     \*     \*     \*

From June onwards the troop convoys (called by the code name "W.S."*) sailed monthly under heavy escort round the Cape to the Middle East and India. At the same time the numerous troop convoys between ports in the Indian Ocean and the continuous stream of Canadian troops reaching this country from across the Atlantic threw the utmost strain on our naval resources. Thus we could not reinstitute the hunting groups which had scoured the seas for the *Graf Spee* in 1939. Our cruisers were disposed in the focal areas near the main shipping routes, and ships sailing independently had to rely on evasive routing and the vastness of the ocean.

On Christmas Day 1940 convoy W.S. 5A, consisting of twenty troopships and supply ships for the Middle East, was approaching the Azores when it was attacked by the cruiser *Hipper*, which had followed the *Scheer* out a month later. Visibility was poor, and the *Hipper* was unpleasantly surprised to find that the escort comprised the cruisers *Berwick*, *Bonaventure*, and *Dunedin*. There was a brief, sharp action between the *Hipper* and the *Berwick*, in which both ships were damaged. The *Hipper* made off, and in the mist succeeded in escaping to Brest, in spite of strenuous efforts by the Home Fleet and by Force H from Gibraltar to catch her; but only one ship of the convoy, which carried over thirty thousand men, the *Empire Trooper*, had to put into Gibraltar for repairs.

We could not regard the state of the outer oceans without uneasiness. We knew that disguised merchant ships in unknown numbers were preying in all the southern waters. The pocket battleship *Scheer* was loose and hidden. The *Hipper* might break out at any moment from Brest, and the two German battle-cruisers *Scharnhorst* and *Gneisenau* must also soon be expected to play their part.

The enormous disproportion between the numbers of the raiders and the forces the Admiralty had to employ to counter them and guard the immense traffic has been explained in Volume I. The Admiralty had to be ready at many points and give protection to thousands of merchant vessels, and could give no guarantee except for troop convoys against occasional lamentable disasters.

\*　　\*　　\*　　\*　　\*

* I have only heard since the war that these initials which I used so often were an Admiralty term signifying "Winston's Specials".

A far graver danger was added to these problems. The only thing that ever really frightened me during the war was the U-boat peril. Invasion, I thought, even before the air battle, would fail. After the air victory it was a good battle for us. We could drown and kill this horrible foe in circumstances favourable to us, and, as he evidently realised, bad for him. It was the kind of battle which, in the cruel conditions of war, one ought to be content to fight. But now our life-line, even across the broad oceans, and especially in the entrances to the Island, was endangered. I was even more anxious about this battle than I had been about the glorious air fight called the Battle of Britain.

The Admiralty, with whom I lived in the closest amity and contact, shared these fears, all the more because it was their prime responsibility to guard our shores from invasion and to keep the life-lines open to the outer world. This had always been accepted by the Navy as their ultimate, sacred, inescapable duty. So we poised and pondered together on this problem. It did not take the form of flaring battles and glittering achievements. It manifested istelf through statistics, diagrams, and curves unknown to the nation, incomprehensible to the public.

How much would the U-boat warfare reduce our imports and shipping? Would it ever reach the point where our life would be destroyed? Here was no field for gestures or sensations; only the slow, cold drawing of lines on charts, which showed potential strangulation. Compared with this there was no value in brave armies ready to leap upon the invader, or in a good plan for desert warfare. The high and faithful spirit of the people counted for nought in this bleak domain. Either the food, supplies, and arms from the New World and from the British Empire arrived across the oceans, or they failed. With the whole French seaboard from Dunkirk to Bordeaux in their hands, the Germans lost no time in making bases for their U-boats and co-operating aircraft in the captured territory. From July onwards we were compelled to divert our shipping from the approaches south of Ireland, where of course we were not allowed to station fighter-aircraft. All had to come in around Northern Ireland. Here, by the grace of God, Ulster stood a faithful sentinel. The Mersey, the Clyde, were the lungs through which we breathed. On the East Coast and in the English Channel small vessels continued to ply under an ever-increasing attack by air, by E-boat,* and by mines. As it was impossible to vary the East Coast route, the passage of each convoy between the Forth and London became almost every day an action in itself. Few large ships were risked on the East Coast and none at all in the Channel.

The losses inflicted on our merchant shipping became most grave during the twelve months from July 1940 to July 1941, when we could claim that the British Battle of the Atlantic was won. Far heavier losses occurred when the United States entered the war before any convoy system was set up along their eastern coast. But then we were no longer alone. The last six months of 1940 showed extremely heavy losses, modified only by the winter gales, and no great slaughter of U-boats. We gained some advantage by larger patterning of depth-charges and by evasive routing, but the invasion threat required strong concentrations in the Narrow Seas and our great volume of anti-U-boat new construction only arrived gradually. This shadow hung over the Admiralty and those who shared their knowledge. The week ending September 22 showed the highest rate of loss since the beginning of the war, and was in fact greater than any we had suffered in a similar period in 1917. Twenty-seven ships, of nearly 160,000 tons, were sunk, many of them in a Halifax convoy. In October, while the *Scheer* was also active, another Atlantic convoy was massacred by U-boats, twenty ships being sunk out of thirty-four.

As November and December drew on, the entrances and estuaries of the Mersey and the Clyde far surpassed in mortal significance all other factors in the war. We could of course at this time have descended upon de Valera's Ireland and regained the southern ports by force of modern arms. I had always declared that nothing but self-preservation would lead me to this. But perhaps the case of self-preservation might come. Then so be it. Even this hard measure would only have given a mitigation. The only sure remedy was to secure free exit and entrance in the Mersey and the Clyde.

Every day when they met, those few who knew looked at one another. One understands the diver deep below the surface of the sea, dependent from minute to minute upon his air-pipe. What would he feel if he could see a growing shoal of sharks

---

* E-boat: the German equivalent of British "light coastal craft".

biting at it? All the more when there was no possibility of his being hauled to the surface! For us there was no surface. The diver was forty-six millions of people in an overcrowded island, carrying on a vast business of war all over the world, anchored by nature and gravity to the bottom of the sea. What could the sharks do to his air-pipe? How could he ward them off or destroy them?

As early as the beginning of August I had been convinced that it would be impossible to control the Western Approaches through the Mersey and Clyde from the Command at Plymouth.

*Prime Minister to First Lord and First Sea Lord*                    4.VIII.40

The repeated severe losses in the North-Western Approaches are most grievous, and I wish to feel assured that they are being grappled with with the same intense energy that marked the Admiralty treatment of the magnetic mine. There seems to have been a great falling off in the control of these Approaches. No doubt this is largely due to the shortage of destroyers through invasion precautions. Let me know at once the whole outfit of destroyers, corvettes, and Asdic trawlers, together with aircraft, available and employed in this area. Who is in charge of their operations? Are they being controlled from Plymouth and Admiral Nasmith's staff? Now that you have shifted the entry from the south to the north, the question arises, is Plymouth the right place for the Command? Ought not a new Command of the first order to be created in the Clyde, or should Admiral Nasmith [C.-in-C. at Plymouth] move thither? Anyhow, we cannot go on like this. How is the southern minefield barrage getting on? Would it not be possible after a while to ring the changes upon it for a short time and bring some convoys in through the gap which has been left? This is only a passing suggestion.

There were always increased dangers to be apprehended from using only one set of Approaches. These dangers cannot be surmounted unless the protective concentration is carried out with vigour superior to that which must be expected from the enemy. He will soon learn to put everything there. It is rather like the early days in the Moray Firth after the East Coast minefield was laid. I am confident the Admiralty will rise to the occasion, but evidently a great new impulse is needed. Pray let me hear from you.

I encountered resistances. The Admiralty accepted my view in September of moving from Plymouth to the North, rightly substituting the Mersey for the Clyde. But several months elapsed before the necessary headquarters organisation, with its operations rooms and elaborate network of communications, could be brought into being, and in the meantime much improvisation was necessary. The new Command was entrusted to Admiral Sir Percy Noble, who, with a large and ever-growing staff, was installed at Liverpool in February 1941. Henceforward this became almost our most important station. The need and advantage of the change was by then recognised by all.

Towards the end of 1940 I became increasingly concerned about the ominous fall in imports. This was another aspect of the U-boat attack. Not only did we lose ships, but the precautions we took to avoid losing them impaired the whole flow of merchant traffic. The few harbours on which we could now rely became congested. The turn-round of all vessels as well as their voyages was lengthened. Imports were the final test. In the week ending June 8, during the height of the battle in France, we had brought into the country 1,201,535 tons of cargo, exclusive of oil. From this peak figure imports had declined at the end of July to less than 750,000 tons a week. Although substantial improvement was made in August, the weekly average again fell, and for the last three months of the year was little more than 800,000 tons.

*Prime Minister to First Lord and First Sea Lord*                    3.XII.40
The new disaster which has overtaken the Halifax convoy requires precise examination. We heard about a week ago that as many as thirteen U-boats were lying in wait on these approaches. Would it not have been well to divert the convoy to the Minches? Would this not have been even more desirable when owing to bad weather the outward-bound convoys were delayed, and consequently the escort for the inward-bound could not reach the dangerous area in time?

*Prime Minister to Chancellor of the Exchequer*                    5.XII.40
Pray convene a meeting to discuss the measures to be taken to reduce the burden on our shipping and finances in consequence of the heavy sinkings off the Irish coast and our inability to use the Irish ports. The following Ministers should be summoned: Trade, Shipping, Agriculture, Food, Dominions. Assuming there is agreement on principle, a general plan should be made for acting as soon as possible, together with a time-table and programme of procedure. It is not necessary to consider either the Foreign Affairs or the Defence aspect at this stage. These will have to be dealt with later. The first step essential is to have a good workable scheme, with as much in it as possible that does not hit us worse than it does the others.

*Prime Minister to Minister of Transport*       13.XII.40

I am obliged to you for your note of December 3 on steel, and I hope that you are pushing forward with the necessary measures to give effect to your proposals.

In present circumstances it seems to me intolerable that firms should hold wagons up by delaying to unload them, and action should certainly be taken to prevent this.

A sample shows that the average time taken by non-tanker cargo ships to turn round at Liverpool rose from 12½ days in February to 15 days in July and 19½ days in October. At Bristol the increase was from 9½ days to 14½ days, but at Glasgow the time remained steady at 12 days. To improve this seems one of the most important aspects of the whole situation.

*Prime Minister to Minister of Transport*       13.XII.40

I see that oil imports during September and October were only half what they were in May and June, and covered only two-thirds of our consumption. I understand that there is no shortage of tankers, that the fall is the result of the partial closing of the South and East Coasts to tankers, and that a large number had to be temporarily laid up in the Clyde and others held at Halifax, Nova Scotia. More recently some tankers have been sent to the South and East Coasts, and oil imports increased during November.

From the reply your predecessor* made to my minute of August 26 I gathered that he was satisfied with the preparations in hand for the importation of oil through the West Coast ports. His expectations do not appear to have been fulfilled.

There are two policies which can be followed to meet this situation. We can either expose oil tankers to additional risk by bringing them to South and East Coast ports, and thus increase our current imports; or we can continue to draw upon our stocks, relying upon being able to replenish them from the West Coast ports when arrangements have been completed for the handling of the cargoes, and accepting the resulting inconvenience. I should be glad if you would consider, in consultation with the First Lord, to what extent each of these two policies should be followed.

I am sending a copy of this letter to the First Lord.

*Prime Minister to First Lord*       14.XII.40

Let me have a full account of the condition of the American destroyers, showing their many defects and the little use we have been

---

* Sir John Reith. He became Lord Reith and Minister of Works and Buildings on October 3, 1940.

able to make of them so far. I should like to have the paper by me for consideration in the near future.

*Prime Minister to First Lord and First Sea Lord*    27.XII.40

What have you done about catapulting expendable aircraft from ships in outgoing convoys? I have heard of a plan to catapult them from tankers, of which there are nearly always some in each convoy. They then attack the Focke-Wulf and land in the sea, where the pilot is picked up, and machines salved or not as convenient.

How is this plan viewed?

As we shall see in the next volume, this project was fruitful. Ships equipped for catapulting fighter aircraft to attack the Focke-Wulf were developed early in 1941.

*Prime Minister to Minister of Transport*    27.XII.40

It is said that two-fifths of the decline in the fertility of our shipping is due to the loss of time in turning round ships in British ports. Now that we are confined so largely to the Mersey and the Clyde, and must expect increasingly severe attacks upon them, it would seem that this problem constitutes the most dangerous part of the whole of our front.

Would you kindly give me a note on:

(a) The facts.
(b) What you are doing, and what you propose to do.
(c) How you can be helped.

*Prime Minister to First Lord*    29.XII.40

These [U-boat decoy ships*] have been a great disappointment so far this war. The question of their alternative uses ought to be considered by the Admiralty. I expect they have a large number of skilled ratings on board. Could I have a list of these ships, their tonnage, speeds, etc. Could they not carry troops or stores while plying on their routes?

\* \* \* \* \*

My indignation at the denial of the Southern Irish ports mounted under these pressures.

*Prime Minister to the Chancellor of the Exchequer*    1.XII.40

The straits to which we are being reduced by Irish action compel a reconsideration of the subsidies [to Ireland]. It can hardly be argued

---

* The modern equivalent of "Q" ships, which had been effectively used in the 1914–18 war to lure the U-boats to their destruction. They were not successful in the changed conditions of this war.

that we can go on paying them till our last gasp. Surely we ought to use this money to build more ships or buy more from the United States in view of the heavy sinkings off the Bloody Foreland.

Pray let me know how these subsidies could be terminated, and what retaliatory measures could be taken in the financial sphere by the Irish, observing that we are not afraid of their cutting off our food, as it would save us the enormous mass of fertilisers and feeding-stuffs we have to carry into Ireland through the de Valera-aided German blockade. Do not assemble all the pros and cons for the moment, but show what we could do financially and what would happen. I should be glad to know about this to-morrow.

*Prime Minister to General Ismay, for C.O.S. Committee* 3.XII.40

I gave you and each of the C.O.S. a copy of the Irish paper. The Chancellor of the Exchequer's comments are also favourable, and there is no doubt that subsidies can be withdrawn at very short notice.

We must now consider the military reaction. Suppose they invited the Germans into their ports, they would divide their people, and we should endeavour to stop the Germans. They would seek to be neutral and would bring the war upon themselves. If they withdrew the various cable and watching facilities they have, what would this amount to, observing that we could suspend all connections between England and Southern Ireland? Suppose they let German U-boats come in to refresh in west coast ports of Ireland, would this be serious, observing that U-boats have a radius of nearly thirty days, and that the limiting factor is desire of crews to get home and need of refit, rather than need of refuelling and provisioning? Pray let me have your observations on these and other points which may occur to you.

I thought it well to try to bring the President along in this policy.

*Former Naval Person to President Roosevelt* 13.XII.40

North Atlantic transport remains the prime anxiety. Undoubtedly Hitler will augment his U-boat and air attack on shipping and operate ever farther into the ocean. Now that we are denied the use of Irish ports and airfields our difficulties strain our flotillas to the utmost limit. We have so far only been able to bring a very few of your fifty destroyers into action, on account of the many defects which they naturally develop when exposed to Atlantic weather after having been laid up so long. I am arranging to have a very full technical account prepared of renovations and improvements that have to be made in the older classes of destroyers to fit them for the present task, and this may be of use to you in regard to your own older flotillas.

In the meanwhile we are so hard pressed at sea that we cannot undertake to carry any longer the 400,000 tons of feeding-stuffs and fertilisers which we have hitherto convoyed to Eire through all the attacks of the enemy. We need this tonnage for our own supply, and we do not need the food which Eire has been sending us. We must now concentrate on essentials, and the Cabinet proposes to let de Valera know that we cannot go on supplying him under present conditions. He will of course have plenty of food for his people, but they will not have the prosperous trading they are making now. I am sorry about this, but we must think of our own self-preservation, and use for vital purposes our own tonnage brought in through so many perils. Perhaps this may loosen things up and make him more ready to consider common interests. I should like to know quite privately what your reactions would be if and when we are forced to concentrate our own tonnage upon the supply of Great Britain. We also do not feel able in present circumstances to continue the heavy subsidies we have hitherto been paying to the Irish agricultural producers. You will realise also that our merchant seamen, as well as public opinion generally, take it much amiss that we should have to carry Irish supplies through air and U-boat attacks and subsidise them handsomely when de Valera is quite content to sit happy and see us strangled.

*　　*　　*　　*　　*

One evening in December I held a meeting in the downstairs War Room with only the Admiralty and the sailors present. All the perils and difficulties, about which the company was well informed, had taken a sharper turn. My mind reverted to February and March 1917, when the curve of U-boat sinkings had mounted so steadily against us that one wondered how many months' more fighting the Allies had in them, in spite of all the Royal Navy could do. One cannot give a more convincing proof of the danger than the project which the Admirals put forward. We must at all costs and with overriding priorities break out to the ocean. For this purpose it was proposed to lay an underwater carpet of dynamite from the seaward end of the North Channel, which gives access to the Mersey and the Clyde, to the 100-fathom line north-west of Ireland. A submerged minefield must be laid three miles broad and sixty miles long from these coastal waters to the open ocean. Even if all the available explosives were monopolised for this task, without much regard to field operations or the proper rearmament of our troops, it seemed vital to make this carpet—assuming there was no other way.

Let me explain the process. Many thousands of contact-mines would have to be anchored to the bottom of the sea, reaching up to within thirty-five feet of its surface. Over this pathway all the ships which fed Britain, or carried on our warfare abroad, could pass and repass without their keels striking the mines. A U-boat, however, venturing into this minefield, would soon be blown up; and after a while they would find it *not good enough* to come. Here was the defensive *in excelsis*. Anyhow, it was better than nothing. It was the last resort. Provisional approval and directions for detailed proposals to be presented were given on this night. Such a policy meant that the diver would in future be thinking about nothing but his air-pipe. But he had other work to do.

At the same time however we gave orders to the R.A.F. Coastal Command to dominate the outlets from the Mersey and Clyde and around Northern Ireland. Nothing must be spared from this task. It had supreme priority. The bombing of Germany took second place. All suitable machines, pilots, and material must be concentrated upon our counter-offensive, by fighters against the enemy bombers, and surface craft assisted by bombers against the U-boats in these narrow vital waters. Many other important projects were brushed aside, delayed, or mauled. At all costs one must breathe.

We shall see the extent to which this counter-offensive by the Navy and by Coastal Command succeeded during the next few months; how we became the masters of the outlets; how the Heinkel 111's were shot down by our fighters, and the U-boats choked in the very seas in which they sought to choke us. Suffice it here to say that the success of Coastal Command overtook the preparations for the dynamite carpet. Before this ever made any appreciable inroad upon our war economy the morbid defensive thoughts and projects faded away, and once again with shining weapons we swept the approaches to the Isle.

# CHAPTER XXXI

# DESERT VICTORY

BEFORE a great enterprise is launched the days pass slowly. The remedy is other urgent business, of which there was at this time certainly no lack. I was myself so pleased that our Generals would take the offensive that I did not worry unduly about the result. I grudged the troops wasted in Kenya and Palestine and on internal security in Egypt; but I trusted in the quality and ascendancy of the famous regiments and long-trained professional officers and soldiers to whom this important matter was confided. Eden also was confident, especially in General Wilson, who was to command the battle; but then they were both "Greenjackets",* and had fought as such in the previous war. Meanwhile, outside the small group who knew what was going to be attempted, there was plenty to talk about and do.

* Rifle Brigade and King's Royal Rifles.

For a month or more all the troops to be used in the offensive practised the special parts they had to play in the extremely complicated attack. The details of the plan were worked out by Lieutenant-General Wilson and Major-General O'Connor, and General Wavell paid frequent visits of inspection. Only a small circle of officers knew the full scope of the plan, and practically nothing was put on paper. To secure surprise, attempts were made to give the enemy the impression that our forces had been seriously weakened by the sending of reinforcements to Greece and that further withdrawals were contemplated. On December 6 our lean, bronzed, desert-hardened, and completely mechanised army of about twenty-five thousand men leaped forward more than forty miles, and all next day lay motionless on the desert sand unseen by the Italian Air Force. They swept forward again on December 8, and that evening, for the first time, the troops were told that this was no desert exercise, but the "real thing". At dawn on the 9th the battle of Sidi Barrani began.

It is not my purpose to describe the complicated and dispersed fighting which occupied the next four days over a region as large as Yorkshire. Everything went smoothly. Nibeiwa was attacked by one brigade at 7 a.m., and in little more than an hour was completely in our hands. At 1.30 p.m. the attack on the Tummar camps opened, and by nightfall practically the whole area and most of its defenders were captured. Meanwhile the 7th Armoured Division had isolated Sidi Barrani by cutting the coast road to the west. Simultaneously the garrison of Mersa Matruh, which included the Coldstream Guards, had also prepared their blow. At first light on the 10th they assaulted the Italian positions on their front, supported by heavy fire from the sea. Fighting continued all day, and by ten o'clock the Coldstream battalion headquarters signalled that it was impossible to count the prisoners on account of their numbers, but that "there were about five acres of officers and two hundred acres of other ranks".

At home in Downing Street they brought me hour-to-hour signals from the battlefield. It was difficult to understand exactly what was happening, but the general impression was favourable, and I remember being struck by a message from a young officer in a tank of the 7th Armoured Division: "Have arrived at the second B in Buq Buq." I was able to inform the House of Commons on the 10th that active fighting was in progress in the desert, that 500 prisoners had been taken and an Italian general killed; and also that our troops had reached the coast. "It is too soon to attempt to forecast either the scope or the result of the considerable operations which are in progress. But we can at any rate say that the preliminary phase has been successful." That afternoon Sidi Barrani was captured.

From December 11 onwards the action consisted of a pursuit of the Italian fugitives by the 7th Armoured Division, followed by the 16th British Infantry Brigade (Motorised) and the 6th Australian Division, which had relieved the 4th Indian Division. On December 12 I could tell the House of Commons that the whole coastal region around Buq Buq and Sidi Barrani was in the hands of British and Imperial troops and that 7,000 prisoners had already reached Mersa Matruh. "We do not yet know how many Italians were caught in the encirclement, but it would not be surprising if at least the best part of three Italian divisions, including numerous Blackshirt formations, have been either destroyed or captured. The pursuit to the westward continues with the greatest vigour. The Air Force are now bombing, the Navy are shelling, the principal road open to the retreating enemy, and considerable additional captures have already been reported.

"While it is still too soon to measure the scale of these operations, it is clear that they constitute a victory which, in this African theatre of war, is of the first order, and reflects the highest credit upon Sir Archibald Wavell, Sir Henry Maitland Wilson, the Staff officers who planned this exceedingly complicated operation, and the troops who performed the remarkable feats of endurance and daring which accomplished it. The whole episode must be judged upon the background of the fact that it is only three or four months ago that our anxieties for the defence of Egypt were acute. Those anxieties are now removed, and the British guarantee and pledge that Egypt would be effectually defended against all comers has been in every way made good."

The moment the victory of Sidi Barrani was assured—indeed, on December 12—General Wavell took on his own direct initiative a wise and daring decision. Instead of holding back in general reserve on the battlefield the 4th British Indian Division, which had just been relieved, he moved it at once to Eritrea to join the 5th British Indian Division for the Abyssinian campaign under General Platt. The division went partly by sea to Port

Soudan, and partly by rail and boat up the Nile. Some of them moved practically straight from the front at Sidi Barrani to their ships, and were in action again in a theatre seven hundred miles away very soon after their arrival. The earliest units arrived at Port Soudan at the end of December, and the movement was completed by January 21. The division joined in the pursuit of the Italians from Kassala, which they had evacuated on January 19, to Keren, where the main Italian resistance was encountered. General Platt had, as we shall see, a hard task at Keren, even with his two British Indian Divisions, the 4th and 5th. Without this farseeing decision of General Wavell's the victory at Keren could not have been achieved and the liberation of Abyssinia would have been subject to indefinite delays. The immediate course of events both on the North African shore and in Abyssinia proved how very justly the Commander-in-Chief had measured the values and circumstances of the situations.

\* \* \* \* \*

I hastened to offer my congratulations to all concerned, and to urge pursuit to the utmost limit of strength.

*Former Naval Person to President Roosevelt* 13.XII.40
I am sure you will be pleased about our victory in Libya. This, coupled with the Albanian reverses, may go hard with Mussolini if we make good use of our success. The full results of the battle are not yet to hand, but if Italy can be broken our affairs will be more hopeful than they were four or five months ago.

*Mr. Churchill to Mr. Menzies, Prime Minister of Australia* 13.XII.40
I am sure you will be heartened by the fine victory the Imperial Armies have gained in Libya. This, coupled with his Albanian disasters, may go hard with Mussolini. Remember that I could not guarantee a few months ago even a successful defence of the Delta and Canal. We ran sharp risks here at home in sending troops, tanks, and cannon all round the Cape while under the threat of imminent invasion, and now there is a reward. We are planning to gather a very large army representing the whole Empire and ample sea-power in the Middle East, which will face a German lurch that way, and at the same time give us a move eastward in your direction, if need be. Success always demands a greater effort. All good wishes.

*Prime Minister to General Wavell* 13.XII.40
I send you my heartfelt congratulations on your splendid victory,

which fulfils our highest hopes. The House of Commons was stirred when I explained the skilful Staff work required, and daring execution by the Army of its arduous task. The King will send you a message as soon as full results are apparent. Meanwhile pray convey my thanks and compliments to Wilson and accept the same yourself.

The poet Walt Whitman says that from every fruition of success, however full, comes forth something to make a greater struggle necessary. Naturally, pursuit will hold the first place in your thoughts. It is at the moment when the victor is most exhausted that the greatest forfeit can be exacted from the vanquished. Nothing would shake Mussolini more than a disaster in Libya itself. No doubt you have considered taking some harbour in Italian territory to which the Fleet can bring all your stuff and which will give you a new jumping-off point to hunt them along the coast until you come up against real resistance. It looks as if these people were corn ripe for the sickle. I shall be glad to hear from you your thoughts and plans at earliest. . . .

As soon as you come to a full-stop along the African coast we can take a new view of our prospects, and several attractive choices will be open.

By December 15 all enemy troops had been driven from Egypt. The greater part of the Italian forces remaining in Cyrenaica had withdrawn within the defences of Bardia, which was now isolated. This ended the first phase of the battle of Sidi Barrani, which resulted in the destruction of the greater part of five enemy divisions. Over 38,000 prisoners were taken. Our own casualties were 133 killed, 387 wounded, and 8 missing.

*Prime Minister to General Wavell*                                    16.XII.40

The Army of the Nile has rendered glorious service to the Empire and to our cause, and we are already reaping rewards in every quarter. We are deeply indebted to you, Wilson, and other commanders, whose fine professional skill and audacious leading have gained us the memorable victory of the Libyan desert. Your first objective now must be to maul the Italian Army and rip them off the African shore to the utmost possible extent. We were very glad to learn your intentions against Bardia and Tobruk, and now to hear of the latest captures of Sollum and Capuzzo. I feel convinced that it is only after you have made sure that you can get no farther that you will relinquish the main hope in favour of secondary action in the Soudan or Dodecanese. The Soudan is of prime importance, and eminently desirable, and it may be that the two Indian brigades [*i.e.*, the 4th British Indian Division] can be spared without prejudice to the Libyan pursuit battle. The Dodecanese will not get harder for a little waiting. But neither of them ought to detract

542

from the supreme task of inflicting further defeats upon the main Italian army. I cannot of course pretend to judge special conditions from here, but Napoleon's maxim, "*Frappez la masse et tout le reste vient par surcroît*", seems to ring in one's ears. I must recur to the suggestion made in my previous telegram about amphibious operations and landings behind the enemy's front to cut off hostile detachments and to carry forward supplies and troops by sea.

Pray convey my compliments and congratulations to Longmore on his magnificent handling of the R.A.F. and fine co-operation with the Army. I hope most of the new Hurricanes have reached him safely. Tell him we are filling up *Furious* again with another even larger packet of flyables from Takoradi. He will also get those that are being carried through in [Operation] "Excess". Both these should arrive early in January.

*Prime Minister to General Wavell*                     18.XII.40
   St. Matthew, chapter vii, verse 7.

"𝔄𝔰𝔨, 𝔞𝔫𝔡 𝔦𝔱 𝔰𝔥𝔞𝔩𝔩 𝔟𝔢 𝔤𝔦𝔳𝔢𝔫 𝔭𝔬𝔲; 𝔰𝔢𝔢𝔨, 𝔞𝔫𝔡 𝔶𝔢 𝔰𝔥𝔞𝔩𝔩 𝔣𝔦𝔫𝔡; 𝔨𝔫𝔬𝔠𝔨, 𝔞𝔫𝔡 𝔦𝔱 𝔰𝔥𝔞𝔩𝔩 𝔟𝔢 𝔬𝔭𝔢𝔫𝔢𝔡 𝔲𝔫𝔱𝔬 𝔶𝔬𝔲."

*General Wavell to Prime Minister*                     19.XII.40
   St. James, chapter i, verse 17.

"𝔈𝔳𝔢𝔯𝔶 𝔤𝔬𝔬𝔡 𝔤𝔦𝔣𝔱 𝔞𝔫𝔡 𝔢𝔳𝔢𝔯𝔶 𝔭𝔢𝔯𝔣𝔢𝔠𝔱 𝔤𝔦𝔣𝔱 𝔦𝔰 𝔣𝔯𝔬𝔪 𝔞𝔟𝔬𝔳𝔢, 𝔞𝔫𝔡 𝔠𝔬𝔪𝔢𝔱𝔥 𝔡𝔬𝔴𝔫 𝔣𝔯𝔬𝔪 𝔱𝔥𝔢 𝔉𝔞𝔱𝔥𝔢𝔯 𝔬𝔣 𝔩𝔦𝔤𝔥𝔱𝔰, 𝔴𝔦𝔱𝔥 𝔴𝔥𝔬𝔪 𝔦𝔰 𝔫𝔬 𝔳𝔞𝔯𝔦𝔞𝔟𝔩𝔢𝔫𝔢𝔰𝔰, 𝔫𝔢𝔦𝔱𝔥𝔢𝔯 𝔰𝔥𝔞𝔡𝔬𝔴 𝔬𝔣 𝔱𝔲𝔯𝔫𝔦𝔫𝔤."

\*    \*    \*    \*    \*

Bardia was our next objective. Within its perimeter, seventeen miles in extent, was the greater part of four more Italian divisions. The defences comprised a continuous anti-tank ditch and wire obstacles with concrete block-houses at intervals, and behind this was a second line of fortifications. The storming of this considerable stronghold required preparation. The 7th Armoured Division prevented all enemy escape to the north and north-west. For the assault there were available the 6th Australian Division, the 16th British Infantry Brigade, the 7th Battalion Royal Tank Regiment (twenty-six tanks), one machine-gun battalion, one Field and one Medium regiment of Corps Artillery.

To complete this episode of desert victory I shall intrude upon the New Year. The attack opened early on January 3. One Australian battalion, covered by a strong artillery concentration, seized and held a lodgment in the western perimeter. Behind them engineers filled in the anti-tank ditch. Two Australian brigades carried on the attack and swept east and south-eastwards. They sang at that time a song from an American film, which soon became popular also in Britain:

"We're off to see the Wizard,
The wonderful Wizard of Oz.
We hear he is a Whiz of a Wiz,
If ever a Wiz there was."

This tune always reminds me of these buoyant days. By the afternoon of the 4th, British tanks—"Matildas", as they were named—supported by infantry, entered Bardia, and by the 5th all the defenders had surrendered. 45,000 prisoners and 462 guns were taken.

By next day, January 6, Tobruk in its turn had been isolated by the 7th Armoured Division, and on the 7th the leading Australian brigade stood before its eastern defences. Here the perimeter was twenty-seven miles long and similar to that of Bardia, except that the anti-tank ditch at many points was not deep enough to be effective. The garrison consisted of one complete infantry division, a corps headquarters, and a mass of remnants from the forward areas. It was not possible to launch the assault till January 21, when, under a strong barrage, another Australian brigade pierced the perimeter on its southern face. The two other brigades of the division entered the bridgehead thus formed, swinging off to left and right. By nightfall one-third of the defended area was in our hands, and early next morning all resistance ceased. The prisoners amounted to nearly 30,000, with 236 guns. The Desert Army had in six weeks advanced over two hundred miles of waterless and foodless space, had taken by assault two strongly fortified seaports with permanent air and marine defences, and captured 113,000 prisoners and over 700 guns. The great Italian army which had invaded and hoped to conquer Egypt scarcely existed as a military force, and only the imperious difficulties of distance and supplies delayed an indefinite British advance to the west.

Throughout these operations vigorous support was provided by the Fleet. Bardia and Tobruk were in turn heavily bombarded from the sea, and the Fleet Air Arm played its part in the battle on land. Above all, the Navy sustained the Army in its advance by handling about 3,000 tons of supplies a day for the

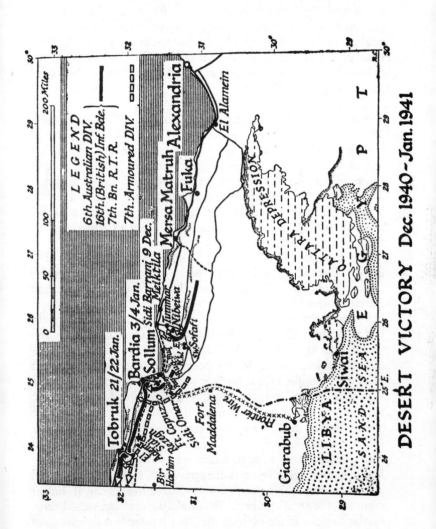

DESERT VICTORY Dec. 1940~Jan. 1941

LEGEND

6th. Australian DIV.
16th. (British) Inf. Bde.
7th. Bn. R.T.R.

7th. Armoured DIV.

0      50   100         200 Miles

Tobruk 21/22 Jan.
Bardia 3/4 Jan.
Sollum Sidi Barrani 9 Dec.
Buq Buq Meiktila
Tummar
Nibeiwa
Sofafi

Mersa Matruh Alexandria
Fuka
El Alamein

EGYPT

QATTARA DEPRESSION

Bir Hacheim Capuzzo
Ft. Omar
Sidi Omar
Fort Maddalena
Frontier Wire

Giarabub
Siwa

LIBYA
SAND SEA

forward troops, besides maintaining an invaluable ferry service for personnel through the captured ports. Our victorious army was also greatly indebted for their success to the mastery which the Royal Air Force gained over the Regia Aeronautica. Although in inferior numbers, the aggressiveness of our pilots soon established a complete moral ascendancy that gave them the freedom of the air. Our attacks on enemy airfields reaped a rich reward, and hundreds of their aircraft were later found wrecked and abandoned.

*　　*　　*　　*　　*

It is always interesting to see the reactions of the other side. The reader is already acquainted with Count Ciano, and should not be too hard on weak people who follow easily into wrong courses the temptations of affluence and office. Those who have successfully resisted all such temptations should form the tribunal. When Ciano faced the firing squad he paid his debts to the full. Villains are made of a different texture. We must not however imagine that it is better to be a rare villain than a Ciano or one of the multitudinous potential Cianos.

We have the Ciano diaries, jotted down each day.* The diary:—December 8: Nothing new. December 9: Intrigues against Badoglio. December 10: "News of the attack on Sidi Barrani comes like a thunderbolt. At first it doesn't seem serious, but subsequent telegrams from Graziani confirm that we have had a licking." Ciano saw his father-in-law twice on this day and found him very calm. "He comments on the event with impersonal objectivity . . . being more preoccupied with Graziani's prestige." On the 11th it was known in the inner circle at Rome that four Italian divisions must be considered destroyed, and, even worse, Graziani dwelt upon the daring and design of the enemy rather than upon any counter-measures of his own. Mussolini maintained his composure. "He maintains that the many painful days through which we are living must be inevitable in the changing fortunes of every war." If the British stopped at the frontier nothing serious would have happened. If, on the contrary, they reach Tobruk, "he thinks the situation would verge on the tragic". In the evening the Duce learned that five divisions had been "pulverised" in two days. Evidently there was something wrong with this army!

* *Ciano's Diaries*, 1939–43, edited by Malcolm Muggeridge, pp. 315–17.

On December 12 a "catastrophic telegram" came from Graziani. He contemplated retiring as far as Tripoli, "in order to keep the flag flying on that fortress at least". He was indignant that he should have been forced into so hazardous an advance upon Egypt by Rommel's undue influence on Mussolini. He complained that he had been forced into a struggle between "a flea and an elephant". Apparently the flea had devoured a large portion of the elephant. On the 15th Ciano was himself by no means sure that the English would be content to stop at the frontier, and records his opinion in that sense. Graziani, in default of military deeds, served up to his master bitter recriminations. Mussolini remarked, perhaps with some justice: "Here is another man with whom I cannot get angry, because I despise him." He still hoped that the British advance would be stopped at least at Derna.

*     *     *     *     *

I had kept the House daily informed of our progress in the desert, and on December 19 I made a long statement on the general war position. I described the improvement of our home defence and urged increasing vigilance. We must expect a continuance of the air attacks, and the organisation of shelters, the improvement of sanitation, and the endeavour to mitigate the extremely bad conditions under which people had to get their night's rest was the first task of the Government at home. "The Air Raid Precautions, the Home Office, and the Ministry of Health are just as much in the front line as are the armoured columns which are chasing the Italians about the Libyan desert." I also thought it necessary to utter a warning about the sinkings in the Atlantic. "They still continue at a very disquieting level; not so bad as in the critical period of 1917, but still we must recognise the recrudescence of the danger, which a year ago we seemed to have mastered. We shall steadily increase, from now on, our resources in flotillas and other methods of defence, but we must regard *the keeping open of this channel to the world against submarines and the long-distance aircraft which are now attacking as the first of all of our military tasks.*"

*     *     *     *     *

I thought it the moment to address the Italian people by the broadcast, and on the night of December 23 I reminded them of the long friendship between Britain and Italy. Now we were at war. ". . . Our armies are tearing and will tear your African Empire to shreds and tatters. . . . How has all this come about, and what is it all for?"

Italians, I will tell you the truth. It is all because of one man. One man and one man alone has ranged the Italian people in deadly struggle against the British Empire, and has deprived Italy of the sympathy and intimacy of the United States of America. That he is a great man I do not deny, but that after eighteen years of unbridled power he has led your country to the horrid verge of ruin can be denied by none. It is one man who, against the Crown and Royal Family of Italy, against the Pope and all the authority of the Vatican and of the Roman Catholic Church, against the wishes of the Italian people, who had no lust for this war, has arrayed the trustees and inheritors of ancient Rome upon the side of the ferocious pagan barbarians.

I read out the message I had sent to Mussolini on becoming Prime Minister and his reply of May 18, 1940, and I continued:

Where is it that the Duce has led his trusting people after eighteen years of dictatorial power? What hard choice is open to them now? It is to stand up to the battery of the whole British Empire on sea, in the air, and in Africa, and the vigorous counter-attack of the Greek nation; or, on the other hand, to call in Attila over the Brenner Pass with his hordes of ravenous soldiery and his gangs of Gestapo policemen to occupy, hold down, and protect the Italian people, for whom he and his Nazi followers cherish the most bitter and outspoken contempt that is on record between races.

There is where one man and one man only has led you; and there I leave this unfolding story until the day comes—as come it will—when the Italian nation will once more take a hand in shaping its own fortunes.

It is curious that on this same day Mussolini, speaking of the morale of the Italian Army, remarked to Ciano,* "I must nevertheless recognise that the Italians of 1914 were better than these. It is not flattering for the *régime*, but that's how it is." And the next day, looking out of the window: "This snow and cold are very good. In this way our good-for-nothing Italians, this mediocre race, will be improved." Such were the bitter and ungrateful reflections which the failure of the Italian Army in Libya and Albania had wrung from the heart of this dark figure

* *Ciano's Diaries*, p. 321.

after six months of aggressive war on what he had thought was the decadent British Empire.

★　　★　　★　　★　　★

This was a time when events were so fluid that every possible stroke had to be studied beforehand, and thus the widest choice of action lay open to us. Our victory in Libya had already stimulated the revolt against Italy in Abyssinia. I was most anxious that the Emperor, Haile Selassie, should re-enter his country as he desired to do. The Foreign Office thought this step premature. I deferred to the judgment of the new Secretary of State, but the delay was short, and the Emperor, eager to run all risks, was soon back on his native soil.

*(Action this Day)*
*Prime Minister to Foreign Secretary and General Ismay,*
*for C.O.S. Committee*          30.XII.40

It would seem that every effort should be made to meet the Emperor of Ethiopia's wishes. We have already, I understand, stopped our officers from entering the Galla country. It seems a pity to employ battalions of Ethiopian deserters, who might inflame the revolt, on mere road-making. We have sixty-four thousand troops in Kenya, where complete passivity reigns, so they surely could spare these road-makers. On the first point, I am strongly in favour of Haile Selassie entering Abyssinia. Whatever differences there may be between the various Abyssinian tribes, there can be no doubt that the return of the Emperor will be taken as a proof that the revolt has greatly increased, and will be linked up with the rumours of our victory in Libya.

I should be glad if a favourable answer could be drafted for me to send to the Emperor.

*Prime Minister to Foreign Secretary*          31.XII.40

One would think the Emperor would be the best judge of when to risk his life for his throne. In your minute you speak of our being "stampeded into premature and possibly catastrophic action". I do not wish at all to be "stampeded", but I should like to know some of the reasons why nothing is to be done for some months yet by the Emperor. I should have hoped the telegram to him could have been more forthcoming, and the one to Sir Miles Lampson rather more positive. These are however only matters of emphasis, and if with your knowledge you are apprehensive of giving more clear guidance I do not press for alteration of the telegrams.

The question of what pledges we give to Haile Selassie about his restoration, and what are our ideas about the Italian position in East

Africa, assuming that our operations prosper, as they may, is one which I was glad to hear from you this morning is receiving Foreign Office attention.

★ ★ ★ ★ ★

Finally, I was most anxious to give Vichy its chance to profit by the favourable turn of events. There is no room in war for pique, spite, or rancour. The main objective must dominate all secondary causes of vexation. For some weeks past the Chiefs of Staff Committee and the General Staff of the War Office had been preparing an Expeditionary Force of six divisions, and making plans, if the French attitude should become favourable, to land in Morocco. We had the advantage of M. Dupuy, the Canadian representative at Vichy, as a channel of communication with Marshal Pétain. It was necessary to keep the United States informed; for I already sensed the President's interest in Tangier, Casablanca, and indeed in the whole Atlantic seaboard of Africa, the German occupation of which by U-boat bases was held by the American military authorities to endanger the security of the United States. Accordingly, with the full approval of the Chiefs of Staff and the War Cabinet, the following message was sent by the hand of M. Dupuy to Vichy and notified by the Foreign Office to our Chargé d'Affaires in Washington.

*Prime Minister to Marshal Pétain*　　　　　　　　　　　31.XII.40
If at any time in the near future the French Government decide to cross to North Africa or resume the war there against Italy and Germany, we should be willing to send a strong and well-equipped Expeditionary Force of up to six divisions to aid the defence of Morocco, Algiers, and Tunis. These divisions could sail as fast as shipping and landing facilities were available. We now have a large, well-equipped army in England, and have considerable spare forces already well trained and rapidly improving, apart from what are needed to repel invasion. The situation in the Middle East is also becoming good.

2. The British Air Force has now begun its expansion, and would also be able to give important assistance.

3. The command of the Mediterranean would be assured by the reunion of the British and French Fleets and by our joint use of Moroccan and North African bases.

4. We are willing to enter into Staff talks of the most secret character with any military representatives nominated by you.

5. On the other hand, delay is dangerous. At any time the Germans

may, by force or favour, come down through Spain, render unusable the anchorage at Gibraltar, take effective charge of the batteries on both sides of the Straits, and also establish their air forces in the aerodromes. It is their habit to strike swiftly, and if they establish themselves on the Moroccan coast the door would be shut on all projects. The situation may deteriorate any day and prospects be ruined unless we are prepared to plan together and act boldly. It is most important that the French Government should realise that we are able and willing to give powerful and growing aid. But this may presently pass beyond our power.

A similar message was sent by another hand to General Weygand, now Commander-in-Chief at Algiers. No answer of any kind was returned from either quarter.

\*     \*     \*     \*     \*

At this stage we may review the numerous tasks and projects for which plans and in most cases preparations had been made, and approval in principle obtained. The first was of course the defence of the Island against invasion. We had now armed and equipped, though not in all cases at the highest standard of modern equipment, nearly thirty high-class mobile divisions, a large proportion of whom were Regulars, and all of whose men had been under intense training for fifteen months. Of these we considered that, apart from the coastal troops, fifteen would be sufficient to deal with oversea invasion. The Home Guard, now more than a million men, had rifles and some cartridges in their hands, apart from our reserve. We therefore had twelve or fifteen divisions available for offensive action overseas as need and opportunity arose. The reinforcement of the Middle East, and especially of the Army of the Nile, from Australia and New Zealand and from India had already been provided for by shipping and by other arrangements. As the Mediterranean was still closed, very long voyages and many weeks were required for all these convoys and their escorts.

Secondly, in case Vichy or the French in North Africa should rally to the common cause, we had prepared an Expeditionary Force of six divisions, with an air component, for an unopposed and assisted landing in Moroccan Atlantic ports, principally Casablanca. Whether we could move this good army to French Morocco or to Ceuta, opposite Gibraltar, more rapidly than the Germans could come in equal numbers and equipment through

Spain depended upon the degree of Spanish resistance. We could however, if invited, and if we liked it, land at Cadiz to support the Spaniards.

Thirdly, in case the Spanish Government yielded to German pressure and became Hitler's ally or co-belligerent, thus making the harbour at Gibraltar unusable, we held ready a strong brigade with four suitable fast transports to seize or occupy some of the Atlantic islands. Alternatively, if the Portuguese Government agreed that we might for this purpose invoke the Anglo-Portuguese Alliance of 1373, "Friends to friends, and foes to foes", we might set up with all speed a base in the Cape Verde Islands. This operation, called "Shrapnel", would secure us the necessary air and refuelling bases to maintain naval control of the critical stretch of the route round the Cape.

Fourthly, a French de Gaullist brigade from England, with West African reinforcements, was to be sent round the Cape to Egypt in order to effect the capture of Jibouti in case conditions there became favourable (Operation "Marie").*

Preparations were also being made to reinforce Malta, particularly in air-power (Operation "Winch"), with the object of regaining control of the passage between Sicily and Tunis. As an important element in this policy plans had been made for the capture by a brigade of commandos, of which Sir Roger Keyes wished to take personal command, of the rocky islet of Pantelleria (Operation "Workshop"). Every effort was ordered to be made to develop a strong naval and air base in Crete at Suda Bay, pending the movement thither of any reinforcements for its garrison which a change in the Greek situation might require.

---

* *Prime Minister to General Ismay, for C.O.S. Committee*                       1.XII.40

General de Gaulle told me that he had in mind an attempt to recover Jibouti—hereinafter to be called "Marie" in all papers and telegrams connected with the operation. He would send three French battalions from Equatorial Africa to Egypt, where General Le Gentilhomme would meet them. These battalions would be for the defence of Egypt, or possibly ostensibly as a symbolic contribution to the defence of Greece. There would be no secret about this. On the contrary, prominence would be given to their arrival. However, when the moment was opportune these battalions would go to Jibouti, being carried and escorted thither by the British Navy. No further assistance would be asked from the British. General de Gaulle believes, and certainly the attached paper favours the idea, that Le Gentilhomme could make himself master of the place, bring over the garrison and rally it, and immediately engage the Italians. This would be a very agreeable development, and is much the best thing de Gaulle could do at the present time. It should be studied attentively, and in conjunction with him. The importance of secrecy, and of never mentioning the name of the place, should be inculcated on all, remembering Dakar. I suppose it would take at least two months for the French battalions to arrive in Egypt.

Kindly let me have a full report.

We were developing airfields in Greece both to aid the Greek Army and to strike at Italy, or if necessary at the Roumanian oilfields. Similarly, the active development of airfields in Turkey and technical assistance to the Turks was in progress.

Finally the revolt in Abyssinia was being fanned by every means, and respectable forces were based on Khartoum to strike in the neighbourhood of Kassala against the menace of the large Italian army in Abyssinia. A movement was planned for a joint military and naval advance from Kenya up the East African coast towards the Red Sea to capture the Italian fortified seaports of Assab and Massawa, with a view to the conquest of the Italian colony of Eritrea.

Thus I was able to lay before the War Cabinet a wide choice of carefully-considered and detailed enterprises which could at very short notice be launched against the enemy, and certainly from among them we could find the means for an active and unceasing overseas offensive warfare, albeit on a secondary scale, with which to relieve and adorn our conduct of the war during the early part of 1941, throughout which the building up of our main war-strength in men and munitions, in aircraft, tanks, and artillery, would be continuously and immensely expanded.

\* \* \* \* \*

As the end of the year approached both its lights and its shadows stood out harshly on the picture. We were alive. We had beaten the German Air Force. There had been no invasion of the Island. The Army at home was now very powerful. London had stood triumphant through all her ordeals. Everything connected with our air mastery over our own Island was improving fast. The smear of Communists who obeyed their Moscow orders gibbered about a Capitalist-Imperialist War. But the factories hummed and the whole British nation toiled night and day, uplifted by a surge of relief and pride. Victory sparkled in the Libyan desert, and across the Atlantic the Great Republic drew ever nearer to her duty and our aid.

At this time I received a very kind letter from the King.

SANDRINGHAM
January 2, 1941

My dear Prime Minister,

I must send you my best wishes for a happier New Year, and may we see the end of this conflict in sight during the coming year. I am already feeling better for my sojourn here; it is doing me good, and the change of scene and outdoor exercise is acting as a good tonic. But I feel that it is wrong for me to be away from my place of duty, when everybody else is carrying on. However, I must look upon it as medicine and hope to come back refreshed in mind and body, for renewed efforts against the enemy.

I do hope and trust you were able to have a little relaxation at Christmas with all your arduous work. I have so much admired all you have done during the last seven months as my Prime Minister, and I have so enjoyed our talks together during our weekly luncheons. I hope they will continue on my return, as I do look forward to them so much.

I hope to pay a visit to Sheffield* next Monday. I can do it from here in the day. . . .

With renewed good wishes,

> I remain,
> Yours very sincerely,
> GEORGE R.I.

I expressed my gratitude, which was heartfelt.

January 5, 1941

Sir,

I am honoured by Your Majesty's most gracious letter. The kindness with which Your Majesty and the Queen have treated me since I became First Lord and still more since I became Prime Minister has been a continuous source of strength and encouragement during the vicissitudes of this fierce struggle for life. I have already served Your Majesty's father and grandfather for a good many years as a Minister of the Crown, and my father and grandfather served Queen Victoria, but Your Majesty's treatment of me has been intimate and generous to a degree that I had never deemed possible.

Indeed, Sir, we have passed through days and weeks as trying and as momentous as any in the history of the English Monarchy, and even now there stretches before us a long, forbidding road. I have been greatly cheered by our weekly luncheons in poor old bomb-battered Buckingham Palace, and to feel that in Your Majesty and the Queen there flames the spirit that will never be daunted by peril, nor wearied by unrelenting toil. This war has drawn the Throne and the people more closely together than was ever before recorded, and Your Majesties are more beloved by all classes and conditions than any of the princes of the past. I am indeed proud that it should have fallen to my

---

* Sheffield had been very heavily bombed.

lot and duty to stand at Your Majesty's side as First Minister in such a climax of the British story, and it is not without good and sure hope and confidence in the future that I sign myself, "on Bardia day", when the gallant Australians are gathering another twenty thousand Italian prisoners,

Your Majesty's faithful and devoted
servant and subject,
WINSTON S. CHURCHILL

\*     \*     \*     \*     \*

We may, I am sure, rate this tremendous year as the most splendid, as it was the most deadly, year in our long English and British story. It was a great, quaintly-organised England that had destroyed the Spanish Armada. A strong flame of conviction and resolve carried us through the twenty-five years' conflict which William III and Marlborough waged against Louis XIV. There was a famous period with Chatham. There was the long struggle against Napoleon, in which our survival was secured through the domination of the seas by the British Navy under the classic leadership of Nelson and his associates. A million Britons died in the first World War. But nothing surpasses 1940. By the end of that year this small and ancient Island, with its devoted Commonwealth, Dominions, and attachments under every sky, had proved itself capable of bearing the whole impact and weight of world destiny. We had not flinched or wavered. We had not failed. The soul of the British people and race had proved invincible. The citadel of the Commonwealth and Empire could not be stormed. Alone, but upborne by every generous heart-beat of mankind, we had defied the tyrant in the height of his triumph.

All our latent strength was now alive. The air terror had been measured. The Island was intangible, inviolate. Henceforward we too would have weapons with which to fight. Henceforward we too would be a highly organised war machine. We had shown the world that we could hold our own. There were two sides to the question of Hitler's world domination. Britain, whom so many had counted out, was still in the ring, far stronger than she had ever been, and gathering strength with every day. Time had once again come over to our side. And not only to our national side. The United States was arming fast and drawing ever nearer to the conflict. Soviet Russia, who with callous miscalculation had adjudged us worthless at the outbreak of the war, and had bought from Germany fleeting immunity and a share of the booty, had also become much stronger and had secured advanced positions for her own defence. Japan seemed for the moment to be overawed by the evident prospect of a prolonged world war, and, anxiously watching Russia and the United States, meditated profoundly what it would be wise and profitable to do.

And now this Britain, and its far-spread association of states and dependencies, which had seemed on the verge of ruin, whose very heart was about to be pierced, had been for fifteen months concentrated upon the war problem, training its men and devoting all its infinitely-varied vitalities to the struggle. With a gasp of astonishment and relief the smaller neutrals and the subjugated states saw that the stars still shone in the sky. Hope, and within it passion, burned anew in the hearts of hundreds of millions of men. The good cause would triumph. Right would not be trampled down. The flag of Freedom, which in this fateful hour was the Union Jack, would still fly in all the winds that blew.

But I and my faithful colleagues who brooded with accurate information at the summit of the scene had no lack of cares. The shadow of the U-boat blockade already cast its chill upon us. All our plans depended upon the defeat of this menace. The Battle of France was lost. The Battle of Britain was won. The Battle of the Atlantic had now to be fought.

# APPENDICES

APPENDICES

# APPENDIX A

## PRIME MINISTER'S PERSONAL MINUTES AND TELEGRAMS

### MAY—DECEMBER 1940

### MAY

*Prime Minister to General Ismay, for those concerned*    18.V.40

The proximity fuze and the necessary rocket projectors have hitherto been treated as important protection for ships, but even larger numbers will be needed, even some perhaps more urgently, for the protection of aircraft factories and other exceptionally important points. What is being done about this? Let proposals be made to-morrow for setting up the necessary manufacture. Are any modifications in the design of the projectors necessary? The Director of Naval Ordnance can go on with the ship side of the business, but be careful no hold-up takes place in the supply for the vulnerable points ashore. Report to-morrow night what organisation or measures are required to procure this production.

*Prime Minister to Secretary of State for the Colonies*    23.V.40

I am in full agreement with the answer you propose to Wedgwood's* Questions, and I do not want Jewish forces raised to serve outside Palestine. The main and almost the sole aim in Palestine at the present time is to liberate the eleven battalions of excellent Regular troops who are now tethered there. For this purpose the Jews should be armed in their own defence, and properly organised as speedily as possible. We can always prevent them from attacking the Arabs by our sea-power, which cuts them off from the outer world, and by other friendly influences. On the other hand, we cannot leave them unarmed when our troops leave, as leave they must at a very early date.

*Prime Minister to Minister of Aircraft Production*    24.V.40

I should be much obliged if you would have a talk with Lindemann, so as to get at some agreed figures upon aircraft outputs, both recent and prospective. I have for a long time been convinced that the Air Ministry do not make enough of the deliveries with which they are

---

* Hon. Josiah Wedgwood, M.P.

supplied, and Lindemann is obtaining for me returns of all aircraft in their hands, so that one can see what use is made of them.

It is of the highest importance that all aircraft in storage and reserve should not only be made available for service, but that these should be organised in squadrons with their pilots. Now that the war is coming so close, the object must be to prepare the largest number of aircraft, even, as you said, training and civil aircraft, to carry bombs to enemy aerodromes on the Dutch, Belgian, and French coasts. I must get a full view of the figures, both of delivery and employment, and this can be kept up to date weekly.

*Prime Minister to Professor Lindemann* 24.V.40

Let me have on one sheet of paper a statement about the tanks. How many have we got with the Army? How many of each kind are being made each month? How many are there with the manufacturers? What are the forecasts? What are the plans for heavier tanks?

NOTE.—The present form of warfare, and the proof that tanks can overrun fortifications, will affect the plans for the "Cultivator", and it seems very likely that only a reduced number will be required.

*Prime Minister to Sir Edward Bridges* 24.V.40

I am sure there are far too many Committees of one kind and another which Ministers have to attend, and which do not yield a sufficient result. These should be reduced by suppression or amalgamation. Secondly, an effort should be made to reduce the returns with which the Cabinet is oppressed to a smaller compass and smaller number. Pray let proposals be made by the Cabinet Office staff for effecting these simplifications.

*Prime Minister to Secretary of State for Air* 27.V.40

In your communiqué to-day you distinguish in several cases between enemy aircraft "put out of action" or "destroyed". Is there any real difference between the two, or is it simply to avoid tautology? If so, this is not in accordance with the best authorities on English. Sense should not be sacrificed to sound.

Will you also report to-day whether you would like the weather to be clear or cloudy for the operations on the Belgian coast.

*Prime Minister to General Ismay and C.I.G.S.* 29.V.40

The change which has come over the war affects decisively the usefulness of "Cultivator No. 6". It may play its part in various operations, defensive and offensive, but it can no longer be considered the only method of breaking a fortified line. I suggest that the Minister of Supply should to-day be instructed to reduce the scheme by one-half. Probably in a few days it will be to one-quarter. The spare available capacity could be turned over to tanks. If the Germans can make tanks in nine months, surely we can do so. Let me have your general proposals for the priority construction of an additional thousand tanks capable of engaging the improved enemy pattern likely to be working in 1941.

There should also be formed, if it does not already exist, an Anti-tank Committee to study and devise all methods of attacking the latest German tanks. Pray let me have suggested list of names.


# JUNE

*Prime Minister to Sir Edward Bridges*                                 3.VI.40

Has anything been done about shipping twenty thousand internees to Newfoundland or St. Helena? Is this one of the matters that the Lord President has in hand? If so would you please ask him about it. I should like to get them on the high seas as soon as possible, but I suppose considerable arrangements have to be made at the other end. Is it all going forward?

*Prime Minister to Secretary of State for Air*                         3.VI.40

The Cabinet were distressed to hear from you that you were now running short of pilots for fighters, and that they had now become the limiting factor.

This is the first time that this particular admission of failure has been made by the Air Ministry. We know that immense masses of aircraft are devoted to the making of pilots, far beyond the proportion adopted by the Germans. We heard some months ago of many thousands of pilots for whom the Air Ministry declared they had no machines, and who consequently had to be "re-mustered": as many as seven thousand were mentioned, all of whom had done many more hours of flying than those done by German pilots now frequently captured. How then therefore is this new shortage to be explained?

Lord Beaverbrook has made a surprising improvement in the supply and repair of aeroplanes, and in clearing up the muddle and scandal of the aircraft production branch. I greatly hope that you will be able to do as much on the personnel side, for it will indeed be lamentable if we have machines standing idle for want of pilots to fly them.

*Prime Minister to Professor Lindemann*                                3.VI.40

You are not presenting me as I should like every few days, or every week, with a short, clear statement of the falling off or improvement in munitions production. I am not able to form a clear view unless you do this.

*Prime Minister to Professor Lindemann*      3.VI.40

See attached paper [Production Programmes: Memorandum by Chiefs of Staff], which seems to contain a lot of loose thinking. Evidently we must "pull forward" everything that can be made effective in the next five months, and accept the consequent retardation of later production, but there is no reason whatever to alter, so far as I can see, the existing approved schemes for a three years' war. Indeed, they will be more necessary than ever if France drops out.

Pray let me have your views.

*Prime Minister to Professor Lindemann*      7.VI.40
(Secret.)

I am much grieved to hear of the further delay in the proximity fuze.

Considering the enormous importance of this, and the directions I have given that all possible pressure should be put behind it, it would surely have been right to have two or three firms simultaneously making the experimental pattern, so that if one failed the other could go on.

Please report to me what has been done.

You have not given me yet either a full statement of the production which is already ordered in rockets for the proximity fuze and in rockets for the ordinary fuze before we get the P.F.

It is of the utmost importance that you should go forward with the stabilising bomb-sight, as we must knock out their aircraft factories at the same rate that they affect ours. If you will gather together (*a*) all the people interested in the P.F. and (*b*) all those interested in the stabilised bomb-sight I will next week receive their reports and urge them on.

*Prime Minister to Minister of Aircraft Production*      11.VI.40

It was decided on December 22 at a conference on bomb-sight design that urgent action should be taken to convert 2,600 A.B.s, Mark II, into stabilised high-altitude bomb-sights, over 90 per cent. of the drawings then being completed. Please let me know exactly what followed. How is it that only one bomb-sight was converted? I should be very glad if you would look at the files and ascertain who was responsible for stifling action.

*Prime Minister to Secretary of State for Air and C.A.S.*      11.VI.40

This report\* is most interesting, and I shall be glad if you will arrange to use the squadron you mentioned yesterday for the purpose of infecting the reaches mentioned, where the traffic is reported to be

---

\* On the "Royal Marine" Operation. See Vol. I.

so heavy. We do not need to ask the French permission for this, but only for the continuous streaming of the naval fluvials. This I am doing. Meanwhile you should act as soon as you can on the lower reaches. Kindly report what you will do.

*Prime Minister to Secretary of State for the Colonies*      16.VI.40

Have you considered the advisability of raising a West Indies Regiment? It might have three battalions, strongly officered by British officers, and be representative of most of the islands; to be available for Imperial service; to give an outlet for the loyalty of the natives, and bring money into these poor islands.

At present we are short of weapons, but these will come along.

*Prime Minister to First Lord of the Admiralty*      17.VI.40

I am content with your proposed disposition of the heavy ships in the West, namely, *Repulse* and *Renown* to maintain the blockade at Scapa; *Rodney*, *Nelson*, and *Valiant* at Rosyth to cover the Island; *Hood* and *Ark Royal* to join *Resolution* at Gibraltar, to watch over the fate of the French Fleet.

It is of the utmost importance that the fleet at Alexandria should remain to cover Egypt from an Italian invasion, which would otherwise destroy prematurely all our position in the East. This fleet is well placed to sustain our interests in Turkey, to guard Egypt and the Canal, and can, if the situation changes, either fight its way westward or go through the Canal to guard the Empire or come round the Cape on to our trade routes.

The position of the Eastern fleet must be constantly watched, and can be reviewed when we know what happens to the French Fleet and whether Spain declares war or not.

Even if Spain declares war it does not follow that we should quit the Eastern Mediterranean. If we have to quit Gibraltar we must immediately take the Canaries, which will serve as a very good base to control the western entrance to the Mediterranean.

*Prime Minister to Minister of Home Security*      20.VI.40

I understand that it was settled last Saturday that your department was to take on the executive control of smoke as a means of hiding factories and similar industrial targets. I should be glad to know whom you have put in charge of this work, which I regard as of the highest importance, and what progress he has made.

*Prime Minister to Admiralty*      23.VI.40

I do not think it would be a good thing to keep *Hood* and *Ark Royal* lolling about in Gibraltar harbour, where they might be bombed at any time from the shore.

Surely when they have fuelled they should go to sea, and come back only unexpectedly and for short visits.

What is being done?

*Prime Minister to General Ismay*                                             24.VI.40

Has any news been received of the German prisoner pilots in France, whose return to this country was solemnly promised by Monsieur Reynaud?

*Prime Minister to Secretary of State for Foreign Affairs*          24.VI.40

It does not seem to be necessary to address the President again upon the subject of destroyers to-day or to-morrow. Evidently he will be influenced by what happens to the French Fleet, about which I am hopeful. I am doubtful about opening Staff talks at the present time. I think they would turn almost entirely from the American side upon the transfer of the British Fleet to transatlantic bases. Any discussion of this is bound to weaken confidence here at the moment when all must brace themselves for the supreme struggle. I will send the President another personal telegram about destroyers and flying-boats a little later on.

*Prime Minister to Secretary of State for the Colonies*            25.VI.40

The cruel penalties imposed by your predecessor upon the Jews in Palestine for arming have made it necessary to tie up needless forces for their protection. Pray let me know exactly what weapons and organisation the Jews have for self-defence.

*Prime Minister to Minister of Supply*                                   25.VI.40

Thank you for your letter of June 22 about increasing the import of steel from the United States. I understand that owing to the transfer of the French contracts to us our volume of purchases for the coming month has more than doubled and that we are now buying at the rate of about six hundred thousand tons a month. This is satisfactory, and we should certainly get as much from the United States as we can while we can.

*Prime Minister to Secretary of State for Foreign Affairs*          26.VI.40

I am sure we shall gain nothing by offering to "discuss" Gibraltar at the end of the war. Spaniards will know that, if we win, discussions would not be fruitful; and if we lose they would not be necessary. I do not believe mere verbiage of this kind will affect the Spanish decision.

*Prime Minister to General Ismay*                                         28.VI.40

Although our policy about the French Navy is clear, I should like to have an appreciation by the Admiralty of the consequences which are likely to follow—namely, a hostile attitude by France, and the seizure by Germany and Italy of any part of the French Navy which we cannot secure. I should like to have this on Sunday next.

*Prime Minister to General Ismay*                                   28.VI.40
This is a very unsatisfactory figure [of civilian labour*]. When I mentioned fifty-seven thousand the other day in the Cabinet I was assured that they represented a very small part of what were actually employed and that a hundred thousand was nearer the mark, and that many more were coming in before the end of the week. Now, instead, we have a figure of only forty thousand. Pray let me have a full explanation of this.
It is very wrong that fighting troops should be kept from their training because of the neglect to employ civilian labour.
The question must be brought up at the Cabinet on Monday.

*Prime Minister to Home Secretary*                                  28.VI.40
Let me see a list of prominent persons you have arrested.

*Prime Minister to Professor Lindemann*                             29.VI.40
If we could have large supplies of multiple projectors and rockets directed by Radar irrespective of cloud or darkness, and also could have the proximity fuze working effectively by day and to a lesser extent in moonlight or starlight, the defence against air attack would become decisive. This combination is therefore the supreme immediate aim. We are not far from it in every respect, yet it seems to baffle us. Assemble your ideas and facts so that I may give extreme priority and impulse to this business.

*Prime Minister to Professor Lindemann*                             29.VI.40
It seems to me that the blockade is largely ruined, in which case the sole decisive weapon in our hands would be overwhelming air attack upon Germany.
We should gain great relief in the immediate future from not having to maintain an army in France or sending supplies of beef, coal, etc., to France. Let me know about this.
How has the question of beef supplies been affected? We are freed from the obligation to supply the French Army with beef. There is really no reason why our Army at home should have rations far exceeding the heavy munitions workers'. The complications about frozen meat and fresh meat ought also to be affected by what has happened, although I am not sure which way.

* Labour for defence works. My former minute, dated 25.VI.40, is recorded in Book I, Chapter VIII, page 151.

# JULY

*Prime Minister to General Ismay*                                    2.VII.40

If it be true that a few hundred German troops have been landed on Jersey or Guernsey by troop-carriers, plans should be studied to land secretly by night on the islands and kill or capture the invaders. This is exactly one of the exploits for which the Commandos would be suited. There ought to be no difficulty in getting all the necessary information from the inhabitants and from those evacuated. The only possible reinforcements which could reach the enemy during the fighting would be by aircraft-carriers, and here would be a good opportunity for the Air Force fighting machines. Pray let me have a plan.

*Prime Minister to Secretary of State for Foreign Affairs*           3.VII.40

I could not reconcile myself to leaving a large number of influential Frenchmen who are the adherents of the Pétain Government free to run an active and effective propaganda in our Service circles and in French circles in this country, against the whole policy of aiding General de Gaulle, to which we are publicly and earnestly committed. The attempt to set up a French Government in Morocco and to obtain control of the *Jean Bart* and other vessels, and to open up a campaign in Morocco, with a base on the Atlantic, is, in my opinion, vital. It was most cordially adopted by the Cabinet in principle, and, apart from technical details, I should find very great difficulty in becoming a party to its abandonment, and to our consequent relegation to the negative defensive, which has so long proved ruinous to our interests.

*(Action this Day)*
*Prime Minister to V.C.N.S. and A.C.N.S.*                           5.VII.40

Could you let me know on one sheet of paper what arrangements you are making about the Channel convoys now that the Germans are all along the French coast? The attacks on the convoy yesterday, both from the air and by E-boats, were very serious, and I should like to be assured this morning that the situation is in hand and that the Air is contributing effectively.

*Prime Minister to First Lord of the Admiralty, Secretary*
*of State for War, and Secretary of State for Air*                   5.VII.40
(Sir E. Bridges to implement)

It has been represented to me that our colleagues not in the War Cabinet but above the "line" are depressed at not knowing more of what is going forward in the military sphere. It would be advantageous if each of the Service Ministers could in rotation have a talk with them, answer questions, and explain the general position. If a weekly meeting were instituted, this would mean that each Service Minister would meet them every three weeks. I trust this would not be too heavy a burden upon you. Nothing must ever be said to anybody about future operations; these must always be kept in the most narrow circles; but explanations of the past and expositions of the present offer a wide field. On the assumption that the above is agreeable to you, I am giving directions through Sir Edward Bridges.

*Prime Minister to Colonel Jacob*                                    6.VII.40

Obtain a most careful report to-day from the Joint Intelligence Staff of any further indication of enemy preparations for raid or invasion. Let me have this to-night.

*Prime Minister to Minister of Aircraft Production*                  8.VII.40

In the fierce light of the present emergency the fighter is the need, and the output of fighters must be the prime consideration till we have broken the enemy's attack. But when I look round to see how we can win the war I see that there is only one sure path. We have no Continental army which can defeat the German military power. The blockade is broken and Hitler has Asia and probably Africa to draw from. Should he be repulsed here or not try invasion, he will recoil eastward, and we have nothing to stop him. But there is one thing that will bring him back and bring him down, and that is an absolutely devastating, exterminating attack by very heavy bombers from this country upon the Nazi homeland. We must be able to overwhelm them by this means, without which I do not see a way through. We cannot accept any lower aim than air mastery. When can it be obtained?

*Prime Minister to Secretary of State for Air*                       11.VII.40

Generally speaking, the losses in the bomber force seem unduly heavy, and the Bremen raid, from which only one out of six returned, is most grievous. At the present time a very heavy price may be paid (*a*) for information by reconnaissance of the conditions in the German ports and German-controlled ports and river-mouths; (*b*) for the bombing of barges or assemblies of ships thus detected. Apart from this, the long-range bombing of Germany should be conducted with a desire to save the machines and personnel as much as possible while keeping up a steady attack. It is most important to build up the numbers of the bomber force, which are very low at the present time.

*Prime Minister to Home Secretary*                                   11.VII.40

You should, I think, prepare a Bill vacating the seat of any Member of Parliament who continues during the present war outside the jurisdiction for more than six months without the leave of the Secretary of State.

*Prime Minister to General Ismay*                              12.VII.40

What is being done to reproduce and install the small circular pill-boxes which can be sunk in the centre of aerodromes, and rise by means of a compressed-air bottle to two or three feet elevation, like a small turret commanding the aerodrome? I saw these for the first time when I visited Langley Aerodrome last week. This appears to afford an admirable means of anti-parachute defence, and it should surely be widely adopted. Let me have a plan.

*Prime Minister to Secretary of State for War*                 12.VII.40

Now is the time to popularise your administration with the troops by giving to all regiments and units the little badges and distinctions they like so much. I saw the London Irish with their green and peacock-blue hackles. We can easily afford the expense of bronze badges, the weight of which is insignificant in metal. All regimental distinctions should be encouraged. The French Army made a great speciality of additional unofficial regimental badges, which they presented to people. I liked this idea, and I am sure it would amuse the troops, who will have to face a long vigil. I am delighted at the action you have taken about bands, but when are we going to hear them playing about the streets? Even quite small parade marches are highly beneficial, especially in towns like Liverpool and Glasgow; in fact, wherever there are troops and leisure for it there should be an attempt at military display.

*Prime Minister to General Ismay, for C.O.S. Committee*      12.VII.40

1. The contacts we have had with the Italians encourage the development of a more aggressive campaign against the Italian homeland by bombardment both from air and sea. It also seems most desirable that the Fleet should be able to use Malta more freely. A plan should be prepared to reinforce the air defences of Malta in the strongest manner with A.A. guns of various types and with aeroplanes. Malta was also the place where it was thought the aerial mine barrage from the "Egg-layer" would be useful. Finally, there are the P.E. fuzes,* which will be coming along at the end of August, which should give very good daylight results. If we could get a stronger air force there we might obtain considerable immunity from annoyance by retaliation.

2. Let a plan for the speediest anti-aircraft reinforcement of Malta be prepared forthwith, and let me have it in three days, with estimates

---

* This was the photo-electric fuze, and, although not very successful, was the forerunner of the later proximity fuze.

in time. It should be possible to inform Malta to prepare emplacements for the guns before they are sent out.

*Prime Minister to General Ismay*                    12.VII.40

Will you bring the following to the notice of the Chiefs of Staff:

It is the settled policy of His Majesty's Government to make good strong French contingents for land, sea, and air service, to encourage these men to volunteer to fight on with us, to look after them well, to indulge their sentiments about the French flag, etc., and to have them as representatives of a France which is continuing the war. It is the duty of the Chiefs of Staff to carry this policy out effectively.

The same principle also applies to Poles, Dutch, Czech, and Belgian contingents in this country, as well as to the Foreign Legion of anti-Nazi Germany. Mere questions of administrative inconvenience must not be allowed to stand in the way of this policy of the State. It is most necessary to give to the war which Great Britain is waging single-handed the broad, international character which will add greatly to our strength and prestige.

I hope I may receive assurances that this policy is being whole-heartedly pursued. I found the conditions at Olympia very bad, and there is no doubt that the French soldiers were discouraged by some officers from volunteering. An opportunity of assisting the French would be to make a great success of their function of July 14, when they are going to lay a wreath on the Foch statue.

*Prime Minister to General Ismay*                    13.VII.40

Draw Admiralty attention to the importance of all these ships, especially *Western Prince*. What is her speed? It would be a disaster if we lost these fifty thousand rifles. Draw attention also to the immense consequence of the convoy which is leaving New York between July 8 and 12. When will these various convoys be in the danger zone? When will they arrive? Let me have a report on the measures to be taken.

*Prime Minister to Sir Edward Bridges*                    13.VII.40

I am receiving from various sources suggestions that there should be another day of prayer and humiliation.

Will you find out privately what is thought about this by the Archbishop.

*Prime Minister to General Ismay*                    14.VII.40

It seems to me very important that everybody should be made to look to their gas-masks now. I expect a great many of them require overhauling, and it may well be Hitler has some gas designs upon us.

Will you consider how the necessary overhauls can be set on foot. Action should be taken at once.

*Prime Minister to General Ismay, for V.C.A.S.*      15.VII.40

I am in full agreement with your proposal for bombing during the present moon-phase. I do not understand however why we have not been able to obtain results in the Kiel Canal. Nothing could be more important than this, as it prevents any movement of prepared shipping and barges from the Baltic for invasion purposes. I heard that you had dropped a number of bombs into this area, but that they did no good. Let me know what you have done about it in the past. How many raids, how many bombs, what kind of bombs, and what is the explanation that the canal still works? Can you make any plans for bettering results in the future? This is surely a matter of the very highest importance, and now is the time when it counts most.

*Prime Minister to General Ismay*      15.VII.40

Make sure that overhead cover against bombing attack is provided for the 14-inch gun. A structure of steel girders should be put up to carry sandbag cover similar to that over the 6-inch guns which are mounted along the coast. All should be camouflaged. You will be told that it will be necessary to change the guns after a hundred and twenty rounds. In that case the structure will have to be taken to pieces and put up again after the gun is changed. There should be no difficulty in this.

*Prime Minister to General Ismay*      17.VII.40

Press the War Office continually to develop the Foreign Legion, either by Pioneer Battalions or otherwise. Let me have weekly reports.

*Prime Minister to Home Secretary*      18.VII.40

I certainly do not propose to send a message by the senior child to Mr. Mackenzie King, or by the junior child either. If I sent any message by anyone, it would be that I entirely deprecate any stampede from this country at the present time.*

*(Action this Day.)*
*Prime Minister to Home Secretary*      19.VII.40

I have noticed lately very many sentences imposed for indiscretion by magistrates' and other courts throughout the country in their execution of recent legislation and regulation. All these cases should be reviewed by the Home Office, and His Majesty moved to remit the

---

* This refers to the Government-sponsored scheme for the evacuation of children to Canada and the U.S.A. The scheme was abandoned after the sinking of the *City of Benares* by a U-boat on September 17, 1940.

sentence where there was no malice or serious injury to the State. By selecting some of those cases which have recently figured in the public eye, and announcing remission publicly, you would give the necessary guidance without which it is difficult for local courts to assess the lead and purpose of Parliament.

*Prime Minister to First Lord and First Sea Lord*   20.VII.40

I have drawn attention to this danger before. I do not think *Hood* should be left lying in Gibraltar harbour at the mercy of a surprise bombardment by heavy howitzers. Both she and *Ark Royal* should go to sea for a cruise, with or without *Valiant* and *Resolution*, as may be thought fit. They could return to fuel or to carry out any operations, provided the Spanish situation has not further deteriorated. Pray let me have your proposals.

*Prime Minister to Foreign Secretary*   20.VII.40

Don't you think we might go very slow on all this general and equitable, fair and honourable peace business between China and Japan? Chiang does not want it; none of the pro-Chinese want it; and so far from helping us round the Burma Road difficulty, it will only make it worse. I am sure that it is not in our interest that the Japanese should be relieved of their preoccupation. Would it not be a good thing to give it a miss for a month or so, and see what happens?

*Prime Minister to Secretary of State for War*   20.VII.40

You may care to see this letter from Colonel Wedgwood on "London Defence". The only scale of attack which it seems to me need be contemplated for the centre of Government is, say, five hundred parachutists or Fifth Columnists. What is the present plan, and what is the scale against which it is being provided?

You might do something for Jos. He is a grand-hearted man.

*Prime Minister to Minister without Portfolio*   20.VII.40

I am rather doubtful, from information which has reached me, whether our home timber resources are being adequately developed.

This, of course, is primarily a matter for the Minister of Supply, who I know has made certain departmental adjustments recently with this particular end in view.

*Prime Minister to General Ismay*   21.VII.40

Let me have a statement showing the scheme of defence for the Central Government, Whitehall, etc. What was the scale of attack prescribed, and who was responsible for taking the measures? What was the reason for attempting to put an anti-tank obstacle across St. James's Park? Who ordered this? When was it counter-ordered?

*Prime Minister to General Ismay*  23.VII.40

I am told that the refuelling of fighter aeroplanes could be much more rapidly achieved if there were more tankers on the aerodromes, and considering that an attack by air would make every minute gained in returning the fighters to the air most precious, I should be glad if measures were taken at once to double or greatly increase the fuelling facilities.

*Prime Minister to Secretary of State for War*  23.VII.40

I do not seem to have had any answer from you to my query about whether the 2nd Canadian Division and all it stands for is being frittered away in Iceland.

*Prime Minister to Secretary of State for War*  23.VII.40

1. It is of course urgent and indispensable that every effort should be made to obtain secretly the best possible information about the German forces in the various countries overrun, and to establish intimate contacts with local people, and to plant agents. This, I hope, is being done on the largest scale, as opportunity serves, by the new organisation under M.E.W. [Ministry of Economic Warfare]. None of this partakes of the nature of military operations.

2. It would be most unwise to disturb the coasts of any of these countries by the kind of silly fiascos which were perpetrated at Boulogne and Guernsey. The idea of working all these coasts up against us by pin-prick raids and fulsome communiqués is one to be strictly avoided.

3. Sir Roger Keyes is now studying the whole subject of medium raids—*i.e.*, by not less than five nor more than ten thousand men. Two or three of these might be brought off on the French coast during the winter. As soon as the invasion danger recedes or is resolved, and Sir R. K.'s paper-work is done, we will consult together and set the Staffs to work upon detailed preparations. After these medium raids have had their chance there will be no objection to stirring up the French coast by minor forays.

4. During the spring and summer of 1941 large armoured irruptions must be contemplated. The material for these is however so far ahead of us that only very general study of their possibilities is now necessary, and no directions need be given to the Staff upon them until the end of August.

*Prime Minister to General Ismay, for Chiefs of Staff*  24.VII.40

Apart from the Anti-Nazi Germans, who can begin by being pioneers, rifles and ammunition should be issued to all foreign corps. Whether this should be from British Service rifles now in the possession of the Home Guard, but in process of being replaced by American rifles, or whether the foreign corps* should be armed with American rifles direct, has no doubt been considered. On the whole I am inclined to the former solution. It is most urgent to rearm the Poles and the French, as we may need them for foreign service in the near future. The armament of these foreign corps ranks *after* the armament of British troops so far as rifles are concerned, but they have priority over the Home Guard. They ought to have a small proportion of Bren guns, etc., even at the expense of our own men. What is being done to furnish them with artillery? Surely some of the 75's can be made to serve the purpose. The Polish unit should be ripened as much as possible. Pray let me have a weekly report of numbers and weapons.

(*Action this Day*)
*Prime Minister to First Lord, First Sea Lord, and*
*V.C.N.S.*                                                                25.VII.40

I cannot help feeling that there is more in the plan of laying mines behind an invader's landing than the Naval Staff felt when I mentioned the matter three weeks ago. In the interval I sent a reminder asking that it should be further considered.

If an invader lands during the night or morning, the flotillas will attack him in rear during the day, and these flotillas will be heavily bombarded from the air, as part of the air battles which will be going on. If however when night falls a curtain or fender of mines can be laid close inshore, so as to cut off the landing-place from reinforcements of any kind, these mines, once laid, will not have to be guarded from air attack, and consequently will relieve the flotilla from the need of coming back on the second day, thus avoiding losses from the air and air protection. At any rate, I think it improvident not to provide for the option whether to seal off the hostile landing by attack of flotillas or mines. There may be several landings, and you may want to leave one sealed off with mines in order to attack another. Of course all the above would apply still more if the landing had got hold of a port instead of merely a beach.

Pray let this matter have further attention, and also say what craft

| * French | .. | .. | .. | .. | 2,000 |
|---|---|---|---|---|---|
| Poles | .. | .. | .. | .. | 14,000 |
| Dutch | .. | .. | .. | .. | 1,000 |
| Czechs | .. | .. | .. | .. | 4,000 |
| Norwegians | .. | .. | .. | .. | 1,000 |
| Belgians | .. | .. | .. | .. | 500 |
| Anti-Nazi Germans | .. | .. | .. | 3,000 |

25,500

are available for the purpose, or how soon they can be provided or adapted.

*Prime Minister to V.C.N.S.*          25.VII.40

Let me have a report on how far the German, Dutch, and Belgian harbours have been sealed up by mines or obstructions.

*Prime Minister to Foreign Secretary*          26.VII.40

I saw Mr. Quo yesterday, at his request, and explained to him frankly the position about the Burma Road. I told him verbally of the message I sent through the Foreign Office to Chiang Kai-shek. He was naturally anxious to extort some promise from me about what would happen when the three months had expired. I said it all depended upon what the situation was then, and that I could make no forecast. I assured him we should put no pressure on General Chiang to consent to terms or negotiations against his will and policy. Mr. Quo seemed fairly satisfied, though rueful.

*Prime Minister to Chancellor of the Exchequer*          28.VII.40

Now that the Roumanian Government are helping themselves to the property of British subjects, ought we not to show the Roumanians that we shall use their frozen fund to compensate our people? I understand that about six weeks ago you blocked Roumanian assets in London. We have been treated odiously by these people.

# AUGUST

*(Action this Day)*
*Prime Minister to First Lord and First Sea Lord*          1.VIII.40

In view of the threatening attitude of Japan, it is vitally important to know about *Bismarck* and *Tirpitz*. Pray let me have your latest information. It seems to me that a great effort will have to be made by the Air Force to disable these ships, as their apparition in the next few months would be most dangerous.

Assuming Japan goes to war with us, or forces us into war, I suppose you would send *Hood*, three 8-inch-gun cruisers, two *Ramillies*, and twelve long-radius destroyers to Singapore.

Let me have the legends [*i.e.*, construction details] of the completed Japanese battle-cruisers.

*(Action this Day)*
*Prime Minister to First Lord and First Sea Lord*          2.VIII.40

I pray that we may never have to make this widespread distribution, but I am in full accord with the principles on which the Admiralty would propose to meet the strain. I should have thought that *Hood* would be a greater deterrent than *Renown*. Please let me have a report of the possibility of air attack on *Bismarck* and *Tirpitz*. This seems to me to be one of the most vital steps to take. Apart from this, there is no need to make any new dispositions at the present time on account of Japanese war-risk.

I was much concerned to hear of the sinking of the three tankers off Tory Island. I should like to see you move some destroyers from the East Coast thither. We had better wait, however, until the August moon-phase is over. During this time also the American guns and rifles will be distributed to the troops.

*(Action this Day)*
*Prime Minister to General Ismay*                                    2.VIII.40
    1. Next week one of my principal tasks must be going through this scheme of the Air Ministry for increasing the pilots and for the training of pilots. Lord Beaverbrook should be asked for his views beforehand.
    2. Let me have a report on the plans for lectures on tactical subjects for the troops in the autumn.
    3. What has been done about the collection of scrap of all kinds? Let me have a short report on one page covering the progress made this year.
    4. When at the Admiralty I took a special interest in the work of the Salvage Department, and held a meeting there four months ago. A naval officer, Captain Dewar, was then in charge. Let me have a report on what has happened to salvage since that day.
    5. I am also expecting this week to reach a settlement about the functions of the A.R.P. and police in the case of invasion. The Lord Privy Seal was dealing with this in the first instance. At the same time we must consider allowing transfers from A.R.P. to the Home Guard, and their being made available for fighting purposes. To what extent has the payment of the A.R.P. personnel been discontinued or restricted? It ought to be continually restricted.
    6. Let me have a report on the progress and future construction of the tank divisions. There should be five armoured divisions by March 31 [1941], and two more by the end of May. Let me know how far the present prospects of men and material allow of this. Let me know also what are the latest ideas for the structure and organisation of an armoured division. This should be prepared on one sheet of paper, showing all the principal elements and accessories.

*Prime Minister to General Ismay*                                    2.VIII.40
    It is very important to get on with the uniforms for the Home Guard. Let me have a forecast of deliveries.

*(Action this Day)*
*Prime Minister to First Lord*           2.VIII.40

My objection was to anything in the nature of sinking at sight or sinking without due provision for the safety of the crews. Provided this is excluded, there can be no reason against sinking a captured ship if, owing to air attack or other military reasons, it is impossible to bring her into port as a prize. The disadvantages of sinking a ship and losing valuable tonnage are obvious, and I do not see why in nineteen cases out of twenty the Admiralty cannot put a prize crew on board and send the ship in, in the ordinary way. I see no objection to the action taken in the *Hermione* case,\* which falls entirely within the general principles set forth above.

*Prime Minister to Sir Edward Bridges*           2.VIII.40

The whole question of holidays and reduced hours should be considered by the Cabinet at an early date. It is far too soon to assume that the danger has passed. It is a great mistake to tell the workpeople that they are tired. On the other hand, certain easements are indispensable. Please communicate with Mr. Bevin, Lord Beaverbrook, and the Minister of Supply, so that their views may be in readiness for Cabinet conversation. I should also like to know what is being done about holidays for the Civil Service and for Ministers, and persons in high Service positions. Something will have to be done about this, but we must be very careful not to be caught while in an August mood.

*Prime Minister to Lord Privy Seal and Home Secretary*     3.VIII.40

The attached memorandum by Lord Mottistone on duties of police in the event of invasion raises a very difficult question, and one that must be speedily settled. We cannot surely make ourselves responsible for a system where the police will prevent the people from resisting the enemy and will lay down their arms and become the enemy's servant in any invaded area. I confess I do not see my way quite clearly to the amendments required in the regulations. In principle however it would seem that the police should withdraw from any invaded area with the last of His Majesty's troops. This would also apply to the A.R.P. and the fire brigades, etc. Their services will be used in other districts. Perhaps on invasion being declared the police, A.R.P., fire brigades, etc., should automatically become a part of the military forces.

---

\* The *Hermione* was a small Greek steamer which was intercepted by our cruisers in the Ægean on July 28, 1940, while carrying a military cargo for Italy. Our ships were attacked by aircraft when making the interception. The *Hermione* was therefore sunk, and her crew left in boats near the land.

*Prime Minister to General Ismay*         3.VIII.40

All secret service reports about affairs in France or other captive countries are to be shown to Major Morton, who is responsible for keeping me informed. Make sure this instruction is obeyed.

*Prime Minister to Secretary of State for War*         3.VIII.40

It seems quite possible that a portion of General de Gaulle's forces will be used in the near future. It therefore becomes of the utmost consequence and urgency to complete the equipment of his three battalions, company of tanks, headquarters, etc. Evidently action is being taken already, but I shall be much obliged if you will accelerate this action by every means in your power, and also if you will let me know in what way the situation has improved since Major Morton's minute of yesterday.

*Prime Minister to Sir Edward Bridges, and others concerned*    3.VIII.40

1. I think the circular about work in the factories and holidays for whole establishments should, whatever the agreement of the Production Council, be brought before the Cabinet on Tuesday by the Minister of Labour. We must give holidays without creating a holiday atmosphere. It would therefore seem desirable to announce only that "such local arrangements as are possible are being made for staggered holidays", or something like that.

2. I approve Sir Horace Wilson's letter to departments. It arose out of my instructions to him.

3. I shall be very glad if you will adjust the holidays of Ministers, and make sure that the Services arrange for similar relief in case of high military officers at the centre of government.

*Prime Minister to Sir Edward Bridges*         4.VIII.40

I circulate to my colleagues the enclosed report on the first use of the U.P. weapon with the wire curtain at Dover. This appears to be of high importance, and may well inaugurate a decisive change in the relations of ground and air, particularly in respect of ships and ports exposed to dive-bombing attack.

*Prime Minister to Professor Lindemann*         4.VIII.40

What are you doing to focus the discussions on food, shipping, and agricultural policy for the second twelve months of the war? I thought it looked like 18,000,000 tons of shipping [for food], plough up 1,500,000 more acres, and instruct the Food Department to submit a plan both for increasing rations and building up further food reserves. This should be possible on the above basis.

*Prime Minister to Secretary of State for Air and C.A.S.*     4.VIII.40

The danger of Japanese hostility makes it all the more important that the German capital ships should be put out of action. I understand that the Air Force intends to make heavy attacks on these ships as soon as there is sufficient moon. *Scharnhorst* and the *Gneisenau*, both in floating docks at Kiel, the *Bismarck* at Hamburg, and the *Tirpitz* at Wilhelmshaven, are all targets of supreme consequence. Even a few months' delay in *Bismarck* will affect the whole balance of sea-power to a serious degree. I shall be glad to hear from you.

*Prime Minister to General Ismay*     5.VIII.40

I am not satisfied with the volume or quality of information received from the unoccupied area of France. We seem to be as much cut off from these territories as from Germany. I do not wish such reports as are received to be sifted and digested by the various Intelligence authorities. For the present Major Morton will inspect them for me and submit what he considers of major interest. He is to see everything and submit *authentic documents for me in their original form.*

Further, I await proposals for improving and extending our information about France and for keeping a continued flow of agents moving to and fro. For this purpose naval facilities can, if necessary, be invoked. So far as the Vichy Government is concerned, it is not creditable that we have so little information. To what extent are American, Swiss, and Spanish agents being used?

*Prime Minister to General Ismay*     5.VIII.40

What orders are extant for the future production of U.P. Multiple Projectors in groups of twenties, tens, fives, and also single projectors?

What amount of ammunition—

        (*a*)  of the ordinary rocket,
        (*b*)  of the aerial mine,
        (*c*)  of the P.E. fuze,
        (*d*)  of the radio fuze,

is on order? What are the forecasts of deliveries in the next six months in all cases?

Presently the P.E. fuze will probably supersede the aerial mine for use in multiple projectors mounted on H.M. ships. This will entail an alteration of the projector tubes. The Admiralty should be asked to study this betimes, so that the new tubes can be fitted on the existing mountings of H.M. ships with the least possible delay from the moment that this change appears desirable.

The Admiralty should also be asked to report whether any progress has been made on firing short aerial mines from ships' guns.

I wish to refresh my memory with what happened about this before I left the Admiralty.

*Prime Minister to Minister of Mines*                                6.VIII.40
I saw it stated that you were piling up large reserves of coal during the summer for use during the winter. I should be glad to know how far this very wise precaution has advanced. We were very short and anxious in January last, and I hope you are taking precautions.

*Prime Minister to Secretary of State for War*                       7.VIII.40
Please let me know what is being done to train men in the use of the sticky bomb, which is now beginning to come through in quantity.

*Prime Minister to General Ismay*                                    9.VIII.40
Ask for a statement of the Ministry of Supply importation programme under various heads. Professor Lindemann should be consulted about these heads. Let me see them.

The programme for the second year of the war has not yet been presented to me in a coherent form.

*Prime Minister to Secretary of State for War and*
*C.I.G.S.*                                                           9.VIII.40
I was much concerned to find that the 1st Division, which has an exceptionally high proportion of equipment, and includes a Brigade of Guards, should be dispersed along the beaches, instead of being held in reserve for counter-attack. What is the number of divisions which are now free and out of the line, and what is the argument for keeping divisions with a high equipment of guns, etc., on the beaches?

*Prime Minister to Lord Beaverbrook*                                 9.VIII.40
If it came to a choice between hampering air production or tank production, I would sacrifice the tank, but I do not think this is the case, as the points of overlap are not numerous and ought to be adjustable. I gathered from you that you thought you could arrange with the Minister of Supply.

*Prime Minister to Minister of Information*                          9.VIII.40
It is important to keep General de Gaulle active in French on the broadcast, and to relay by every possible means our French propaganda to Africa. I am told the Belgians will help from the Congo.

Have we any means of repeating to the West African stations the agreement made between us and de Gaulle?

*Prime Minister to General Ismay*                                    10.VIII.40
Let me have a weekly return of the deliveries to troops of the

American 75's and the .300 rifles to Home Guard, with consequent liberation by them of Lee-Metfords. Begin at once.

*Prime Minister to General Ismay, for C.O.S. Committee* 10.VIII.40
The Prime Minister would be glad to have a report from the C.O.S. Committee, after conference with the C.-in-C. Home Forces, upon the small arms ammunition position on the beaches and with the Reserves.

*Prime Minister to Minister of Mines* 11.VIII.40
I felt sure you would take advantage of the breakdown of the export market to increase our stocks all over the country. I hope you will press on with this, especially as regards our essential gas, water, and electricity works. I note that the gas and electricity supplies are about 20 per cent. up. We cannot go wrong in piling up such well-distributed stocks, which are sure to be used sooner or later.

I am sending a note to the Minister of Transport to call his attention to the position of the railways.

The tremendous upset in your plans due to the collapse of France and the loss of three-quarters of our export markets must have put a great strain on your department. It must be very difficult after all your efforts to increase production to explain the sudden slump, but I have no doubt the men will understand. Indeed, what you tell me about the fortitude of the Kent miners is an encouraging sign of the spirit which I believe informs all the working men in the country.

*Prime Minister to Minister of Information* 11.VIII.40
In view of certain activities we are planning for General de Gaulle, it is of the highest importance that the broadcasting of French news in North and West Africa should be carried to the highest point. Please make sure that the B.B.C. conform to this requirement, and let me have a report on Monday to the effect that all is satisfactorily arranged.

I cannot emphasise too strongly that you have full authority to make the B.B.C. obey.

*(Action this Day)*
*Prime Minister to Minister of Transport* 11.VIII.40
I should be grateful for a full report on the steps taken by your department to deal with the difficulties which may arise from the bombing and closing of ports.

One-quarter of our imports, it seems, normally comes in through the Port of London, and one-fifth through the Mersey, with a tenth each through Southampton, the Bristol Channel, and the Humber.

We must envisage these entrances being wholly or partially closed, either one at a time or even several at a time, but I have no doubt you have worked out plans to take account of the various contingencies.

In view of our large accretions of shipping it may well be that port facilities and roadway facilities may be a more stringent bottleneck than shortage of tonnage, so that the preparations you make to meet the various possible eventualities may be of the greatest importance.

*Prime Minister to Sir Edward Bridges*                                12.VIII.40

How does the position now stand about a Timber Controller under the Ministry of Supply?

Ask for a short summary from the Ministry of Supply of the present timber position and policy.

*Prime Minister to Lord Privy Seal and*
*Captain Margesson*                                                   12.VIII.40

It would probably be convenient for me to make a general statement on the war, covering the first year and also the first quarter of the new Government, before the House rises. This would be expected, and I suppose Tuesday, the 20th, would be the best day. This should, of course, be in public session. Perhaps you will let me know what you wish. An announcement could be made in good time this week.

It would save me a lot of trouble if a record could be taken at the time, so that the speech could be repeated over the wireless in the evening, or such parts of it as are of general interest. Can this be arranged without a Resolution? If not, could a Resolution be passed this week? I do not think the House would object.

*Prime Minister to Home Secretary*                                   12.VIII.40

The drafts [about instructions to police in case of invasion] submitted do not correspond with my view of the recent Cabinet decision. We do not contemplate or encourage fighting by persons not in the armed forces, but we do not forbid it. The police, and as soon as possible the A.R.P. services, are to be divided into combatant and non-combatant, armed and unarmed. The armed will co-operate actively in fighting with the Home Guard and Regulars in their neighbourhood, and will withdraw with them if necessary; the unarmed will actively assist in the "stay put" policy for civilians. Should they fall into an area effectively occupied by the enemy, they may surrender and submit with the rest of the inhabitants, but must not in those circumstances give any aid to the enemy in maintaining order, or in any other way. They may assist the civil population as far as possible.

*Prime Minister to Minister of Transport*                    13.VIII.40

I should be glad to know what stocks of coal are now held by the railways, and how they compare with those normally held. With the stoppage of our export trade to Europe there should be a great surplus just now, and no doubt you are taking advantage of this to fill up every available dump so that we shall have a well-distributed stock for the railways in case of any interruptions, or even in case of another very hard winter. Negotiations about price should not be allowed to hold up the process of re-stocking. If necessary, some form of arbitration will have to be employed to make sure that the prices paid are fair.

*Prime Minister to Secretary of State for War*                    13.VIII.40

If, owing to lack of equipment and other facilities, it is necessary to limit the numbers of the active Home Guard, would it not be possible to recruit a Home Guard Reserve, members of which would for the time being be provided with no weapons and no uniform other than arm-bands? Their only duties would be to attend such courses of instruction as could be organised locally in the use of simple weapons like the "Molotov cocktail", and to report for orders in the event of invasion.

Unless some such step is taken, those who are refused enlistment will be bewildered and disappointed, and one of the primary objects of the Home Guard, which was to provide for the people as a whole an opportunity of helping to defend their homes, will be lost. I am anxious to avoid the disappointment and frustration which the stoppage of recruiting for the Home Guard is likely to cause to many people.

Please let me know what you think of this proposal.

*Prime Minister to General Ismay*                    19.VIII.40

Is it true that Admiral [John] Cunningham says that the only suitable day for "Menace" [Dakar] is September 12, and that if this day is missed owing to storm no other days will be open till the 27th or 28th, when tide and moon will again be satisfactory? All this raises most grave questions. The Admiral cannot take up a position that only in ideal conditions of tide and moon can the operation be begun. It has got to be begun as soon as possible, as long as conditions are practicable, even though they be not the best. People have to fight in war on all sorts of days, and under all sorts of conditions. It will be a great misfortune if there is any delay beyond the 8th. Pray report to me on this to-day.

*Prime Minister to General Ismay*                    21.VIII.40

I am not convinced by these arguments about flame-throwers. The question is one to be settled relatively to other forms of war effort.

The prospects of invasion are rapidly receding. The likelihood in an invasion of a column of troops marching up the very defile in which these installations have been laid on appears remote. The idea of setting up a Petroleum Warfare Executive is a needless multiplication of our apparatus. I have no doubt whatever that the method would be very effective if ever the occasion arose, but will it arise, and in this case would it be at the point expected? Troops do not march along roads without having first cleared their way by small parties and guarded their flanks on each side of defiles.

*Prime Minister to First Lord*                                          22.VIII.40

I await your proposals about the resumption of the capital ship programme, which was approved by the late Cabinet on my initiative. This cannot be settled apart from the general demand upon steel and labour, but in principle I favour its resumption.

I hope opportunity will now be taken to repair the disastrous neglect to convert the *Royal Sovereign* class into properly armoured and bulged bombarding vessels with heavy deck armour. These will be needed next year for the attack on Italy. It is lamentable that we have not got them now. They should certainly take precedence over the resumption of battleship construction.

*Prime Minister to General Ismay*                                       24.VIII.40

Report to me on the position of Major Jefferis. By whom is he employed? Whom is he under? I regard this officer as a singularly capable and forceful man, who should be brought forward to a higher position. He ought certainly to be promoted Lieutenant-Colonel, as it will give him more authority.

*Prime Minister to C.A.S. and V.C.A.S.*                                 24.VIII.40

It is of high importance to increase both the number of squadrons and the number of aircraft and crews immediately available. After a year of war we have only operationally fit about 1,750, of which again only three-quarters are immediately available. You cannot rest satisfied with this, which is less than the number we were supposed to have available before the war.

*Prime Minister to Minister of Transport*                               25.VIII.40

I have read with interest your memorandum on Port Clearance.

I note that the Minister of Shipping doubts whether the country could be supplied through the West Coast ports on the scale you envisage. I should be glad to have your views on this.

Does not the widespread dislocation caused by the cold spell last winter raise some doubts as to the ready adaptability of the railway system in case of sudden emergency?

No doubt arrangements have been made for the importation of oil, which is not included in the food or supply programme. It appears that over two-fifths of our oil imports come through London and Southampton in peace-time. Our stocks are high, but if road transport had to be used more fully to relieve the railways our consumption would of course increase.

I presume that you have discussed their import programmes with the Ministers of Food and of Supply, so that alternative schemes will be ready in case of great diversions.

*Prime Minister to Secretary of State for War*          25.VIII.40

I have been following with much interest the growth and development of the new guerrilla formations of the Home Guard known as "Auxiliary Units".

From what I hear these units are being organised with thoroughness and imagination, and should, in the event of invasion, prove a useful addition to the regular forces.

Perhaps you will keep me informed of progress.

*Prime Minister to First Lord and First Sea Lord*          25.VIII.40

The enclosed returns show losses of over forty thousand tons reported in a single day. I regard this matter as so serious as to require special consideration by the War Cabinet. Will you therefore have prepared a statement showing the recent losses, their cause, the measures which have been taken by the Admiralty to cope with the danger, any further measures which you feel it necessary to propose, and whether there is any way in which the War Cabinet can assist the Admiralty.

I should be glad if you would make this report to the War Cabinet on Thursday next.

*(Action this Day)*
*Prime Minister to General Ismay*          25.VIII.40

Address the War Office forthwith upon the situation disclosed at Slough. Point out the danger of this large concentration of vehicles; the desirability of dispersing and concealing the vehicles. Ask that a plan should be examined for de-centralising this depot as far as possible. We should also make sure that no sediment or surplus accumulates in the depot. It would be a great pity if a thousand valuable vehicles were ruined by an air attack.

*Prime Minister to Secretary of State for Air*          25.VIII.40

I visited Kenley [Air Station] on Thursday, saw the gunner in question, and had a rocket fired off. Moreover, it was the Admiralty Committee over which I presided early in the year which produced the idea of using these distress rockets. I am therefore well acquainted with the subject. The Air Ministry, not for the first time, spread itself into very large demands, and, using its priority, barged in heavily into other forms of not less important production. I agree that P.A.C.* rockets may be a good interim defence against low-flying attack, but they have to take their place in the general scheme. I thought myself about five thousand a month would be sufficient, but I am willing to agree to one thousand five hundred a week, or six thousand a month. This figure could be somewhat extended if the wire-recovery projects you mention were further developed and proved an effective economy.

(*Action this Day*)
*Prime Minister to Secretary of State for War*　　　　　　25.VIII.40

War Office have accepted from the War Cabinet the responsibility for dealing with delayed-action bombs. This may become a feature of the enemy attack. A number were thrown last night into the City, causing obstruction. They may even try them on Whitehall! It seems to me that energetic effort should be made to provide sufficient squads to deal with this form of attack in the large centres. The squads must be highly mobile, so as not to waste men and material. They must move in motor-lorries quickly from one point to another. I presume a careful system of reporting all unexploded bombs and the time at which they fell is in operation, and that this information will be sent immediately to the delayed-action section of Home Defence, which has no doubt already been established, and also the various local branches. The service, which is highly dangerous, must be considered particularly honourable, and rewards should follow its successful discharge.

I should be very glad to see your plans for the new section, together with numbers, and it will also be interesting to have a short account of the work done up to date and the methods employed. I presume you are in touch with all the scientific authorities you need.

On the other hand, I am asking the Air Ministry for information as to their reciprocating this process on the enemy.

(General Ismay to see.)

*Prime Minister to Secretary of State for Air*　　　　　　25.VIII.40

I cannot feel you are justified in maintaining the present scale of communication squadrons when we are fighting so heavily. The sole end should surely be to increase the reserve and operational strength of our fighting squadrons and to meet the problem of trainer aircraft. Surely your dominant idea should be "Strength for battle". Every-

---

* The letters stand for "Parachute and Cable". The P.A.C. rocket was one form of the U.P. weapon. A description of the latter is given in a minute dated January 13, 1940. See Volume I, p. 594, first edition; New Edition, p. 674.

thing should be keyed on to this, and administrative convenience or local vested interests must be made to give way. In your place I should comb and re-comb. I have been shocked to see the enormous numbers at Hendon, and I would far rather give up flying on inspections altogether for members of the Government than that this should be made a reason for keeping these forces out of the fight.

I should have thought that Hendon could provide at least two good squadrons of fighter or bomber aircraft of the reserve category, and that they should have the machines issued to them and practise on them as occasion serves. Then they could be thrown in when an emergency came.

Ought you not every day to call in question in your own mind every non-military aspect of the Air Force? The tendency of every Station Commander is naturally to keep as much in his hands as possible. The Admirals do exactly the same. Even when you have had a thorough search, if you look around a few weeks later you will see more fat has been gathered.

I hope you will feel able to give some consideration to these views of your old friend.

*Prime Minister to First Lord and First Sea Lord*          27.VIII.40

Will you kindly send the following to Admiral Cunningham, Commander-in-Chief Mediterranean:

Following from Prime Minister, Minister of Defence:

Main object of directive was to safeguard Alexandria. Only a limited number of troops can be maintained Mersa Matruh, as G.O.C. Middle East will inform you. Every effort is to be made to defend this position. If however it and intermediate positions are forced or turned, it will be necessary to hold the line from Alexandria southwards along the cultivated area [of the Delta]. Air attack on the Fleet at Alexandria is not necessarily less effective from a hundred and twenty miles distance than from twenty miles, since aeroplanes often fly at three hundred miles per hour and have ample endurance. In practice it is usually thought better to hold aerodromes a little back of the actual fighting line. They do not move forward concurrently with the fronts of armies. Everyone here understands the grievous consequence of the fall of Alexandria, and that it would probably entail the Fleet leaving the Mediterranean. If however you have any helpful suggestion to make for the more effective defence of Mersa Matruh or of any positions in advance of it I should be obliged if you would tell me.

*Prime Minister to General Ismay, for Joint Planning Staff*          28.VIII.40

Now that the long nights are approaching the question of the black-out must be reviewed. I am in favour of a policy not of *black-out* but of *blackable-out*. For this purpose a considerable system of auxiliary electric street-lighting must be worked out. The whole of the centre of London, now lighted by incandescent gas, must be given priority. The best methods in the centres of other great cities must also be studied and local schemes must be examined. Thus the lights can be switched down and up and finally out on an air-raid warning being given. The lights themselves should not be of a too brilliant character. The subdued lighting of shop windows must also be studied with a view to extending the facilities given last Christmas on a permanent basis. Where factories are allowed to continue working at night in spite of the black-out there can be no objection to extending blackable-out lighting to the surrounding districts, thus tending to make the target less defined. Consideration should also be given to decoy lighting and baffle lighting in open spaces at suitable distances from vulnerable points.

*Prime Minister to Secretary of State for Air, C.A.S.,*
*and General Ismay* 29.VIII.40

I was much concerned on visiting Manston Aerodrome yesterday to find that although more than four clear days have passed since it was last raided the greater part of the craters on the landing ground remained unfilled and the aerodrome was barely serviceable. When you remember what the Germans did at the Stavanger aerodrome and the enormous rapidity with which craters were filled I must protest emphatically against this feeble method of repairing damage. Altogether there were one hundred and fifty people available to work, including those that could be provided from the Air Force personnel. These were doing their best. No effective appliances were available, and the whole process appeared disproportionate to the value of maintaining this fighting vantage ground.

All craters should be filled in within twenty-four hours at most, and every case where a crater is unfilled for a longer period should be reported to higher authorities. In order to secure this better service it will be necessary to form some crater-filling companies. You might begin with, say, two of two hundred and fifty each for the South of England, which is under this intensive attack. These companies should be equipped with all helpful appliances and be highly mobile, so that in a few hours they can be at work on any site which has been cratered. Meanwhile, at every aerodrome in the attack area, and later elsewhere, there must be accumulated by local contractors stocks of gravel, rubble, and other appropriate materials sufficient to fill without replenishment at least one hundred craters. Thus the mobile air-field repair companies would arrive to find all the material all ready on the spot.

I saw some time ago that the Germans filled in the shell holes by some process of having the gravel in wooden frameworks. The

V.C.N.S. drew my attention to it during the Norwegian operation, and he could perhaps put you on to the telegram referred to.

In what department of the Air Ministry does this process now fall?

After the craters had been refilled camouflage effort might be made to pretend they had not been, but this is a refinement.

*Prime Minister to General Ismay*
*(For all departments concerned, including Service Dept.,*
*Home Security, M.A.P., and Supply)*                     30.VIII.40

We must expect that many windows will be broken in the bombing raids, and during the winter glass may become scarce, with serious resultant damage to buildings if not replaced.

The utmost economy is to be practised in the use of glass. Where windows are broken they should, if possible, be boarded up, except for one or two panes. We cannot afford the full-sized windows in glass. All glass not needed for hot-houses should be stored if the hot-houses are empty. I saw at Manston a large hot-house with a great quantity of glass; enough was broken to make it useless, and I directed that the rest should be carefully stored.

What is the condition of glass supply? It would seem necessary to press the manufacturers.

Government buildings should all be fitted with emergency windows, containing only one or two glass panes, which, when the existing framework is blown in, can be substituted. Let me have a full report on the position.

*Prime Minister to General Ismay*                     31.VIII.40

If French India wish for trade they should be made to signify association with General de Gaulle. Otherwise no trade! This is not a matter upon which to be easy-going. Secretary of State for India to be informed.

The accession of any French possessions now is of importance.

*Prime Minister to General Ismay*                     31.VIII.40

I have not approved of any further cruiser tanks being dispatched to the Middle East beyond those which have already gone. Although in principle it is desirable to complete the dispatch of a full armoured division, further movements from this country can only be decided in relation to situation of home forces. No decision of this importance must be taken without reference to me, and in this case I should have to consult the Cabinet.

*Prime Minister to Minister of Supply*                     31.VIII.40

I am very glad to know that the chemical warfare stocks are piling up in this country. Let me know what the total now amounts to. The necessary containers should be brought level with supply. Do these stocks keep? Press on.


# SEPTEMBER

*Prime Minister to General Ismay, for C.O.S. Committee*       1.IX.40

Of course if the Glider scheme is better than parachutes we should pursue it, but is it being seriously taken up? Are we not in danger of being fobbed off with one doubtful and experimental policy and losing the other which has already been proved? Let me have a full report of what has been done about the Gliders.

*Prime Minister to First Lord and First Sea Lord*       1.IX.40

I am deeply concerned at your news that you cannot attack these batteries of German long-range guns until the 16th. You are allowing an artillery concentration to be developed day after day which presently will forbid the entry of all British ships into the Straits of Dover, and will prepare the way for an attack on Dover itself. Pray let me know what you propose to do about this.

Surely while the big guns are actually being hoisted into position and cannot fire back is the time for action. The general weakness of the defences of Dover itself in heavy guns is also a matter of great seriousness. We must not simply look at dangers piling up without any attempt to forestall them. *Erebus* will have to face double the fire on the 16th that she or any other ship would have to face in the next week.

I remember well that it was customary to bombard the Knocke and other German batteries on the Belgian coast very frequently during the late war. It was possible to fire most accurately by night after a buoy had been fixed and sound-ranging used. I ask for proposals for action this week. Look at the photographs attached.

*Prime Minister to General Ismay, for C.O.S. Committee*       1.IX.40

I presume you will be thinking about what is to happen should "Menace" succeed, with little or no bloodshed. It would seem that as soon as de Gaulle has established himself there and in the place a little to the north he should try to get a footing in Morocco, and our ships and troops could be used to repeat the process of "Menace", if it has been found to work, immediately and in a more important theatre. This operation may be called "Threat".

*Prime Minister to Secretary of State for War*       1.IX.40

I should be glad to have a full report of the arrangements being made to provide educational and recreational facilities for the troops during the coming winter. Who will be responsible for this important branch of work? ·

*Prime Minister to Secretary of State for India*                    1.IX.40
1. I am sorry to say that I cannot see my way to diverting aeroplanes or A.A. guns from the battle now raging here for the defence of India, which is in no way pressing; neither is it possible to divert American supplies for the building of an aircraft industry in India. We are already running risks which many might question in the reinforcement and re-equipment of the Middle East, and when the battle at home dies down this theatre will absorb all our surplus for a long time to come.
2. It is very important that India should be a help and not on the balance a burden at the present time. The debit balance is heavy when you consider the number of British troops and batteries locked up there, and the very exiguous Indian forces which, after a year of war, have reached the field. I am glad you are making increased efforts to form Indian divisions for the very large important operations which seem likely to develop in the Middle East in 1941.

*Prime Minister to First Lord, First Sea Lord, and Controller*     5.IX.40
I continue to be extremely anxious for *King George V* to get away to the north. It would be disastrous if *Bismarck* were finished and something happened to *King George V*. Surely the electricians, etc., can go north in her and finish up at Scapa. It would be most painful if you lost this ship now, after all these long, vexatious delays, just at the moment when she is finished and most needed. The Tyne is very ill-defended compared with Scapa.

(*Action this Day*)
*Prime Minister to Foreign Secretary*                              5.IX.40
Would it not be well to send a telegram to Lord Lothian expressing War Cabinet approval of the manner in which he handled the whole destroyer question, and paying him a compliment?
At the same time, what is being done about getting our twenty motor torpedo-boats, the five P.B.Y. [flying-boats], the one hundred and fifty to two hundred aircraft, and the two hundred and fifty thousand rifles, also anything else that is going? I consider we were promised all the above, and more too. Not an hour should be lost in raising these questions. "Beg while the iron is hot."

(*Action this Day*)
*Prime Minister to Secretary of State for War and C.I.G.S.*        8.IX.40
I am very pleased with this telegram [about the Cavalry Division in Palestine]. It has been heartbreaking to me to watch these splendid

units fooled away for a whole year. The sooner they form machine-gun battalions, which can subsequently be converted into motor battalions, and finally into armoured units, the better. Please let nothing stand in the way. It is an insult to the Scots Greys and House-hold Cavalry to tether them to horses at the present time. There might be something to be said for a few battalions of infantry or cavalrymen mounted on ponies for the rocky hills of Palestine, but these historic Regular regiments have a right to play a man's part in the war. I hope I may see your telegram approving this course before it goes.

*Prime Minister to First Lord* 9.IX.40

I have read your papers on the new programme. I understand you are going to redraft your memorandum after reading the one I presented to the Cabinet in March. I am not content at all with the refusal to reconstruct the *Royal Sovereign* class.* I think these should have precedence over all battleships, except those which can finish by the end of 1942. This would mean that you could get on with the *Howe*, the position of the other five capital ships being considered next year when the time for presenting the Navy Estimates comes. I see no reason why work should not proceed on the aircraft-carrier *Indefatigable*, and on the eight suspended cruisers. I am quite ready to approve the refilling of all slips vacated by anti-submarine craft, provided that a maximum limit of fifteen months is assigned to the completion of all new craft. All very large-size destroyers taking over this period to build must be excluded from the emergency war-time programme.

After your final proposals are ready we can have a conference.

*Prime Minister to General Ismay* 10.IX.40

1. The prime defence of Singapore is the Fleet. The protective effect of the Fleet is exercised to a large extent whether it is on the spot or not. For instance, the present Middle Eastern Fleet, which we have just powerfully reinforced, could in a very short time, if ordered, reach Singapore. It could, if necessary, fight an action before reaching Singapore, because it would find in that fortress fuel, ammunition, and repair facilities. The fact that the Japanese had made landings in Malaya and had even begun the siege of the fortress would not deprive a superior relieving fleet of its power. On the contrary, the plight of the besiegers, cut off from home while installing themselves in the swamps and jungle, would be all the more forlorn.

2. The defence of Singapore must therefore be based upon a strong

* See note under my minute of 7.IX.40, Book II, Chapter XXII, and also my minutes of 15.IX.40 and 26.XII.40, below.

*local* garrison and the general potentialities of sea-power. The idea of trying to defend the Malay peninsula and of holding the whole of Malaya, a large country four hundred by two hundred miles at its widest part, cannot be entertained. A single division, however well supplied with signals, etc., could make no impression upon such a task. What could a single division do for the defence of a country nearly as large as England?

3. The danger of a rupture with Japan is no worse than it was. The probabilities of the Japanese undertaking an attack upon Singapore, which would involve so large a proportion of their fleet far outside the Yellow Sea, are remote; in fact, nothing could be more foolish from their point of view. Far more attractive to them are the Dutch East Indies. The presence of the United States Fleet in the Pacific must always be a main preoccupation to Japan. They are not at all likely to gamble. They are usually most cautious, and now have real need to be, since they are involved in China so deeply.

4. I should have preferred the Australian Brigade to go to India rather than Malaya, but only because their training in India will fit them more readily for the Middle East. I am delighted to know they can be trained in the Middle East.

5. I do not therefore consider that the political situation is such as to require the withholding of the 7th Australian Division from its best station strategically and administratively. A telegram should be drafted to the Commonwealth Government in this sense.

*Prime Minister to the Mayor of Tel Aviv, Palestine*          15.IX.40

Please accept my deep sympathy in losses sustained by Tel Aviv in recent air attack. This act of senseless brutality will only strengthen our united resolve.

*Prime Minister to First Lord*          15.IX.40

1. Your new programme. I am very doubtful whether the Japanese figures are correct. The Naval Intelligence Branch are very much inclined to exaggerate Japanese strength and efficiency. I am not however opposed to the resumption of the battleship programme, provided it can be fitted in with more immediate war-time needs. Much of the battleship plant and labour would not be useful for other purposes. Pray let me have a paper showing the demands these ships would make in each year they are under construction, in money, steel, and labour. Every effort must be concentrated upon *Howe*.

2. I should be content if two R. [*Royal Sovereign*] class vessels were taken in hand as soon as the invasion situation has cleared and we get *King George V* in commission. Meanwhile material can be collected

and preparations made. This should enable them to be ready in eighteen months from now—*i.e.*, the summer of 1942.

3. You should press on with *Indefatigable*, but we need not consider an additional aircraft-carrier until early next year. The drawings can however be completed.

4. I suppose you realise that the Belfast type take over three years to build. Considering a large programme of cruisers is already under construction, I hope you will not press for these four to be added to the programme of this year.

5. I am all for building destroyers, and I do not mind how large they are, or how great their endurance, *provided* that they can be constructed in fifteen months. This should be taken as the absolute limit, to which everything else must be made to conform. We were making destroyers which took three years to build, everyone thinking himself very clever in adding one improvement after another. I should like to discuss the destroyer designs with the Controller and the Director of Naval Construction. They must be built only for this war, and have good protection from aircraft. Extreme speed is not so important. What you say about the U-boats working continually farther west is no doubt true, but the corvettes, formerly called whalers, have very fine endurance and range.

6. The submarine programme is already very large, and makes inroads on other forms of war requirements. I think you would be wise to re-examine the demand for the fourteen additional to the twenty-four to which the Treasury have agreed.

7. Great efforts should be made to produce the landing-craft as soon as possible. Are the Joint Planning Committee satisfied that these numbers are sufficient?

8. I am surprised you ask for only fifty anti-E-boats. Unless this is the utmost limit of your capacity, one hundred would be more appropriate.

9. Speaking generally, the speed of construction and early dates of completion must at this time be considered the greatest virtues in new building. It is no use crowding up the order books of firms and filling the yards with shipping orders which everyone knows cannot be completed. You have, I presume, consulted Sir James Lithgow about this programme, and have heard his views upon the consequences it will have upon merchant shipping building and our already reduced steel output. It is very wrong to trench too deeply upon the needs of other services in time of war.

10. What has happened to the armoured torpedo ram which I asked the D.N.C. to design?

*Prime Minister to Colonel Jacob*          15.IX.40

1. More than a year ago it was considered possible that we should soon be able to develop Radar inland. Since then however we have relied entirely on the Observer Corps. These have done splendid work; but in cloudy weather like yesterday and to-day they have the greatest difficulty in functioning accurately. If we could have even half a dozen stations which could work inland I am assured that very great advantages would be reaped in interception. This is especially important over the Sheerness–Isle of Wight promontory, which is likely to be the main line of air attack on London. I am told that there are duplicate installations already at some of the stations on this sector of the coast as an insurance against bombing. These might be turned round and put in action. In other cases new stations could be made. I regard this matter as of the highest urgency.

2. To-morrow, Monday, Air Marshal Joubert de la Ferté will assemble all necessary scientific authorities and make a report that day to me on—

    (*a*) the desirability of the above,

    (*b*) its practicability and the time it will take to get even a few stations into action.

He should make proposals for putting into service at the earliest moment six or twelve stations, and for rebuilding their reserves.

3. Should a feasible scheme emerge, I will myself bring it before the Minister of Aircraft Production.

*Prime Minister to General Sikorski*          17.IX.40

I deeply appreciated your telegram of September 14 conveying the relief felt by the Polish Government, the Polish armed forces, and the Polish people at the fortunate escape of the King and Queen from the recent German bombing of Buckingham Palace. As their Majesties stated, these dastardly attacks have only strengthened the resolution of all of us to fight through to final victory.

*Prime Minister to Home Secretary*          18.IX.40

The enemy will try by magnetic mines and other devices to smash as much glass as possible, and the winter is coming on. We must immediately revert to more primitive conditions in regard to daylight in dwellings. All glass in the country should be held, and every effort made to increase the supply. Everyone should be encouraged or pressed to reduce window-glass to at least one-quarter of its present compass, keeping the rest as spare. Windows should be filled as may be most convenient with plywood or other fabric, and the spare panes kept to replace breakages. The quicker this can be done in the target centres the better. Will you convene a meeting of the departments concerned and reach decisions for action of a violent character and on the broadest lines, inviting me to assist you in suppressing obstruction.

*Prime Minister to Home Secretary*                                        19.IX.40
I sent you a minute on this subject last night, and you were going to look into it for me.

How many square feet of glass have been destroyed up to date? Can any estimate be formed? If of course our monthly production is ahead of the damage there is no need to worry.

Let me have the best estimate possible.

*Prime Minister to Postmaster-General*                                   19.IX.40
There are considerable complaints about the Post Office service during air raids. Perhaps you will give me a report on what you are doing.

*Prime Minister to C.I.G.S.*                                              21.IX.40
I understood that all brigades from India consisted of one British and three Indian battalions, which would be the normal and desirable formation. But this telegram seems to suggest that Indian brigades have only Indian troops. If so the change made by C.-in-C. Middle East is most desirable.

*Prime Minister to First Sea Lord and Controller*                        21.IX.40
How is the expenditure of naval ammunition proceeding in the Middle East, as well as in the North Sea and Channel? Let me know of any weak points in the supplies which are emerging. Have you got over the difficulty of the 4.7 ammunition? Let me have a short note.

*Prime Minister to Secretary of State for Air*                           21.IX.40
Pray have a look at the Air Ministry communiqué issued in this morning's papers. It includes the following:

"The enemy formations were engaged by our fighters, but cloud conditions made interception difficult. Reports so far received show that four enemy aircraft were shot down. Seven of our fighters have been lost, the pilots of three being safe."

It is very unwise to let the Germans know that their new tactics have been successful and that they resulted in our losing seven fighters as against four.

We do not of course want to conceal our losses, at the present time, when we are prospering, but surely there is no need to relate them to any particular action.

*Prime Minister to General Ismay*                                        22.IX.40

Make sure through every channel that all arrangements are made to bring these rifles [from U.S.A.] over at full speed. They must be distributed in at least four fast ships. Could not some of them come by passenger liner? Let me know what Admiralty can do. Make sure there is no delay at Purco's [Purchasing Commission] end through repacking as described by General Strong, U.S.A.

*Former Naval Person to President Roosevelt*                            22.IX.40

I asked Lord Lothian to speak to you about our remaining desiderata. The 250,000 rifles are most urgently needed, as I have 250,000 trained and uniformed men into whose hands they can be put. I should be most grateful if you could arrange the necessary release. Every arrangement will be made to transport them with the utmost speed. They will enable us to take 250,000 .303 rifles from the Home Guard and transfer them to the Regular Army, leaving the Home Guard armed with about 800,000 American rifles. Even if no ammunition is available these rifles will be none the less useful, as they can draw upon the stock which has already reached us.

*Prime Minister to General de Gaulle*                                   22.IX.40

From every quarter the presence of General Catroux was demanded in Syria. I therefore took the responsibility in your name of inviting the General to go there. It is of course perfectly understood that he holds his position only from you, and I shall make this clear to him again. Sometimes one has to take decisions on the spot because of their urgency and the difficulty of explaining to others at a distance. There is time to stop him still if you desire it, but I should consider this was a very unreasonable act.

All good fortune in your enterprise to-morrow morning.

*Prime Minister to Minister of Supply*                                  23.IX.40

I regard the production of G.L. sets* as of prime importance, and every step should be taken to accelerate output. I understand that the chief difficulty at the moment is that of obtaining the skilled labour required, and I wish everything possible to be done to meet this requirement. Speed is vital.

*(Action this Day)*
*Prime Minister to Secretary of State for War and C.I.G.S.*            23.IX.40

There is not much in the report referred to, and what there is applies equally to the Soudan. We are piling up troops and artillery in Kenya which are urgently needed in the Soudan.

With regard to what you say about the vast strategical front of the

---

* A Radar set for anti-aircraft gun control.

Kenya operation: if we lie back on the railway from Mombasa to the lake we have a lateral line of communication incomparably superior to any line by which we can be approached, and it should be possible to move our forces so as to have sudden superior strength at the point where the enemy advance develops. Although no one can say for certain where the enemy's blow will fall, I am convinced that the true disposition would economise [troops] to the utmost in Kenya in order to reinforce the Soudan. The one concession which is needed for Kenya is about ten cruiser tanks. If these were put on suitable vehicles on the railway they could strike with deadly effect, and with *surprise*, at any Italian movement. But the mere piling up of guns and brigades is a most painful process to watch.

In order to raise these points, I must ask that the move of the Mountain Battery from Aden to Kenya shall be held up, and that instead the question of moving it or another battery to the Soudan shall be considered. Please let me have a statement showing ration, rifle, machine-gun, and artillery strength of all troops in Kenya.

*Prime Minister to Captain, H.M. destroyer "Churchill"*     25.IX.40
   Am delighted that your ship should be named after the great Duke of Marlborough, and I am sending you one of his handwritten letters for your Ward Room for luck. Thank you so much for your kind message.

*Prime Minister to Foreign Secretary*     25.IX.40
   Lord Lothian's proposal to return for a flying visit appeals to me. Pray authorise it, and arrange it as you think most serviceable and convenient.

*Prime Minister to General Ismay, for C.O.S. Committee*     26.IX.40
   If these facts [about the use of the blind beam for bombing] are accurate they constitute a deadly danger, and one of the first magnitude. I expect the Chiefs of Staff to use all the resources at their disposal and to give me a report by to-morrow night—

   (a) upon the reality of the danger,
   (b) upon the measures to counter it.

In making any recommendation for action the Chiefs of Staff may be sure that the highest priorities and all other resources will be at their disposal.

*Prime Minister to Home Secretary*     26.IX.40
   The composition hat for air raids which Mr. Bevin is promoting seems to me of the utmost importance, and if it gives a measure of protection against falling splinters, etc., it should certainly be mass-produced on a great scale, and eventually made a full issue.

Pray let me have a report to-day on the experimental aspect, and in conjunction with the Minister of Supply let me have estimates for production.

*Prime Minister to Minister of Labour*         26.IX.40

I was delighted with your hat, and something on these lines should certainly be mass-produced as soon as possible for issue pending steel hats. I think it is a mistake to call it a "rag hat", as I see is done in some of the papers to-day. I hope you will think of some better name.

I am calling for a full report to-day from the Home Secretary.

*Prime Minister to Secretary of State for Air and C.A.S.*      26.IX.40

Considering that everything depends upon Lord Beaverbrook's success in obtaining the supply of aircraft, and the heavy blows he is receiving at Bristol, Southampton, and elsewhere, I earnestly trust you will see that his wishes are met fully and immediately in the matter of these spares.

*Prime Minister to Minister of Agriculture*        26.IX.40

I am far from satisfied at the proposal to reduce pigs to one-third of their present number by the middle of the autumn. This is certainly not what was understood by the Cabinet. Why do you not ask for a greater proportion of feeding-stuffs in the imports? We could then see what, if anything, had to give way to it. Meanwhile, what arrangements are you making for curing the surplus bacon that will come upon the market through the massacre of pigs? What increases have you been able to establish in the pig population by encouraging people to feed individual pigs from household refuse?

*Prime Minister to Minister of Supply*        28.IX.40

Recent air raids have shown that the production of certain vital munitions, and particularly De Wilde ammunition, has been concentrated in one factory, with the result that output has been seriously curtailed by one successful raid. Pray let me have a report on the distribution of the production of every important key munition. It will then be possible to assess the danger of serious reductions in output and to consider what can be done to distribute the risk more widely.

*Prime Minister to General Ismay, for C.O.S. Committee*     28.IX.40

1. These two papers [about the supply of material for chemical warfare] cause me great anxiety. I had understood that Randle [factory] had been working at full capacity as a result of the orders given by the War Cabinet on October 13, 1939—*i.e.*, almost exactly

a year ago. What is the explanation of the neglect to fulfil these orders, and who is responsible for it?

2. Secondly, it appears that practically no steps have been taken to make projectiles or containers either for air or artillery to discharge these various forms of gas. The programme now set out would clearly take many months before any results are realised. Let me have an immediate report on this. The highest priority must be given. I regard the danger as very great.

3. Thirdly, the possibility of our having to retaliate on the German civil population must be studied, and on the largest scale possible. We should never begin, but we must be able to reply. Speed is vital here.

4. Fourthly, instant measures should be taken to raise Randle to full production, and above all to disperse the existing stock.

5. What are the actual amounts in stock?

*Prime Minister to General Ismay*                                    29.IX.40

These figures [about A.A. fire, first year of war] are encouraging. You should ask General Pile however to send in the account for September.

I should like to see a return of the ammunition fired every twenty-four hours during September as soon as possible.

*(Action this Day)*
*Prime Minister to Minister of Supply and President of the*
*Board of Trade*                                                      30.IX.40

I am sure we ought to increase our steel purchases from the United States so as to save tonnage on ore. I should like to buy another couple of million tons, in various stages of manufacture. Then we should be able to resume the plan of the Anderson shelters, and various other steel requirements which press upon us. I would if necessary telegraph to the President.


# OCTOBER

*Prime Minister to Foreign Secretary*                                 4.X.40

This shows the very serious misconception which has grown up in this Ambassador's mind about the consequences of the United States entering the war. He should surely be told forthwith that the entry of the United States into war, either with Germany and Italy or with Japan, is fully conformable with British interests; that nothing in the munitions sphere can compare with the importance of the British Empire and the United States being co-belligerents; that if Japan attacked the United States without declaring war on us we should at once range ourselves at the side of the United States and declare war upon Japan.

It is astonishing how this misleading Kennedy* stuff that we should do better with a neutral United States than with her warring at our side should have travelled so far. A clear directive is required to all our Ambassadors in countries concerned.

*Prime Minister to Secretary of State for War*       9.X.40
  . . . Anyone can see that aircraft are needed in the Middle East. What is not so easy is whether they can be spared here. Remember that we are still vastly inferior in numbers, both of fighters and bombers, to the German Air Forces, and that heavy losses have been sustained by our air production. The Chief of the Air Staff and Secretary of State must be asked for a precise recommendation.

*Prime Minister to General de Gaulle*       10.X.40
  I have received your telegram with great pleasure, and I send my best wishes to you and to all other Frenchmen who are resolved to fight on with us. We shall stand resolutely together until all obstacles have been overcome and we share in the triumph of our cause.

*Prime Minister to General Ismay, for C.O.S. Committee*     12.X.40
  This development of Radar with German long-range coastal batteries is serious. We have for a long time been on the track of this device, and I drew attention to it some weeks ago. I was then told that it had to have a low priority because of other even more urgent needs. Perhaps it may now be possible to bring it forward. Evidently it will turn night into day so far as defence against sea bombardment is concerned.

  Pray see if some proposals can be made without injury to other radio projects.

*Prime Minister to C.I.G.S.*       13.X.40
  There are great disadvantages in stationing many British troops on the West African coast. In view of the altered situation, pray consider bringing one of the West African brigades back from Kenya by one of the convoys returning empty. This should not add in any way to shipping burdens.

*Prime Minister to Sir James Grigg*       13.X.40
  A hot discussion is raging in the A.T.S. about whether members who marry should, if they wish, be allowed to quit. Nearly everyone is in favour of this. It seems futile to forbid them, and if they desert

* United States Ambassador to Britain.

there is no means of punishing them. Only the most honourable are therefore impeded. Pray let me have, on one sheet of paper, a note on this showing the pros and cons.

*Prime Minister to General Ismay*         14.X.40

Let a report be prepared on two sheets only showing what are the possibilities of Germany developing the munitions industries, especially aircraft, of the countries she has overrun, and when these evil effects are likely to become manifest.

*Prime Minister to First Lord*         15.X.40

1. If you wish to circulate the Naval Staff paper\* of October 13, which I have now read, I do not demur. It is of course a most pessimistic and nervous paper, which it is very depressing to receive from the Admiralty. Instances of the overdrawn character of the paper are found in para. 3, which claims that we must maintain "general control in every sea", whereas effective power of passage is all we require in many cases. And in para. 5, "German strength, in which from *now onwards* [October 15] must be counted the *Tirpitz* and the *Bismarck*." This is not true, as even the *Bismarck* has, I suppose, to work up, like the *King George V*, which should be ready as soon, or earlier. The *Tirpitz* is three months behind the *Bismarck*, according to every statement I have received, and it is hoped by that time we shall have the *Prince of Wales* and *Queen Elizabeth*. If such statements are made to the Cabinet I should be forced to challenge them.

2. The whole argument is meant to lead up to the idea that we must submit to the wishes of Vichy because they have the power to drive us out of Gibraltar by bombing. I fully share the desire of the Naval Staff not to be molested in Gibraltar, but I do not think that the enforcement of the blockade will lead the French to do this, still less to declare war upon us. I do not believe the Vichy Government has the power to wage war against us, as the whole French nation is coming more and more on to our side. I have dealt with this in a minute on general policy which is being circulated, and of which I enclose you the relevant extract.

3. The redeeming point in this paper is the suggestion that we should tell the Vichy Government that if they bomb Gibraltar we shall retaliate, not against, say, Casablanca, but Vichy, to which I would add, or any other place occupied by the Vichy Government. This is the proper note to strike, and it is also important to bear in mind that while humbleness to Vichy will not necessarily prevent them being ordered to make war upon us by their German masters, a firm attitude will not necessarily deter them from coming over to our side.

\* On the naval aspect of our policy towards the Vichy Government.

These questions are not urgent because of the failure to intercept *Primoguet*.*

*Prime Minister to C.A.S.*

18.X.40

What arrangements have we got for blind landings for aircraft? How many aircraft are so fitted? It ought to be possible to guide them down quite safely, as commercial craft were done before the war in spite of fog. Let me have full particulars. The accidents last night are very serious.

*Prime Minister to C.I.G.S.*

19.X.40

I was very much pleased last week when you told me you proposed to give an armoured division to Major-General Hobart.† I think very highly of this officer, and I am not at all impressed by the prejudices against him in certain quarters. Such prejudices attach frequently to persons of strong personality and original view. In this case General Hobart's original views have been only too tragically borne out. The neglect by the General Staff even to devise proper patterns of tanks before the war has robbed us of all the fruits of this invention. These fruits have been reaped by the enemy, with terrible consequences. We should therefore remember that this was an officer who had the root of the matter in him, and also vision.

In my minute last week to you I said I hoped you would propose to me the appointment that day, *i.e.*, Tuesday, but at the latest this week. Will you very kindly make sure that the appointment is made at the earliest moment.

Since making this minute I have carefully read your note to me and the summary of the case for and against General Hobart. We are now at war, fighting for our lives, and we cannot afford to confine Army appointments to persons who have excited no hostile comment in their career. The catalogue of General Hobart's qualities and defects might almost exactly have been attributed to most of the great commanders of British history. Marlborough was very much the conventional soldier, carrying with him the goodwill of the Service. Cromwell, Wolfe, Clive, Gordon, and in a different sphere Lawrence, all had very close resemblance to the characteristics set down as defects. They had other qualities as well, and so I am led to believe has General Hobart. This is a time to try men of force and vision and not to be exclusively confined to those who are judged thoroughly safe by conventional standards.

---

* A French merchant ship.
† General Hobart, at this time a corporal in the Home Guard, was accordingly appointed to command an armoured division, and in that capacity rendered distinguished service to the very end of the war. I had a pleasant talk with him on the day we first crossed the Rhine in 1945. His work was then highly esteemed by General Montgomery.

I hope therefore you will not recoil from your proposal to me of a week ago, for I think your instinct in this matter was sound and true.

*Prime Minister to C.I.G.S.* 19.X.40

Are there no younger men available for this strenuous administrative appointment [Director-General Home Guard]? The bringing back of retired officers for posts like these causes much criticism, both in and out of Service circles. Why not try to find a man still in the forties, and give him temporary rank?

*Prime Minister to General Ismay, for C.O.S. Committee* 19.X.40

In view of the forecasts of small arms ammunition, and the very great improvement in our position which will be effected from the factories coming into bearing in October, and the expanding output expected before March 31, 1941, and having regard to the fact that unless there is an invasion no operations are possible except in the Middle East, and then only on a comparatively moderate scale, I am of opinion that a very much larger issue may be made now to the Commander-in-Chief Home Forces for practice. I understand he has only two million rounds a week for this purpose, and that training is grievously hampered in consequence. Although it seems a risk to deplete our small War Office reserve, I think it should be considered whether from November 1 onwards the amount issued for practice should not be doubled—*i.e.*, four million a week. I shall be glad if you will consult the Chiefs of Staff immediately.*

*Prime Minister to General Ismay* 20.X.40

1. When was the last meeting of the Commanders-in-Chief, Naval, Air, and Military? Was it not found very useful? Who attended it?

I should be willing to preside over such a meeting in the course of the next week or so.

2. Let me have a plan for the imparting of more information about our war policy to these very high officers.

*Prime Minister to Secretary of State for Air and C.A.S.* 20.X.40

I am deeply concerned with the non-expansion, and indeed contraction, of our bomber force which must be expected between now and April or May next, according to present policy. Surely an effort should be made to increase our bomb-dropping capacity during this period. In moonlight periods the present arrangements for bombing are the best possible, and the only difficulty is our small numbers compared to the many attractive military targets. On no account should the limited bomber force be diverted from accurate bombing of military objectives reaching far into Germany. But is it not possible

* It was decided to provide the increased amount.

603

to organise a Second Line Bomber Force, which, especially in the dark of the moon, would discharge bombs from a considerable and safe height upon the nearest large built-up areas of Germany, which contain military targets in abundance? The Ruhr, of course, is obviously indicated. The object would be to find easy targets, short runs, and safe conditions.

How is such a Second Line or Auxiliary Bomber Force to be improvised during the winter months? Could not crews from the training schools do occasional runs? Are none of the Lysander and Reconnaissance pilots capable of doing some of this simpler bombing, observing that the Army is not likely to be in action unless invasion occurs? I ask that a wholehearted effort shall be made to cart a large number of bombs into Germany by a second line organisation such as I have suggested, and under conditions in which admittedly no special accuracy would be obtained. Pray let me have the best suggestions possible, and we can then see whether they are practical or not.

How is it that so few of our bombers are fitted with blind landing appliances? M.A.P. tells me that a number of Lorenz equipments are available. The grievous losses which occurred one day last week ought not to be repeated. Not only do the bombers need the blind landing facilities (which have been used in commercial aviation for years), but also if fighter aircraft are to operate by night, as they must increasingly, such aircraft must also be furnished with the means of making safe landings. Pray let me have your observations.

*Prime Minister to Secretary of State for Air and C.A.S.*     20.X.40

In connection with the plans now being developed for night fighting, not only by individual Aircraft Interception fitted machines, but by 8-gun-fighter squadrons, it is worth considering whether in any area where our fighters are operating and the guns have to remain out of action these guns should not fire blank charges. This would (*a*) confuse the enemy by the flashing on the ground, and tend to make him less aware of the impending fighter attacks—it would thus have a strictly military reason; (*b*) it would make a noise to drown the approach of our attacking fighters, and also to avoid discouraging silence for the population. It would not be legitimate to fire blank merely for the second purpose, but if there is a military reason the objection disappears.

*Prime Minister to C.I.G.S.*     20.X.40

I am concerned by the very low state of equipment of the Polish troops, whose military qualities have been proved so high. I hope to inspect them on Wednesday this week.

Pray let me have during Monday the best proposals possible for equipping them. I am most anxious they should not become disheartened.

*(Action this Day)*
*Prime Minister to Secretary, War Office*                    20.X.40

It is impossible to take away steel helmets from "the Home Guard in Government offices". Four were killed outside Downing Street on Thursday night. Whitehall is as heavily bombed as any part of the country. It will be difficult to take helmets away from anyone to whom they have been issued. I am astonished to see that the Army is aiming at three million helmets. I was not aware that we had three million men. Let me have a full return of all steel helmets in possession of the Regular Army, showing the different branches, *i.e.*, whether field army or training or holding battalions, etc., or in store. . . .

*Prime Minister to C.I.G.S. and Sir James Grigg*                 21.X.40

This very lengthy report by General Irwin[*] on how he was carried out to Freetown and back emphasises all the difficulties of the operation in which he was concerned. He foresaw all the difficulties beforehand, and the many shortcomings in the preparations. He certainly felt throughout that he was plunged into the midst of a grave and hazardous undertaking on political rather than military grounds. All this makes it the more surprising that he should have wished to persist in this operation, with all its defects and dangers, of which he was so acutely conscious, after these had been so formidably aggravated by the arrival through a naval failure of the French cruisers and reinforcements in Dakar, and in the teeth of the considered opinion of the War Cabinet and the Chiefs of Staff that conditions had now so changed as to make the original plan impossible. However, any error towards the enemy and any evidence of a sincere desire to engage must always be generously judged. This officer was commanding a division very ably before he was selected for the expedition, and I see no reason why he should not resume these duties now that he has returned. He would make a mistake, however, if he assumed either (*a*) that no enterprise should be launched in war for which lengthy preparation has not been made, observing that even in this connection twenty-five Frenchmen took Duala, and with it the Cameroons, or (*b*) that ships can in no circumstances engage forts with success. This might well be true in the fog conditions which so unexpectedly and unnaturally descended upon Dakar; but it would not necessarily be true of the case where the ships' guns could engage the forts at ranges to which the forts could not reply, or where the gunners in the forts were frightened, inefficient, or friendly to the attacking force.

[*] See Book II, Chapter XXIV.

*Prime Minister to Secretary of State for the Colonies*
*[Lord Lloyd]* 21.X.40

I am afraid I have been some time in studying your notes on the African continent, and its strategic and political dangers in the present war. I should deprecate setting up a special committee. We are overrun by them, like the Australians were by the rabbits. I see no reason to assume that we shall be at war with Vichy France or Spain, or that the South African position will develop dangerously. I should have thought that you would be able, with your own military experience and political knowledge, to gather such officials of the Colonial Office as you may need around you, and prepare yourself any reports you may think it right to present to the Defence Committee or the War Cabinet. If however you feel the need of being associated with a committee, I suggest that the Middle East Ministerial Committee takes on the agenda you have outlined as an addition to their present sphere.

P.S.—I am trying to move one of the West African brigades back from Kenya to the West Coast.

*(Action this Day)*
*Prime Minister to Minister of Information and*
*Sir Alexander Cadogan* 24.X.40

Sir Walter Citrine leaves this country shortly for the United States on a mission from the Trades Union Congress to American labour. He is a man of exceptional qualities and consequence, and is a Privy Councillor. He should certainly have a diplomatic status conferred upon him which will facilitate his movements. The T.U.C. are paying all his expenses in connection with the purely Labour side of the business, but I think that any expenses he may incur in work useful in the national interest should be defrayed by the Ministry of Information. Perhaps the Minister would look into this and see what can be done. In any case, Sir Walter should be treated with the greatest consideration, as I am sure we can count on his entire loyalty and discretion.

# NOVEMBER

*Prime Minister to C.A.S.* 1.XI.40

How is it that when we have 520 crews available for bombing operations and only 507 aircraft similarly available we do not draw on the aircraft storage units, where a large number are awaiting use?

*Prime Minister to Secretary of State for Air* 1.XI.40

Let me have, on not more than two sheets of paper, an analysis of the German aviators taken prisoners of war since July 1, showing numbers, ages, amount of training, etc., distinguishing between bomber and fighter prisoners. Any other information about them would be welcome.

*Prime Minister to First Sea Lord*            6.XI.40

Although I feel sceptical about the pocket-battleship going to Lorient, the Air Force should be thinking of attacking him there at the earliest moment, and should be warned *now*. If he goes to Lorient he runs a chance of being caught by you on the way in, bombed while he is there, and caught again on the way out. There is only one way in and out of Lorient. Very different is his position at Kiel, where he can come out via the Heligoland Bight or through the Skagerrak or sneak up the Norwegian Corridor to Trondheim. I would much rather see him go to Lorient than break south or stay out on the Atlantic route or go back one side or the other of Iceland.

If he continues preying on the trade you ought to be able to bring him to action.

On further reflection I agree it is better our two heavy ships should stay in the north.

These notes are only for your consideration.

*Prime Minister to C.I.G.S.*            6.XI.40

You impressed upon me how important it was to have a first-rate man in charge of the Home Guard, and what a compliment it would be if the former Chief of Staff in France was chosen; so General Pownall was appointed. But a few weeks later I was astonished to learn he was to go to America on the mission now discharged by General Pakenham-Walsh. With some difficulty I stopped this change. However, a little later Pownall was sent to Ireland. Whereas I suppose he would have done very well for the Home Guard, just as he got to know his job and men were beginning to look to him he was whisked off to something else, and General Eastwood took his place. This is, I think, only a month ago. However, I dutifully set myself to work to make General Eastwood's acquaintance, and I suppose so did the principal officers of the Home Guard. I formed a favourable opinion of him, particularly on account of his age, which is under fifty. I suppose he has been working very hard for the month, trying to learn his immense new task, and he certainly had begun to speak about it with knowledge. Now you propose to me to send him away, and to appoint a third new figure, all in four months.

All these rapid changes are contrary to the interests of the Service, and open to the most severe criticism. I am not prepared to agree to dismiss General Eastwood from the Home Guard command. If you wish to set up this Directorate-General, he must have it, so far as I am concerned. However, the Secretary of State will be back in two days, if all goes well, and I am sending a copy of this minute to him. I shall still expect to be consulted.

*Prime Minister to C.A.S.*          6.XI.40

Last night at least seven of our planes crashed on landing or were lost. The slow expansion of the bomber force is, as you know, a great anxiety to me. If bombing in this bad weather is imposing altogether undue risks and losses on the pilots, the numbers might be slacked down in order to accumulate our strength while at the same time keeping various objectives alive.

*Prime Minister to Sir Edward Bridges*          8.XI.40

Many of the executive departments naturally have set up and developed their own statistical branches, but there appears to be a separate statistical branch attached to the Ministerial Committee on Production, and naturally the Ministry of Supply's statistical branch covers a very wide field. I have my own statistical branch under Professor Lindemann.

It is essential to consolidate and make sure that agreed figures only are used. The utmost confusion is caused when people argue on different statistical data. I wish all statistics to be concentrated in my own branch as Prime Minister and Minister of Defence, from which alone the final authoritative working statistics will issue. The various departmental statistical branches will, of course, continue as at present, but agreement must be reached between them and the Central Statistical Office.

Pray look into this, and advise me how my wish can be most speedily and effectively achieved.

*Prime Minister to Minister of Transport*          8.XI.40

Let me know what progress has been made in breaking up the queues, and in bringing vehicles into service. With the earlier black-out it must be very hard on many.

*Prime Minister to First Sea Lord*          9.XI.40

Please let me have a report on the improvements in the Asdic and hydrophone technique which have been made in the last year.

*Prime Minister to Minister of Transport*          9.XI.40

Preliminary inspection seems to indicate that the time of turn-round in ports has increased in recent months rather than the reverse. This is probably due to the concentration of traffic on a few West Coast ports. Are the delays caused by inadequate port facilities or by difficulties in clearing the goods from the docks? Have you a scheme to exploit to the full our large resources of road transport if the railways prove inadequate to deal with these special problems?

*Prime Minister to C.A.S.*                                   10.XI.40

Altogether, broadly speaking, 1,000 aircraft and 17,000 air personnel in the Middle East provide 30½ squadrons, with a total initial equipment of 395 operational types, of which it is presumed 300 are ready for action on any date. Unhappily, out of 65 Hurricanes only 2 squadrons (apart from Malta) are available. These are the only modern aircraft, unless you count the Blenheim IVs. All the rest of this enormous force is armed with obsolete or feeble machines. The process of replacement should therefore be pressed to the utmost, and surely it should be possible to utilise all this skilled personnel of pilots and ground staff to handle the new machines. Therefore "remounting" the Eastern Air Force ought not in principle to require more personnel, except where new types are more complicated. However, as part of the reinforcements now being sent—*i.e.*, four Wellington and four Hurricane squadrons—we are sending over 3,000 additional personnel.

In the disparity between the great mass of men and numbers of aircraft on charge, and the fighting product constantly available, which is painfully marked both here and at home, lies the waste of R.A.F. resources. What is the use of the 600 machines which are not even included in the initial equipment of the 30 squadrons? No doubt some can be explained as training, communication, and transport. But how is it that out of 732 operational types only 395 play any part in the fighting?

I hope that a most earnest effort will be made to get full value for men, material, and money out of this very large force, first, by remounting, second, by making more squadrons out of the large surplus of machines not formed in squadrons, third, by developing local O.T.U.s [Operational Training Units] or other training establishments.

*Prime Minister to Minister of Health*                       10.XI.40

I see your total of homeless is down by 1,500 this week to about 10,000. Please let me know how many new you had in, and how many former went out. With such a small number as 10,000, you ought to be able to clean this up if you have another light week.

What is the average time that a homeless person remains at a rest centre?

*Prime Minister to Secretary of State for Air\**              10.XI.40

There is a shelter at Chequers which gives good protection from

---

\* The Air Ministry began making proposals for the greater protection of Chequers by sending Bofors guns.

lateral damage. There is the household to consider. Perhaps you will have the accommodation inspected.

The carriage drive is being turfed.

I cannot bear to divert Bofors from the fighting positions. What about trying a few rockets, which are at present only in an experimental stage?

I am trying to vary my movements a little during the moonlight intervals. It is very good of you and your Ministry to concern yourselves with my safety.

*Prime Minister to Secretary of State for War*                    10.XI.40

I hope you will look into this yourself. We had the greatest difficulty in carrying these sticky bombs through, and there was every evidence they would not have received fair play had I not gone down myself to see the experiment. Now is the chance to let the Greeks try this method out, and it would seem that it might be very helpful to them.

What is this tale that they are dangerous to pack and handle? They are of course dispatched without their detonators, and therefore cannot explode.

*Prime Minister to Air C.-in-C., Middle East*                    12.XI.40

I am trying every day to speed up the arrivals in your command of Hurricanes, etc. This is especially important in the next three weeks. Pray report daily what you actually receive, and how many you are able to put into action.

I was astonished to find that you have nearly 1,000 aircraft and 1,000 pilots and 16,000 air personnel in the Middle East, excluding Kenya. I am most anxious to re-equip you with modern machines at the earliest moment; but surely out of all this establishment you ought to be able, if the machines are forthcoming, to produce a substantially larger number of modern aircraft operationally fit? Pray report through the Air Ministry any steps you may be able to take to obtain more fighting value from the immense mass of material and men under your command.

I am grieved that the imperative demands of the Greek situation and its vital importance to the Middle East should have disturbed your arrangements at this exceptionally critical time. All good wishes.

*Prime Minister to Sir Edward Bridges and*
*General Ismay*                    12.XI.40

The Prime Minister has noticed that the habit of private secretaries and others of addressing each other by their Christian names about matters of an official character is increasing, and ought to be stopped. The use of Christian names in inter-departmental correspondence should be confined only to brief explanatory covering notes or to purely personal and private explanations.

It is hard enough to follow people by their surnames.

*Prime Minister to Home Secretary*                                    12.XI.40

How are you getting on with the comfort of the shelters in the winter—flooring, drainage, and the like? What is being done to bring them inside the houses? I attach the greatest importance to gramophones and wireless in the shelters. How is that going forward? Would not this perhaps be a very good subject for the Lord Mayor's Fund? I should not be surprised if the improved lighting comes up again before many weeks are out, and I hope that the preparations for it will go forward.

*Prime Minister to Foreign Secretary*                                  12.XI.40

We shall most certainly have to obtain control of Syria by one means or another in the next few months. The best way would be by a Weygand or a de Gaullist movement, but this cannot be counted on, and until we have dealt with the Italians in Libya we have no troops to spare for a northern venture. On no account must Italian or Caitiff-Vichy influences become or remain paramount in Syria.

*Prime Minister to Lord Beaverbrook*                                   15.XI.40

I do not think this could be said without the approval of the Air Ministry, and indeed of the C.O.S. Committee. My own feeling would be against giving these actual figures.* They tell the enemy too much. It is like getting one of the tail bones of the ichthyosaurus from which a naturalist can reconstruct the entire animal. The more I think about it the more I am against it.

*Prime Minister to Secretary of State for Air and C.A.S.*             15.XI.40

This amounts to a loss of eleven of our bombers in one night. I said the other day by minute that the operations were not to be pressed unduly during these very adverse weather conditions. We cannot afford to have losses of this kind in view of your very slow replacements. If you go on like this you will break the bomber force down to below a minimum for grave emergencies. No results have been achieved which would in any way justify or compensate for these losses. I consider the loss of eleven aircraft out of one hundred and thirty-nine—*i.e.*, about 8 per cent.—a very grievous disaster at this stage of our bomber development.

Let me have the losses during the first half of November.

* Figures of aircraft strength proposed to be used by Lord Beaverbrook in a broadcast.

*Prime Minister to C.A.S.* 17.XI.40

1. I watch these figures every day with much concern. My diagrams show that we are now not even keeping level, and there is a marked downward turn this week, especially in the Bomber Command. Painful as it is not to be able to strike heavy blows after an event like Coventry, yet I feel we should for the present *nurse* Bomber Command a little more. This can be done (1) by not sending so many to each of the necessary objectives, (2) by not coming down too low in the face of heavy prepared batteries and being content with somewhat less accuracy, and (3) by picking out soft spots where there is not too much organised protection, so as to keep up our deliveries of bomb content. There must be unexpecting towns in Germany where very little has been done in Air Raid Precautions and yet where there are military objectives of a minor order. Some of these could be struck at in the meanwhile.

2. I should feel differently about this if our bomber force were above five hundred, and if it were expanding. But, having regard to the uncertainties of war, we must be very careful not to let routine bombing and our own high standards proceed without constant attention to our resources. These remarks do not apply, of course, to Italy, against which the full-scale risk should be run. The wounded *Littorio* is a fine target.

*(Action this Day)*
*Prime Minister to First Lord and First Sea Lord* 18.XI.40

I was assured that 64 destroyers would be available for the North-Western Approaches by November 15. This return of Asdic-fitted ships, which goes to November 16, shows 60. But what is disconcerting is that out of 151 destroyers only 84 are available for service, and out of 60 for the North-Western Approaches only 33 are available for service. When we held our conference more than a month ago the Admiral was found with only 24 destroyers available, and all that has happened in the month that has passed is that another 9 have been added to his available strength. But meanwhile you have had the American destroyers streaming into service, and I was assured that there was a steady output from our own yards. I cannot understand why there has been this serious frustration of decisions so unitedly arrived at, nor why such an immense proportion of destroyers are laid up from one cause or another. Are the repairs falling behind? What has happened to the American destroyers? Are we failing in repairs and new construction?

I should be glad to have a special conference at 10 a.m. on Tuesday at the Admiralty War Room.

*Prime Minister to General Ismay, for Chiefs of Staff*  18.XI.40

I am informed that on the night of November 6-7 one of the German K.G. 100 Squadron* came down in the sea near Bridport. This squadron is the one known to be fitted with the special apparatus with which the Germans hope to do accurate night-bombing, using their very fine beams. Vital time was lost during which this aircraft or its equipment might have been salvaged because the Army claimed that it came under their jurisdiction, made no attempt to secure it, and refused to permit the naval authorities to do so.

Pray make proposals to ensure that in future immediate steps are taken to secure all possible information and equipment from German aircraft which come down in this country or near our coasts, and that these rare opportunities are not squandered through departmental differences.

*Prime Minister to Prime Minister of New Zealand*  18.XI.40

Your telegram is being dealt with departmentally. We dwell under a drizzle of carping criticism from a few Members and from writers in certain organs of the Press. This has an irritating effect, and would not be tolerated in any other country exposed to our present stresses. On the other hand, it is a good thing that any Government should be kept keen and made aware of any shortcomings in time to remedy them. You must not suppose everything is perfect, but we are all trying our best, and the war effort is enormous and morale admirable. All good wishes.

*Prime Minister to Prime Minister of Canada*  20.XI.40

1. I am most grateful to you for your message and for your very generous offer to afford facilities for a further expansion of the Joint Air Training Plan. I am confident that we shall be able to make excellent use of it.

2. A review of the air training requirements in the light of the latest developments is at present in progress, and it is of the utmost value to the War Cabinet in this connection to know that in such further measures as prove to be necessary they can rely on the continuance of the wholehearted assistance of the Canadian Government, which has already made such a notable contribution to our common effort.

3. As soon as our review is completed I will let you know, for your consideration, what we think would be the best direction for our further joint efforts.

4. As you mention in your message, any measures for the extension of the Joint Training Plan must form the subject of discussion and

* See p. 339.

agreement between all the Governments concerned. Would you agree to my repeating to the Prime Ministers of the Governments of Australia and New Zealand the text of your message and of this reply, or would you prefer to take this action yourself?

5. Subject to your agreement, we should like to offer a cordial invitation to Air Vice-Marshal Breadner to pay a short visit to this country. Such a visit would be most valuable for the purpose of consultation on many training questions, and would give Air Vice-Marshal Breadner the fullest and most up-to-date information on our plans for the future development of the Air Force.

*Prime Minister to Secretary of State for the Dominions*   22.XI.40
I think it would be better to let de Valera stew in his own juice for a while. Nothing could be more harmless or more just than the remarks in the *Economist*. The claim now put forward on behalf of de Valera is that we are not only to be strangled by them, but to suffer our fate without making any complaint.

Sir John Maffey should be made aware of the rising anger in England and Scotland, and especially among the merchant seamen, and he should not be encouraged to think that his only task is to mollify de Valera and make everything, including our ruin, pass off pleasantly. Apart from this, the less we say to de Valera at this juncture the better, and certainly nothing must be said to reassure him.

Let me see the Parliamentary questions as they come in.

*Prime Minister to Secretary of State for the Colonies*   22.XI.40
As the action has been announced, it must proceed, but the conditions in Mauritius must not involve these people being caged up for the duration of the war. The Cabinet will require to be satisfied about this. Pray make me your proposals.

[Ref.: Proposal to ship to Mauritius Jewish refugees who had illegally emigrated to Palestine.]

*Prime Minister to First Lord and First Sea Lord*
*(General Ismay to see)*   22.XI.40
1. In my view Admiral Stark is right, and Plan D* is strategically sound and also most highly adapted to our interests. We should therefore, so far as opportunity serves, in every way contribute to

---

* Plan D: Provision of all possible naval and military aid in the European field to the exclusion of any other interest. This would involve the adoption of a strictly defensive plan in the Pacific and abandonment of any attempt seriously to reinforce the Far East, with accepted consequences. On the other hand, by full-scale concentration in the European area the defeat of Germany was ensured with certainty, and if subsequently it was in the American interest to deal with Japan requisite steps would be possible.

strengthen the policy of Admiral Stark, and should not use arguments inconsistent with it.

2. Should Japan enter the war on one side and the United States on ours, ample naval forces will be available to contain Japan by long-range controls in the Pacific. The Japanese Navy is not likely to venture far from its home bases so long as a superior battle-fleet is maintained at Singapore or at Honolulu. The Japanese would never attempt a siege of Singapore with a hostile, superior American fleet in the Pacific. The balance of the American fleet, after providing the necessary force for the Pacific, would be sufficient, with our Navy, to exercise in a very high degree the command of all the seas and oceans except those within the immediate Japanese regions. A strict defensive in the Far East and the acceptance of its consequences is also our policy. Once the Germans are beaten the Japanese would be at the mercy of the combined fleets.

3. I am much encouraged by the American naval view.

*Prime Minister to Home Secretary*                        23.XI.40

There seems to be great disparity in these sentences [on A.F.S. men for looting], and I wonder whether any attempt is being made to standardise the punishments inflicted for this very odious crime. Five years' penal servitude for stealing whisky for immediate consumption seems out of proportion when compared with sentences of three or six months for stealing valuables. Exemplary discipline is no doubt necessary, as people must be made to feel that looting is stealing. Still, I should be glad to know that such cases are being reviewed and levelled out.

*Prime Minister to C.I.G.S.*                        24.XI.40

I sent you to-day two Foreign Office telegrams from Bucharest and Sofia respectively, which concur in an estimate of thirty thousand Germans, or one full division, as the maximum in Roumania at the present time. In view of this your Intelligence Branch should carefully review the advice they gave to the effect that there were five divisions in Roumania and that these could be assembled on the Bulgarian-Greek frontier in three or four days. I thought myself that this estimate was altogether too pessimistic, and credited the enemy with a rapidity of movement and a degree of preparedness which were perhaps more serious than the facts. Will you have the whole problem examined most carefully again? I had thought myself that it would be a fortnight before anything serious could happen on the Greek frontier, and that perhaps it might be a month. The great thing is to get the true picture, whatever it is.

*Prime Minister to General Ismay, and others concerned*        24.XI.40

This paper shows that we have completely failed to make cruiser tanks, and that there is no prospect of the present deficiency being made up in the next year. We must therefore equip our armoured divisions in the best possible way open to us in these melancholy circumstances. At this stage in tank production numbers count above everything else. It is better to have any serviceable tank than none at all. The formation and training of the divisions can proceed, and the quality and character of the vehicles be improved later on. The I tank should not be disdained because of its slow speed, and in default of cruisers must be looked upon as our staple for fighting. *We must adapt our tactics for the time being to this weapon as we have no other.* Meanwhile the production of cruiser tanks and of A.22 [a new model] must be driven forward to the utmost limit.


*Prime Minister to General Ismay*        24.XI.40

The full order for the thirty-five thousand vehicles should be placed in the United States without further delay. Meanwhile the inquiry into the scale required by the War Office is to proceed.


*Prime Minister to Foreign Secretary*        27.XI.40

The Greek complication seems to me serious. It will be of enormous advantage to us if Germany delays or shrinks from an attack on Greece through Bulgaria. I should not like those people in Greece to feel that, for the sake of what is after all only a parade, we had pressed them into action which could be cited by Germany as a justification for marching. The only thing to do is to put the meeting off until we can see a little more clearly on this very confused chessboard of Eastern Europe.

I think the Dominions should be told that we are waiting for the Greek situation to define itself more clearly, and that this ought not to take more than a fortnight. I do not think it is necessary to give any reasons to the Allied Governments, except to assure them that the delay will be short.


*Prime Minister to General Ismay*        28.XI.40

It is of no use giving me these reports five days late. The Admiralty know every day exactly the state of the flotillas. I do not know why this matter should go through the War Cabinet or Defence Ministry. Pray tell the Admiralty to send direct to me, every week, the state of their flotillas.

I am much concerned that the patrols on the Western Approaches should only have gone up to thirty effective. Let me see the chart showing previous weeks to-morrow.

*Prime Minister to Minister of Labour*                           28.XI.40

I shall be obliged if you will let me know the present unemployment figures, divided into as many categories as is convenient, and compared with

> (*a*) how they stood at the outbreak of war, and
> (*b*) when the new Government was formed.

*Prime Minister to First Sea Lord*                               30.XI.40

It is to me incomprehensible that with the 50 American destroyers coming into service we should not have been able to raise the total serviceable to above 77 by November 23, when they stood at 106 on October 16. What happened between October 16 and October 26 to beat down serviceable destroyers by 28 vessels, and why did they go down from 84 to 77 between November 16 and November 23—just at the very time when another dozen Americans were coming into service?

*Prime Minister to C.-in-C. Home Forces*                       30.XI.40

I have authorised the ringing of church bells on Christmas Day, as the imminence of invasion has greatly receded. Perhaps, however, you will let me know what alternative methods of giving the alarm you would propose to use on that day, and, secondly, what steps would be taken to ensure that the ringing of the bells for church services and without any invasion does not in fact lead to an alarm. There must certainly be no relaxation of vigilance.

# DECEMBER

*Prime Minister to Secretary of State for the Dominions*
*(General Ismay to see for C.O.S. Committee)*                       1.XII.40

All this talk about Atlantic Operations and Atlantic Islands is most dangerous, and is contrary to the decision to describe such operations as "Shrapnel". I see no need for these long and pointless telegrams, and it is becoming quite impossible to conduct military operations when everything has to be spread about the departments and around the world like this.

Kindly give me the assurance that there will be no further discussion of these matters by telegram without my seeing the messages before they are multiplied.

Let me also know exactly the lists of officials and departments to whom these telegrams have been distributed.

APPENDIX A

*(Action this Day)*
*Prime Minister to C.-in-C. Mediterranean*      3.XII.40
(Personal and most secret.)

1. Your 270. We considered whole matter this morning with the Director of Combined Operations, Sir Roger Keyes, who will execute it with full control of all forces employed, and final plans are now being prepared by him. His appointment will not be naval, but limited to these combined operations. If necessary he will waive his naval rank. Cannot feel that air counter-attack will be serious having regard to size island, broken character, many houses and detached forts, in which comparatively small attacking force will be intermingled with defenders. Enemy aircraft will not know who holds what till all is over, and even then Italian flags may be displayed on soft spots.

2. Capture of "Workshop"* no doubt a hazard, but Zeebrugge would never have got past scrutiny bestowed on this. Commandos very highly trained, carefully picked volunteers for this kind of work. Weather and fixed date of convoy may of course prevent attempt, in which case whole outfit will go to Malta or Suda for other enterprises. If conditions favourable nothing will be stinted.

3. Apprehensions you have that A.A. guns, etc., will be diverted from Eastern Mediterranean and new commitment created may be mitigated by capture of enemy A.A., which are numerous. Enemy unlikely attempt recapture, even though garrison left will be small Commandos will come away after handing over to Regular troops, and be available for further operations.

4. Comparing "Workshop" with other operation you mention, in future called "Mandibles"† (*repeat* "Mandibles"), kindly weigh following considerations:

"Mandibles" requires ten or twelve thousand men and is far larger affair if the two big ones are to be taken. Little ones you mention would stir up all this area without any important reward unless process continued. Secondly, captures in "Mandibles" area would excite keen rivalry of Greeks and Turks, which above all we don't want now. Thirdly, our reports show "Mandibles" slowly starving, and perhaps we shall get them cheaper later. Apart from the above, trying "Workshop" does not rule out "Mandibles" afterwards, unless ships and landing-craft are lost, which they may be. Also perhaps operations on enemy's land communications along North African shore may present opportunities.

5. On strategic grounds "Workshop" gives good air command of most used line of enemy communications with Libyan army, and also

* Capture of Pantelleria.
† Operations against the Dodecanese.

618

increased measure air protection for our convoys and transports passing so-called Narrows. Joint Staffs here consider very high value attaches to removal of this obstruction to our East and West communications. Besides all this, we need to show ourselves capable of vehement offensive amphibious action. I call upon you therefore to use your utmost endeavours to procure success should conditions be favourable at zero hour.

*Prime Minister to Minister of Aircraft Production*          3.XII.40

The King asked me to-day whether there was any shortage of instruments for aircraft.

*Prime Minister to General Ismay*          4.XII.40

1. Two searchlights [at Suda Bay] seem very insufficient. What is going to be done to increase them?

2. In view of the torpedoing of the *Glasgow* by a seaplane while at anchor, ought not ships at anchor to be protected by nets at short range? I gather this was the Italian method at Taranto, but at the moment of the attack they had taken them off. Pray let me have a note on this.

*Prime Minister to Secretary of State for War*          9.XII.40

### ARMY ORGANISATION

1. I understand that you are asking for another big call-up shortly. The papers talk about a million men. This forces me to examine the distribution of the men you have. According to your paper, 27 British divisions are credited to Expeditionary Force and Middle East. These divisions are accounted for at 35,000 men each, to cover corps, army, and line of communications troops, etc., plus 70,000 security troops in M.E. [Middle East].

2. The approved establishment of a British division at the present time is 15,500 men. It comprises only 9 battalions with an establishment of 850, *i.e.*, about 7,500. The establishment of all battalions comprises a considerable proportion of servicing elements, and I doubt whether the rifle and machine-gun strength—*i.e.*, fighting strength—amounts to more than 750. Thus the total number of men who actually fight in the infantry of a British division is 6,750. This makes the fighting infantry of 27 divisions, in what used to be called bayonet or rifle strength, 182,250. It used to be said that the infantry was "the staple of the Army", to which all other branches were ancillary. This has certainly undergone some modification under new conditions, but none the less it remains broadly true. The structure of a division is built round its infantry of 9 battalions, with a battery to each battalion, the necessary proportion of signallers and sappers, the battalion,

brigade, and divisional transport, and some additional elements, the whole being constituted as an integral and self-contained unit of 15,500 men.

3. When we look at the division as a unit, we find that 27 divisions at 15,500 official establishment require no less than 1,015,000 men. This gives an actual burden of 35,000 men for every divisional unit of 15,500 men, the units themselves being already fully self-contained. Nearly 20,000 men have therefore to be accounted for for each division of the E.F. [Expeditionary Force] or M.E. over and above the full approved establishment of 15,500.

This great mass, amounting to 540,000, has now to be explained. We are assured that the corps, army, L. of C. [line of communications] troops, etc., plus the 70,000 security troops in the M.E., justify this enormous demand upon the manhood of the nation.

4. One would have thought, if this were conceded, that the process was at an end. On the contrary, it is only just beginning. There still remain nearly two million men to be accounted for, as are set out on the attached table and graph. No one can complain of 7 divisions for the Home Field Force, though it is surprising that they should require 24,000 men for divisional establishments of 15,500. This accounts for 170,000 men.

5. A.D.G.B. [Air Defence of Great Britain] 500,000 must be submitted to for the present, pending improved methods of dealing with the night-bomber and increased British ascendancy in the air.

6. 200,000 men for the permanent staffs and "unavailable" at training and holding units is a distressing figure, having regard to the great margins already provided. Staffs, static and miscellaneous units, "Y" list, etc., require 150,000 after all the 27 divisions and the 7 home divisions have been fully supplied with corps and army troops. Apart from everything necessary to handle an army of 27 divisions and 7 home divisions, there is this mass of 350,000 staffs and statics, living well off the nation as heroes in khaki.

7. Compared with the above, overseas garrisons, other than M.E., of 75,000 seems moderate. India and Burma at 35,000 is slender.

8. 150,000 men for the corps, army, and L. of C. troops for divisions other than British requires to be explained in detail. I understand the Australian and New Zealand Forces had supplied a great many of their rearward services. At any rate, I should like to see the exact distribution of this 150,000 in every category behind the divisions which they are expected to serve.

9. The net wastage, 330,000, is of course a purely speculative figure. But it might well be supplied from the 350,000 permanent staffs, static and other non-availables already referred to.

10. Deducting for the moment the 330,000 men for wastage, which deals with the future up till March 1942, and 110,000 required for overseas garrisons other than M.E., India, and Burma, we face a total of 2,505,000 required for the aforesaid 27 divisions plus 7 home divisions, equal to about 74,000 per division. If the 500,000 for A.D.G.B. is omitted we still have over 2,000,000 men—*i.e.*, nearly 60,000 men mobilised for each of 34 divisions.

Before I can ask the Cabinet to assent to any further call-up from the public, it is necessary that this whole subject shall be thrashed out, and that at least a million are combed out of the fluff and flummery behind the fighting troops, and made to serve effective military purposes. We are not doing our duty in letting these great numbers be taken from our civil life and kept at the public expense to make such inconceivably small results in the fighting line.

*Prime Minister to General Ismay*                                           9.XII.40

Let me have a report on the development of the Salvage Section of the Admiralty, showing the work that has been done and what expansions, if any, are contemplated to meet the ever-growing need of repairs rapidly.

*Prime Minister to General Ismay*                                          11.XII.40

Let models be made of Rhodes and Leros. Report when they will be ready.

*(Action this Day)*
*Prime Minister to Secretary of State for Air*                        14.XII.40

There is one thing about the warfare between the Air Ministry and M.A.P. which is helpful to the public interest, namely, that I get a fine view of what is going on and hear both sides of the case argued with spirit. Will you very kindly address yourself to the various statements made in this letter attached [from Lord Beaverbrook], and especially to the one that on September 1 you had over 1,000 unserviceable trainer aircraft? I have long suspected that the inefficiency which formerly ruled in the A.S.U.s and left us with only 45 aeroplanes when the new Government was formed, as against about 1,200 now, was reproduced in all the trainer establishments and communication flights and that a great mass of aeroplanes were kept in an unserviceable state, and I remember particularly the statement of one of your high officers that the Training Command worked on a basis of 50 per cent. unserviceable. Who is responsible for repair and training establishments? If I were you, I should throw the whole business of repair on to M.A.P., and then you would be able to criticise them for any shortcomings.

See also the figures of how repaired aircraft and engines have increased since the change was made.

I recur to the point I made to you yesterday when you sent me your letter to M.A.P. The Air Ministry's view is that the Germans have nearly 6,000 aeroplanes in front-line action, and we have about 2,000. Air Ministry also believe that the German output is 1,800 a month, out of which they provide only 400 for training establishments, while we, out of 1,400 output, provide also 400. How do you, then, explain that the Germans are able to keep three times our establishment in front-line action with only an equal monthly subscription of trainer aircraft? Apparently, on your figures, which I may say I do not accept (except for controversial purposes), the Germans can keep three times as large a force in action as you can for the same number of trainer planes. I know that you will rightly say you are preparing for the expansion of the future, but they have to keep going on a threefold scale, and expand as well.

I await with keen interest further developments of your controversy.

*Prime Minister to Lord Beaverbrook*                                    15.XII.40

It is a magnificent achievement* in the teeth of the bombing. Quite apart from new production, the repaired aircraft has been your own creation. We now have 1,200 in the A.S.U.s, which is a great comfort. Dispersion has greatly hampered you, but was absolutely necessary as an insurance to spread the risk.

In addition, you have not confined yourself to mere numbers, but, on the contrary, have pushed hard into quality.

The reason why there is this crabbing, as at A,† is of course the warfare which proceeds between A.M. and M.A.P. They regard you as a merciless critic, and even enemy. They resent having had the M.A.P. functions carved out of their show, and I have no doubt they pour out their detraction by every channel open. I am definitely of opinion that it is more in the public interest that there should be sharp criticism and counter-criticism between the two departments than that they should be handing each other out ceremonious bouquets. One must therefore accept the stimulating but disagreeable conditions of war.

---

* Table from Lord Beaverbrook giving comparison of actual output of aircraft with programme.

† Paragraph in Lord Beaverbrook's minute of 14.XII.40 to the effect that it is sometimes said that the output of the Ministry of Aircraft Production would have been equalled by the Air Ministry if there had not been any change in May 1940.

*Prime Minister to Secretary of State for the Dominions*     15.XII.40

You will see from my telegram to Mr. Menzies that I do not view the situation in the Far East as immediately dangerous. The victory in Libya has reinforced, nay redoubled, the argument there set forth. I do not wish to commit myself to any serious dispersion of our forces in the Malay peninsula and at Singapore. On the contrary, I wish to build up as large as possible a Fleet, Army, and Air Force in the Middle East, and keep this in a fluid condition either to prosecute war in Greece and presently Thrace, or reinforce Singapore should the Japanese attitude change. I could not commit myself to the dispatch of many of the aircraft mentioned, certainly not the P.B.Y.s [flying-boats] at this juncture, when we have a major peril to face on the North-Western Approaches. I could not therefore agree to your telegram, and I should have thought my own (as amended in red) was quite sufficient at the present time.

*Prime Minister to C.A.S.*     15.XII.40

How are you getting on with the development on a large scale of aerodromes in Greece to take modern bombers and fighters, and with the movement of skeleton personnel, spare parts, etc., there?

It is quite clear to me this is going to be most important in the near future, and we must try not to be taken by surprise by events.

I should be glad to have a fortnightly report.

*Prime Minister to C.I.G.S.*     20.XII.40

Please let me know the earliest date when the 2nd Armoured Division—

(a) will land at Suez, and
(b) can be available for action in the Western Desert.

*Prime Minister to C.A.S.*     20.XII.40

I hope you will try to take a few days' rest, and seize every opportunity of going to bed early. The fight is going to be a long one, and so much depends upon you. Do not hesitate to send your deputy to any meetings I may call.

Pray forgive my giving you these hints, but several people have mentioned to me that you are working too hard.

[The possible use and counter-use of poison gas, should invasion come in the New Year, rested heavily upon me. Our progress in this sphere was however considerable.]

*Prime Minister to Minister of Supply*     21.XII.40

You will remember that the War Cabinet ordered an inquiry into the fact that bulk storage for 2,000 tons of mustard gas which had been ordered by the Cabinet in October 1938 was still not ready in October 1940.

The latest information which I have received from your Ministry shows that the bulk stock of mustard gas on December 9 was 1,485 tons. I was also informed through your Ministry that 650 tons of additional new storage was to have become available last week, and that production was being increased accordingly. Was this promise fulfilled?

Meanwhile I note that the filling of the new 25-pdr. base-ejection shell has at last begun in earnest, and that 7,812 of this type of shell had been filled by December 9. I should be glad to know how this figure compares with the total reserve of this type of shell required by the Army, and when this reserve is likely to be attained.

None of the new 6-inch base-ejection shells has yet been filled. What reserve does the Army require of this type of shell, and when is this reserve expected to be ready?

I am sending a copy of this minute to the Secretary of State for War.

*Prime Minister to Minister of Supply*                    22.XII.40

I learn that the Central Priority Department has been conducting a special investigation into the requirements of materials likely to be short.

I am told that much the most serious case is that of drop-forgings, on which the production of aeroplanes, tanks, guns, and transport all depend. Requirements for 1941 are estimated at 441,000 tons. Home production is now at the rate of 208,000. I am informed that there are orders in the United States for 7,000 tons, and that these are likely to rise to an annual rate of 25,000 by the end of 1941. Even if the requirements are considerably overstated, the deficiency is very serious.

Some moderate rate of expansion at home is expected, but we need to double the output. There are 14,000 workers in the industry, but it is reported that only 300 recruits have been received since August, that the industry alleges that it cannot absorb more than 1,000 new workers in each quarter, and that it is difficult to get recruits. All this needs looking into.

Meanwhile the only possible immediate action seems to be to increase purchases of drop-forgings in America, if necessary sending a special expert there for this purpose.

*Prime Minister to Minister of Works and Buildings*          22.XII.40

I understand that there is a serious shortage of accommodation for welfare services of all kinds to meet the needs of the homeless as well as of the evacuation schemes, and that you, in conjunction with the Minister of Health, have undertaken to seek for premises.

I hope that you will use your utmost endeavours to press on with this work.

I should be glad if you would let me have a return of commandeered premises which have not yet been used for war purposes, and which might be suitable for use in this way.

*Prime Minister to Lord Chatfield*                                         22.XII.40

I am grieved to find how very few George Medals have been issued. I had hoped there would be ten times as many. The idea was that you would go about and get into touch with local authorities where there had been heavy bombing, and make sure that recommendations were sent forward which could be sifted, and that you would stir the departments on the subject. Can you not do something more in this direction? You ought by now to have a number of typical cases which could be circulated to the authorities and departments concerned, who would thereafter be asked to match them from their experience.

Let me know if I can be of any assistance.

*Prime Minister to First Sea Lord*                                         22.XII.40

Very soon the Baltic will be frozen. Let me know its state and future prospects.

What has been happening to the Swedish ore during this summer? The Naval Staff should make the necessary inquiries.

What traffic has been moving down the Leads?

How has the position of German ore supplies been affected by the events of the last eight months? Is there any reason why we should not sow magnetic mines in the Leads, even if we do not lay a regular minefield? We seem to have forgotten all about this story.

I should be glad to have a note upon this, and to know whether anything can be done.

*Prime Minister to General Ismay*                                         22.XII.40

The work of the Joint Planners divides itself naturally into two parts:

    (a) all the current work they do for the C.O.S. Committee, and
    (b) the long-term future projects which are indicated to them, and on which they are already at work.

It is to these latter that I now turn. I think it would be well to appoint a Director of Future Schemes, or some other suitable title, who would guide and concert the preparation of the special schemes, preside over any meetings of the Joint Planners engaged upon them, and have direct access to me as Minister of Defence. I think Major Oliver Stanley [the former Secretary of State for War], with his experience of foreign politics and Cabinet government, would be able to impart to all this work a liveliness which I cannot supply except at rare intervals. He would have to be given a temporary Army rank to make him senior.

Pray make me proposals for giving effect to this idea.

*Prime Minister to Minister of Aircraft Production*                     22.XII.40

I am disturbed to see from reports sent to me by the Minister of Supply that deliveries to the Royal Air Force of bombs and containers charged with gas have dropped very noticeably during the past month, the total during the four weeks from November 11 to December 9 being:

| | | | | |
|---|---|---|---|---|
| 30-lb. bombs | .. | .. | .. | Nil |
| 250-lb. bombs | .. | .. | .. | 18 |
| 250-lb. containers | .. | .. | .. | Nil |
| 500-lb. containers | .. | .. | .. | 25 |
| 1,000-lb. containers | | .. | .. | 9 |

I understand that the reason for this decline is that factories have been bombed, and that difficulties have been encountered in the supply of certain component parts.

Nevertheless it is of vital importance that we should have the largest possible supply of aircraft gas-containers for immediate retaliation if need be, and I should be glad to know what steps are being taken to improve the delivery of these containers and what is the forecast of these deliveries over the next three months.

[I was concerned at the grave affronts to the rights and liberties of the individual which the safety of the State had required. Having been brought up on the Bill of Rights, *habeas corpus*, and trial by jury conceptions, I grieved to become responsible, even with the constant assent of Parliament, for their breach. In June, July, August, and September our plight had seemed so grievous that no limits could be put upon the action of the State. Now that we had for the time being got our heads again above water a further refinement in the treatment of internees seemed obligatory. We had already set up an elaborate sifting process, and many persons who had been arrested in the crisis were released by the Home Secretary, who presided over this field.]

*Prime Minister to Home Secretary*                                     22.XII.40

It must be remembered that these political *détenus* are not persons against whom any offence is alleged, or who are awaiting trial or are on remand. They are persons who cannot be proved to have committed any offence known to the law, but who because of the public danger and the conditions of war have to be held in custody. Naturally I feel distressed at having to be responsible for action so utterly at variance with all the fundamental principles of British liberty, *habeas corpus*, and the like. The public danger justifies the action taken, but that danger is now receding.

In the case of Mosley and his wife there is much prejudice from the Left, and in the case of the Pandit Nehru from the Right. I particularly asked that the rigorous character of the latter's imprisonment should be removed. In foreign countries such people are confined in fortresses—at least, they used to be when the world was still civilised.

These reflections led me to look into the details of Mosley's present confinement, as well as others of that category. Does a bath every week mean a hot bath, and would it be very wrong to allow a bath every day? What facilities are there for regular outdoor exercise and games and recreation under Rule 8? If the correspondence is censored, as it must be, I do not see any reason why it should be limited to two letters a week. What literature is allowed? Is it limited to the prison libraries? Are newspapers allowed? What are the regulations about paper and ink for writing books or studying particular questions? Are they allowed to have a wireless set? What arrangements are permitted to husbands and wives to see each other, and what arrangements have been made for Mosley's wife to see her baby, from whom she was taken before it was weaned?

I should be grateful if you would let me know your own view upon these matters.

*Prime Minister to Prime Minister of Australia*                23.XII.40

1. I am most grateful for your promised help at Singapore in respect both of troops and of equipment and ammunition, and hope that you will make these available as proposed. If so we will arrange to relieve your troops in May by the equivalent of a division from India.

2. The danger of Japan going to war with the British Empire is in my opinion definitely less than it was in June after the collapse of France. Since then we have beaten off the attacks of the German Air Force, deterred the invader by our ever-growing land strength, and gained a decisive victory in Libya. Since then the Italians have shown their weakness by sea, land, and air, and we no longer doubt our ability to defend the Delta and the Canal until or unless Germany makes her way through Turkey, Syria, and Palestine. This would be a long-term affair. Our position in the Eastern Mediterranean is enormously improved by the possession of Crete, where we are making at Suda Bay a second Scapa, and also by our victories and those of the Greeks, and the facilities we now have for building up strong air bases in Greece from which Italy can be attacked.

3. The naval and military successes in the Mediterranean and our

growing advantage there by land, sea, and air will not be lost upon Japan. It is quite impossible for our Fleet to leave the Mediterranean at the present juncture without throwing away irretrievably all that has been gained there and all the prospects of the future. On the other hand, with every weakening of the Italian naval power the mobility of our Mediterranean Fleet becomes potentially greater, and should the Italian Fleet be knocked out as a factor, and Italy herself broken as a combatant, as she may be, we could send strong naval forces to Singapore without suffering any serious disadvantage. We must try to bear our Eastern anxieties patiently and doggedly until this result is achieved, it always being understood that if Australia is seriously threatened by invasion we should not hesitate to compromise or sacrifice the Mediterranean position for the sake of our kith and kin.

4. Apart from the Mediterranean, the naval strain has considerably increased. When *Bismarck* and *Tirpitz* join the German Fleet, which they may have done already, the Germans will once again be able to form a line of battle. The *King George V* is ready, but we do not get *Prince of Wales* for several months, nor *Duke of York* till midsummer, nor *Anson* till the end of the year 1941. For the next six months we must keep more concentrated at Scapa Flow than has been necessary so far. The appearance of a raiding pocket-battleship in the Atlantic has forced us to provide battleship escort again for our convoys, and we are forming hunting-groups for the raiders in the South Atlantic, and if necessary in the Indian Ocean. We have always to consider the possibility of the undamaged portion of the French Fleet being betrayed by Darlan to Germany.

5. For all these reasons we are at the fullest naval strain I have seen either in this or the former war. The only way in which a naval squadron could be found for Singapore would be by ruining the Mediterranean situation. This I am sure you would not wish us to do unless or until the Japanese danger becomes far more menacing than at present. I am also persuaded that if Japan should enter the war the United States will come in on our side, which will put the naval boot very much on the other leg, and be a deliverance from many perils.

6. As regards air reinforcements for Malaya, the Conference at Singapore recommended the urgent dispatch of considerable numbers of aircraft. With the ever-changing situation it is difficult to commit ourselves to the precise number of aircraft which we can make available for Singapore, and we certainly could not spare the flying-boats to lie about idle there on the remote chance of a Japanese attack when they ought to be playing their part in the deadly struggle on the North-Western Approaches. Broadly speaking, our policy is to build up as large as possible a Fleet, Army, and Air Force in the Middle East, and

keep this in a fluid condition, either to prosecute war in Libya, Greece, and presently Thrace, or reinforce Singapore should the Japanese attitude change for the worse. In this way dispersion of forces will be avoided and victory will give its own far-reaching protections in many directions.

7. I must tell you finally that we are sending enormous convoys of troops and munitions to the Middle East, and we shall have nearly three hundred thousand men there by February. This again entails heavy escort duties. But great objects are at stake, and risks must be run in every quarter of the globe, if we are to emerge from all our dangers, as I am sure we shall.

8. I am arranging for details as regards shipping and equipment, etc., to be taken up direct between the War Office and Army Headquarters, Melbourne.

With all good wishes.

*Prime Minister to General Ismay*                                    23.XII.40
Please see that I have a good supply of photographs of war places; for instance, Sollum, Bardia, etc.

One of your staff might be told off to give some attention to this.

*Prime Minister to General Ismay, for C.O.S. Committee*
*Note for M. Dupuy (travelling to North Africa)*                     23.XII.40
Should you see Generals Weygand or Noguès you should explain that we now have a large, well-equipped army in England, and have considerable spare forces already well trained and rapidly improving, apart from what are needed to repel invasion.

The situation in the Middle East is also becoming good. If at any time in the near future the French Government decide to resume the war in Africa against Italy and Germany we would send a strong and well-equipped Expeditionary Force to aid the defence of Morocco, Algiers, and Tunis. These divisions could sail as fast as shipping and landing facilities were available. The British Air Force has now begun its expansion, and would also be able to give important assistance. The command of the Mediterranean would be assured by the reunion of the British and French Fleets, and by our joint use of Moroccan and North African bases. We are willing to enter into Staff talks of the most secret character with General Weygand, or any officers nominated by him.

On the other hand, delay is dangerous. At any time the Germans may, by force or favour, come down through Spain, render unusable the anchorage at Gibraltar, take effective charge of the batteries on both sides of the Straits, and also establish their Air Forces in the aerodromes. It is their habit to strike swiftly, and if they establish themselves at Casablanca the door would be shut on all projects. We are quite ready to wait for a certain time, provided that there is a good hope of bold action, and that plans are being made. But the situation may deteriorate any day and prospects be ruined. It is most important that the Government of Marshal Pétain should realise that we are able and willing to give powerful and growing aid. But this may presently pass beyond our power.

*Prime Minister to Minister of Shipping*                     24.XII.40
I see you made a speech about the Americans taking foreign ships. Could you let me have the text of it, together with any reactions you may have noticed in the American Press? I have the impression that the Americans were not quite pleased with the request addressed to them, as they do not consider that sufficient use is being made of British tonnage at the present time. In this connection you will remember my repeated inquiries as to the amount of British tonnage now plying exclusively between ports not in the United Kingdom.

According to the latest monthly report of the Ministry of Shipping, two and one-third million tons of British non-tanker shipping of over 1,600 tons is trading between overseas countries. Pray let me have a full explanation of this. About two million tons of Norwegian, Belgian, Polish shipping, excluding tankers, is also trading abroad.

*Prime Minister to Sir Edward Bridges and General Ismay*     25.XII.40
With the new year a fresh effort must be made to restrict the circulation of secret matters in Service and other departments. All the markings of papers in the Service departments, Foreign Office, Colonial and Dominions Offices, etc., should be reviewed with a view to striking off as many recipients as possible.

The officials concerned in roneo-ing the various circulations should be consulted, and a return made for me showing how many copies are made of different secret documents.

Pray report to me how this object can be achieved.

*Prime Minister to Secretary of State for the Dominions*     25.XII.40
No departure in principle is contemplated from the practice of keeping the Dominions informed fully of the progress of the war. Specially full information must necessarily be given in respect of theatres where Dominion troops are serving, but it is not necessary to circulate this to the other Dominions not affected. Anyhow, on the whole an effort should be made not to scatter so much deadly and secret information over this very large circle. . . . There is a danger that the Dominions Office staff get into the habit of running a kind of newspaper full of deadly secrets, which are circularised to the four

principal Governments with whom they deal. The idea is that the more they circulate the better they are serving the State. Many other departments fall into the same groove, loving to collect as much secret information as possible and feeling proud to circulate it conscientiously through all official circles. I am trying steadily to restrict and counter-act these tendencies, which, if unchecked, would make the conduct of war impossible.

While therefore there is no change in principle, there should be considerable soft-pedalling in practice.

I wish to be consulted before anything of a very secret nature, especially anything referring to operations or current movements, is sent out.

*Prime Minister to Minister of Health and Minister
of Home Security*                                          25.XII.40
I enclose minutes of our meeting yesterday, on which action is being taken.

I am convinced there should be only one authority inside the shelters, who should be responsible for everything pertaining to the health and comfort of the inmates. This authority should be charged with sanitation and storing of the bedding, etc. I cannot feel that the Home Security and Home Office, with all its burdens and duties under the enemy attack, ought to be concerned with questions affecting vermin and sanitation. These ought to be in the province of the Ministry of Health, who should be made responsible for the whole interior life of the shelters, big or small.

*Prime Minister to Sir Edward Bridges and Professor
Lindemann*                                                 26.XII.40
I must examine the Import Programme for 1941 next week. 5 p.m. in the Lower War Room, Monday, Tuesday, and Wednesday. Agenda to be drawn up by you and Professor Lindemann. Let me see by Saturday night here the immediate lay-out of the shipping programme in relation to food and supply and the demand for the Services in the face of present losses. Professor Lindemann will present me by Saturday night with the salient facts and graphs. To be summoned to the meeting the following:

> Lord President,
> Lord Privy Seal,
> Minister without Portfolio,
> Minister of Aircraft Production,
> Minister of Supply,
> Ministers of Food, Transport, and Shipping.

(Ministers only.)

*Prime Minister to Ministry of Supply*                           26.XII.40

The discrepancy between weapons and ammunition is terrible in the case of the anti-tank rifles, 2-inch and 3-inch mortars, the climax being reached with the 3-inch mortars. We have enough A.T. rifles to equip 23½ divisions, but only enough ammunition at 32,000 rounds per month to equip 5½. We have enough 2-inch mortars at 108 per division to equip 33 divisions, but ammunition at 32,400 rounds per month suffices only for 4½ divisions. The worst of all is the 3-inch mortar, where, oddly enough, we have at 18 per division enough to equip nearly 40 divisions, but at 14,000 rounds per month only enough ammunition for 1½ divisions.

*Prime Minister to First Lord*                                   26.XII.40

Provided that it can be arranged that four of the 15-inch can be cocked up within six months from now, and all other repairs be completed, I agree to abandon my long-cherished hope, in which I have been so continuously frustrated, of making *Resolution* into an effective fighting ship for inshore action.

The story of these four ships since the war began ranks with the story of the two-gun turret of the *K.G.V.* class in the most melancholy pages of the Admiralty annals.

I hope I may have your positive assurance that the six months condition will be fulfilled, barring enemy action of course.*

*Prime Minister to First Sea Lord*                               26.XII.40

I consider a greater effort should be made to interrupt the ore traffic through the Leads during January and onwards. This should certainly come before the Iceland–Faroes channel, which is a vast operation undertaken chiefly to use mines made for quite a different purpose, in conditions which have passed away. Now that we have not to give notice, and can lay secretly anywhere conditions are much more favourable for mining the Norwegian coast than they were last year, but the need to act seems to be almost as great.

Pray let me have a further report.

*Prime Minister to General Ismay, for C.O.S.*
*Committee and others concerned*                                 26.XII.40

Tactical requirements must be paramount during invasion. I am deeply anxious that gas warfare should not be adopted at the present time. For this very reason I fear the enemy may have it in mind, and perhaps it may be imminent. Every precaution must be kept in order, and every effort made to increase retaliatory power.

Sometimes I have wondered whether it would be any deterrent on

---

* See also my minute of 15.IX.40.

the enemy if I were to say that we should never use gas ourselves unless it had first been used against us, but that we had actually in store many thousands of tons of various types of deadly gas with their necessary containers, and that we should immediately retaliate upon Germany. On the whole, I think it is perhaps better to say nothing unless or until we have evidence that the attack is imminent. After all, they can make the calculations to which Professor Lindemann refers for themselves. They would certainly say we had threatened them with gas warfare, and would soon invent a pretext. Thirdly, there would be too much bluff in any such statement. If anyone is of a different opinion I shall be glad to know. The subject causes me much anxiety.

*Prime Minister to Home Secretary*        26.XII.40

I read in the papers of many people being sentenced for various offences against war regulations and for doing things which would not arise in peace-time. I am curious to know how the prison population compares with pre-war, both for imprisonment and penal servitude cases.

I should be much obliged if you could give me a few very simple figures. Are there a great many more now in gaol?*

*Prime Minister to Minister of Shipping*        27.XII.40

Let me have on one sheet of paper the main heads of your pro-ramme as at present settled of imports

    (a) in the next four months,
    (b) for the year 1941.

I should be glad to have this during to-morrow (Saturday).

*(Action this Day)*
*Prime Minister to General Ismay, for C.O.S. Committee*        27.XII.40

1. I do not recognise at all the account of my views given about "Marie".† I was under the impression that I had given a written minute. Pray let this be sought for. It is very unusual for me to give any directions other than in writing. To avoid further misunderstanding, the following is set forth:

2. The operation "Marie" has been regarded by the Chiefs of Staff, and is considered by me, to be valuable and important. For this purpose not only the Foreign Legion battalion but two other French battalions should be sailed in the January 4 convoy, and deposited at Port Soudan, where they can either intervene in "Marie" or in Egypt. There is no use sending only the Foreign Legion without any other troops of the French forces. Therefore I have asked for proposals to sail transports capable of taking the other two battalions empty from

---

* The figures were reassuring.
† Occupation of Jibouti.

here to Freetown, so that the whole French force can go round together.

Pray let me have to-day the proposals for giving effect to this.

There will be plenty of time to consider the political aspects when these troops have arrived at Port Soudan.

*Prime Minister to Lord Privy Seal*                                              27.XII.40

You very kindly sent me a report about cold storage of meat, dated November 14, and I wonder whether you would care to bring it up to date in the light of later happenings. I am very much concerned about the meat position.

*Prime Minister to Secretary of State for War and C.I.G.S.*          27.XII.40

1. Hitherto the production of anti-tank rifles has been a bright spot, and we have nearly thirty thousand already made. On the other hand, the ammunition for this weapon is deplorably in arrear, being in fact less than one-fifth of the proper proportion. The failure to "marry" the ammunition and the A.T. rifle is one of the worst blots on our present munitions programme. It is little less than a fraud on the troops to issue these large quantities of A.T. rifles, which would quickly become useless and worth no more than old iron through ammunition shortage. In many cases it has not been possible to allow any rounds for practice at all, these having to be saved for actual use against the enemy.

2. In these circumstances one would expect that the War Office would have concentrated their desires on ammunition, instead of increasing the already gigantic disproportion of A.T. rifles to ammunition. On the contrary however, for reasons which I have never heard mentioned, the Army requirement of A.T. rifles is suddenly raised from thirty-one thousand to seventy-one thousand for the same number of divisions. When was this decision taken? by whom? and what were the arguments? Was any attempt made at the time to make sure that the ammunition, already lagging so far behind, could catch up this enormous increase in rifles? Let me have a full report on this transaction.

3. However, the Germans have now twice bombed the Small Heath factory and checked the output of A.T. rifles in a most decisive manner. There can be no possibility of fulfilling the increased War Office demand of seventy-one thousand at the date desired. On the other hand, it is to be hoped that the ammunition supply will now have a chance of overtaking the weapons. It would therefore appear that a valuable and necessary readjustment of our programme has resulted from enemy action.

4. Arising out of the above, I wish to be informed when any large

changes are made in the existing programmes for the Army, particularly when these necessitate setting up new plants which can only be set up at the expense of other urgent work. All important modifications of the equipment tables set out in my diagrams are to be reported to me before action is taken.

*Prime Minister to C.A.S. and Air Ministry*      29.XII.40

It seems odd that only one machine should have been dispatched from Takoradi during the week ending December 27, when no fewer than forty-four are piled up there waiting. Is there a breakdown in the handling work at Takoradi? Could we have a special report on conditions there? Quite soon they will have the second instalment from the *Furious* upon them.

*(Action this Day)*
*Prime Minister to Secretary of State for Air, C.A.S., and*
*Minister of Aircraft Production*      30.XII.40
   (Secret)

1. I am deeply concerned at the stagnant condition of our bomber force. The fighters are going ahead well, but the bomber force, particularly crews, is not making the progress hoped for. I consider the rapid expansion of the bomber force one of the greatest military objectives now before us. We are of course drawing upon the bomber force for the Coastal Command and for the Middle East. If the bottleneck is, as I am told, crews, we must either have the pilots and personnel we are sending out to the Middle East returned to us after they have delivered their machines, or, what would be less injurious to formed squadrons, have other pilots and personnel sent back from the Middle East in their place. The policy is to remount the Middle East, and this must be achieved before reinforcements of a permanent character can be indulged in. Even before the recent reinforcements there were one thousand pilots in the Middle East. Air Marshal Longmore must be told to send back an equal number of good men of the various classes, and not add to his already grossly distended personnel.

2. In order to increase the number of crews available the training must be speeded up and a certain measure of dilution accepted.

3. The figures placed before me each day are deplorable. Moreover, I have been told on high authority that a substantial increase in numbers available for operations against Germany must not be expected for many months. I cannot agree to this without far greater assurance than I have now that everything in human wit and power has been done to avert such a complete failure in our air expansion programme.

4. So far as aircraft are concerned, the question arises, from constant study of the returns, whether sufficient emphasis is put upon bomber

production. The fighters are streaking ahead, and it is a great comfort that we have so good a position in them. We must however increase our bomb deliveries on Germany, and it appears that some of the types and patterns most adapted to this are not coming forward as we had hoped. I am well aware of the damage done by enemy action, but I ask whether it cannot be remedied, and what further steps are possible.

5. I wish to receive a programme of expansion week by week, and also a plan set forth showing what measures can be taken to improve the position, which at present is most distressing and black.

# APPENDIX B

The first table contains the figures which were given to the President in my letter of December 8, 1940.*

The second table gives the final assessment in the light of post-war knowledge.

\* See Book II, Chapter XXVIII, p. 494.

# TABLE I

## WEEKLY LOSSES AT SEA

| Week ended | British | | Allied | | Neutral | | Total: British, Allied, and Neutral | |
|---|---|---|---|---|---|---|---|---|
| | No. | Gross tons | No. | Gross tons | No. | Gross tons | No. | Gross tons |
| **1940** | | | | | | | | |
| June 2 .. | 28 | 79,415 | 5 | 25,137 | 2 | 4,375 | 35 | 108,927 |
| June 9 .. | 13 | 49,762 | 8 | 22,253 | 4 | 14,750 | 25 | 86,765 |
| June 16 .. | 15 | 60,006 | 10 | 40,216 | 6 | 23,170 | 31 | 123,392 |
| June 23 .. | 16 | 91,373 | 12 | 81,742 | 12 | 39,159 | 40 | 212,274 |
| June 30 .. | 6 | 30,377 | 4 | 13,626 | 5 | 19,332 | 15 | 63,335 |
| | 78 | 310,933 | 39 | 182,974 | 29 | 100,786 | 146 | 594,694 |
| July 7 .. | 14 | 75,888 | 4 | 18,924 | 5 | 21,968 | 23 | 116,780 |
| July 14 .. | 10 | 40,469 | 5 | 13,159 | 7 | 24,945 | 22 | 78,573 |
| July 21 .. | 12 | 42,463 | 2 | 3,679 | 7 | 13,723 | 21 | 59,865 |
| July 28 .. | 18 | 65,601 | 2 | 7,090 | .. | .. | 20 | 72,691 |
| | 54 | 224,421 | 13 | 42,852 | 19 | 60,636 | 86 | 327,909 |
| August 4 .. | 14 | 67,827 | 2 | 7,412 | 5 | 13,768 | 21 | 89,007 |
| August 11 .. | 9 | 32,257 | 2 | 7,674 | 2 | 6,708 | 13 | 46,639 |
| August 18 .. | 10 | 41,175 | 1 | 7,590 | 2 | 4,134 | 13 | 52,899 |
| August 25 .. | 20 | 108,404 | 1 | 1,718 | 2 | 8,692 | 23 | 118,814 |
| September 1 | 12 | 62,921 | 5 | 15,038 | 5 | 18,460 | 22 | 96,419 |
| | 65 | 312,584 | 11 | 39,432 | 16 | 51,762 | 92 | 403,778 |
| September 8 | 13 | 44,975 | 4 | 18,499 | 3 | 13,715 | 20 | 77,189 |
| September 15 | 13 | 55,153 | 4 | 12,575 | 3 | 7,379 | 20 | 75,107 |
| September 22 | 22 | 148,704 | 3 | 13,006 | 5 | 14,425 | 30 | 176,135 |
| September 29 | 11 | 56,096 | 4 | 12,119 | 2 | 7,351 | 17 | 75,566 |
| | 59 | 304,928 | 15 | 56,199 | 13 | 42,870 | 87 | 403,997 |
| October 6 | 8 | 30,886 | 3 | 5,742 | 1 | 3,687 | 12 | 40,315 |
| October 13 | 10 | 52,668 | 3 | 17,537 | 4 | 14,544 | 17 | 84,749 |
| October 20 | 34 | 154,279 | 7 | 24,686 | 6 | 26,816 | 47 | 205,781 |
| October 27 | 6 | 9,986 | 2 | 6,874 | 1 | 1,583 | 9 | 18,443 |
| November 3 | 13 | 65,609 | 4 | 5,403 | .. | .. | 17 | 71,012 |
| | 71 | 313,428 | 19 | 60,242 | 12 | 46,630 | 102 | 420,300 |
| November 10 | 11 | 69,110 | 2 | 10,236 | 2 | 8,617 | 15 | 87,963 |
| November 17 | 15 | 57,977 | 3 | 15,383 | 1 | 1,316 | 19 | 74,676 |
| November 24 | 20 | 80,426 | 3 | 12,415 | .. | .. | 23 | 92,841 |
| December 1 | 9 | 41,360 | 3 | 5,734 | 1 | 5,135 | 13 | 52,229 |
| | 55 | 248,873 | 11 | 43,768 | 4 | 15,068 | 70 | 307,709 |
| Grand Total, May 27 to Dec. 1, 1940 | 382 | 1,715,167 | 108 | 425,467 | 93 | 317,752 | 583 | 2,458,386 |

NOTES.—Week ended December 1 is the last full week for which details are available, and from the nature of the circumstances must be considered provisional.

For the whole period the following commissioned ships of 500 gross tons and over (formerly merchant vessels) have been lost by enemy action: 20 vessels, of 183,000 gross tons approximately.

# TABLE II

## MONTHLY TOTALS OF SHIPPING LOSSES, BRITISH, ALLIED, AND NEUTRAL

May 1940 to December 1940

| Months | British | | Allied | | Neutral | | Total: British, Allied, and Neutral | |
|---|---|---|---|---|---|---|---|---|
| | No. | Gross Tons | No. | Gross Tons | No. | Gross Tons | No. | Gross Tons |
| May 1940 | 31 | 82,429 | 26 | 134,078 | 20 | 56,712 | 77 | 273,219 |
| June 1940 | 61 | 282,560 | 37 | 187,128 | 27 | 101,808 | 125 | 571,496 |
| July 1940 | 64 | 271,056 | 14 | 48,239 | 20 | 62,672 | 98 | 381,967 |
| August 1940 | 56 | 278,323 | 13 | 55,817 | 19 | 59,870 | 88 | 394,010 |
| September 1940 | 62 | 324,030 | 19 | 79,181 | 9 | 39,423 | 90 | 442,634 |
| October 1940 | 63 | 301,892 | 17 | 73,885 | 17 | 66,675 | 97 | 442,452 |
| November 1940 | 73 | 303,682 | 13 | 47,685 | 5 | 24,731 | 91 | 376,098 |
| December 1940 | 61 | 265,314 | 11 | 70,916 | 7 | 21,084 | 79 | 357,314 |
| Totals | 471 | 2,109,286 | 150 | 696,929 | 124 | 432,975 | 745 | 3,239,190 |

# APPENDIX C

## AIRCRAFT STRENGTH
## DURING THE BATTLE OF BRITAIN
### 1940*

### 1.—AIRCRAFT PRODUCTION IN 1940

|  | | | Gross Production | Fighter Production |
|---|---|---|---|---|
| January | .. | .. | 802 | 157 |
| February | .. | .. | 719 | 143 |
| March | .. | .. | 860 | 177 |
| April | .. | .. | 1,081 | 256 |
| May | .. | .. | 1,279 | 325 |
| June | .. | .. | 1,591 | 446 |
| July | .. | .. | 1,665 | 496 |
| August | .. | .. | 1,601 | 476 |

* See Book II, Chapter XVI.

## 2.—OPERATIONAL STRENGTHS OF BOMBER COMMAND WEEK BY WEEK DURING BATTLE OF BRITAIN

### SUMMARISED ORDER OF BATTLE (BOMBER COMMAND) AND BOMBER AIRCRAFT IN AIRCRAFT STORAGE UNITS

#### BOMBER COMMAND

| Date | Total No. of Squadrons | No. of Squadrons Operationally Fit | Total I.E. Operational Squadrons ★ | Total Aircraft Available for Operations |
|---|---|---|---|---|
| July 11, 1940 .. | 40 | 35 | 560 | 467 |
| July 18, 1940 .. | 40 | 35 | 560 | 510 |
| July 25, 1940 .. | 40 | 35 | 554 | 517 |
| August 1, 1940 .. | 40 | 35 | 560 | 501 |
| August 8, 1940 .. | 41 | 36 | 576 | 471 |
| August 15, 1940 .. | 37 | 31 | 496 | 436 |
| August 22, 1940 .. | 37 | 31 | 496 | 491 |
| August 29, 1940 .. | 38 | 32 | 512 | 482 |
| September 5, 1940 | 39 | 36 | 576 | 505 |
| September 12, 1940 | 41 | 38 | 608 | 547 |
| September 19, 1940 | 42 | 38 | 608 | 573 |
| September 26, 1940 | 42 | 38 | 608 | 569 |

★ "I.E.", abbreviation for "Initial Equipment".

## AIR SUPPLY UNITS

### NUMBER OF AIRCRAFT EQUIPPED TO CURRENT OPERATIONAL STANDARDS, READY FOR DISPATCH

| Date | Within 48 Hours | Additionally within 4 Days |
| --- | --- | --- |
| July 11, 1940 .. | 285 | 128 |
| July 18, 1940 .. | 272 | 111 |
| July 25, 1940 .. | 251 | 111 |
| August 1, 1940 .. | 249 | 111 |
| August 8, 1940 .. | 191 | 111 |
| August 15, 1940 .. | 210 | 111 |
| August 22, 1940 .. | 152 | 116 |
| August 29, 1940 .. | 145 | 124 |
| September 5, 1940 | 103 | 124 |
| September 12, 1940 | 113 | 123 |
| September 19, 1940 | 107 | 121 |
| September 26, 1940 | 165 | 109 |

## 3.—OPERATIONAL STRENGTHS OF FIGHTER COMMAND WEEK BY WEEK

| Date | Total Number of Squadrons in Fighter Command | Number of Squadrons Available for Operations | Number of Aircraft Serviceable and Available for Operation |
|------|------|------|------|
| July 10 .. | 57 | 54 | 656 |
| July 17 .. | 57 | 52 | 659 |
| July 24 .. | 60 | 50 | 603 |
| July 31 .. | 61 | 54 | 675 |
| Aug. 7 .. | 61 | 56 | 714 |
| Aug. 14 .. | 61 | 54 | 645 |
| Aug. 21 .. | 61 | 57 | 722 |
| Aug. 28 .. | 61 | 58 | 716 |
| Sept. 4 .. | 63 | 58 | 706 |
| Sept. 11 .. | 63 | 60 | 683 |
| Sept. 18 .. | 64 | 61 | 647 |
| Sept. 25 .. | 64 | 61 | 665 |

## 4.—COMPARISON OF BRITISH AND GERMAN FIGHTER STRENGTH DURING THE BATTLE OF BRITAIN

The preceding table gives the overall strength of Fighter Command, including Blenheims and Defiants. But these cannot be reckoned, for the purposes of comparison, as part of the day fighting force, which consisted of Hurricanes and Spitfires.

After taking representative dates in the period July 10–October 31, the approximate daily average of squadrons *available for operations* of these two latter types combined is:

| | |
|---|---|
| Squadrons     .. .. | 49 |
| Aircraft and crews ready | 608 |

On the German side figures of serviceability are not at present available; comparison can therefore be made only in terms of Initial Equipment. The German I.E. was:

| | | |
|---|---|---|
| Single engine | approximately | 850 |
| Twin engine (Me. 110) | " | 350 |
| | TOTAL | 1,200 |

The comparable Initial Equipment figures for British squadrons available for operations, averaged over the 12 weeks, was 980.

# APPENDIX D

CORRESPONDENCE RELATING TO DAKAR
BETWEEN MR. CHURCHILL AND MR. MENZIES*

*Mr. Menzies to the Prime Minister*                                    29.IX.40

We are very disturbed in regard to Dakar incident, which has had unfortunate effect in Australia. First, as to matter of substance:

It is difficult to understand why attempt was made unless overwhelming chances of success. To make what appears at this distance to be a half-hearted attack is to incur a damaging loss of prestige.

Second, as to matter of procedure:

It is absolutely wrong that Australian Government should know practically nothing of details of engagement and nothing at all of decision to abandon it until after newspaper publication. I have refrained from any public criticism, but privately can tell you that absence of real official information from Great Britain has frequently proved humiliating. Finally, I must say frankly that Australian Government profoundly hopes difficulties have not been underestimated in the Middle East, where clear-cut victory is essential.

*Prime Minister to Mr. Menzies*                                        2.X.40

I am very sorry to receive your message of September 29, because I feel that the great exertions we have made deserve a broad and generous measure of indulgence should any particular minor operation miscarry. . . . The situation at Dakar was revolutionised by arrival of French ships from Toulon with Vichy personnel and the manning of the batteries by the hostile French Navy. Although every effort was made, the British Navy was not able to stop these ships on their way. After strongly testing the defences, and sustaining the losses I have already reported to you, the naval and military Commanders did not consider they had the strength to effect and support a landing, and I think they were quite right not to get us committed to a shore operation, which could not, like the naval attack, be broken off at any moment, and might have become a serious entanglement.

With regard to your criticisms, if it is to be laid down that no attempt is to be made which has not "overwhelming chances of success", you will find that a complete defensive would be imposed upon us. In dealing with unknown factors like the degree of French resistance it is impossible to avoid uncertainty and hazard. For in-

* See Book II, Chapter XXIV.

stance, Duala, and with it the Cameroons, were taken by twenty-five Frenchmen after their Senegalese troops had refused to march. Ought we to have moved in this case without having overwhelming force at hand? Secondly, I cannot accept the reproach of making "a half-hearted attack". I hoped that you had not sustained the impression from these last five months of struggle which has excited the admiration of the whole world that we were "a half-hearted Government" or that I am half-hearted in the endeavours it is my duty to make. I thought indeed that from the way my name was used in the election quite a good opinion was entertained in Australia of these efforts.

Every care will always be taken to keep you informed before news is published, but we could not prevent the German and Vichy wireless from proclaiming the course of events as they occurred at Dakar before we had received any information from our Commanders.

With regard to what you say about the Middle East, I do not think the difficulties have been under-estimated, but of course our forces are much smaller than those which the Italians have in Libya and Abyssinia, and the Germans may always help them. The defection of France has thrown the whole Middle East into jeopardy and severed our communications through the Mediterranean. We have had to face the threat of invasion here and the full strength of Germany's air bombing attack on our cities, factories, and harbours. Nevertheless we have steadfastly reinforced the Middle East, and in spite of all our perils at home and scanty resources have sent over thirty thousand men, nearly half our best tanks, many anti-aircraft guns needed to protect our vital aircraft factories, two of the finest units in the Fleet, the *Illustrious* and *Valiant*, and a considerable number of Hurricane fighters and Wellington bombers. We have done this in the face of an accumulation across the Channel and the North Sea of barges and shipping sufficient to carry half a million men to these shores at a single voyage and in a single night. Therefore, if the Middle East difficulties and dangers have not been fully met, it is not because the Mother Country has shirked her share of perils and sacrifice. At present the situation in Egypt and the Soudan looks better than we feared some time ago. Still, my dear Prime Minister and friend, as you have allowed me to deem you, I cannot guarantee "clear-cut victory" in the Middle East, or that Cairo, Khartoum, the Suez Canal, and Palestine may not fall into Italian or German hands. We do not think they will, and we are trying our utmost to resist the attacks which are massing against us. But I can make no promises at all of victory, nor can I make any promises that regrettable and lamentable incidents will not occur, or that there will not be disappointments and blunders. On the contrary, I think the only certainty is that we have very bad times indeed to go through before we emerge from the mortal perils by which we are surrounded.

I felt it due to your great position and the extremely severe tone of your message to reply with equal frankness.

*Mr. Menzies to Mr. Churchill* 4.X.40

I have received your message of October 2 and am very disturbed by some of its contents.

We were, and are, concerned about the failure at Dakar. My telegram concerning it was somewhat crudely expressed, as I can see on perusing it again. But I still do not understand how it can be construed as containing even the faintest suggestion that you or the British Government are half-hearted in policy, spirit, or achievement.

As the recent election here has left my own position extremely precarious and I may therefore soon go out of office, I would like to take the opportunity of saying to you that I have been very proud on behalf of Australia to be associated, even though at a distance, with the efforts of Winston Churchill and the British people. Such machinery as I possess in my own country has at all times been exercised so as to encourage the Australian people to realise that Great Britain is fighting our battles and that her heroism and superhuman cheerfulness and philosophy must be for us not only a shield but an inspiration.

As for yourself, praise from me would be an impertinence, but what I cabled you on September 3, the anniversary of the war, represented my whole heart and mind. I am indeed grieved to think that you should have felt my recent telegram to be either carping or discouraging.

I say no more about Dakar because it, no doubt, has lessons which it is not necessary for me to underline. Real point I make is that we, at this distance, will learn the lessons of events the more rapidly if information about those events can come to us as promptly and as fully as possible.

As to the Middle East, I have not sought or intended to seek guarantees. All that we ask—and I am sure it is granted before the asking—is that the Middle East should be as fully reinforced and equipped as is humanly possible. Your telegram has given me great satisfaction on this point.

You point out that if the Middle East difficulties and dangers have not been fully met it is not because the Mother Country has shirked her share of the perils or sacrifice: this is, of course, splendidly true. But I hope that you do not entertain any idea that Australia is shirking her share. We have many thousands of men in the Middle East, as many as shipping has been able to take. We have in camp in Australia further Expeditionary Force approximating eighty-five thousand men, many of whom will shortly be moving to the Middle East.

In spite of much public doubt caused by a real fear of what Japan may do, my Government has raised naval, air, and military forces and pledged our resources to munitions production on a scale previously unknown and regarded only a year ago as impossible.

We have done this notwithstanding the parochial interests and issues which in the recent elections succeeded in defeating us in the all-important State of New South Wales. We have set no limit to our contribution, because we know that there is no limit to the total British risk.

I mention these matters because I desire to make it clear that our anxiety about our main overseas theatre of actual participation in the war is not only intelligible but acute.

Please, my dear Prime Minister, do not interpret anxieties arising from these facts as either fearful, selfish, or unduly wrongheaded. And, above all, please understand that whatever interrogative or even critical telegrams I may send to you in secret Australia knows courage when it sees it and will follow you to a finish, as to the best of my abilities I certainly shall.

*Prime Minister to Mr. Menzies*                                    6.x.40

I am deeply grateful for your generous message. Forgive me if I responded too controversially to what I thought was somewhat severe criticism. I am having an account prepared of the Dakar incident, in all its stages, which I will send for the confidential information of yourself and your colleagues. I do not propose to defend myself at any length in Parliament, as such a spectacle would only gratify the enemy. I am deeply grateful for all that Australia has done under your leadership for the Common Cause. It has been a great comfort having some of the Australians here during these anxious months. I greatly admired their bearing and spirit when I inspected them. They had just received twenty-four good field guns. They are soon going to join the rest of the Australian Army in the Middle East, where they will probably be in the forefront of the fighting next year. We shall do everything in our power to equip them as they deserve. For the moment it seems that the situation in the Middle East is steady. Should the armies engage near Mersa Matruh the forces available during the next month or six weeks would not appear to be ill-matched in numbers. This should give a good chance to General Wilson, who is reputed a fine tactician, and the excellent troops he has. The Londoners are standing up magnificently to the bombing, but you can imagine the numerous problems which a ruthless attack like this upon a community of eight million people creates for the administration. We are getting the better of our difficulties, and I feel confident that the act of mass terror which Hitler has attempted will fail, like his magnetic mines and other deadly schemes. All good wishes personally for yourself.

# APPENDIX E

## LIST OF OPERATIONAL CODE NAMES

CATAPULT: Seizure, control, or effective disablement or destruction of all the accessible French Fleet.

COMPASS: Offensive operations in the Western Desert.

CROMWELL: Alarm word to be used if invasion of Britain imminent.

DYNAMO: Naval evacuation of the B.E.F., May 1940.

EXCESS: Aircraft reinforcements to the Middle East, January 1941.

HATS: Passing of Fleet reinforcements through Mediterranean and running of supply convoy to Malta.

MANDIBLES: Operations against the Dodecanese.

MARIE: Occupation of Jibouti.

MENACE: Occupation of Dakar.

MULBERRY: Artificial harbours.

OVERLORD: Liberation of France.

SEA LION: German plan for the invasion of Britain.

SHRAPNEL: Occupation of Cape Verde Islands.

TORCH: Anglo-American invasion of North Africa.

WINCH: Fighter reinforcement to Malta.

WORKSHOP: Capture of Pantelleria.

# APPENDIX F

## LIST OF ABBREVIATIONS

| | |
|---|---|
| A.A. guns | Anti-aircraft guns, or ack-ack guns. |
| A.D.G.B. | Air Defence of Great Britain. |
| A.F.S. | Auxiliary Fire Service. |
| A.F.V.s | Armoured fighting vehicles. |
| A.G.R.M. | Adjutant-General Royal Marines. |
| A.R.P. | Air Raid Precautions. |
| A.S.U. | Air Supply Units. |
| A.T. rifles | Anti-tank rifles. |
| A.T.S. | (Women's) Auxiliary Territorial Service. |
| B.E.F. | British Expeditionary Force. |
| C.A.S. | Chief of the Air Staff. |
| C.I.G.S. | Chief of the Imperial General Staff. |
| C.-in-C. | Commander-in-Chief. |
| CONTROLLER | Third Sea Lord and Chief of Material. |
| C.N.S. | Chief of the Naval Staff (First Sea Lord). |
| C.O.S. | Chiefs of Staff. |
| D.N.C. | Director of Naval Construction. |
| F.O. | Foreign Office. |
| G.H.Q. | General Headquarters. |
| G.O.C. | General Officer Commanding. |
| G.Q.G. | Grand-Quartier-Général. |
| H.F. | Home Forces. |
| H.M.G. | His Majesty's Government. |
| L.D.V. | Local Defence Volunteers (renamed Home Guard). |
| M.A.P. | Ministry of Aircraft Production. |
| M.E.W. | Ministry of Economic Warfare. |
| M.O.I. | Ministry of Information. |
| M. OF L. | Ministry of Labour. |
| M. OF S. | Ministry of Supply. |

| O.K.H. | Oberkommando des Heeres. Supreme Command of the German Army. |
| O.T.U. | Operational Training Unit. |
| P.M. | Prime Minister. |
| U.P. | Unrotated projectiles—*i.e.*, code name for rockets. |
| V.C.A.S. | Vice-Chief of the Air Staff. |
| V.C.I.G.S. | Vice-Chief of the Imperial General Staff. |
| V.C.N.S. | Vice-Chief of the Naval Staff. |
| W.A.A.F. | Women's Auxiliary Air Force. |
| W.R.N.S. | Women's Royal Naval Service ("Wrens"). |

# INDEX

# INDEX